Ulrich Kück, Nicole Frankenberg-Dinkel (Eds.)
Biotechnology
De Gruyter Textbook

I0031434

Also of interest

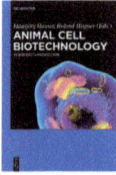

Animal Cell Biotechnology
Hansjörg Hauser, Roland Wagner (Eds.), 2014
ISBN 978-3-11-027886-6, e-ISBN (PDF) 978-3-11-027896-5,
e-ISBN (EPUB) 978-3-11-038142-9, Set-ISBN 978-3-11-027897-2

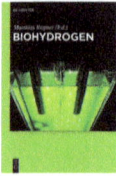

Biohydrogen
Matthias Rögner (Ed.), 2015
ISBN 978-3-11-033645-0, e-ISBN (PDF) 978-3-11-033673-3,
e-ISBN (EPUB) 978-3-11-038934-0, Set-ISBN 978-3-11-033674-0

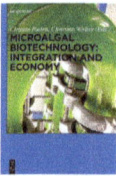

Microalgal Biotechnology: Integration and Economy
Clemens Posten, Christian Walter (Eds.), 2012
ISBN 978-3-11-029827-7, e-ISBN 978-3-11-029832-1,
Set-ISBN 978-3-11-029833-8

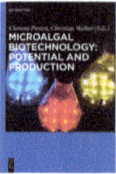

Microalgal Biotechnology: Potential and Production
Clemens Posten, Christian Walter (Eds.), 2012
ISBN 978-3-11-022501-3, e-ISBN 978-3-11-022502-0,
Set-ISBN 978-3-11-174200-7

Biomedical Engineering / Biomedizinische Technik
Olaf Dössel (Editor in Chief), 6 Issues per year
ISSN 0013-5585, e-ISSN 1862-278X

Ethik und Recht in Medizin und Biowissenschaften
Bernice S. Elger, Nikola Biller-Andorno, Bernhard Rütsche (Eds.), 2014
ISBN 978-3-11-028462-1, e-ISBN 978-3-11-028551-2,
e-ISBN (EPUB) 978-3-11-037394-3

Trennungsmethoden der Analytischen Chemie
Rudolf Bock, Reinhard Nießen, 2014
ISBN 978-3-11-026544-6, e-ISBN (PDF) 978-3-11-026637-5,
e-ISBN (EPUB) 978-3-11-037049-2

Biotechnology

Edited by
Ulrich Kück and Nicole Frankenberg-Dinkel

DE GRUYTER

Editors

Prof. Dr. Ulrich Kück
Ruhr-Universität Bochum
Lehrstuhl für Allgemeine und Molekulare Botanik
Fakultät für Biologie und Biotechnologie
ND 7/130 Süd
Universitätsstraße 150
44780 Bochum, Germany
ulrich.kueck@rub.de

Prof. Dr. Nicole Frankenberg-Dinkel
Technische Universität Kaiserslautern
Abteilung Mikrobiologie
P.O. Box 3049
67653 Kaiserlautern, Germany
Nfranken@rhrk.uni-kl.de

ISBN 978-3-11-034110-2
e-ISBN (PDF) 978-3-11-034263-5
e-ISBN (EPUB) 978-3-11-038360-7

Library of Congress Cataloging-in-Publication Data
A CIP catalog record for this book has been applied for at the Library of Congress.

Bibliographic information published by the Deutsche Nationalbibliothek
The Deutsche Nationalbibliothek lists this publication in the Deutsche Nationalbibliografie;
detailed bibliographic data are available on the Internet at http://dnb.dnb.de.

© 2015 Walter de Gruyter GmbH, Berlin/Boston
Typesetting: Meta Systems Publishing Printservices GmbH, Wustermark
Printing and binding: CPI books GmbH, Leck
Cover image: Prof. Dr. Ulrich Kück
♾ Printed on acid-free paper
Printed in Germany

www.degruyter.com

Contents

Mathias Lübben and Raphael Gasper

Ulrich Kück and Ines Teichert

Ansgar Poetsch and Andreas Harst

Markus Piotrowski

Stefan Wiese, Dennis Stern and Alice Klausmeyer

Andreas Faissner, Jacqueline Reinhard and Ursula Theocharidis

Olivia A. Masseck, Katharina Spoida and Stefan Herlitze

Hermann Lübbert

List of contributing authors

Prof. Dr. Andreas Faissner
Ruhr Universität Bochum, Lehrstuhl für Zellmorphologie und Molekulare Neurobiologie,
Universitätsstraße 150, NDEF 05/593, 44801 Bochum, **Andreas.Faissner@rub.de**

Dr. Raphael Gasper-Schönenbrücher
Ruhr-Universität Bochum, AG Proteinkristallographie, Fakultät für Biologie und Biotechnologie,
Universitätsstr. 150, NDEF 04/316, 44801 Bochum, **Raphael.Gasper@bph.rub.de**

Sarah Gaßmeyer
Ruhr-Universität Bochum, NG Mikrobielle Biotechnologie, Fakultät für Biologie und
Biotechnologie, Universitätsstr. 150, ND 2/29, 44801 Bochum, **Sarah.Gassmeyer@rub.de**

Prof. Dr. Thomas Happe
Ruhr-Universität Bochum, AG Photobiotechnologie, Fakultät für Biologie und Biotechnologie
Universitätsstraße 150, ND 2/169, 44801 Bochum, **Thomas.Happe@rub.de**

Andreas Harst
Ruhr-Universität Bochum, Lehrstuhl Biochemie der Pflanzen, Fakultät für Biologie und
Biotechnologie, Universitätsstraße 150, ND 3/128, 44801 Bochum, **Andreas.Harst@rub.de**

Dr. Anja Hemschemeier
Ruhr-Universität Bochum, AG Photobiotechnologie, Fakultät für Biologie und Biotechnologie,
Universitätsstraße 150, ND 2/134, 44801 Bochum, **Anja.Hemschemeier@rub.de**

Prof. Dr. Stefan Herlitze
Ruhr-Universität Bochum, Lehrstuhl für Allgemeine Zoologie und Neurobiologie, Fakultät für
Biologie und Biotechnologie Universitätsstraße 150, ND 7/32, 44801 Bochum,
Stefan.Herlitze@rub.de

Dr. Alice Klausmeyer
Ruhr Universität Bochum, AG Molekulare Zellbiologie, Fakultät für Biologie und Biotechnologie,
Universitätsstraße 150, NDEF 05/588, 44801 Bochum, **Alice.Klausmeyer@rub.de**

Jun.Prof. Dr. Robert Kourist
Ruhr-Universität Bochum, NG Mikrobielle Biotechnologie, Fakultät für Biologie und
Biotechnologie, Universitätsstr. 150, ND 1/130, 44801 Bochum, **Robert.Kourist@rub.de**

PD Dr. Mathias Lübben
Ruhr-Universität Bochum, Molekulare Biologie von Membranproteinen, Fakultät für Biologie und
Biotechnologie, Universitätsstr. 150, ND 04/398, 44801 Bochum,
Mathias.Luebben@rub.de

Prof. Dr. Hermann Lübbert
Ruhr-Universität Bochum, Lehrstuhl für Tierphysiologie, Fakultät für Biologie und Biotechnologie,
Universitätsstr. 150, ND 04/398, 44801 Bochum, **Hermann.Luebbert@rub.de**

Dr. Olivia Masseck
Ruhr-Universität Bochum, Lehrstuhl für Allgemeine Zoologie und Neurobiologie, Fakultät für
Biologie und Biotechnologie, Universitätsstraße 150, ND 7/34, 44801 Bochum,
Masseck@neurobiologie.rub.de

Prof. Dr. Axel Mosig
Ruhr-Universität Bochum, AG Bioinformatik, Fakultät für Biologie und Biotechnologie,
Universitätsstr. 150, ND 04/173, 44801 Bochum, **Axel.Mosig@rub.de**

Prof. Dr. Franz Narberhaus
Ruhr-Universität Bochum, Lehrstuhl für Biologie der Mikroorganismen, Fakultät für Biologie und
Biotechnologie, Universitätsstr. 150, NDEF 06/783, 44801 Bochum, **Franz.Narberhaus@rub.de**

Dr. Marc Nowaczyk
Ruhr-Universität Bochum, Lehrstuhl Biochemie der Pflanzen, Fakultät für Biologie und
Biotechnologie, Universitätsstraße 150, ND 2/150, 44801 Bochum,
Marc.M.Nowaczyk@rub.de

PD Dr. Markus Piotrowski
Ruhr-Universität Bochum, Lehrstuhl für Pflanzenphysiologie, Fakultät für Biologie und
Biotechnologie, Universitätsstraße 150, ND 3/49, 44801 Bochum, **Markus.Piotrowski@rub.de**

PD Dr. Ansgar Poetsch
Ruhr-Universität Bochum, Lehrstuhl Biochemie der Pflanzen, Fakultät für Biologie und
Biotechnologie, Universitätsstraße 150, ND 3/130, 44801 Bochum, **Ansgar.Poetsch@rub.de**

Dr. Jacqueline Reinhard
Ruhr Universität Bochum, Lehrstuhl für Zellmorphologie und Molekulare Neurobiologie,
Universitätsstraße 150, NDEF 05/342, 44801 Bochum, **Jacqueline.Reinhard@rub.de**

Dr. Sascha Rexroth
Ruhr-Universität Bochum, Lehrstuhl Biochemie der Pflanzen, Fakultät für Biologie und
Biotechnologie, Universitätsstraße 150, ND 3/133, 44801 Bochum, **Sascha.Rexroth@rub.de**

Prof. Dr. Matthias Rögner
Ruhr-Universität Bochum, Lehrstuhl Biochemie der Pflanzen, Fakultät für Biologie und
Biotechnologie, Universitätsstraße 150, ND 3/125, 44801 Bochum, **Matthias.Roegner@rub.de**

Dr. Katharina Spoida
Ruhr-Universität Bochum, Lehrstuhl für Allgemeine Zoologie und Neurobiologie, Fakultät für
Biologie und Biotechnologie, Universitätsstraße 150, ND 7/34, 44801 Bochum,
Katharina.Spoida@rub.de

Dennis Stern
Ruhr Universität Bochum, AG Molekulare Zellbiologie, Fakultät für Biologie und Biotechnologie,
Universitätsstraße 150, NDEF 05/598, 44801 Bochum, **Dennis.Stern@rub.de**

Dr. Ines Teichert
Ruhr Universität Bochum, Lehrstuhl für Allgemeine und Molekulare Botanik, Fakultät für Biologie
und Biotechnologie, Universitätsstraße 150, ND 7/176, 44801 Bochum, **Ines.Teichert@rub.de**

Dr. Ursula Theocharidis
Ruhr Universität Bochum, Lehrstuhl für Zellmorphologie und Molekulare Neurobiologie,
Universitätsstraße 150, NDEF 05/342, 44801 Bochum, **Ursula.Theocharidis@rub.de**

Prof. Dr. Stefan Wiese
Ruhr Universität Bochum, AG Molekulare Zellbiologie, Fakultät für Biologie und Biotechnologie,
Universitätsstraße 150, NDEF 05/598, 44801 Bochum, **Stefan.Wiese@rub.de**

Introduction

The term "biotechnology" connects traditional as well as novel biological techniques and applications. Although biotechnology is an old (ancient, traditional) technology and has been used for over 6.000 years by mankind, it rejuvenated during the last decades as new tools in molecular biology became available. Biotechnology is a very broad discipline, which employs biological processes, organisms, cells or cellular systems to develop new tools, products and technologies that improve our daily live. As such, these technologies are valuable in biological research, agriculture, industry and medicine. Based on its broad orientation, the different areas of biotechnology have been organized in a color coded classification with the major disciplines being the **white, green** and **red** biotechnology. However, due to constant technological advancement the diversification proceeded, leading to **blue, yellow, grey, black** and even **gold** biotechnology.

Table 1: Colors of biotechnology

Color classification	Description
white	Industrial processes involving microorganisms
green	Processes improving agriculture
red	Medicine, health, diagnostics
blue	Marine and aquatic systems
yellow	Food biotechnology and nutrition
grey	Environmental biotechnology: removal of pollutants, bioremediation
black	Biowarfare, bioterrorism
gold	Bioinformatics

This textbook is based on the lecture series "Biotechnology", which is held at the Ruhr-University Bochum, Germany. It is devoted to Bachelor and Master students with a focus on biotechnology, but also to graduate students and postdocs seeking for new facets in their research. 14 lecturers of this series have provided individual chapters to this book that are grouped according to the color coded classification of biotechnology. These are:

White biotechnology includes technologies that employ microorganisms in chemical production. Here, gene-based technologies are used to generate efficient production strains. In addition, white biotechnology aims in designing processes and products that consume little resources and energy compared to older, traditional methods.

Blue biotechnology refers to the application of molecular biological methods to marine and freshwater organisms. This includes the use of marine organisms (animals, plants, algae, and bacteria) or their derivatives to isolate and/or to develop new products in the fields of medicine, cosmetics, food and feed supplements. Involved are environmental as well as industrial applications.

Green biotechnology includes technologies that have a positive impact on agriculture. This includes the generation of new crop plants by means of genetic manipulation (transgenic plants) and traditional crossing as well as the production of biofertilizers and biopesticides. In general, green biotechnology has several aims: A) create new crop varieties resistant to disease and pests; B) to produce crop varieties with improved nutritional properties (i.e. elevated vitamin content); C) to modify plants in such a way that they can be used as bio-factories to produce bioactive/biomedical substances.

Red biotechnology refers to the areas of biotechnology that are related to medicine. It includes the production of vaccines and antibodies, new tools in molecular diagnostics, drug development and genetic engineering for disease treatment.

All chapters provide in the beginning a short introduction, followed by specific chapters on varies aspects of biotechnology. At the end, representative questions and literature for further readings are given for students with an interest in biotechnology.

This textbook would not have been possible without the support of the **de Gruyter** Publisher. In particular we appreciate the help by Mr. Leonardo Milla, Mrs. Karola Seitz and Mrs. Simone Witzel, who supported planning and production of this book.

Ulrich Kück and Nicole Frankenberg-Dinkel,   February 2015

Nicole Frankenberg-Dinkel

1 History of biotechnology and classical applications in food biotechnology

Biotechnology is defined as the use of biological processes, organisms, or systems to manufacture products intended to improve the quality of human life. Derived from the greek word βίος (bios-life) and combined with the word technology, it represents an interdisciplinary science which employs enzymes, cells or whole organisms for technical use. Biotechnology combines knowledge from a wide variety of disciplines including microbiology, biochemistry, genetics, bioinformatics but also process technology and chemical engineering.

1.1 History of food biotechnology (a brief overview)

Biotechnology is not new. Very early examples date back 10,000 years and include in a very broad-ranging definition the domestication of animals, the planting of crops and the use of microorganisms to produce alcoholic beverages (wine, beer), dairy products (yoghurt, cheese), bread but also Sauerkraut and silage (Figure 1.1). Dogs, sheep and cows were among the first domesticated animals and used for

Fig. 1.1: Various food products made with the help of microorganisms. Most of the product displayed are produced by the action of yeasts and/or lactic acid bacteria.

meat and milk production. For plants, rice, wheat, and barley are thought to be the first crops to be farmed to obtain a reliable source of food. The main purpose to employ fermenting microorganisms was to preserve and refine food. The most relevant organisms in these processes even until to date are yeasts and lactic acid bacteria.

1.1.1 Bread-baking in ancient Egypt

Bread and beer was central to the ancient Egyptian diet and consumed with every meal. Early descriptions, mainly from artistic scenes, date back 4000 years BC. Bread was mainly made from emmer wheat (*Triticum dicoccum*) and in the early days only sometimes included yeast for leavening the dough.

1.1.2 Ancient beer brewing

Several recipes for beer have been derived from inscriptions. One includes the use of a yeasty bread ("beer bread"), which was lightly baked and crumbled into water. Dates were added for flavor and the mixture was fermented in a tub before it was sieved and filled into jars for storage. However, there is also evidence for brewing beer from barley and emmer wheat by heating and subsequent mixing with yeast. The existence of yeast was not yet known to the brewer but based on experience they knew that the addition of the sediment from a well-done beer enhanced the fermentation effect. Traditionally, beer brewing in ancient Egypt was a female activity.

The Celts and Teutons preferred a slightly acidic beer (mead) which they stored in large jars underground to keep the temperature low. Honey was added for sweetness and flavor. Much later, in the 6th century, monks developed a more sophisticated method for brewing beer. Still until nowadays the "monastery beers" are known for their richness and high alcohol content and served the monks as a nutritious drink during Lenten season.

1.1.3 Winemaking (vinification)

Like for beer brewing there is also evidence for ancient winemaking not only on the European continent but also in Asia (China). Traditionally, wine was made by fermenting grape juice employing yeasts which were already present on the grape skin. These natural yeasts can be used for the primary fermentation, which converts the sugars of the grape juice into alcohol (details see 1.3). However, in order to avoid unpredicted results from natural yeast variations on the grapes, addition

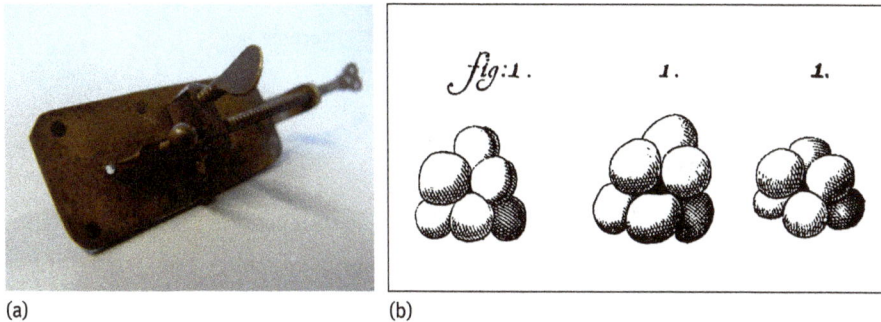

Fig. 1.2: Early microscopy. (a) Replica of a Leeuwenhoek microscope from Museum Boerhaave Leiden (photograph courtesy of medizinhistorische Sammlung Ruhr University Bochum). (b) Yeast (nl. "gist") as drawn schematically by Antonie van Leeuwenhoek in his letter to Thomas Gale of June 14th 1680, pp. 6–10.

of cultured yeasts in nowadays winemaking is a must. Some wines require a secondary, malolactic fermentation in order to obtain a softer taste. Malic acid which is mostly present in red wine must has a tart-like taste and is converted by certain strains of lactic acid bacteria to lactic acid which has a more softer taste (see also 1.5.6 of this chapter).

1.1.4 The first description of bacteria and yeast

It was not until the 17th century that the Dutch merchant Antonie van Leeuwenhoek (1632–1723) observed the first bacteria and yeast, which he investigated with his home-made microscope. This first microscope had a single glass lens with a magnification of about 200-fold (Figure 1.2) which enabled van Leeuwenhoek to observe yeast and bacteria at the cellular level. Although yeast was already used in concentrated and enriched form for bead baking, beer brewing, and winemaking, it was still unknown that it was a living creature. Due to van Leeuwenhoek's keen interest in nature, he was inspired to investigate sheep hair, water droplets from a rain barrel but also beer samples with his microscope, which resulted in numerous drawings. He was the first to draw images of bacteria and yeasts in their morphologic diverse from (Figure 1.2) which he later sent to the Royal Society in London. Although van Leeuwenhoek never attended a University, he was elected to the Royal Society in 1680 due to the discovery of his "little animals".

1.2 A work horse of food biotechnology: Yeast

Microorganisms can be divided in eukaryotic (greek: *karyon*, nucleus) and prokaryotic ones. While eukaryotic microorganisms have a complex cell architecture, prokaryotes are rather simple. The yeast *Saccharomyces cerevisiae* (baker's yeast)

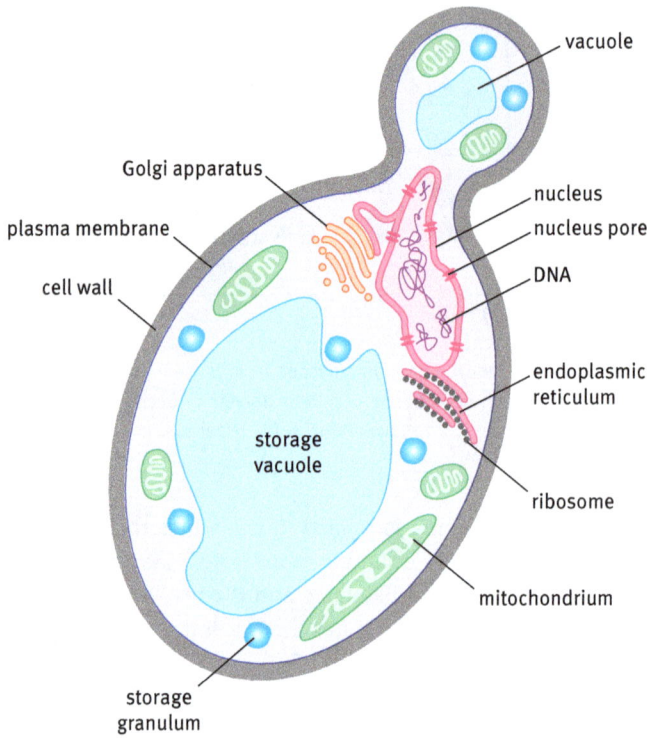

Fig. 1.3: Schematic drawing of a yeast cell.

is a eukaryotic model organism (see also chapter 4). It belongs to the fungi, specifically to the group of Ascomycetes. This class is characterized by typical sporangia, called asci (singular ascus), which are generated during the sexual life cycle. Like all other eukaryotes its cell structure is complex including compartmentalization (nucleus, mitochondria, Golgi apparatus etc.) (Figure 1.3). Yeast consists of only a single cell which propagates by budding and each daughter cell itself can give rise to new cells (Herskowitz 1988). This asexual propagation involves mitosis and results in the duplication of a haploid or diploid genome that is then evenly distributed among mother and daughter cells. In nature, *S. cerevisiae* exists in the diploid state. Under certain growth conditions (e.g. nitrogen starvation), yeast cells are able to undergo a life cycle with meiotic divisions, which result in the generation of asci with four haploid ascospores that carry one of the mating type loci ("alpha" or "a"). In the laboratory, these haploid cells can be separated to receive haploid strains, while in nature two haploid strains of the opposite mating type will immediately fuse after germination to generate diploid strains. This life cycle of fungal yeast is known as the sexual life cycle. In biotechnological applications however, the asexual cycle is of major importance. Notably, in some applications, e.g. during breeding of brewer's yeasts, the sexual life cycle is used to generate recombinant strains with newly acquired properties.

Yeasts grow heterotrophically, meaning they require an external carbon source (glucose, maltose, trehalose) with the central metabolic pathways in action (glycolysis, citric acid cycle (TCA cycle; Krebs cycle) and oxidative phosphorylation) under aerobic conditions. As yeast is a facultative anaerobe, it can also grow under fermenting conditions. The biotechnological use mainly relies on the metabolic pathways being active under anaerobic conditions. Here, yeast ferments the sugar to alcohol (ethanol) and carbon dioxide, a process that takes place in the cytosol of the cell.

1.2.1 Glycolysis

During glycolysis, a series of reactions transform glucose into pyruvate with the formation of two molecules of adenosine triphosphate (ATP) and the reducing equivalents nicotinamide adenine dinucleotide (NADH). Glycolysis is operative under oxic and anoxic conditions and starts with the activation of glucose by two molecules of ATP and a subsequent isomerization. The resulting fructose 1,6-bisphosphate is cleaved by aldolase to yield the triosephosphate isomers dihydroxyacetone phosphate and glyceraldehyde-3-phosphate. Energy extraction occurs through substrate level phosphorylation in two reactions. First, phosphoglycerate kinase converts 1,3-bisphosphoglycerate to 3-phosphoglycerate under the generation of ATP. The second reaction is the final step of glycolysis where phosphoenolpyruvate is transformed to pyruvate by pyruvate kinase (Figure 1.4). All glycolytic enzymes are located in the cytoplasm of the yeast cell. In case other sugars instead of glucose are present they are either converted to glucose or intermediates of glycolysis and then metabolized.

Under aerobic conditions pyruvate is transported into the mitochondrium and subsequently converted by the pyruvate dehydrogenase complex to acetyl-CoA. The latter enters the TCA cycle in which it is fully oxidized to carbon dioxide and electrons in form of reducing equivalents. During oxidative phosphorylation electrons are transported in a series of redox-reaction to the final oxidase which reduces oxygen to water. During this process, protons are translocated over the inner mitochondrial membrane into the matrix. The generated proton gradient is then used by ATP synthase to generate ATP. Overall, this process yields a total of 36–38 molecules of ATP per molecule of glucose (for further details see standard biochemistry text books).

1.3 Alcohol fermentation by yeast

Under anaerobic conditions, when the terminal electron acceptor for oxidative phosphorylation, oxygen, is missing, many microorganisms including yeast can undergo fermentation. The general principle of a fermentation involves an oxidative and a reductive part. In the oxidative part the substrate is oxidized to interme-

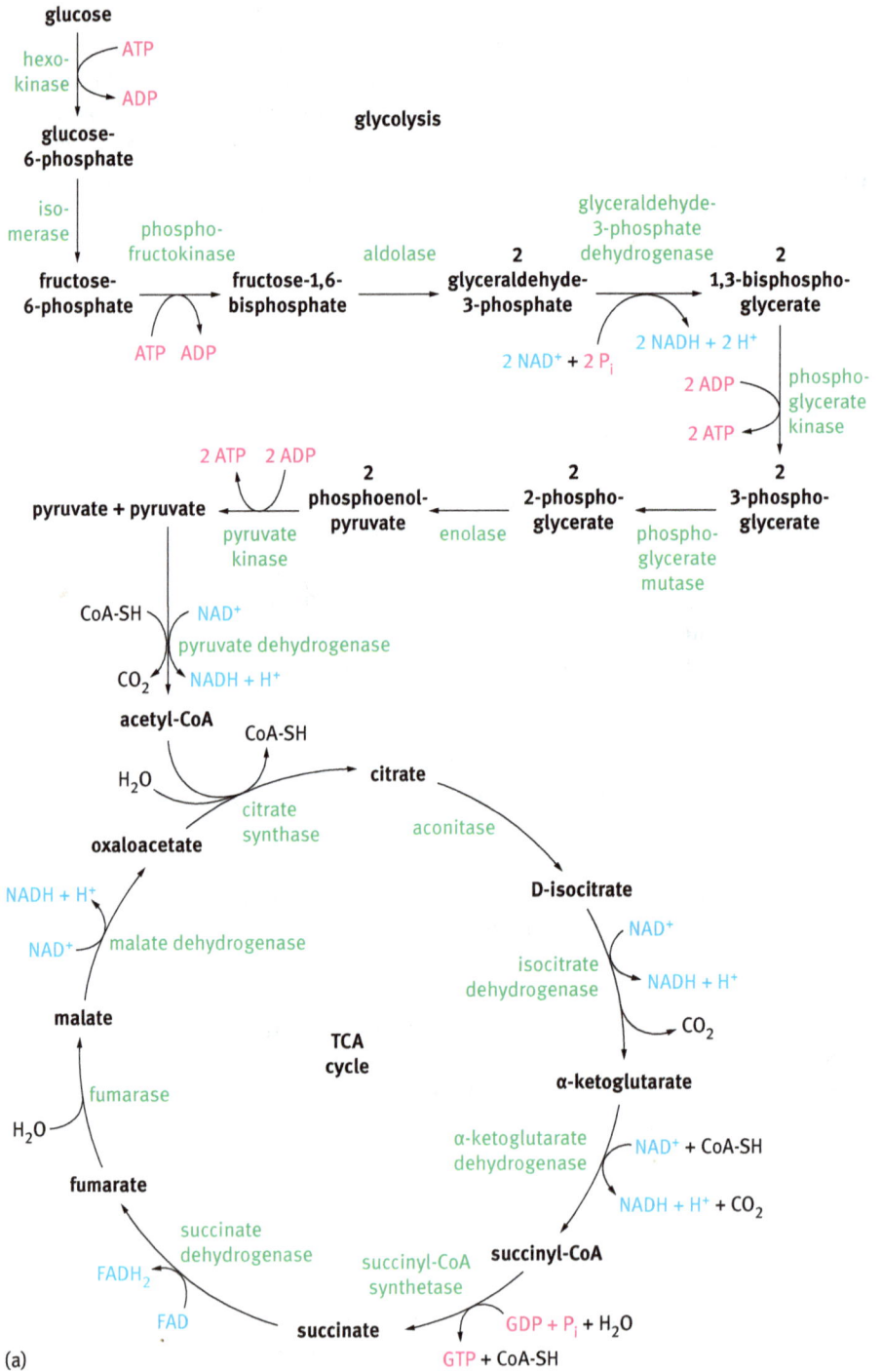

glucose

hexo-
kinase

ATP

ADP

glycolysis

**glucose-
6-phosphate**

iso-
merase

phospho-
fructokinase

aldolase

glyceraldehyde-
3-phosphate
dehydrogenase

**fructose-
6-phosphate**

ATP ADP

**fructose-1,6-
bisphosphate**

**2
glyceraldehyde-
3-phosphate**

2 NAD$^+$ + 2 P$_i$

**2
1,3-bisphospho-
glycerate**

2 NADH + 2 H$^+$

2 ADP

2 ATP

phospho-
glycerate
kinase

2 ATP   2 ADP

pyruvate + pyruvate

pyruvate
kinase

**2
phosphoenol-
pyruvate**

enolase

**2
2-phospho-
glycerate**

phospho-
glycerate
mutase

**2
3-phospho-
glycerate**

CoA-SH

NAD$^+$

pyruvate dehydrogenase

CO$_2$

NADH + H$^+$

acetyl-CoA

CoA-SH

H$_2$O

citrate
synthase

citrate

aconitase

oxaloacetate

NADH + H$^+$

NAD$^+$

malate dehydrogenase

D-isocitrate

NAD$^+$

isocitrate
dehydrogenase

NADH + H$^+$

CO$_2$

malate

**TCA
cycle**

fumarase

H$_2$O

α-ketoglutarate

α-ketoglutarate
dehydrogenase

NAD$^+$ + CoA-SH

NADH + H$^+$ + CO$_2$

fumarate

succinate
dehydrogenase

FADH$_2$

FAD

succinate

succinyl-CoA
synthetase

succinyl-CoA

GDP + P$_i$ + H$_2$O

GTP + CoA-SH

(a)

diates, which are subsequently reduced to give the fermentation product. During this reaction reducing equivalents (i.e. nicotinamide adenosine dinucleotide, NADH) are reoxidized for reuse in the oxidative part of the fermentation. ATP is only generated by substrate level phosphorylation without the involvement of ATP synthase. Therefore, the overall energy yield of fermentations is rather low and the end products are often still rich in energy.

During alcohol fermentation, yeast converts glucose to pyruvate in the oxidative part (i.e. glycolysis). This is also the moment when ATP is generated (Figure 1.4 and 1.5; phosphoglycerate kinase: 1,3-bisphosphoglycerate + ADP → 3-phosphoglycerate + ATP and pyruvate kinase: phosphoenolpyruvate + ADP → pyruvate + ATP). Next, instead of converting pyruvate into acetyl-CoA, pyruvate is decarboxylated by pyruvate decarboxylase to yield acetaldehyde (Figure 1.5). The last and reductive step of alcohol fermentation involves the reoxidation of NADH and the subsequent conversion of acetaldehyde to ethanol by alcohol dehydrogenase. Overall, alcohol fermentation of one molecule of glucose yields a total of each two molecules of ATP, carbon dioxide and ethanol. The regeneration of NADH is important to keep glycolysis running as this is the only way the organism is able to generate energy (ATP).

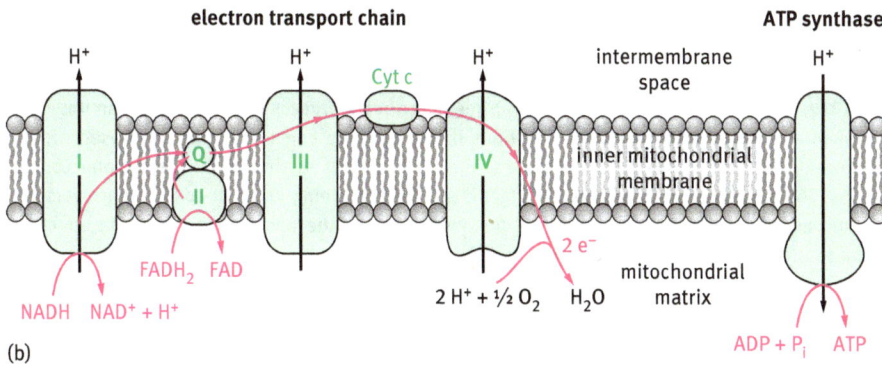

electron transport chain　　　　　　　**ATP synthase**

H^+　　　　H^+　　　Cyt c　　　H^+　　intermembrane　　H^+
　　　　　　　　　　　　　　　　　　　　space

I　　　Q　　III　　　　IV　　inner mitochondrial
　　　　II　　　　　　　　　　membrane

FADH$_2$　FAD　　　　　　　　　　2 e$^-$

NADH　NAD$^+$ + H$^+$　　　2 H$^+$ + ½ O$_2$　H$_2$O　mitochondrial
　　　　　　　　　　　　　　　　　　　matrix

(b)　　　　　　　　　　　　　　　ADP + P$_i$　ATP

Fig. 1.4: The central metabolic pathways for glucose utilization in yeast. (a) Glucose is first oxidized during glycolysis to the central metabolite pyruvate which is decarboxylated to acetyl-CoA. During the following citric acid (TCA cycle), acetyl-CoA is fully oxidized and the generated reducing equivalent shunt into the electron transport chain. (b) In a series of redox-reactions electrons are transported to the final oxidase which reduces oxygen. The concomitantly transported protons generate a proton gradient across the inner mitochondrial membrane which is used by ATP synthase to generate ATP. Glycolysis takes place in the cytoplasm of the yeast cell, pyruvate is subsequently transported into the mitochondrial matrix. The final electron transport chain is located in the inner mitochondrial membrane.

glucose

2 ADP + P_i

2 ATP

2 ethanol

2 NAD⁺

alcohol
dehydro-
genase

2 NADH

2 CO_2

2 pyruvate ——→ 2 acetaldehyde

pyruvate
decarboxylase

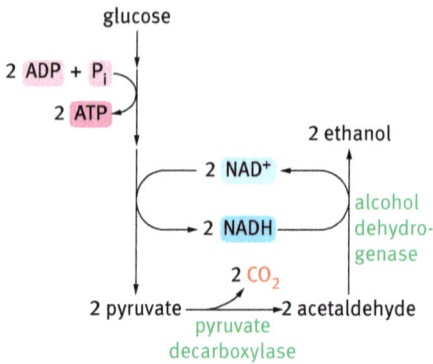

Fig. 1.5: Alcohol fermentation by yeast. In the oxidative part of the fermentation pathway, glucose is oxidized via glycolysis to pyruvate which is subsequently decarboxylated by pyruvate decarb-oxylase to acetaldehyde. In the following reductive part of the pathway, reducing equivalents are reoxidized through the reduction of acetaldehyde to ethanol by alcohol dehydrogenase.

Box 1.1: Pasteur effect.

In 1861, the French microbiologist Louis Pasteur discovered that yeasts grown under anaerobic conditions consumed more glucose than under aerobic conditions. The higher turnover of glucose under these conditions can be explained by the low energy yield under anaerobic conditions. While fermentation of glucose only yields two molecules of ATP, under aerobic condition and full oxidation, the yield is almost 13-fold higher. Higher anaerobic turnover of glucose is used to generate more ATP and in turn also yields more ethanol.

Box 1.2: Crabtree effect.

In 1922, the biochemist Herbert Grace Crabtree described a process by which yeast can undergo fermentation even under aerobic conditions if high glucose concentrations are present. High glucose concentrations (~ 100 mg/L) lead to the production of ethanol even under aerobic conditions. The mechanism of the Crabtree effect is still far from being understood and controversially discussed. In this regards, it is still unknown whether the onset of the Crabtree effect is due to the saturation of respiratory capacity or is the result of glucose-mediated repression of respiration (Vemuri et al., 2007).

1.3.1 Winemaking

The first step during winemaking or vinification is the crushing of the harvested grapes (Figure 1.6). For white wine, the crushed grapes are pressed and separated from stems, skin and pits (this residue is called pomace) yielding grape must. In contrast, crushed red grapes are directly fermented with only the stems being removed. One reason for this is that the red colored pigments (anthocyanins) are located in the skin of the grapes and are only solubilized once alcohol is produced.

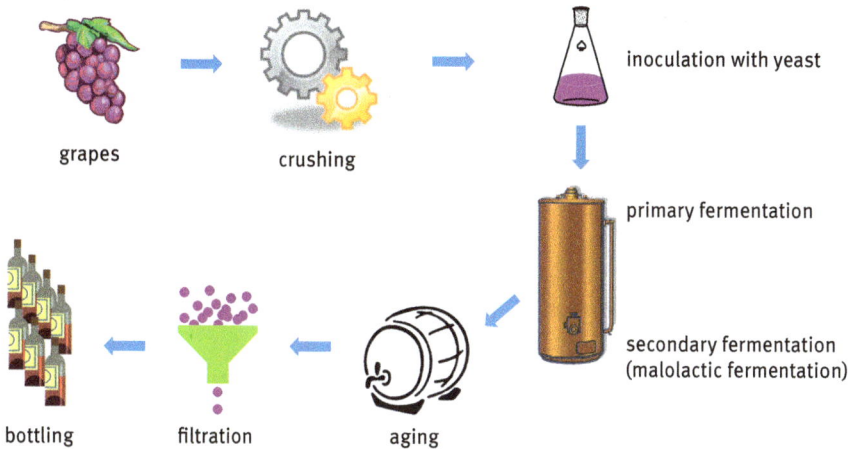

Fig. 1.6: The major steps of winemaking. See text for details.

The next step often involves the addition of sulfur dioxide. In commercial wine-making, this chemical is used as an antioxidant and antimicrobial agent to preserve wine quality and freshness. It can be added during various stages of vinification. Most commonly, it is first added to the crushed grapes to prevent oxidation of sensitive wine ingredients, enzymatic browning (oxidation of phenolic compounds by oxygen) and production of a sherry-like, nutty aroma which is due to the production of acetaldehyde. In addition, at this stage, sulfur dioxide has an inhibitory, antimicrobial effect on microorganisms present on the grape skins. It therefore, serves to reduce spoilage bacteria and yeast (*Acetobacter*, *Lactobacillus*, *Pediococcus* and *Brettanomyces*) and natural yeasts.

Primary fermentation is initiated by the addition of cultured yeast which ferments the grape sugars into ethanol and carbon dioxide. Initially, grapes store sugar in form of sucrose. During ripening however, the sucrose molecules are hydrolyzed by the enzyme invertase (β-fructosidase) into glucose and fructose. At the end, 15–25 % of the grape will be composed of simple sugars. For primary fermentation, glucose is the preferred substrate and is consumed first. Not all sugars present in the grape must can be fermented by yeast but will add to the flavor of the wine. Those sugars include arabinose, rhamnose and xylose. Up to 400 additional products are, furthermore, known to contribute to the flavor and taste of a wine.

For every gram of sugar that is converted, about half a gram of ethanol is produced. To achieve a 12 % alcohol concentration, the must should contain about 24 % sugars. In case the sugar content of the must is too low to achieve the desired alcohol concentration, sugar can be added. However, depending on the country several governmental restrictions apply on the amount of sugar allowed to be added.

The primary fermentation takes about 6–8 days and can be followed by a secondary, bacterial fermentation: the malolactic fermentation (see 1.5.6). Once the fermentations are completed, yeast is removed by filtering and the wine is transferred to (oak) barrels or steals tanks for maturation. This process can take anything from six months to three years and is especially important for the maturation of red wine. In a last step, the wine is bottled.

1.3.2 Beer brewing

Beer brewing involves microbial activity in every step of the process with the fermentation of cereal derived sugars being the major contributor (for an overview of beer brewing see Figure 1.7). According to the German beer purity law of 1516, beer may only contain four ingredients: water, hops, barley and yeast. Each brewery has its specific yeast strains which are employed in brewing. Furthermore, there are differences depending which type of beer is brewed. Bottom-fermenting yeasts are used for brewing Lager and Pilsener beer, while the German "Hefeweizen" (wheat beer) employs top-fermenting strains. Bottom-fermenting strains are exclusively pure culture breeding yeasts. Initially, the classification of top- and bottom-fermenters was introduced because in older days those where the regions in the brewing tank from where the yeasts were collected for the next brew. It then became clear that top- and bottom-fermenting yeasts are different species that prefer different temperatures (warmer vs. cooler temperatures). Brewing yeast strains are polyploid and contain 3–4 copies of each chromosome. The significance of this is still unknown but might generate more stable organisms and boost enzyme production, which lead to a more rapid metabolism.

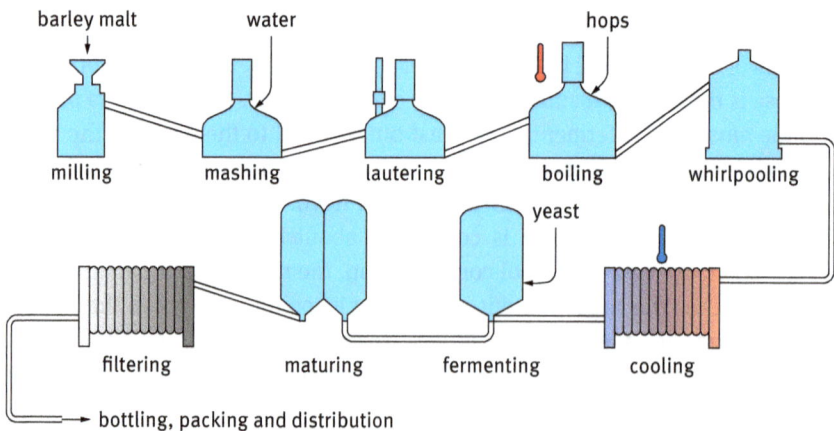

Fig. 1.7: Overview of the beer brewing process. See text for details.

Malting

In order to prepare barley grains for beer brewing they have to be converted into barley malt in a process called malting. Barley grains are soaked for two days in water to initiate germination. This induces the production of amylases which start to degrade the starch present in the grain into simple sugar molecules (mostly maltose; disaccharide composed of two glucose units). Malting is terminated by boiling to stop the germination process. Afterwards, the grains are dried in a process called kiln-drying.

Box 1.3: Starch.

Starch is a polysaccharide used as carbohydrate storage in plants. Chemically, starch is a "poly-glucose" composed of α-D-glucose molecules which are connected via α-1,4-glycosidic bonds in amylose and more branched via α-1,6-glycosidic and α-1,4-glycosidic bonds in amylopectin. Roughly, starch consists of 20 to 25 % amylose and 75 to 80 % amylopectin.

Mashing

In the following mashing procedure, the dried malt is crushed into fine powder and mixed with hot water. Within this process, the starch is further broken down and the resulting products (glucose, maltose, dextrines) are brought into solution.

Therefore, the mixture is heated to different temperatures to allow certain enzymes to work (process also known as enzymatic rest). The progress of the mashing is monitored by the brewer using the iodine test.

– 40 °C: β-glucanase (cellulose) breaks down β-glycosidic bonds of cellulose from the bran of the grain
– 50 °C: proteases break down proteins of grain into amino acids for yeast nutrition
– 62 °C: β-amylase break down starch from the end liberating maltose molecules
– 72 °C: α-amylase break down starch by breaking α-1,4-glycosidic bonds (brach points of starch)
– 78 °C: mash-out: Release of additional 2 % starch which makes the mash less viscous.

Lautering

During the process of lautering, the mash is filtered, resulting in a clear liquid called wort, and the residual grains.

Wort boiling

In the subsequent step, the wort is brought to a boil to sterilize the liquid and ensure the killing of unwanted contaminating bacteria. In addition, all enzymatic activities are terminated (enzymes are denatured) at this point and thereby the carbohydrate composition of the wort is fixed. The boiling liquid is then supplemented with flowers of the female hops plant. During the boiling process, alpha acids contained in the hops plant are solubilized and isomerized to iso-alpha acids, which are responsible for the bitter flavor of the beer. The bitterness of the beer can be adjusted through the boiling time as longer boiling times results in more iso-alpha acid production. One common iso-alpha acid found in beer is humulone. Iso-alpha acids not only add to the flavor of the beer but also possess bacteriostatic activity against contaminating Gram-positive bacteria (e.g. lactic acid bacteria, *Pediococcus* sp.).

Before fermentation can be initiated, solid particles from the hopped wort are separated in a process called whirlpooling. The solid particles within the wort will settle into a cone at the center of the whirlpool vessel (see also Figure 1.7).

Fermentation
Before fermentation can be initiated, the wort has to be cooled down to fermenting temperature and transferred to a fermentation vessel. Cultured yeast is added and fermentation starts. Initially, the yeast also consumes oxygen for cell proliferation and growth. Bottom-fermented beers are fermented at cooler temperature (7–15 °C) and fermentation takes between 8 and 10 days. Top-fermenting yeasts prefer somewhat warmer fermenting temperatures (20–26 °C) and fermentation takes only 4–6 days. Finally, after filtering, the beer is transferred into a new container and matured at low temperatures.

Filtering
After a final filtration step, the beer is bottled and distributed.

1.3.3 Side products during yeast fermentation

The major products during alcohol fermentation are ethanol and carbon dioxide. However, during fermentation other compounds are being generated which mainly contribute to the flavor of the fermented product. These include esters, fusel alcohols, ketones, various phenolics and fatty acids. Fruity notes in beer are due to esters while phenols cause the spicy and sometimes medicinal notes. The keton diacetyl gives a butter or butterscotch note. Since diacetyl is unstable and easily oxidizes, it produces stale, raunchy flavors as the must ages. Fusel alcohols are heavier molecular weight alcohols and are a main cause for hangovers. Fatty acids also tend to oxidize in old must and produce off-flavors.

1.4 Alcohol fermentation by bacteria

Although mainly restricted to yeast, alcohol fermentation has also been described for bacteria. The most prominent bacterium undergoing alcohol fermentation is *Zymomonas mobilis*. *Z. mobilis* is a Gram-negative, rod shaped bacterium, which is a facultative anaerobe. Therefore, *Z. mobilis* posseses both, respiratory and fermenting potential. *Zymomonas* species can easily be isolated from plant saps which are also the major nutrient source for the biotechnological use of *Zymomonas*. Together with *Saccharomyces cerevisiae*, *Z. mobilis* is involved in the fermentation of sugar of the Agave plant used for the production of Tequila. In contrast to *S. cerevisiae*, *Z. mobilis* does not use glycolysis for glucose oxidation but rather a pathway known as the Entner-Doudoroff-pathway (Figure 1.8). This pathway is also highly distributed among Pseudomonas species. Like in glycolysis, glucose is first activated by one molecule of ATP to generate glucose-6-phosphate. Glucose-6-phosphate is further oxidized to 6-phosphogluconate with the generation of NADPH; phosphogluconate is dehydrated to the characteristic intermediate 2-keto-3-deoxy-6-phosphogluconate (KDPG) and finally split KDPG-aldolase into glyceraldehyde-3-phosphate and pyruvate. The former is a key intermediate of glycolysis and further catabolized via glycolysis to generate NADH and two molecule of ATP. Due to the direct formation of pyruvate in the Entner-Doudoroff-pathway, the yield of ATP

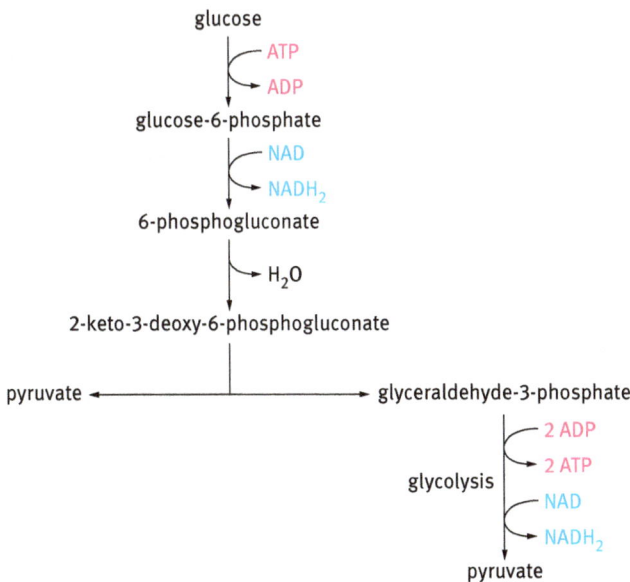

Fig. 1.8: The Entner-Doudoroff-pathway is an alternative route for glycose oxidation used by various bacteria. During alcohol fermentation this pathway is active in the bacterium *Zymomonas mobilis* which is employed for making Tequila. Sometimes the pathway is also designated KDPG-pathway after its intermediate 2-keto-3-deoxy-6-phosphogluconate.

is lower than in glycolysis. Pulque is an old mexican beverage obtained through fermentation of agave juice by a mixture of yeast and bacteria and the precursor of Tequila.

For real Tequila, only the agave species *Agave tequilana* may be used (Cruz and Alvarez-Jacobs 1999). Only alcoholic beverages made with Blue Agave or Agave Azul (*Agave tequilana* Weber blue variety) grown in the states of Jalisco, Minoacán, Nayarit, Guanajuato, and Tamaulipas are allowed to be labeled Tequila. The Blue Agave is high in inulin (see box 1.4) which serves as the carbon source during fermentation. However, in order to degrade this polysaccharide into smaller sugar units, the agave plant has to be cut and cooked to hydrolyze the inulin. The cooking of the agave has several reasons: At a low pH (pH 4.5) and high temperature, the inulin and other compounds are hydrolyzed. Furthermore, cooking makes the agave softer which facilitates the following milling step. The fermentation of the Agave wort is comparable to the fermentation of beer. In order to obtain the high alcohol content of the final beverage (> 38 % (v/v) and < 55 % (v/v)), the fermented Agave wort is distilled. During this process, the alcohol is separated from the wort and concentrated. In a first distillation, the alcohol concentration of the wort is increased to 20–30 % (v/v). Here, the first fraction called heads and the last fraction called tails are separated out. The head fractions are usually enriched in low boiling point compounds like acetaldehyde, ethylacetate, methanol, propanol, butanol and 2-methyl propanol. These compounds are responsible for the pleasant flavor and taste of the Tequila. The tails, on the other hand, contain high boiling point compounds which are strong in taste and flavor. Therefore, this fraction is not used further. Next, the liquid from the first distillation step is re-distilled. The final product contains between 40 % (v/v) and 55 % (v/v) ethanol. For silver or white Tequila, this is the final production step. Other Tequilas, like rested or aged Tequila, are aged in oak barrels for a minimum of 2 month. In a final step, the Tequila is filtered prior to bottling.

Box 1.4: Inulin.

Inulin is a polysaccharide found in many plant types and belongs to the dietary fibers and the class of fructanes. Structurally inulin is heterogeneous and consists of glucose and fructose moieties connected via β-2,1 glycosidic bonds. The degree of polymerization ranges from 2 to 60. Since the human digestive system does not possess enzymes to break down β-2,1 glycosidic bonds, inulin can be used to reduce caloric intake or as a dietary fiber.

1.5 Lactic acid fermentation

1.5.1 Yoghurt

Fermented dairy products have been popular for a very long time in Mediterranean countries but also in central and southeast Asia and central Europe. One of the major products obtained from fermenting milk is yoghurt. Thermophilic lactic acid bacteria of the genera *Streptococcus salivarius subsp. thermophilus* (*S. thermophilus*) and *Lactobacillus delbrueckii subsp. bulgaricus* (*L. bulgaricus*) are among the traditional bacteria used in homofermentative lactic acid fermentation. Both have an optimal growth temperature around 50 °C (Zourari et al., 1992).

Box 1.5: Probiotic drinks and yoghurt.

In the early 1990s, propiotic drinks (Actimel®, Yakult®) and yoghurt (Activia®) entered the European market. In addition to the traditional lactic acid bacteria, these probiotic products employ special strains: *Lactobacillus casei* DN-114001 (Actimel®), *Lactobacillus casei* Shirota (Yakult®) and *Bifidobacterium animalis* DN 173 010 (Activia®).

Consumption of these probiotic dairy products is claimed to have several beneficial effects on human health, including maintenance of gut flora, modulation of the immune system, regulation of bowel habits, and constipation, and also effects on some gastrointestinal infections.

In 2010, the European Food Safety Authority (EFSA) determined that a cause and effect relationship has not been established between the consumption of *Lactobacillus casei* strain Shirota and maintenance of the upper respiratory tract defense against pathogens by maintaining immune defenses.

During milk fermentation, large amounts of lactic acid (lactate) are produced from lactose. The increase in lactic acid leads to the acidification of milk. Subsequently, the reduced pH causes the progressive solubilization of micellar calcium phosphate, resulting in demineralization of casein micelles and destabilization, which finally leads to the precipitation of casein. The acidification, furthermore, prevents the growth of harmful bacteria.

Box 1.6: Casein.

Milk protein and a mixture of different proteins: αS1, αS2, β and κ-casein. Casein forms micelles together with calcium phosphate and is the major nitrogen source for lactic acid bacteria during lactic acid fermentation. To metabolize casein, the bacteria excrete extracellular proteinases, membrane bound aminopeptidases and intracellular proteinases and peptidases. Cow milk contains ~2.6 % (w/v) casein.

Lactic acid bacteria are Gram-positive cocci and facultative anaerobes. As they are unable to synthesize porphyrines (i.e. heme), they neither do possess cytochrome c nor catalase, both enzymes critical for an aerobic lifestyle. Lactic acid bacteria utilized in yoghurt production perform a homofermentative lactic acid fermentation which leads to lactic acid (lactate) as the major product. Acetate, diacetyl, and acetoin are produced in small amounts as side products and give the specific flavor of the dairy product. The major carbon source utilized in milk fermentation is lactose (IUPAC name: 4-O-(β-D-Galactopyranosyl)-D-glucopyranose) which is a disaccharide composed of D-glucose and D-galactose with a β-1,4 glycosidic bond. Metabolism of lactose requires two things: 1. The lactose has to be transport-

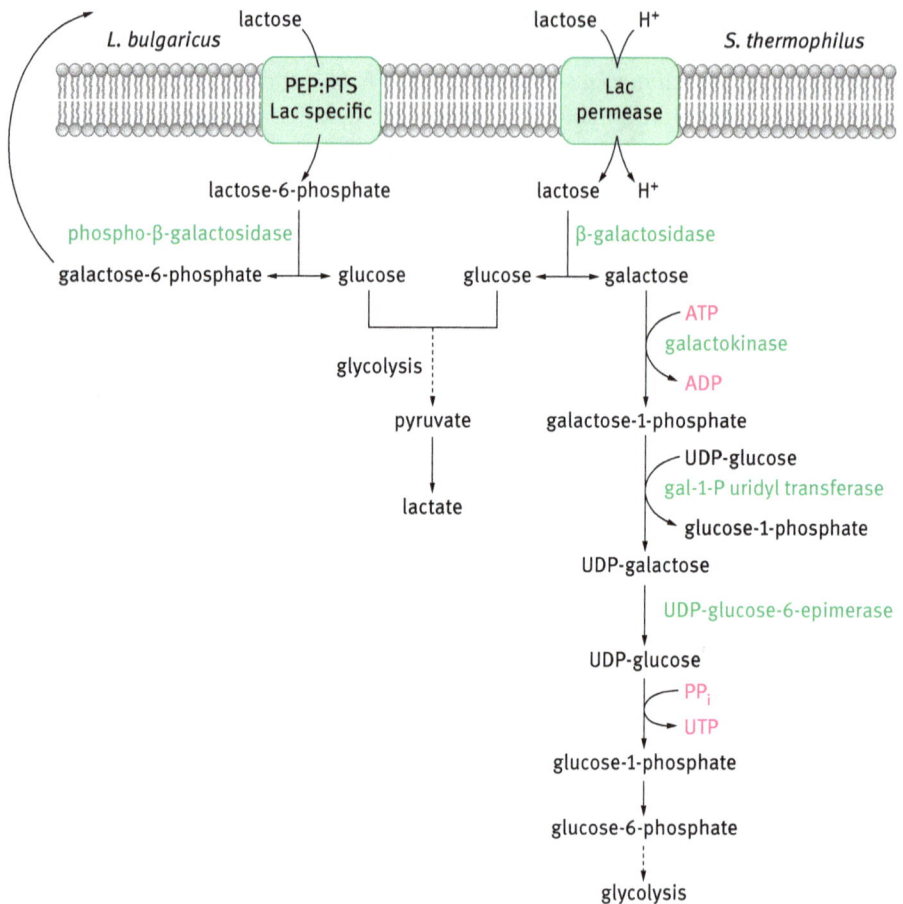

Fig. 1.9: Utilization of lactose by lactic acid bacteria. *L. bulgaricus* uses a phosphoenolpyruvate-dependent phosphotransferase (PEP:PTS) importer which phosphorylates lactose upon import. *S. thermophilus* on the other hand employs lactose permease to import lactose. Both lactose-6-P and lactose are cleaved by the respective β-galactosidase, and the glucose moiety is fermented to lactate. The galactose is either exported as galactose-6-P (*L. bulgaricus*) or converted to glucose-6-P and the channeled into glycolysis and lactic acid fermentation (*S. thermophilus*).

ed into the bacterial cell and 2. The lactose has to be cleaved into the monosaccharides to enter the central metabolic pathways. Lactococci and Streptococci use different transport systems to move lactose into the cell. Lactococci utilize a phosphoenolpyruvate (PEP)-dependent phosphotransferase (PTS) system by which lactose is phosphorylated to lactose-6-phosphate during transport (Figure 1.9). Lactose-6-phosphate is intracellularly hydrolyzed to glucose and galactose-6-phosphate by phospho-β-galactosidase. Interestingly, *L. bulgaricus* uses only the glucose moiety of lactose in lactic acid fermentation while galactose-6-phosphate is released into the growth medium. The glucose moiety of lactose directly enters glycolysis (Figure 1.4). Other strains also metabolize the galactose moiety, which is first converted via galactose-1-phosphate and UDP-galactose to the respective glucose derivatives which then enter glycolysis as glucose-6-phosphate. Finally, the reducing equivalents NADH which are produced in glycolysis by glyceraldehyde-3-phosphate dehydrogenase are reoxidized through the reduction of pyruvate to lactate by the enzyme lactate dehydrogenase. Fermentation of one hexose to two molecules of lactate yields 2 molecules of ATP which are produced in the oxidative part of the pathway (i.e. glycolysis). Depending on the stereospecificity of lactate dehydrogenase, L(+)- or D(−)-lactic acid or a mixture of both stereoisomers is produced. D(−)-lactic acid is only slowly metabolized in humans and may cause metabolic disorders if consumed in excess. Therefore, the world health organization recommends the daily consumption of D(−)-lactic acid not to exceed 100 mg/kg body weight. Regular yoghurt usually contains both stereoisomers (Table 1.1).

1.5.2 Kefir

Kefir is a traditional milk-based beverage originating from the north Caucasus mountains and Tibet. Heterofermentative lactic acid bacteria like *Lactococcus lactis* and *Lactobacillus acidophilus*, together with the yeast *Kluyveromyces marxianus* and some acetic acid bacteria, ferment milk to lactate, ethanol, and carbon dioxide. The microorganisms form so-called Kefir grains which contain a complex mixture of the microorganisms in a matrix with an appearance of cauliflower buds. At low temperatures, the fermentation process is dominated by the yeast, producing ethanol and carbon dioxide. At higher temperatures, the lactic acid bacteria are more active, leading to a higher lactate content of the final product. Heterofermentative lactic acid fermentation differs from the homofermentative pathway in the way sugars are oxidized. Heterofermentative lactic acid bacteria, including some species of *Lactobacillus* and *Leuconostoc*, metabolize sugars via the pentose phosphate pathway in which glucose is oxidized to 6-phosphogluconate (Figure 1.10). Subsequently, this intermediate is decarboxylated to a pentose (ribulose-5-phosphate). The key enzyme phosphoketolase then cleaves pentose phosphate into glyceraldehyde-3-phosphate and acetyl phosphate. While glyceraldehyde-3-phosphate is further metabolized via glycolysis to pyruvate and subsequently to lactate, acetyl phosphate is reduced to ethanol in a two-step reaction. This part of the pathway reoxidizes NADH with no additional yield of ATP. To netto yield from one molecule of glucose is one molecule each of lactate, ethanol, CO_2 and ATP. Pentoses on the other hand will be fermented via xylose-5-phosphate to acetyl phosphate and glyceraldehyde-3-phosphate. Acetyl phosphate is converted to acetate with the yield of an additional molecule of ATP.

glucose

$\quad$ ATP

$\quad$ ADP

glucose-6-phosphate

$\quad$ NAD

$\quad$ NADH$_2$

6-phosphogluconate

$\quad$ NAD

$\quad$ NADH$_2$

CO_2 ←———————→ pentose phosphate

$\quad$ P$_i$

glyceraldehyde-3-phosphate ←———→ acetyl phosphate

P$_i$ — NAD $\qquad$ NADH$_2$

$\quad$ NADH$_2$ $\qquad$ P$_i$ — NAD

1,3-bisphosphoglycerate $\qquad$ acetaldehyde

$\quad$ ADP $\qquad$ NADH$_2$

$\quad$ ATP $\qquad$ NAD

glycolysis $\quad$ 3-phosphoglycerate $\qquad$ ethanol

$\quad$ H$_2$O

phosphoenolpyruvate

$\quad$ ADP

$\quad$ ATP

pyruvate

$\quad$ NADH$_2$

$\quad$ NAD

lactate

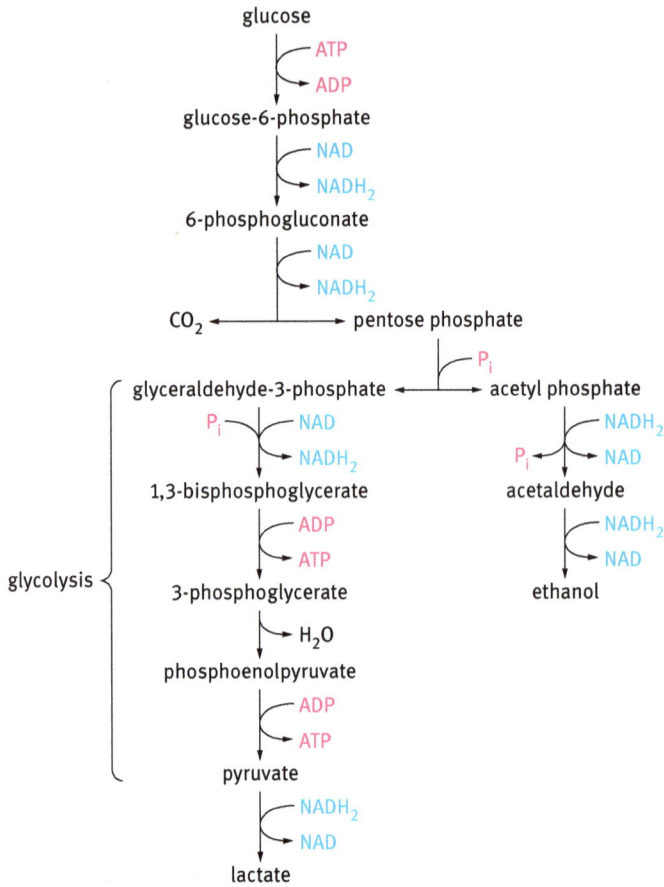

Fig 1.10: Heterofermentative lactic acid fermentation. This pathways utilizes the pentose phosphate pathway to oxidize glucose. The key enzyme phospho-ketolase cleaves the pentose phosphate to glycealdehyde-3-phosphate and acetyl phosphate. Glyceraldehyde-3-phosphate enters the last steps of glycolysis where it is oxidized to pyruvate with the generation of ATP (substrate level phosphorylation). In the final reductive step, NADH is reoxidized with the generation of lactate. The second branch of the pathway reduces acetyl phosphate via acetaldehyde to ethanol. It is only regenerating NAD$^+$ with no additional yield of ATP. However, additional ATP is generated when pentoses are metabolized and acetate is formed.

1.5.3 Cheese

Either homo- or heterofermantative lactic acid bacteria are used for the production of cheese depending on the type of cheese to be made. Homofermentaive lactic acid bacteria are used for cheese like cheddar where a clean acid like flavor is required. To obtain more fruity flavors like in Emmental cheese, heterofermentative lactic acid bacteria are employed. In case a mould-ripened cheese is made (Camembert, Roquefort), fungal spores are added (either to the milk or later to the cheese curd) (see also chapter 4).

After some time of lactic acid fermentation when sufficient lactate has been produced, rennet is added to cause the casein to precipitate. This is due to the presence of the enzyme chymosin which converts κ-casein to para-kappa-caseinate (the main component of cheese curd) and glycomacropeptide. The latter is lost in the cheese whey. During this process, cheese curd is formed and the milk fat is trapped in a casein matrix. The next step in cheese making involves drainage of the "cheese water" (whey) and ripening. During this process, microorganisms produce metabolites which are responsible for the flavor of the final cheese product.

Box 1.7: Chymosin or rennin.

Chymosin/rennin is an aspartate endopeptidase that is commercially isolated from stomachs of newborn calves. Attempts to produce recombinant bovine chymosin in bacteria or molds (genera *Aspergillus*) were already successful, however, applications to use it in cheese production failed due to consumer concerns.

One example is Emmental cheese, which gets its additional flavor and "eyes" during a ripening process where propionic acid bacteria (for example *Propionibacterium freudenreichii*) ferment lactose in a secondary fermentation to propionate, acetate, and carbon dioxide. At the beginning, the produced carbon dioxide is bound by water. However, once the water is saturated with carbon dioxide, gaseous carbon dioxide is released. As the cheese has already developed a rind, the gas can no longer diffuse and accumulates which in turn generates the cavities (eyes).

1.5.4 Sourdough

Sourdough likely originated in ancient Egyptian times around 1,500 BC and was likely the first form of leavening available to bakers. Sourdough remained the usual form of leavening down into the European Middle Ages until being replaced by barm (yeast) from the beer brewing process, and then later purpose-cultured yeast.

Various lactic acid bacteria (*Lactobacillus sanfranciscensis, L. plantarum, L. brevis* and *L. fermentum*) together with yeasts (*Saccharomyces cerevisiae, S. exiguus Candida humilis*), build the microbiological flora of sourdough. They constitute a stable symbiotic culture in which the lactic acid bacteria are responsible for fermenting the sugars originating from the starch of the flour producing lactate for flavor. The yeast, on the other hand, ferments the sugar to produce high amounts of carbon dioxide which leavens the dough.

Maltose (degradation product of starch; glucose-glucose (α-1,4)) is the preferred carbon source for *L. sanfranciscensis* and metabolized in a heterofermentative lac-

Table 1.1: Lactic acid bacteria used in food biotechnology.

Species	used in	product
Streptococcus salivarius subsp. *thermophilus*	yoghurt	homofermentative L(+)-lactic acid
Lactobacillus delbrueckii subsp. *bulgaricus*	yoghurt	D(−)-lactic acid
Lactococcus lactis	kefir, buttermilk	homofermentative L(+)-lactic acid
Lactobacillus acidophilus	kefir	homofermentative L(+)-lactic acid
Lactobacillus plantarum	sourdough	homofermentative
L. brevis	sourdough	heterofermentative
L. sanfranciscensis	sourdough	heterofermentative
Oenococcus oeni	winemaking malolactic fermentation	heterofermentative L(+)-lactic acid
Leuconostoc mesenteroides	sauerkraut	Heterofermentative D(−)-lactic acid

tic acid fermentation yielding lactate, acetate, ethanol and CO_2 (Figure 1.10, Table 1.1).

Obligate and facultative heterofermentative lactic acid bacteria can switch between the two pathways depending which sugars are available. Pentoses are metabolized via the heterofermentative pathway, while hexoses enter glycolysis and the homofermentative pathway.

1.5.5 Sauerkraut

Sauerkraut as a traditional German food is made from white cabbage. The cabbage is planed and layered together with salt (sodium chloride) into a clay pot which is sealed to generate anaerobic conditions. The planning of the cabbage releases plant saps and naturally occurring lactic acid bacteria (*Leuconostoc* species) will ferment the sugars therein to D(−)-lactate via a heterofermentaive lactic acid fermentation. The high salt content inhibits the growth of unwanted spoilage bacteria. In a similar way, grass is conserved for cattle feeding (silage).

1.5.6 Malolactic fermentation by lactic acid bacteria during winemaking

Malolactic fermentation in winemaking is a process to deacidify wine to obtain a softer taste. Usually, at the end of the alcohol fermentation, the wine is inoculated with cultures of lactic acid bacteria of the genus *Oenococcus oeni*. Alternatively, the reaction can also be catalyzed by certain strains of *Lactobacillus* and *Pediococ-*

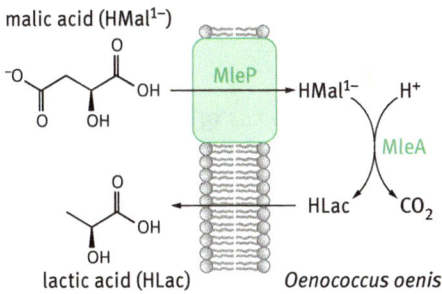

malic acid (HMal^{1-})

lactic acid (HLac)

Oenococcus oenis

Fig. 1.11: Secondary proton motive force-generating system via malate/lactate exchange during malolactic acid fermentation. At low pH (pH 4), malic acid is present in its monoanionic malic acid (HMal^{1-}) form. Upon transport through the malate transporter (MleP) into the higher pH of the bacterial cytoplasm, a net-negative charge is generated, which creates an electrical potential. The decarboxylation of malate into L-lactic acid by malolactic enzyme (MleA) leads to the release of carbon dioxide and a proton, which generates the pH gradient to produce ATP.

cus. During malolactic fermentation, one of the major wine acids L-malic acid (Grape must contains approximately 5 g/L of malic acid) is decarboxylated to L(+)-lactic acid and CO$_2$ is released. By definition, this process is not a fermentation but rather a decarboxylation, catalyzed by soluble decarboxylases. However, the bacteria are able to conserve energy during this process by chemiosmosis. At the low pH of wine (~ pH 4), malic acid is present in its anionic form which is transported into the cell where it is decarboxylated and concomitantly protonated to lactic acid (due to the higher pH in the cytoplasm) (Figure 1.11).

The latter is the exported, which results in the generation of an electrochemical gradient (negative inside) and a netto loss of protons on the inside. This electrochemical gradient can subsequently be used by F$_1$F$_0$-ATPase to generate ATP. However, the ATP produced during this process is not sufficient for their growth. Therefore, the heterofermentative lactic acid bacteria still require the fermentation of pentoses and hexoses.

At the end of the alcohol fermentation, yeasts are also able to convert malic acid to lactic acid in a NADH-dependent reaction. Lactic acid is, furthermore, decarboxylated to ethanol and subsequently reduced to ethanol.

Although the presented examples for food biotechnology represent only a selection and are far from being complete, it was intended to give the reader an overview of the different processes employed by microorganisms for the production of food.

Key-terms

Alcohol fermentation, beer, central metabolic pathways, cheese, Entner-Doudoroff-pathway, food biotechnology, lactic acid fermentation, malolactic fermentation, Sauerkraut, wine, yeast, yoghurt

Questions

- Who built the first microscope?
- Which metabolic pathway is used by *Zymomonas mobilis* to oxidize glucose?
- Which are the ingredients of beer according to the German beer purity law?
- What happens during malolactic fermentation and where is it important?
- What is the difference between homofermentative and heterofermentative lactic acid fermentation?
- What causes the cavities in Emmental cheese?
- Explain the Crabtree effect.

Further readings

Renneberg, R. 2008. Biotechnology for Beginners. *Academic Press, Waltham, Massachusetts.*
Berg, J., Tymoczko, J. L., Gatto, G. J. 2015. Biochemistry, 8th ed, *Macmillan Higher Education, London.*
Slonczewski, J. L., Foster, J. W. 2011. Microbiology: An Evolving Science, 2nd ed, *W. W. Norton & Company, New York.*
For german speaking students also recommended:
Fuchs, G. 2014. Allgemeine Mikrobiologie, 9th ed, *Thieme Verlag, Stuttgart, New York.*

References

Cruz, M. C. & Alvarez-Jacobs J. 1999. Production of tequila from agave: historical influences and contemporary processes. The alcohol textbook: *a reference for the beverage, fuel and industrial alcohol industries* (3rd ed), Nottingham University Press, Alltech Inc.
Herskowitz, I. 1988. Life cycle of the budding yeast *Saccharomyces cerevisiae. Microbiol Rev* 52(4): 536–553.
Vemuri, G. N., Eiteman, M. A., McEwen, J. E., et al. 2007. Increasing NADH oxidation reduces overflow metabolism in *Saccharomyces cerevisiae. Proc Natl Acad Sci* 104(7): 2402–2407.
Zourari, A., Accolas, J. P. & Desmazeaud, M. J. 1992. Metabolism and biochemical characteristics of yogurt bacteria. A review. *Lait* 72(1): 1–34.

Robert Kourist and Sarah Gaßmeyer

2 Biocatalysis and enzyme engineering

This chapter gives an introduction to the industrial application of enzymes. The first part shows with the example of lipases how the selectivity of enzymes makes them attractive catalysts, and how this can be exploited for sustainable syntheses. As enzymes often do not fulfill all requirements for technical applications, they often have to be optimized. The second part is dedicated to protein engineering, a method for the generation of enzyme variants with tailor-made catalytic properties for technical applications.

Enzymes are the catalysts of nature. Evolution optimized them over millions of years to mediate a wide diversity of reactions in a multitude of biochemical pathways. More than a hundred years ago, it was discovered that enzymes outside of the living cell can catalyze chemical reactions. Since then, enzymes have been applied for a number of industrial processes. The fact that enzymes are active in aqueous solution and at mild reaction conditions makes them highly desirable catalysts for environmentally benign processes. Due to their excellent selectivity, biocatalysts are widely applied for the synthesis of fine chemicals. Most interestingly, enzymes can catalyze many reactions that are very challenging for traditional chemistry. Wide application of enzymes as catalysts in organic chemistry started in the 1980s and has resulted so far in various industrial applications. Nonetheless, the industrial application of enzymes has some limitations, which are related to insufficient activity, stability, and selectivity under process conditions. An impressive development of new methods in biocatalysis, together with considerable advances in molecular biology, may help to overcome these limitations: The cloning and heterologous production of enzymes make the redesign of the biocatalysts feasible and practicable. Protein engineering techniques allow to modify features like temperature- and pH-stability and selectivity and make the enzymes fit for industrial applications. In addition, the progress in genomics in combination with efficient high-throughput-screening facilitates the identification of new biocatalysts. The easy access to a wide diversity of biocatalysts shortens the time that is necessary to develop a new process, which is essential in order to make biocatalysis a profitable method to conduct industrial chemical reactions.

2.1 Introduction to biocatalysis

Several enzymes are produced in industrial scale, with proteases, amylases, and lipases exceeding the 10^4 t scale. These enzymes are used as industrial catalysts, but also in detergents, procession of leather, and food production. Table 2.1 shows a selection of industrial biocatalytic processes. Isolated enzymes are usually immo-

Table 2.1: Selection of Large scale industrial processes (Breuer and Hauer 2011, DiCosimo, McAuliffe et al. 2013).

Product	Biocatalyst	Scale (tons per year)
a) Immobilized enzymes		
High fructose corn syrup from corn syrup	Glucose isomerase	10^7
Acrylamide from acrylonitrile	Nitrile hydratase	10^5
Transesterification of food oils	Lipase	10^5
Lactose hydrolysis, GOS synthesis	Lactase	10^5
Biodiesel from triglycerides	Lipase	10^4
Antibiotic modification	Penicillin G acylase	10^4
L-Aspartic acid from Fumaric acid	Aspartase	10^4
Aspartame synthesis	Thermolysin	10^4
Chiral resolution of alcohols and amines	Lipase	10^3
b) fermentative processes		
L-glutamic acid	*Corynebacterium glutamicum*	10^6
L-lysine	*Corynebacterium glutamicum*	10^5
Vitamin C	*Aspergillus niger*	10^4
Vitamin B2	*Ashbya gossypii*	10^3
Antibiotics	*Penicillium chrysogenum and others*	$>10^4$

bilized on carriers. This improves the stability under process conditions and allows reuse of the enzymes. Box 2.1 gives an overview on important parameters for a successful biocatalytic process.

Box 2.1: Important parameters for technical enzymes.

The molar yield of a reaction is defined as moles product per moles substrate:

$$\% \ yield = \frac{moles_{product}}{moles_{substrate}}$$

This parameter indicates if the conversion of the substrate to the product was complete, but does not give any information on the amount of catalyst and the demand for reactors, which causes cost for installation, maintenance, and purification. The space time yield (STY) indicates these parameters and is, therefore, preferred to judge the usefulness of a process. Obviously, it is desirable to conduct a process fast and with a high product concentration.

$$STY = \frac{product}{reaction \ volume \times reactiontime} = \left[\frac{g}{L \times h}\right]$$

Increasing the amount of catalyst accelerates the reaction and improves STY, but causes higher cost for the catalyst. The productivity indicates product formation per amount of catalyst. Obviously, immobilization and 10-fold reuse without activity loss results in a 10-fold improved productivity. Consequently, most industrial enzymes are used in immobilized form.

$$productivity = \frac{product}{catalyst \ amount} = \left[\frac{g}{mg}\right]$$

In practice, enzymes are often sold as cell powders or in immobilized formulation and it is often difficult to determine the exact molar concentration of an enzyme. The number of molar conversions serves to quantify the amount of enzyme. An enzyme unit is defined as the molar product formation per min.

Enzyme unit U: $U = \dfrac{\mu mol}{min}$

Specific activity: $\dfrac{U}{mg} = \dfrac{\mu mol}{min \times mg}$

2.1.1 Lipases – robust enzymes for industrial biotechnology

Lipases belong to the most-used industrial enzymes (Table 2.1). Vegetable oils and animal grease mainly contain triglycerides that consist of three fatty acids bound as carboxylesters to a glycerol molecule (Figure 2.1a). Lipases are hydrolytic enzymes with the ability to cleave triacylglycerols, the main component of natural lipids. Lipolytic enzymes play an important role in the degradation of fats. Because lipid droplets cannot be imported into the cell, many bacteria and fungi excrete lipases to degrade insoluble lipids (compare also section 4.3.2). The released fatty acids are then imported and metabolized. As exoenzymes, lipases have outstanding stability and are even active in organic solvents or a water-lipid interfaces. Consequently, the most widely used lipases are lipase A from the yeast *Pseudozyma antarctica* (CAL-A) and the lipase from the bacterium *Burkholderia cepacia*. Immobilization on carriers can further increase this stability.

The α/β-hydrolase-fold
Lipases are characterized by a common structural motif. This so-called α/β-hydrolase-fold consists of a central β-sheet surrounded by several α-helices. Interestingly, lipases share this structural fold with several other enzyme classes that have very different catalytic mechanisms, including nitrilases, dehalogenases, and epoxid-hydrolases. The α/β-hydrolase-fold shows how evolution used a structural scaffold for the creation of different biocatalytic activities. All lipases have a catalytic triad consisting of a serine, a histidine, and an aspartate or a glutamate. This catalytic motif is extremely conserved throughout the vast and highly diverse α/β-hydrolase-fold superfamily. Because the serine plays a decisive role in the mechanism, lipases belong to the class of serine-hydrolases.

Mechanism of lipases and other serine-hydrolases
Comparison of the lipase mechanism to chemical hydrolysis of esters shows how skillful nature combines different catalytic elements to accelerate a reaction that

(a) natural function of lipases

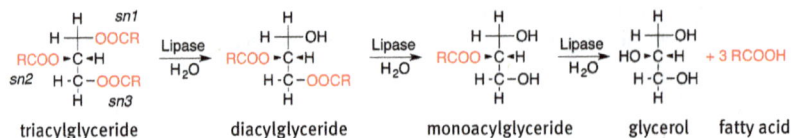

triacylglyceride   diacylglyceride   monoacylglyceride   glycerol   fatty acid

(b) mechanisms of acid- and base-catalyzed chemical hydrolysis

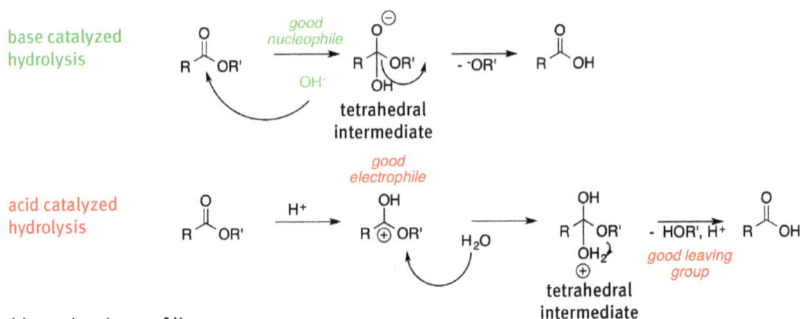

(c) mechanisms of lipases

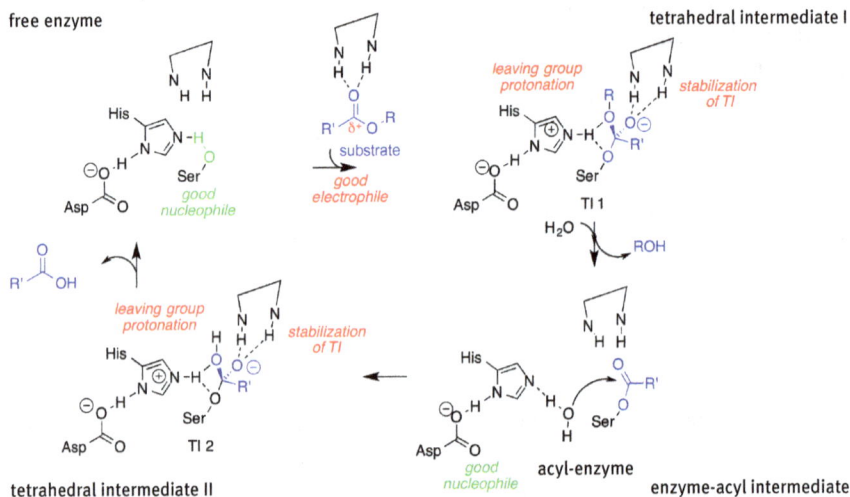

Fig. 2.1: (a) The natural reaction of lipases; (b) chemical mechanisms of ester hydrolysis; (c) mechanism of lipase-catalyzed ester hydrolysis. Base catalysis (green) and acid catalysis (red) are highlighted, (blue) substrate.

otherwise would occur extremely slowly: Under very basic or very acidic conditions, carboxyl esters hydrolyze rapidly (Figure 2.1b). Because lipids were traditionally hydrolyzed using plant ashes (with a high content of potassium hydroxide) as natural catalysts for soap manufacture, this process is often called saponification. At neutral pH, however, hydrolysis occurs extremely slowly. Interestingly, lipases catalyze a fast hydrolysis in neutral solutions. Under basic conditions, the concentration of hydroxyl ions (OH⁻) is high. OH⁻ ions are very good nucleophiles, meaning that they can donate easily an electron pair to form a covalent bond. The nucleophilic attack leads to the formation a tetrahedral intermediate. Collapse of this tetrahedral intermediate releases the alcohol anion (RO⁻). Because of the reduced stability due to the charge on the oxygen atom, this release is rather slow.

The alcohol anion is thus considered a bad leaving group. In acid catalysis, a water molecule performs the nucleophilic attack on the carboxyl ester. Neutral water is a much weaker electron donor than the hydroxyl anion. Nevertheless, the carboxylic oxygen atom is protonated, which increases its electrophilicity, i.e. the ability to accept electrons to form a bond. In the resulting tetrahedral intermediate, a proton migrates from the water molecule to the alcohol moiety of the carboxylic ester. After collapse of the tetrahedral intermediate, a neutral alcohol is released. Obviously, a neutral alcohol is a much better leaving group than the alcoholate anion that was the leaving group in base catalysis. In conclusion, the underlying principle of base-catalyzed saponification is nucleophile activation, while acid-catalyzed ester hydrolysis occurs via electrophile activation and leaving group protonation.

A closer look on the mechanism of serine-hydrolases (Figure 2.1c) shows how lipases combine all these mechanisms in order to accomplish ester hydrolysis in neutral solution: The three amino acids of the catalytic triad are connected by hydrogen bonds. This results in a shift of electrons from the acidic residue to the histidine, thus increasing its strength as base. Hydrogen bonding between the catalytic serine and the histidine leads to a nucleophile activation similar to that in base-catalyzed hydrolysis. Lipases contain a group of two or three hydrogen bond donors in their active center. The hydrogen bond donors of this so-called oxyanion hole are either nitrogen atoms of the peptide backbone or amino acid side chains such as threonine or aspartate.

Hydrogen bonds between the oxyanion hole and the carboxyl group of the substrate lead to a simultaneous electrophile activation similar to that of the acid-catalyzed hydrolysis. The resulting tetrahedral intermediate bears a negative charge: an oxyanion. This is stabilized by hydrogen bonding with the oxyanion hole. The histidine binds the hydrogen atom from the serine and transfers it to the alcohol moiety of the ester. The protonated alcohol can leave a neutral alcohol, which is a much better leaving group than the charged alcoholate anion in base catalysis. Leaving group protonation resembles acid-catalyzed ester hydrolysis. After elimination of the alcohol, the acid is still bound to the catalytic serine. Nucleophilic attack of a water molecule, activated by the histidine, initiates the second part of the catalytic cycle, which results in the release of the free acid and the free enzyme. Comparison of the mechanism of the lipase with acid- and base-catalyzed hydrolysis shows that lipases apply simultaneously catalytic elements from both. This makes ester hydrolysis at neutral pH possible.

2.1.2 Regioselective enzymatic interesterification of lipids

In contrast to chemical hydrolysis, lipases show a high selectivity in substrate recognition. This selectivity is used for the processing of nutritional lipids. As explained above, fats and oils are composed of triacylglycerids. The length, the degree of unsaturation, and the position of the fatty acid on the glycerol determine the physical properties of the fat such as the melting point. Most food products require a melting point where the fat melts in the mouth, not in the hands of the customer. Unfortunately, the amount of natural fats with this melting point is limited. The mixture of two fats with different melting points (so-called 'blending') is not feasible as the different triacylglycerids separate and form distinct phases. Traditionally, controlled reduction of the double bonds of a fat with a high content of unsaturated fatty acids was a cheap and efficient alternative to control the melting point. However, this hydrogenation with H_2 and metal catalysts leads to the accumulation of *trans*-fatty acids. *Trans*-fatty acids pose a high risk for coronary heart diseases and have to be avoided. A recently established process for the synthesis of *trans*-free fats is the interesterification. Figure 2.2 shows the possible combination of the chemical interesterification of two fats with different fatty acids. While the first hast palmitic acid in position 1 and 3 and oleic acid in the second, the second fat contains stearic and oleic acid. Chemical interesterification bases on the fact that the hydrolysis of fatty acid esters is reversible. In reaction conditions with a catalyst and a minute amount of water, fatty acids are cleaved and re-esterified, resulting in a random distribution of the fatty acids. The chemical reaction does not distinguish between the different positions of the fatty acids on the glycerol molecule, and the interesterification results in 18 possible products. All of them have different physical properties. Several lipases, however, show a high regioselectivity towards triacylglycerids (see Box 2.2). They can discriminate between fatty acids bound to the 1,3 position of glycerol and the 2 position, respectively. In the example shown in Figure 2.2, the lipase cleaves and re-esterifies only stearic and palmitic acid and leave oleic acid unreacted. The resulting product mixture contains only three possible variants. The reaction is conducted in the presence of only small amounts of water, as water would shift the equilibrium towards hydrolysis of the fats. This example shows that the use of lipases allows to avoid the accumulation of harmful *trans*-fatty acids.

Box 2.2: Regioselectivity.

Regioselectivity is the ability of a catalyst to distinguish between chemically similar functional groups that are on different positions of a molecule. An example is the selectivity of lipases towards the three fatty acids in triglycerides. Many lipases only hydrolyze the first and third fatty acid, while the second does not react. In contrast, chemical hydrolysis or non-selective lipase cleave all three fatty acids. This property is used for the selective interesterification of lipids (Figure 2.2).

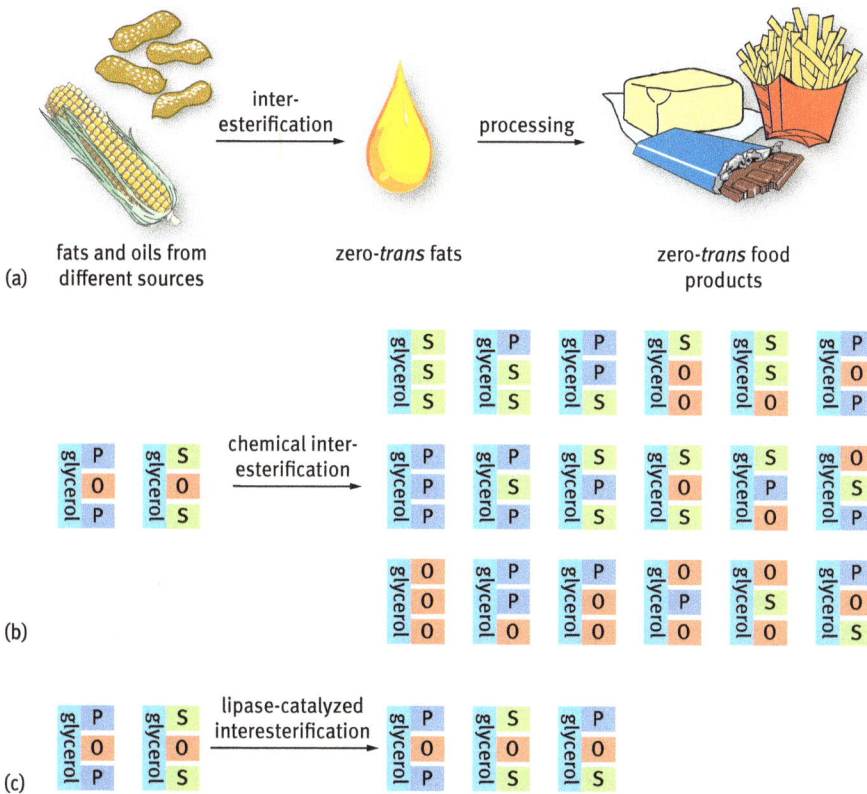

Fig. 2.2: Application of lipases for food processing leads to healthier products. (a) Industrial inter-esterification for the production of nutritional lipids results in a complex mixture of lipids with different fatty acids to control the melting point for the production of margarine and other nutritional fats; (b) Chemical interesterification (above) of a mixture of two fats with palmitic (P), oleic (O), and stearic (S) acid results in over 18 possible fat variants. (c) A lipase with specificity for the 1 and 3-position of the triglyceride, however, selectively produces only three fat variants, which results in improved properties of the fat for nutritional applications. Adapted from (DiCosimo, McAuliffe et al. 2013).

The final price of the products depends on the price of the natural starting material (often a vegetable oil) and the cost of the process, mostly the cost of the enzymes. Application of lipases for lipid processing has thus been limited to cases where the enzyme offers a clear advantage, as the controlled product distribution in enzymatic interesterification. Nevertheless, commercial application of lipases poses strong requirements on the manufacture of lipases and their stability. In enzymatic esterification, the lipase has to be active in a slurry of fats, fatty acids, and water. Lipase stability can be much increased by immobilization of the enzyme on inorganic carrier materials such as diatomaceous earth. This example underlines that a successful technical application of enzymes requires a close interdisciplinary collaboration of scientists from biological sciences, chemistry, and process engineering.

2.1.3 Enantioselectivity of enzymes

In a reaction with an achiral reagent or catalyst, the two enantiomers of a chiral molecule (see Box 2.3) react with the same reaction rate. The reason for this lies in the equal covalent and non-covalent interactions with the achiral reagent, similar to a can of beer in the left or right hand. If chiral molecules interact with other chiral entities, both enantiomer undergo different interactions. This can easily be imagined with a right-handed person shaking first a right, then a left hand. The different complexes are called 'diastereomers'. This results in different transition state energies and in different reaction rates.

A chiral catalyst can, thus, form a diastereomeric complex and discriminate between enantiomers. Enzymes are composed of optically pure molecules, the L-enantiomers of the proteinogenic amino acids. They are optically pure. Thus, enzymes often convert one enantiomer much faster than the other enantiomer. If one would imagine a hypothetical enzyme composed of D-amino acids, it would be the mirror-image and consequently prefer the other enantiomer. The enantioselectivity, also termed enantiomeric ratio or E-value of an enzyme towards a racemic substrate, is defined by how much an enantiomer is faster converted in comparison to the other.

Box 2.3: Chiral molecules.

Chirality describes the fact that a molecule is non-superposable (not identical) on its mirror image. Molecules that contain a chiral carbon atom usually have two non-superposable structures. These two mirror structures of a chiral molecule are called enantiomers. A pairs of enantiomers are often designated as "right-" and "left-handed". The most common naming convention is by configuration (R- and S-). It labels each chiral atom center R or S according to which its substituents are each assigned a priority, according to the Cahn–Ingold–Prelog priority rules (CIP). Other conventions to name enantiomers is the sign of the rotation of plane-polarized light, which is either positive (+) or negative (–) or the Fisher system using L- and D- for the enantiomers.

Even though they share the same structure, both enantiomers often have different biologic activity, and in most chiral pharmaceutical chiral molecules, usually only one enantiomer has the desired effect.

Enantioselectivity is an intrinsic value of the enzyme towards a substrate in specific reaction conditions. To determine it, the kinetic parameters according to Michaelis-Menten kinetics of both enantiomers are determined experimentally. It

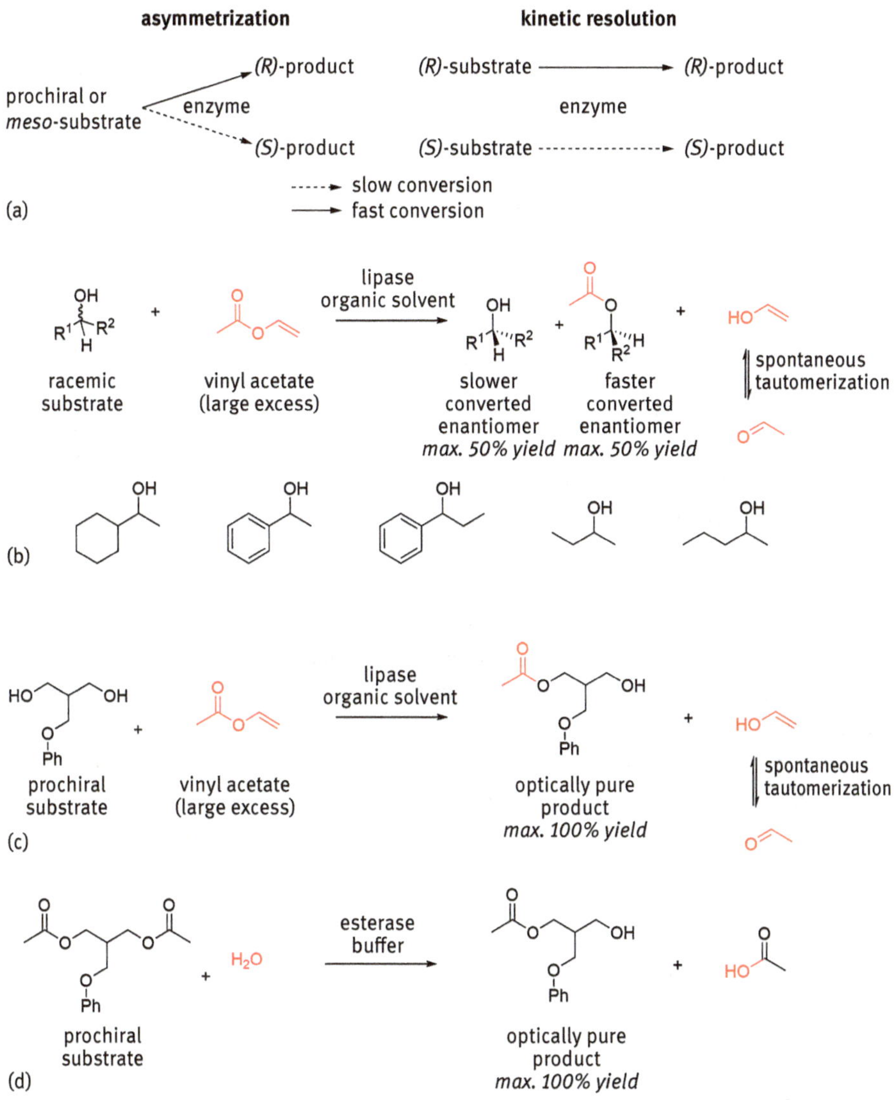

asymmetrization   **kinetic resolution**

(a)

Fig. 2.3: Above: Schematic representation of enzymatic asymmetrization (left) and kinetic resolution (right) with an (R)-selective biocatalyst. Both reactions are based on the different rates in the enzymatic formation of the two enantiomers of the product. (b) lipase-catalyzed alcohol-synthesis by *kinetic resolution* (Patel, Banerjee et al. 2000). The lipase is enantioselective for one enantiomer and leaves the other mostly unreacted. (c) and (d) Asymmetric syntheses using hydrolases.

is assumed that the conversion of the two enantiomers are two distinct reactions, where both enantiomers compete for the active site of the enzyme. The reaction rate of an enzyme towards a given substrate depends on the catalytic efficiency, which is the ratio of the k_{cat} and K_M value. The enantioselectivity is the ratio of the specificity constants k_{cat}/K_M towards both enantiomers (2):

$$E = \frac{v_R}{v_S} \tag{1}$$

$$E = \frac{(k_{cat}/K_M)_R}{(k_{cat}/K_M)_S} \tag{2}$$

Two main types of enantioselective reactions can be used for the synthesis of optically pure compounds. In the first, more commonly applied process, only one enantiomer of a racemic mixture is converted to the product while the other reacts much slower. If the reaction is stopped at 50 % conversion, one enantiomer is almost completely converted, while the other remains mostly unreacted. By this means, a mixture of compounds with different chemical and physical properties is obtained. Substrate and product can easily be separated afterwards. As the process is based on the different reaction rates of the enzymatic conversion of the two enantiomers, it is referred to as 'kinetic resolution'. Kinetic resolution is a separation of two enantiomers by a selectivity catalyzed reaction (Figure 2.3a). The maximal yield is, therefore, limited to 50 %. In the second, the so-called asymmetrization, a prochiral substrate is converted selectively to one of the two possible enantiomers of a chiral product. Asymmetrizations yield one product enantiomer with high optical purity and with a theoretical yield of 100 % (Figure 2.3b, c).

2.1.4 Lipase-catalyzed kinetic resolution of alcohols

The high enantioselectivity of lipases is industrially applied for the production of pure alcohols (Patel, Banerjee et al. 2000). These simple molecules are used as building blocks for the generation of highly complex molecules for the pharmaceutical industry. The lipase-catalyzed ester hydrolysis (clockwise in Figure 2.2) is a reversible reaction. By chosing appropriate reaction conditions, the reaction can be inversed (to counter-clockwise) towards esterification. A special case is the transesterification, in which lipases transfer an acyl group from an alcohol to another. Transesterification reactions require low water concentrations to avoid the hydrolysis of the esters and are usually conducted in organic solvents. A typical acyl donor is vinyl acetate. Transfer of the acetate group to the chiral alcohol releases ethanol. Fast tautomerization of ethanol to acetaldehyde removes this by-product and pushes the reaction equilibrium to the product side. Despite these advantages, the limited yield of 50 % in kinetic resolutions make them rather expensive, and asymmetric synthesis are the preferred reaction form for the production of chiral alcohols.

2.1.5 Lipase-catalyzed kinetic resolution of amines

The enantioselectivity of lipases can be used to produce optically pure amines in the so-called acylation of amines. This is one of the most important reactions for the industrial synthesis of amines. The company BASF uses this process for the synthesis of amines in 1000-ton scale. Application of vinyl esters is not possible because the released acetaldehyde reacts with the amine group of the substrate. The reaction equilibrium is shifted by using a large excess of acetic acid esters. Interestingly, 2-methoxyethyl esters have more than 100-fold increased activity to other esters. The molecular effect of the oxygen atom during catalysis has not been clarified yet. The synthesis of optically pure amines is often very challenging for (chemical) homogenous and heterogenous catalysis, and, lipase-catalyzed kinetic resolution is still in use today. In the conversion of prochiral molecules with two alcohol functions, lipases and esterases often convert preferentially on of them, resulting in an asymmetric synthesis with 100 % maximal yield. This approach is possible either as transesterification or as hydrolysis (Figure 2.4a).

In contrast to nutritional lipids, fine chemicals such as the starting material for highly valuable pharmaceutical ingredients achieve much higher prices. The cost for the manufacturing of the enzyme is less important in relation to the gains, resulting in a larger number of successful enzymatic processes.

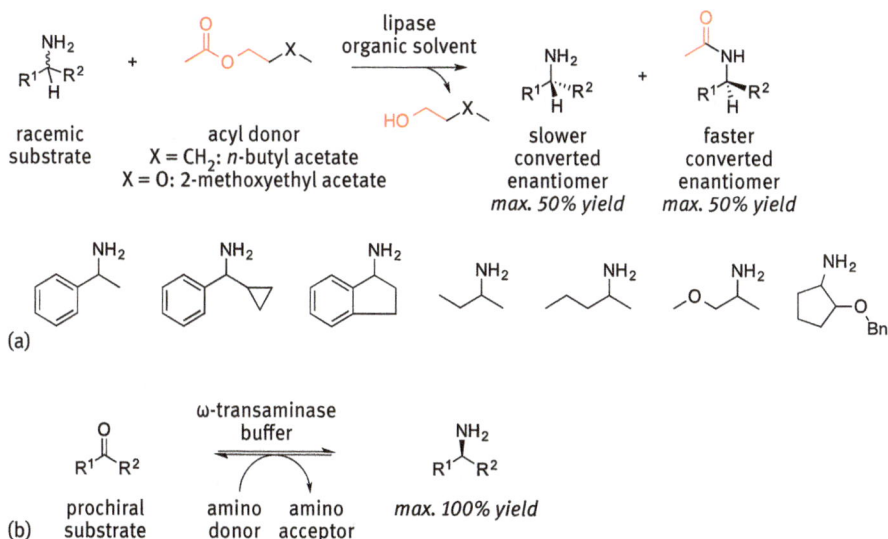

Fig. 2.4: Examples for the biocatalytic preparation of optically pure amines. The lipase-catalyzed kinetic reaction (a) is a robust and well-established reaction but is limited to 50 % maximal yield. In contrast, transaminase-catalyzed asymmetric synthesis allows a full conversion of the starting material with 100 % maximal yield (b) and is getting increasing significance for the synthesis of amines.

2.1.6 Transaminase-catalyzed asymmetric synthesis of amines

Figure 2.4 shows a comparison of two alternative enzymatic strategies for the synthesis of chiral amines. The lipase-catalyzed kinetic resolution (Figure 2.4a) is a robust and commercially successful synthesis. However, the catalytic separation of two enantiomers suffers from an intrinsic limitation of maximal 50 % yield. Recently, aminotransferases (or transaminases) have been established as alternative catalysts for amine synthesis. Transaminases use the cofactor pyridoxal-phosphate to transfer an amino group from a donor (an amine) to an acceptor (a ketone or aldehyde). Usually, inexpensive glutamate or alanine serves as amino donor. Transamination of a prochiral ketone gives rise to an optically pure amine (Figure 2.4b). Similar to the reactions shown in Figure 2.3b, this reaction is an asymmetric synthesis and allows a complete conversion with maximal 100 % yield. Another advantage is the possibility to use water as solvent instead of organic compounds. Despite these advantages, the application of transaminases meets several challenges among them. The equilibrium of the reaction must be shifted to the product side, and transaminases often suffer from product and substrate inhibitions, which makes the technical application difficult. Nevertheless, most of these issues could be solved by reaction engineering, and transaminases are gaining increasing importance for industrial amine synthesis. The following section presents a case where an industrial chemocatalytic reaction was replaced by a biocatalytic, transaminase-based process. Despite the obvious advantages of transaminases for this reaction, no suitable enzymes for this particular reaction were available. The company, therefore, used protein engineering to extend the substrate spectrum of a transaminase to catalyze the desired reaction and generated, thus, a useful enzyme. The following section will firstly introduce to structure-function relationships in enzymes, and how protein engineering can be used to modify the catalytic properties of biocatalysts.

2.2 Introduction to enzyme engineering

Enzymes as biocatalyst have already many applications in industrial processes. However, biocatalysis is mostly limited to reactions that are found in nature. Even if a suitable enzyme for a desired reaction is available, often much is left for improvement. This regards the stability under operational conditions, as well as catalytic properties like regio-, enantio-, and substrate-specificity. The second part of this chapter will give an introduction on how enzymes can be adapted for catalytic applications. It starts with a short review of protein biosynthesis and structure of enzymes. Then two examples will outline the strengths and weaknesses of two alternative methods for the engineering of enzymes; rational protein design and directed evolution.

Box 2.4: Basic priciples of protein biosynthesis.

To understand complex biotechnological tools like enzyme engineering, it is essential to fully understand basic principles of protein biosynthesis. The first step of protein synthesis is the transcription. A defined DNA segment is copied by an enzyme into a defined mRNA sequence. The genetic information leads always from a gene to mRNA and to the enzyme. During transcription, a RNA polymerase reads the DNA sequence, and produces a complementary, antiparallel RNA strand, the so-called transcript. For this, the DNA double helix is opened and the RNA polymerase binds to the template strand for the synthesis of the complementary RNA strand. The coding strand, thus, has the same sequence than the mRNA strand.

Coding strand	5'-ATG TGC GGC GAG CGA TAA-3'
Template strand	3'-TAC ACG CCG CTC GCT ATT-5'
mRNA	5'-AUG UGC GGC GAG CGA UAA-3'
Amino acid sequence	N-term-M-C-G-E-R-Stop-C-term

The bacterial RNA-Polymerase is a multi enzyme complex, which consists of enzymes and sigma factors. These sigma factors start the transcription. They recognize the promotor area of a gene, and initiate transcription by enabling binding of RNA-polymerase to promotor DNA. The promotor of *E. coli* consists of two hexameric sequence motives, which lie 10 and 35 base pairs upstream from the coding gene. The consensus sequences are TTGACA TATAAT. After binding the RNA polymerase moves a transcription bubble, which splits the double helix DNA into two strands of unpaired DNA nucleotides, by breaking the hydrogen bonds between complementary DNA nucleotides. Matching RNA nucleotides are added by the RNA polymerase. They are paired with complementary DNA nucleotides of one DNA strand. The RNA strand is formed by connecting the sugar-phosphate backbone. The hydrogen bonds between RNA and DNA break, and release the newly synthesized RNA strand (Figure 2.5).

Box 2.5: Protein translation.

The nucleotide sequence of the transcript m-RNA is translated by a universal genetic code into an amino acid sequence. This process is called protein translation. Distinct RNAs (t-RNAs), which have an specific anticodon, arrange the m-RNA and the amino acids to form the protein. Three nucleotides code for one amino acid. With this triplett code out of 4 bases, 64 combinations are possible. As there are just 20 proteinogenic amino acids, they are coded by more than one triplett. Translation starts always with the starting codon AUG, which codes for methionin. The correct starting point is important for the correct protein. By a shift of one or two bases, a so called frame shift, the sequence and the, therefore, also the protein structure is changed drastically. The starting methionin is often cleaved by a specific protease after translation.

The protein biosynthesis is done by ribosomes. In *E. coli* and other prokaryotes, the ribosome consists of two subunits with a size of 50 and 30 s. Ribosomes are multi enzyme complexes with three different aminoacyl-t-RNA binding sites. With respect to the m-RNA, the three sites are oriented 5' to 3'. They are the aminoacyl site (A); this site binds the incoming t-RNA with the complementary codon on the m-RNA. The peptidyl site (P) holds the t-RNA with the growing polypeptide chain, and the exit site (E) holds the t-RNA without its amino acid. When a Stop-codon occurs in the m-RNA sequence, the translation stops by freeing the new synthesized peptide chain.

2.2.1 Structure-function relationships

The catalytic properties of enzymes depend from their amino acid sequences, which are encoded in the nucleotide sequence of their genes. This linear sequence of amino acids is the primary structure of a protein. As peptides are formed by the coupling of the amino acids, proteins start with a free amino group of the first amino acids (the N-terminus), and terminate with a free carboxy group (the C-terminus). The sequence is written from N-terminus to C-terminus. It is unique for each protein and defines the structure and function of the protein. Post-translational modification, such as disulfide bond formation, phosphorylations and glycosylations are usually considered as a part of the primary structure. The two main types of secondary structures are α-helices and β-sheets. They are defined by patterns of hydrogen bonds between the main-chain peptide groups. The tertiary structure folding is driven by the non-specific hydrophobic interactions like the burial of hydrophobic residues from water. The α-helices and β-sheets are folded into a compact globular structure (Figure 2.5). The structure is stabilized by specific tertiary interactions, such as salt bridges, hydrogen bonds, and the tight packing of side chains. Tertiary structure refers to the three-dimensional structure of a single, double, or triple bonded protein molecule. The quaternary structure is the three-dimensional structure of a multi-subunit protein, and defines how the subunits fit together.

This connection between the primary sequence and the function has several consequences. Firstly, the structure is directly determining the function, which is subject to natural selection and evolution. As a result, the structure of analogous enzymes sharing similar function is usually highly conserved, while their sequence may differ considerably. Consequently, enzyme structure is more conserved than amino acid sequence. Secondly, changes in the amino acid sequence may change

primary structure	secondary structure	tertiary structure	quartenary structure
sequence of a chain of amino acids	amino acids in a chain are linked by hydrogen bonds	protein folding by interaction of α-helices and β-sheets	protein consisting of more than one amino acid chain

-Ala-Ser-Cys-Gly-Phe-

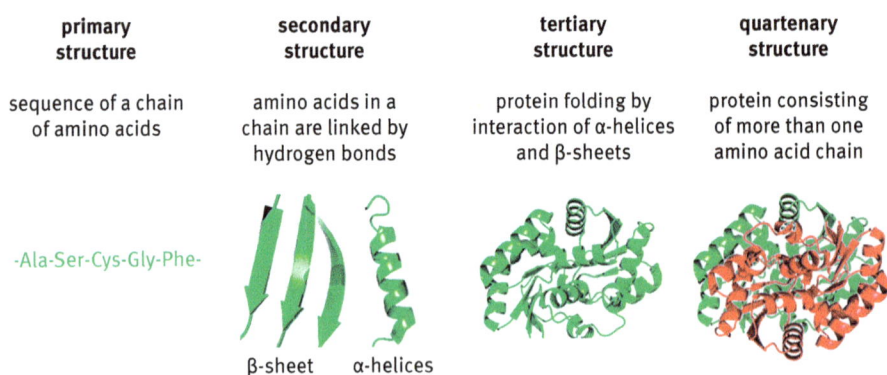

β-sheet   α-helices

Fig. 2.5: Schematic representation of protein folding. Highlighted are primary, secondary, tertiary, and quaternary structures.

the structure and, thus, alter the three-dimensional structure, which in turn will influence the catalytic properties of an enzyme. This can be exploited for a targeted modification of the catalytic properties. As each amino acid can be replaced with 19 other amino acids, the number of theoretically possible enzymes is extremely high. All variants of this vast diversity encode for enzymes with different catalytic properties. The number of possible variants has, therefore, been referred to as the "sequence space". Because it is impossible to explore the whole sequence space in experiments, the key question for a successful engineering of enzymes is how to navigate in sequence space, i.e. how to identify positions whose exchange has the desired effect.

2.2.2 Enzyme engineering

Two main strategies have been developed for enzyme engineering: Rational protein design and directed evolution (Figure 2.6). In rational protein design, the researchers formulate hypotheses on the enzymatic mechanism. A very important prerequisite for rational design is the availability and quality of structural information. Determination of structures by X-ray diffraction and NMR-spectroscopy made almost 100,000 enzyme structures available. In cases where an enzyme structure has not been resolved yet, often structural data from similar enzymes is available. Comparison of the sequences allows to generate a homology-model of the enzyme of interest, which gives some but often unsatisfactory information. Based on structural and mechanistic considerations, researchers try to predict the outcome of amino acid exchanges. Then mutations are introduced to the gene by site-directed mutagenesis (Figure 2.6). In turn, the characterization of these new mutants helps the

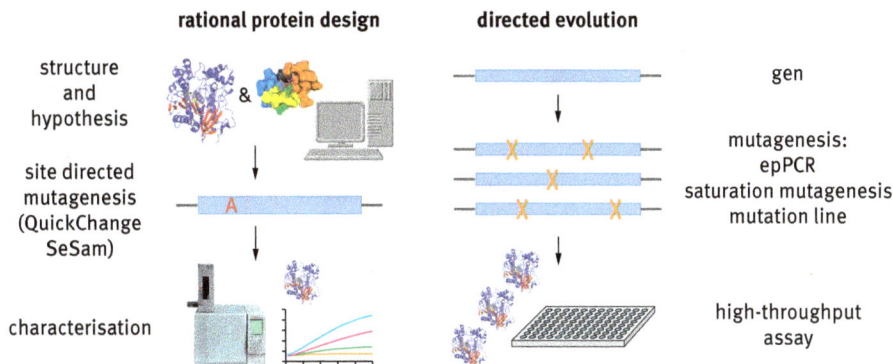

Fig. 2.6: Comparison scheme of rational protein design and directed evolution for protein optimization.

researcher to increase the understanding on the relationship between structure and function. Rational design has the advantage that it increases our understanding of enzymatic mechanisms and is based on a rational basis, which allows a certain predictability. However, the current understanding of enzymes is rather poor, and successful rational protein design usually requires long research experience with a biocatalyst. Unfortunately, the available knowledge on the molecular mechanisms of enzymes is still quite incomplete, and the main challenge of rational protein design is that it still often fails. Moreover, it is very difficult to predict the chances of success.

In lieu of a successful and, more importantly, reliable method for protein engineering, researchers followed the approach of natural evolution by random mutation and selection of the best variants. This approaches uses molecular biology methods to introduce random mutations into the gene of an enzyme. For the introduction of mutations into genes, several methods are available. Treatment of bacterial strains with UV light or mutagenic chemicals leads to a random accumulation of mutants. Similarly, a polymerase chain reaction (PCR) using a polymerase with high error rate introduces random mutations in the PCR product. These procedures are rather unspecific and address the whole gene of interest. Other methods use primers leading to mutations on specific sites, and allow to randomize selected regions of the gene encoding for specific amino acids in the protein. The result of all these mutagenesis methods is a collection of DNA-molecules with randomized mutants, which are unknown to the experimenter. After transfer of these DNA-molecules into bacterial or yeast cells, the DNA is translated into the corresponding protein, and the experimentator characterizes the variant by selection or screening. A typical screening is a colorimetric reaction in which product formation can be done easily by spectroscopy. In these experiments, a mutant library is cultivated in microtiter plates. After cell disruption, the test reaction is performed in each well (Figure 2.6). Variants with improved activity are isolated and serve as basis for the next round of improvement. In natural evolution, the criteria for the selection of a variant, the so-called fitness function, is the survival of the organism. However, in enzyme evolution, the experimentation defines the fitness function by the choice of screening reaction. *in vitro* enzyme evolution is, therefore, referred to as 'directed evolution'. Directed evolution allows the identification of beneficial amino acid substitutions that cannot be predicted by our current knowledge. The proceeding is very similar to computational optimization algorithms such as 'genetic algorithms'. However, while directed evolution works well even without mechanistic knowledge, the resulting variants increase our understanding of the enzymatic mechanism and may inspire rational experiments. Very often, studies on enzyme engineering do not rely purely on rational design or directed evolution but rather try to combine both methods. The following section outlines such an example, in which the substrate spectrum of a transaminase was expanded using directed evolution inspired by molecular modeling.

Box 2.6: Mutagenesis.

Mutagenesis is the generation of mutations within the genetic information of an organism. It may occur spontaneously in nature or as a result of exposure to mutagens. In nature mutagenesis can lead to cancer and various heritable diseases, but it is also the base for biologic diversity. In biological or medical research, mutagenesis as well as breeding is used to create desired features.

2.2.3 Protein engineering of a biocatalyst for the synthesis of the blockbuster drug Sitagliptin

A recent example outlines the potential of protein engineering for the generation of useful catalysts, and the advantages of biocatalysts regarding cost and sustainability. The American pharmaceutical Merck manufactures sitagliptin for the treatment of type II diabetes mellitus. Due to the prevalence of this ailment in industrial societies, sitagliptin belongs to the top blockbuster drugs and is produced in tonscale. As the structure of the drug contains a chiral amine, the introduction of chirality is an important step in the overall synthesis (Savile, Janey et al. 2010). Merck used an asymmetric synthesis for the synthesis of the (R)-amine (Figure 2.7). The process suffered from several drawbacks, mostly an inedequate optical purity of the product (max. 97 % ee). Regulations require that the Rh-catalyst should be completely removed lest the heavy metal contaminates the pharmaceutical ingredients. A complete removal by carbon treatment is quite expensive.

The company reasoned that a stereoselective amination using a transaminase will avoid the use of heavy metals and save a reaction step. This reaction is an asymmetric synthesis, as shown in Figure 2.4 and, thus, has 100 % maximal yield. However, no transaminase was available that catalyzed the conversion of the bulky starting compound to the desired product. Most available ω-transaminases were (S)-specific, i.e. they created the (S)-enantiomers of the products and were, thus, not useful for these reactions. At this time, only one bacterial (R)-selective transaminase was known. Unfortunately, this enzyme did not accept the bulky starting compounds from Figure 2.7. Merck chose to use protein engineering to extend the substrate spectrum of the (R)-selective transaminase. As the three-dimensional structure of the transaminase had not been determined yet, they used the crystal structure of a similar enzyme to generate a so-called homology model. By computer-based molecular modeling, they generated a model of the substrate in the active site and were able to identify the amino acids that were likely to determine the substrate binding. Despite its usefulness, the available information from the homology model was not accurate enough to make predictions on the outcome of amino acid substitutions. The most promising strategy was to combine the available knowledge with random mutagenesis. They postulated three criteria for a successful biocatalyst: Firstly, the new variant should convert the starting compound.

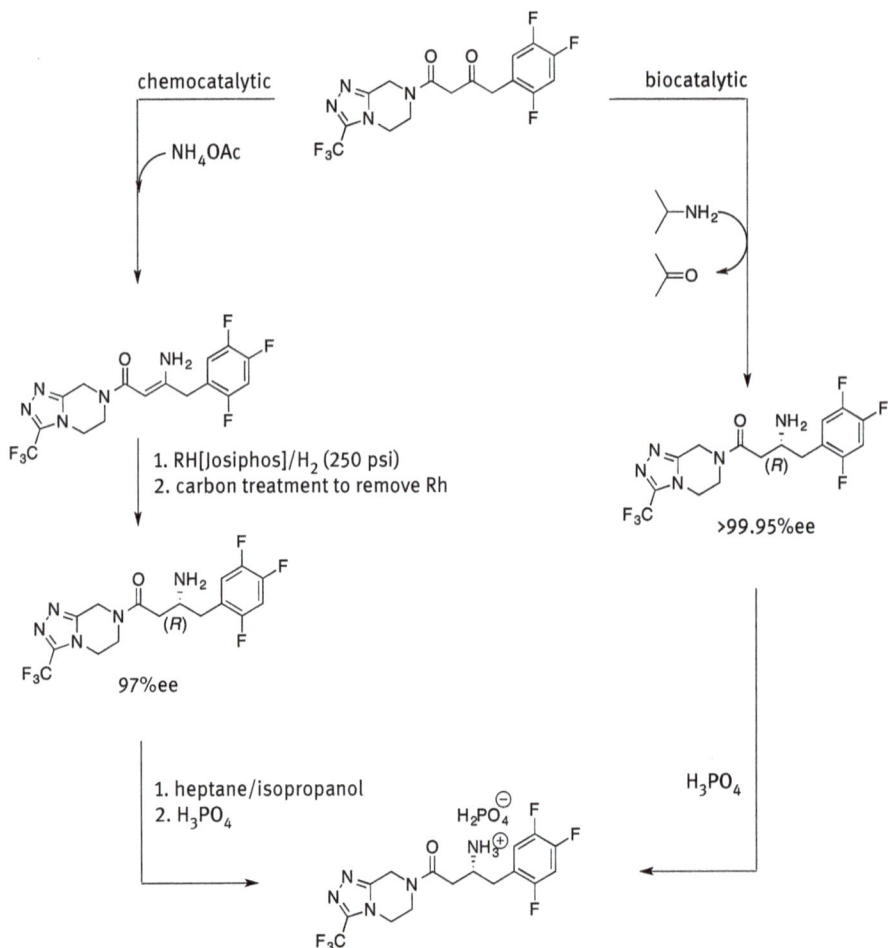

Fig. 2.7: Comparison of chemocatalytic and biocatalytic strategies for the synthesis of the block-buster drug sitagliptin. The chemocatalytic route has limited stereoselectivity and requires use of harmful heavy metals and high pressure. An alternative enzymatic process using a ω-transaminase offers several advantages, among them a shorter route, higher selectivity, and sustainable reaction conditions (adapted from Savile et al. 2010).

Secondly, the variant should be able to work at high substrate concentrations. This meant that substrate and product inhibition should be prevented. Thirdly, the low solubility of the starting material required the addition the organic co-solvent dimethyl sulfoxide (DMSO). The new variant should thus tolerate a high concentration of DMSO. Figure 2.8 shows the final strategy basing of a simultaneous increase of the three parameters. The idea was to use random mutagenesis for the identification of improved variants. Those would be used as basis for another screening round under 'tighter' screening conditions. As the enzyme did not convert the starting compound, they designed a surrogate substrate with a smaller structure and

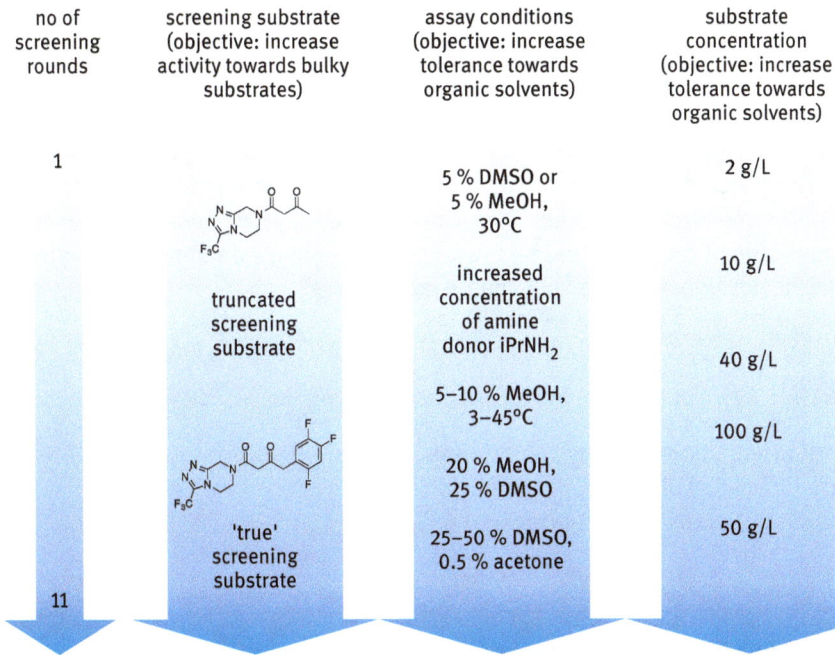

no of screening rounds	screening substrate (objective: increase activity towards bulky substrates)	assay conditions (objective: increase tolerance towards organic solvents)	substrate concentration (objective: increase tolerance towards organic solvents)
1		5 % DMSO or 5 % MeOH, 30°C	2 g/L
	truncated screening substrate	increased concentration of amine donor iPrNH₂	10 g/L
			40 g/L
		5–10 % MeOH, 3–45°C	100 g/L
		20 % MeOH, 25 % DMSO	
	'true' screening substrate	25–50 % DMSO, 0.5 % acetone	50 g/L
11			

Fig. 2.8: Strategy for the engineering of the (R)-selective transaminase for the conversion of sitagliptin (adapted from Savile et al. 2010). In order to combine improvement of the activity towards large substrates with an increased tolerance towards high concentrations of co-solvent and substrate, all three parameters were simultaneously altered.

switched later to the 'true' substrate. Similarly, screening started at low concentrations of substrate and co-solvent. The final round was done using a concentration of 50 % DMSO and a substrate concentration of 50 g/L. The wildtype enzyme was not active under these conditions!

The final variant had 27 amino acid exchanges. 10 of these were predicted initially on basis of the model, while the others were on different positions. By choosing a purely rational approach, the latter positions would have escaped the experimentators. This shows the value of random mutagenesis for the improvement of enzymes.

After implementation of the process using the new transaminase variant, the biocatalytic process showed several advantages (Figure 2.7). The optical purity of the product of the transaminase (99.95 % ee) is higher than that of the chemobiocatalytic product (97 % ee). This might seem a small difference, but regulations require an optical purity higher than 98 % ee for pharmaceutical applications. The enzyme reaction saves one reaction step. Each reaction step leads to a loss of material and to cost for product purification and donwnstream processing. The biocatalytic reaction improved the productivity (in g/L) of the enantioselective reaction by 10 % and increased the overall yield of the total synthesis by 10 %. This leads to a 19% reduction in waste accumulations and eliminates harmful heavy metals. Moreover, the first chemical step requires high pressure, which is difficult to realize in industrial scale and needs costly facilities and training of the co-workers.

In conclusion, this example underlines the value of protein engineering for the development of tailor-made biocatalyst variants for sustainable chemical reactions. Biocatalysis already assumes an important role in industrial production of chemicals, pharmaceuticals, and food. It is a tool for the improvement of the environmental profile of processes and is, thus, expected to make an important contribution to the development of a sustainable industry.

Key-terms

Regio and enantio-selectivity, biocatalysis, protein engineering, rational protein design, directed evolution

Questions

- Draw the mechanism of the hydrolysis of a carboxylamide by a lipase and indicate base and acidic catalysis.
- Why is enzymatic interesterification of lipids superior to chemical interesterification?
- What is the enantioselectivity of an enzyme?
- A bacterial lipase is applied for the synthesis of the chiral building block 1-phenyl ethanol. In order to investigate the enantioselectivity, a student determines the kinetic parameters of the enzyme. For the (R)-enantiomer, the value for k_{cat} is 10 s^{-1}, the K_M-value is 5 mM. For the (S)-enantiomer, the value for k_{cat} is 0.02 s^{-1}, the K_M-value is 4 mM. (a) draw the structure of 1-phenyl ethanol and the reaction equation; (b) calculate the E-value.
- In a similar example, the value for k_{cat} of the (R)-enantiomer is 8 s^{-1}, the K_M-value is 0.04 mM. For the (S)-enantiomer, the value for k_{cat} is 0.9 s^{-1}, the K_M-value is 200 mM. Calculate the E-value.
- Additional question for interested readers: Compare the two examples. Keep in mind that the two enantiomers compete for binding in the active site of the enzyme. In which case, is the binding of the substrate decisive for the enantioselectivity, in which case the k_{cat} (and hence the transition state energy of during the enzymatic reaction)?

Further readings

Faber, K. 2011. Biotransformations in Organic Chemistry, 6th ed, *Springer Verlag*, Berlin, Heidelberg, New York.

Arnold, F. H., Georgiou, G. (Eds) 2003. Directed Enzyme Evolution: Screening and Selection Methods, Vol 230, *Humana Press,* Totowa, USA.

Arnold, F. H., Georgiou, G. (Eds) 2003. Directed Enzyme Evolution Library Creation, Vol 231, *Humana Press,* Totowa, USA.

Lutz, S., Bornscheuer, U. T. (Eds) 2012. Protein Engineering Handbook, Vol 3, *Wiley-VCH,* Weinheim.

References

Breuer, M., Hauer, B. 2011. Biokatalyse in der chemischen Industrie. Molekulare Biotechnologie. M. Wink (Ed), *Wiley-VCH,* Weinheim.

DiCosimo, R., McAuliffe, J., Poulose, A. J., et al. 2013. Industrial use of immobilized enzymes. *Chem Soc Rev* 42: 6437–6474.

Patel, R., Banerjee, A., Nanduri, V., et al. 2000. Enzymatic resolution of racemic secondary alcohols by lipase B from Candida antarctica. *J Am Oil Chem Soc* 77: 1015–1019.

Savile, C. K., Janey, J. M., Mundorff, E. C., et al. 2010. Biocatalytic asymmetric synthesis of chiral amines from ketones applied to sitagliptin manufacture. *Science* 329: 305.

Mathias Lübben and Raphael Gasper

3 Prokaryotes as protein production facilities

Often, the term "biotechnology" is associated with the making and refinement of low molecular weight chemicals. However, another all-important facet is the production of biological macromolecules, in particular the versatile compound class of proteins. These days, bacteria are the most important production plants for many proteins. They have a strong potency to synthesize high levels of specific proteins which are needed for a countless variety of technological, medicinal, and scientific purposes. This huge potential relies on two facts: First, a vast amount of knowledge on the genetics of these microorganisms has been accumulated and the methods of their genetic manipulation are highly developed. Second, bacteria are readily grown at low cost to large quantities upon easy handling. Currently, only a few of the roughly 80,000 species known from rRNA cataloging can be applied in the laboratory, but the estimated total of 10^7 to 10^9 undiscovered bacterial species indicates the rich potential slumbering in this biological diversity. As will be seen, in most cases, the particular need for a certain protein determines the choice of the microorganism. One has to take into account a number of considerations and special conditions before the pipette is grabbed to start the task.

This chapter will focus on general considerations regarding prokaryotic protein production before going into detail and discussing the utilization of different prokaryotic organisms, in particular *Escherichia coli*.

3.1 General considerations of prokaryotic protein production

3.1.1 Purpose of the protein – homologous vs. heterologous expression strategies

Beyond all following considerations, the purpose of protein synthesis has to be clearly defined. The task could be very easy if the polypeptides were simply needed without biological activity, as may be the case for animal immunization to raise antibodies. In contrast, a huge number of active proteins have been bacterially produced and biochemically purified at relatively large amounts (up to the milligram scale) to homogeneity. The workflow and possible experimental strategies are schematically summarized in Figure 3.1.

If the protein should be obtained in folded, fully functional state, much higher efforts would have to be undertaken. In general, prokaryotes of different species are not at all capable of synthesizing any specific kind of proteins as might by envisioned by the concept of the universal genetic code alone. The individual potencies are dependent on the required cofactors, such as specific metal centers or particular organic prosthetic groups which are being made only by highly special-

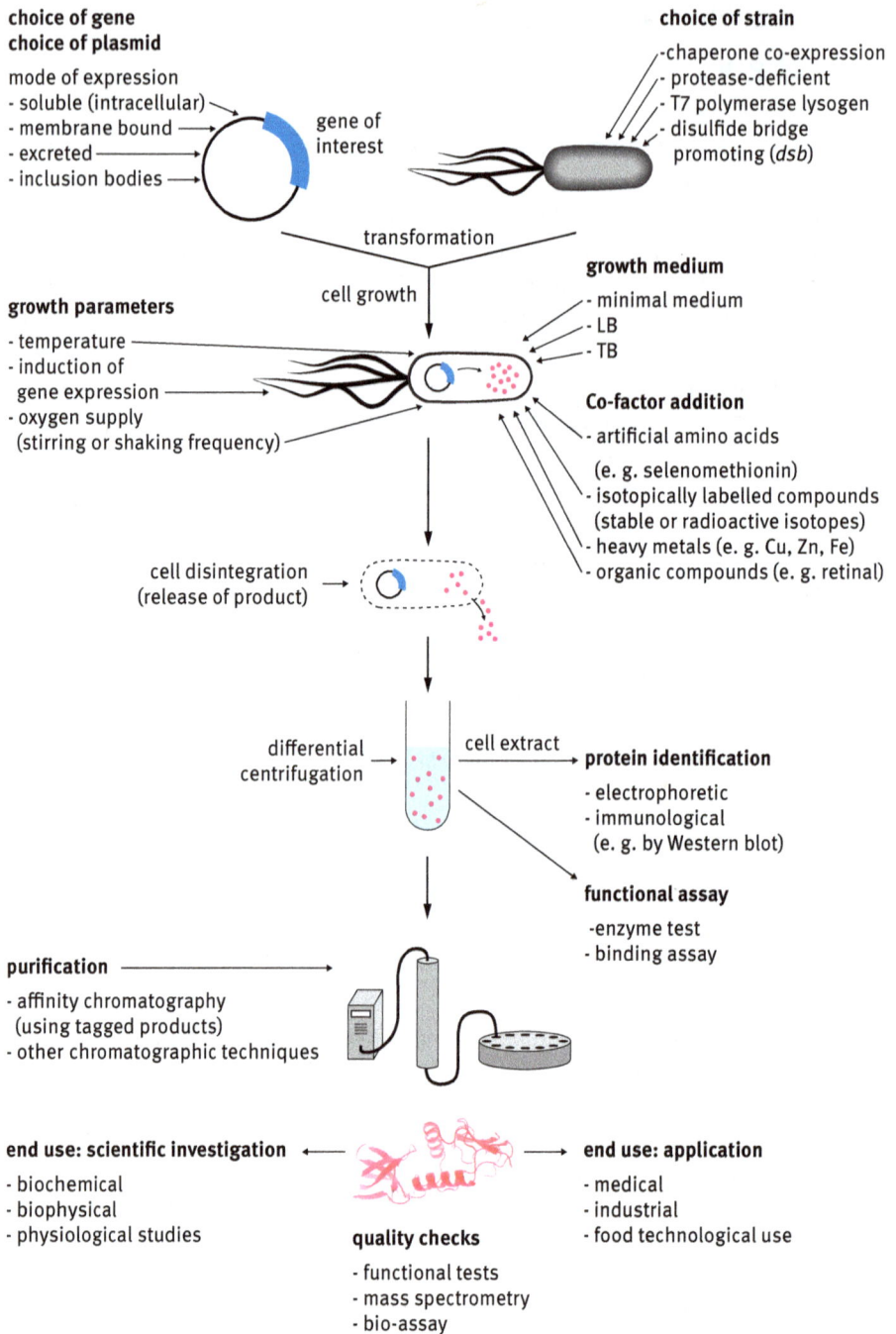

choice of gene
choice of plasmid

mode of expression
- soluble (intracellular)
- membrane bound
- excreted
- inclusion bodies

gene of interest

choice of strain
- chaperone co-expression
- protease-deficient
- T7 polymerase lysogen
- disulfide bridge
  promoting (*dsb*)

transformation

cell growth

growth medium
- minimal medium
- LB
- TB

growth parameters
- temperature
- induction of
  gene expression
- oxygen supply
  (stirring or shaking frequency)

Co-factor addition
- artificial amino acids
  (e. g. selenomethionin)
- isotopically labelled compounds
  (stable or radioactive isotopes)
- heavy metals (e. g. Cu, Zn, Fe)
- organic compounds (e. g. retinal)

cell disintegration
(release of product)

differential
centrifugation

cell extract

protein identification
- electrophoretic
- immunological
  (e. g. by Western blot)

functional assay
-enzyme test
- binding assay

purification
- affinity chromatography
  (using tagged products)
- other chromatographic techniques

end use: scientific investigation
- biochemical
- biophysical
- physiological studies

end use: application
- medical
- industrial
- food technological use

quality checks
- functional tests
- mass spectrometry
- bio-assay

Table 3.1: Use of bacteria or bacterial components for vaccination.

Vaccine	Application (prevention of)	Bacterial source
BCG ("Bacille Calmette-Guérin")	Tuberculosis	Attenuated *Mycobacterium bovis*
Cholera	Cholera	Attenuated *Vibrio cholerae*
Haemophilus	Meningitis	Isolated polysaccharides from *Haemophilus influenzae*
Tetanus toxoid	Tetanus	Cross-linked toxin of *Clostridium tetani*

Box 3.1: Expression and purification of Botulinum toxin in *Clostridium botulinum*.

The potent Botulinum neurotoxin type A (acting at a minimum lethal dose of 30 pg/kg body weight, measured in mice) is produced from culture fluids of *Clostridium botulinum*, strain Hall A grown to the stationary phase. Being liberated from autolyzed cells, the crude toxin is repeatedly precipitated at pH 3.5 and extracted by neutral buffer. After chromatography, the final product, called "crystalline complex", is obtained after precipitation with ammonium sulfate. It consists of a noncovalent association of the 150 kDa neurotoxin with 6–7 nontoxic proteins (total molecular mass 900 kDa). The specific need for the natural host strain is due (i) to proteolytic cleavage of the single-chain protoxin by endogenous proteases to the mature disulfide-bridged double chain form and (ii) to the toxin stabilization by the nontoxic components. The crystalline Botulinum toxin is therapeutically applied against focal dystonias and involuntary movement disorders, and is also used in cosmetics ("Botox").

ized microorganisms under exactly defined growth conditions. If the protein of interest is representative for a particular organism, it is obvious to select the organism itself for homologous expression or, at least, a close relative thriving under identical living conditions. To give examples, specific neurotoxins were industrially produced from strains of *Clostridium tetani* (Table 3.1) and *Clostridium botulinum* (Box 3.1).

Another important issue arises from the presence of possible folding helpers or chaperones, which may be requested for cofactor incorporation and/or correct membrane insertion. For example, this occurs during biosynthesis of cytochrome *c* oxidases in *Rhodobacter sphaeroides,* which depends on the electron transporting

◄ **Fig. 3.1:** Workflow of protein production in a prokaryote. The basic strategy for prokaryotic protein production using the Gram-negative eubacterium *Escherichia coli* as an example is shown schematically, starting from selection of plasmid and expression strain and ending with the final use of the purified protein. After selection of the target gene (blue), the main principles affecting construct engineering and setup of expression conditions the transformed bacteria are grown, the transcription is induced and the protein of interest (red) is synthesized. By disintegration of the bacteria the protein is extracted into solution, from which it is purified by means of designed affinity tags by chromatographic methods. Many aspects to be considered for successful operation are listed as keywords and will be discussed in the text.

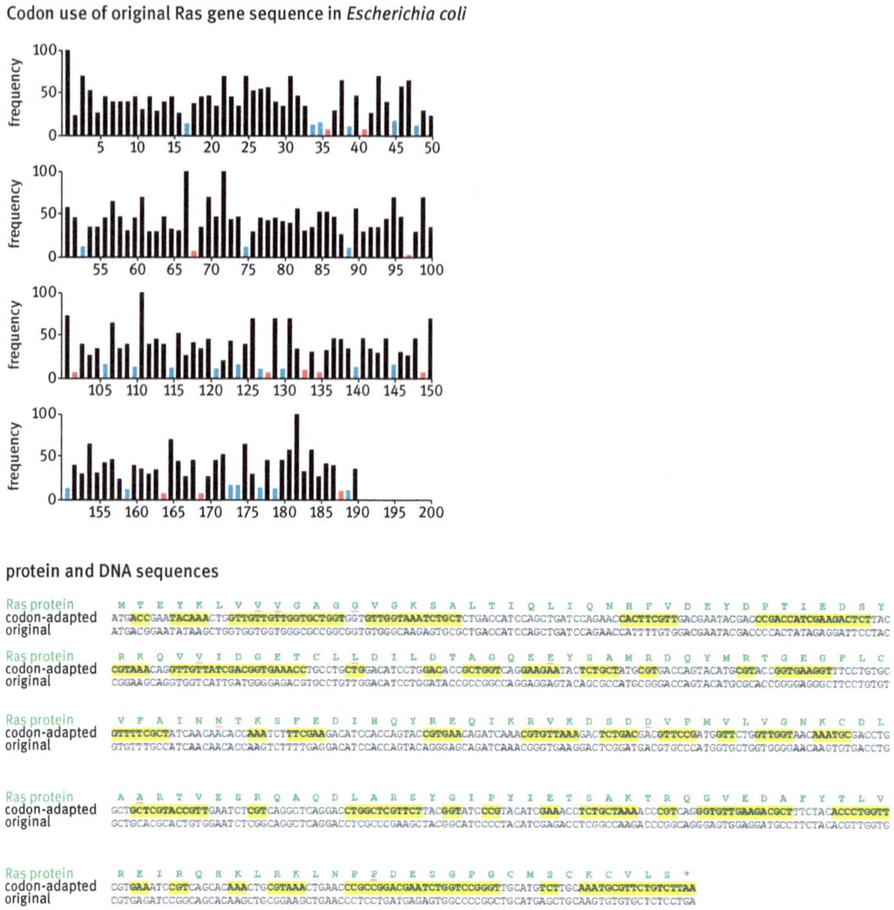

Codon use of original Ras gene sequence in *Escherichia coli*

protein and DNA sequences

Ras protein · codon-adapted · original

M T E Y K L V V V G A G G V G K S A L T I Q L I Q N H F V D E Y D P T I E D S Y

R K Q V V I D G E T C L L D I L D T A G Q E E Y S A M R D Q Y M R T G E G F L C

V F A I N N T K S F E D I H Q Y R E Q I K R V K D S D D V P M V L V G N K C D L

A A R T V E S R Q A Q D L A R S Y G I P Y I E T S A K T R Q G V E D A F Y T L V

R E I R Q H K L R K L N P P D E S G P G C M S C K C V L S *

Fig. 3.2: Adaptation of "rare codons" to improve protein synthesis. Eukaryotic proteins such as the human proto-oncogene encoded Ras could be expressed in prokaryotes. The gene sequence contains codons which are underrepresented in bacterial hosts. Without proper adaptation of rare codons according to the prokaryotes' codon usage the expression rates could be significantly diminished. The computer program GCUA (http://gcua.schoedl.de/) calculates the frequency of codon use if the human gene is expressed in the prokaryote *Escherichia coli*. Codons used at low abundance are colored in blue, and those at less than 10 % in *E. coli* in red. The computer program JAVA CODON ADAPTATION TOOL (http://www.jcat.de/) helps to adjust the human gene sequence to the codon usage needs of *E. coli*, which is necessary to achieve optimized synthesis of the protein (green sequence). The alignment of both original and codon optimized gene sequences (bottom) demonstrates the number of base changes required to fully convert the original human into a codon-adapted, referred to as "bacterialized" gene (highlighted in yellow).

prosthetic groups hemes A and B, and the activity of the copper center assembling gene products PCu$_A$C, Sco and Cox11 (Thompson et al. 2012). For the synthesis of bacteriorhodopsin, highest yields are gained by means of the natural host *Halobac-*

terium salinarum (see below), whereas other microbial rhodopsins were obtained by heterologous expression in *Escherichia coli,* after supplying the growth medium with the cofactor *all trans*-retinal (Fu et al. 2010). Irrespective of specific cofactor needs, it has to be kept in mind that bacteria, in general, are incapable of carrying out posttranslational protein modifications, such as glycosylation or lipidation, which widely occur in eukaryotes. The industrially produced posttranslationally unsubstituted polypeptides of insulin, diverse interferones, certain cytokines such as Granulocyte-Colony Stimulating Factor (G-CSF), and Granulocyte Macrophage Colony-Stimulating Factor (GM-CSF), for example, are produced in the bacterial host *Escherichia coli* in pharmacologically active form. In contrast, recombinant erythropoietin ("EPO") requires heterologous production in mammalian CHO cells (Chinese hamster ovary), as its therapeutic activity is strictly dependent on the correct glycosylation pattern. The use of prokaryotic expression systems to gain functional proteins is anyway precluded in these cases, for which alternative expression systems are available.

Another serious concern of heterologous expression may arise from codon constraints of the target gene, often occurring in eukaryotic genes that frequently have "rare codons" for arginine, lysine, leucine, and others. The biosynthetic rates could be drastically reduced or even stalled when the frequencies of synonymous codons differ too much between source gene and host organism. Modifications of individual codons could be even used to restrict otherwise overshooting protein production rates (Goodman et al. 2013). Potential shortages of specifically encoded activated amino acids could be complemented by co-transfection with plasmids bearing the respective cognate tRNAs (see below). Alternatively, these defects could be compensated for by limited site-directed codon adaptation (Figure 3.2) or by chemical *de novo* synthesis of the entirely tailored gene.

3.1.2 Choice of a suitable expression host

Subsequent to the selection of a suitable host for protein production, the bacteria can be ordered from one of the big biological culture collections, such as the American Type Culture Collection (ATCC), the Japanese RIKEN collection, or the German Collection of Microorganisms and Cell Cultures (DSMZ). Furthermore, specialized strain collections are available for certain microorganisms. These allow for selection of the proper host for each individual task, e.g. the *Escherichia coli* stock center (CGSC), http://cgsc.biology.yale.edu/, offers a huge variety of strains with defined genetic backgrounds. Such specific genetic configurations may be indispensable, if intended group-selective amino acid labelling of a protein requests the use of an auxotrophic strain. The genetic background of a bacterial strain is characterized by a standardized code, which often specifies the affected gene with a four-letter annotation, as shown for a few examples in Table 3.2.

Table 3.2: Genetic terminology, relevant for partly understanding the background of specific expression strains. In these examples, defect genes are denoted by a four letter code (written in italics), sometimes additionally specified by numbers or indices. Many more genes are relevant to fully characterize the genetic background of prokaryotic strains. Also, many rules apply to this genetic nomenclature, but a detailed description is beyond the subject of this chapter.

Genotype	Phenotype
mcrA	No restriction of DNA having methylated cytosine at the sequence ^mCG
mcrB	No restriction of DNA having methylated cytosine at the sequence R^mC
ompT	Outer membrane protease
recA1	Reduced occurrence of unwanted recombination, DNA repair deficiency
thi-1	Defect in thiamine biosynthesis; thiamine requirement
lacI^q	Overproduction of LacI repressor protein
F$^+$, F$^-$	Donor, acceptor of F-plasmid
DE3	λ DE3 lysogen
Rif^R	Resistance to the antibiotic Rifampicin
dnaK	Dysfunction of gene product encoding the HSP-70 type chaperone
groEL	Dysfunction of gene product encoding Chaperone HSP-60
lon	ATP-dependent protease
grpE	Nucleotide exchange factor
leuB	Leucine requirement

Another strain library is the KEIO collection, http://www.thermoscientificbio.com/non-mammalian-cdna-and-orf/e.-coli-keio-knockouts/, consisting of almost 4000 constructed *Escherichia coli* mutants, in which each individual gene has been systematically knocked out by transposon mutagenesis. Among other applications, these strains would be required if a clean genetic background should provide the expression of a site-specifically altered gene free of any wild-type contamination. The genetic background is characterized by a nomenclature that describes the specific changes of the bacterial strain in relation to the wild-type in a typical four letter code.

3.1.3 Adjustment of growth conditions

Growth rates (measured by photometric recording of turbidity in the longwave VIS range) and growth efficiency (wet mass after centrifugation) of prokaryotes are strictly dependent on the media composition (Figure 3.3). In most cases, complex media providing carbon, nitrogen sources (e.g. from digested meat and yeast extracts), and buffered salt will do their task (see Table 3.3). Other instances may depend on defined media, e.g. minimal salt medium, supplemented with a single carbon source such as glucose. In particular, specific additives must be included to the defined media, if the host strain is incapable to synthesize the respective components due to genetic changes.

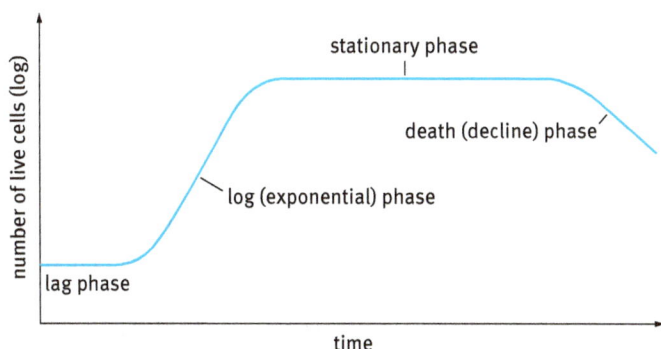

Fig. 3.3: Bacterial growth curve. Bacterial growth follows three different stages. The initial 'lag phase', followed by an exponential 'log phase' that finally leads to the 'stationary phase', where no net increase in cell density can be observed.

Table 3.3: Commonly bacterial growth used media.

Name	Ingredients	Properties	max OD
Minimal Media (M9)	Na_2HPO_4 / KH_2PO_4 / NaCl / NH_4Cl / Glucose / vitamins / trace elements	Low growth rate Possibility of better expression of toxic or membrane proteins	~1
LB	1 % Peptone / 0.5 % yeast extract / NaCl	Typical standard media	2–3
2xYT	1.6 % Peptone / 1 % yeast extract / NaCl	Rich media, originally for M13 phage production	3–5
TB	1.2 % Peptone / 2.4 % yeast extract / K_2HPO_4 / KH_2PO_4 / glycerol	Rich media with additional phosphate and glycerol	5–8
SB (Super Broth)	3.2 % Peptone / 2 % yeast extract / NaCl	Very rich media	~10
SOB (Super Optimal Broth)	2 % Peptone / 0.5 % yeast extract / NaCl / KCl / $MgCl_2$ / $MgSO_4$	Very rich media with additional Mg^{2+} for high density cell growth	10–15
Autoinduction media	Different types of rich media for high cell density	Supplemented with glucose, glycerol and lactose	~10

A critical factor in bacterial protein production is the adjustment and mainte-nance of adequate growth conditions. Cell mass and product yield depend on the energy supply of bacteria, in form of substrates and of appropriate electron accept-ors. Because of its high redox potential, the best results are obtained with molecu-lar oxygen. Cultivation in bioreactors (= fermenters) (see chapters 8 and 9) allows the precise regulation of pH values, and enables a reproducible control of oxygen supply.

3.1.4 Choice of a suitable vector system and appropriate mode of expression (control of gene dosis)

The intended objectives for use of the desired product strictly determine the strategies and methods applied for expression. Consequently, they demand a careful planning of the target gene, and vector construct which defines the nature of the desired product, and the quantity and kinetics by which the mRNA is being presented to the translational apparatus. Compared with eukaryotes, some highly advanced bacterial expression systems (such as the one from *Escherichia coli*, see below) offer the advantage of enormous versatility. The gene of interest could be encoded on the bacterial genome, on episomal plasmids present in various copy numbers per cell, or on bacteriophage DNA. Besides many other factors, the amount of the target gene ("gene dosis") strongly controls extent and course of protein expression. Although the expression levels are modulated by individual promoter activities and mRNA stabilities, the transcriptional and translational productivities are strictly dependent on the number of DNA templates, i.e. the single genomic copy would be expected to yield lower amounts than the multiple ones provided by vectors that occur in numbers of up to 500 per bacterial cell. However, the magnitudes of expressed protein do not relate proportionately to the gene copy number. Some expressed genes would kill the host bacteria already at low production rates. The expression of such harmful proteins has to be regulated with extreme caution (see below).

3.1.5 Multicistronic gene expression

It is often easier to express single proteins in bacteria, although large modular proteins consisting of multiple individual domains may cause problems in folding. In nature, these proteins occur predominantly in higher eukaryotes, indicating that utilization of bacterial systems may be impractical for complex polypeptide chains made up by multiple domains. Although the majority of proteins are translated from a singly encoded gene, bi- or even multicistronic expression in bacteria may be requested to deliver larger entities formed by different polypeptides stabilized by protein-protein interactions. A striking example of a multi-subunit complex with an $\alpha_3\beta_3\gamma\delta\varepsilon$ protomer composition is the production of the human F_1-ATPase in *Escherichia coli* (Yoshida 2014).

3.1.6 Measures in case of toxic target genes

Severe problems in protein production may occur when the product is so toxic to the prokaryotic expression host that the metabolic charges cause dramatic growth reduction or even kill the cells. These conditions exert a strong pressure for recombinative rearrangements, and may lead to inactivation of the target gene and culture overgrowth of non-expressing cells, or even an entire inhibition of growth.

These drawbacks are partly overcome by starting the protein expression after the cell density reached a certain size (for example at a sample turbidity matching to an absorbance of 0.5 at the wavelength 600 nm, which corresponds to a number of 4×10^8 cells per ml of culture volume). Popular measures taken are the controlled onset by use of a temperature-sensitive transcription system or the external addition of repressor binding inducers, such as a galactoside, or arabinose in *E. coli* recombinant systems (see below). Artful tricks applied to fight the challenges of so-called leaky expression consist of promoter modification, vector-mediated overexpression of a repressor, or co-expression of an inhibitory gene product (see below).

No additional vector manipulations would be needed if the growth media composition governs the gene expression. This is controlled either by use of the "autoinducer" media (Table 3.3), or by inclusion of glucose, which is intended to metabolically lower the intracellular cAMP level, and the dissociation of the catabolite activator protein needed for the efficient target gene transcription (see below). Alternatively, the so-called plating method takes advantage of the effective bacterial growth behaviour and stable enforcement of antibiotic resistance on solid agar: Starter cultures are scraped off from plates and suspended in liquid medium for protein production.

3.1.7 Tagging of expressed protein to meet special needs

Another way of counteracting the problem of intracellularly accumulated unfolded proteins is the approach to make the products soluble by genetic fusion of special ready-folding domains to the N- or C-terminus of the target protein. Coupling partners that have been successfully used to this end are the glutathione-S-transferase (GST), the maltose binding protein (MBP), Green fluorescent protein (GFP), the NusA domain, or the thioredoxin protein (Trx) (Table 3.7). In addition to folding support, oligopeptide tags could be attached to the terminal protein positions for tag-specific immunological verification of the product in crude cell extracts. Several fluorescent proteins, e.g. GFP or mCherry and their derivatives fulfill a double role: They can be attached to ease the expression, and their chromophores also act as sensitive optical probes for the proteins of interest. Oligohistidine-, Strep-, Flag- and many other antigenic epitope- or protein domain-based tags are frequently used for the efficient product purification by affinity chromatography. The attached protein domains and tags can be readily removed by proteases at predesigned cleavage sites.

3.1.8 Handling of inclusion bodies

Folded polypeptides exhibiting biological activity are routinely targeted to the cytoplasm or cytoplasmic membrane. However, according to a recent review (Structural Genomics Consortium et al. 2008), only about 10 % of all tested eukaryotic proteins could be produced in soluble form in *E. coli*, whereas a fraction of about 50 % archaeal and eubacterial proteins were successfully made. The native folding sequence involves a build-up of disoriented secondary structure elements as first step, followed by a volume shrinking event through hydrophobic collapse, producing the "molten globule intermediate", which rearranges to the finally folded state (Burgess 2009). If this process is perturbed, especially in combination with high synthetic rates, nascent polypeptides often aggregate to form the so-called "inclusion bodies", consisting of entangled and partly unstructured fibers, having pronounced particulate features.

This occurs, because in the cellular environment with ~200 mg/ml protein concentration, unfolded proteins bear a threat to all cellular proteins due to their exposed hydrophobic cores. Inclusion bodies are strongly enriched in recombinant protein, which makes them highly attractive for purification (Ramon et al. 2014). There are instances where the production of inclusion bodies is desired, e.g. for raising sequence-specific antibodies. To obtain biological activity, the protein needs to be refolded, for which numerous protocols exist (see REFOLD database, http://refold.med.monash.edu.au/, Box 3.2). As a last measure, if the protein is only partly unfolded, or threatens the viability of the cell, it is subjected to degradation in the proteasome.

The formation of inclusion bodies can be avoided by slowing down biosynthetic rates to minimize levels of unfolded translation product, and to provide more time for *in vivo* folding processes. To this end, the gene dose is reduced by low copy number vectors and/or the gene expression rates are decreased by lowering the temperature to 20 °C or less and/or harnessing weaker promoters. Co-expres-

Box 3.2: Refolding of inclusion bodies.

Inclusion bodies of aggregated protein can make up more than 50 % of the prokaryotic cell protein. Subsequent to cell disruption, inclusion bodies are readily collected from the bacterial homogenates by centrifugation. Often, the particles consist of more than 90 % of pure target protein. It is, thus, worthwhile trying to refold the protein into its native structural and functional state by means of a standard protocol. To this end, the particulate material is solubilized with high concentrations of the denaturants urea or guanidium hydrochloride, or with low concentrations of the detergent Sarkosyl, performed under reducing conditions to avoid the unwanted formation of possible unnatural disulfide bonds. The refolding takes place by dilution into the denaturant-free solvent at very low protein concentrations to prevent reaggregation in presence of a redox buffer, which enables the controlled formation of disulfide bonds. The monodisperse target protein is separated from unfolded material or soluble multimers by ion exchange chromatography. Many variants of the basic protocol are in use, which often include low concentrations of naturally binding ligands to stabilize the target protein. The efficiency of refolding is measured by appropriate bioassays.

sion of chaperone genes is another strategy intended to improve the yield of soluble, natively folded protein (see below).

A possible detrimental accumulation of intracellular protein could be avoided by targeted excretion of products, which is controlled by the fusion of a so-called leader sequence to the N-terminus of the desired protein. Dependent on the cellular architecture, the two different cases have to be discerned: The excreted proteins are enriched in the periplasmic space bounded by the inner and outer membranes of Gram-negative bacteria (e.g. in *E. coli*), whereas Gram-positives release their products directly into the media (e.g. *Bacillus subtilis*, see below).

3.1.9 Protein stability – folding aided by chaperones

In addition to high levels of gene expression, the most important factor for successful protein production is adequate folding of the protein after translation at the ribosome, together with high stability within the bacterial cell.

Bacterial organisms exhibit a huge set of diverse chaperones (Baneyx et al. 2004). These are protein entities acting as folding helpers that are, on one hand, responsible for accurate folding of polypeptides that emerge from the ribosome. On the other hand, chaperones are caretakers of a viable metabolism: If a large amount of unfolded protein accumulates, the unfolded protein response induces further chaperone expression. If this fails, *E. coli* starts to encapsulate unfolded/aggregated protein in the inclusion bodies (see above).

In biotechnological terms, the use of strong promoters and optimization of protein production increase the likelihood of producing unfolded protein. This means that the major strategy is to keep all recombinant protein properly folded by either giving the cells enough time, or by reducing the rate at which recombinant proteins are produced. Often, production of less total protein leads to more high quality product.

Chaperones, so called folding-helpers, are significantly upregulated upon any kind of stress, such as elevated temperatures. Hence, they are also classified as "heat shock proteins" (HSPs).

In bacteria, a sequential pathway ensures correct protein folding. Upon release from the ribosome, Trigger factor (TF) binds to the nascent polypeptide, catalysing peptidyl-prolyl *cis/trans* isomerisation. Instantly, HSP70-like DnaK together with DnaJ/GrpE bind to short, hydrophobic patches of (partly) unfolded proteins, thereby shielding them from the surrounding and increasing the time frame within proteins have to fold. Next in line, if necessary (10 % of native proteins), is the GroEL/GroES (Hsp60/Hsp40-like) system. It forms an 800 kDa barrel-like complex, with a cap (GroES) and a large chamber that receives and retains unfolded proteins, which again enhances the time period for the protein to reach a folded state. This process is also governed by ATP-dependent conformational changes of the GroEL structure.

Apart from this *de novo* folding pathway, other mechanisms exist that assist in recovery of misfolded proteins, which is of particular importance for cell stress and recombinantly expressed proteins. Holdases or holding chaperones (e.g. small Hsp family, Hsp31, Hsp33) are produced under cell stress, and bind folding intermediates on their surface until DnaK, GroEL/ES molecules are free to accept new substrates. Disaggregating chaperones, e.g. ClpB (Hsp100-like) are produced in severe cellular stress, when protein aggregates occur. ClpB belongs to the superfamily of AAA-ATPases that form hexameric rings. Under ATP consumption, this ring slides along unfolded protein, like a pearl on a string, thereby leaving behind free, unfolded polypeptides to the DnaKJE and GroEL/GroES systems.

High success has been observed by introducing additional genes that code for chaperones, and coexpressing them with the gene of interest (Georgiou et al. 1996). An increase in the intracellular concentration of early (DnaK/DnaJ) or late (GroEL/GroES) seems to be sufficient for many difficult target proteins. For example, this method was crucial in producing eukaryotic kinases in *E. coli*, thereby allowing for easier biochemical assays that led to improved development of kinase inhibitors.

Chaperones differ between bacterial organisms and, to an even higher extent, between different kingdoms. Eukaryotic proteins often need specialized chaperones that are not present in prokaryotic organisms, e.g. particular folding helpers or special assembly proteins aiding the insertion of co-factors or metal centres. Hence, these proteins can be difficult to produce.

3.1.10 Membrane insertion and folding of membrane proteins in prokaryotes

Membrane proteins constitute 20 % of *E.coli*'s proteome, 31 % in *Caenorhabditis elegans* and an estimated 15 %–39 % in the human proteome (Almen et al. 2009). Despite the high number of genes and their functional relevance, membrane protein research has been a niche discipline. Over the past 1–2 decades, improved technologies of membrane protein production and purification have been developed, which have led to an innovative push forward to revealing the function and elucidating the structure of membrane proteins. To this end, it has been mandatory to understand how membrane protein biosynthesis is performed in pro- and eukaryotes. The following section focusses on the bacterial pathways.

Membrane topology of proteins
Knowledge about the membrane protein topology is essential for a sensible design of experiments. Numerous programs for predicting membrane protein topology are available now, in addition to many experimental techniques to determine protein folds. Soluble proteins orient most of their hydrophobic amino acids into their core,

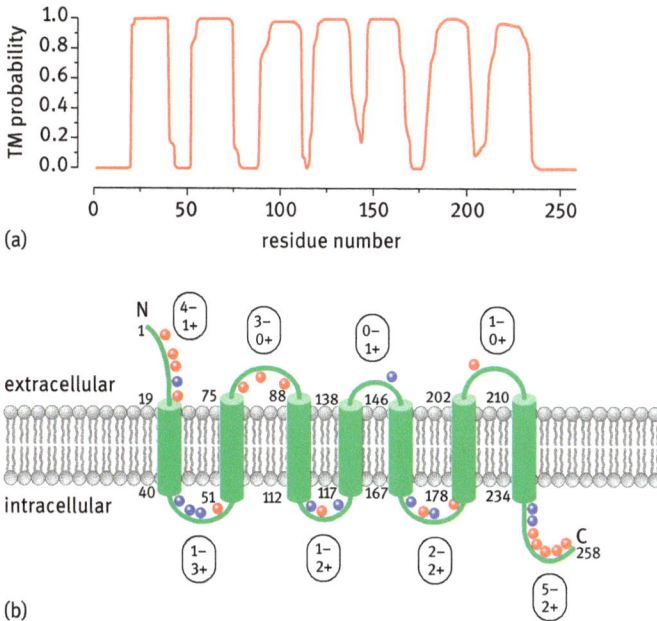

Fig. 3.4: Topology and structure prediction of an integral membrane protein. (a) Hydrophobicity profile of archaeorhodopsin3 of the archaeon *Halorubrum sodomense*, obtained from the prediction program TMHMM (http://www.cbs.dtu.dk/services/TMHMM/). The ordinate scale is from 0 to 1, corresponding to low and high probability for a residue being located in a hydrophobic environment. This plot marks the sequence regions that form transmembrane helices. (b) Topology of the transmembrane segments of archaerhodopsin3 within the membrane, determined from the hydrophobicity profile. The orientation of the transmembrane segments is assigned by the "inside-positive rule". Red dots correspond to negative charges from aspartic and glutamic acid side chains, blue dots show positive charges from lysine and arginine side chains.

which creates the driving force for folding in solution. In contrast, the hydrophobic parts of membrane proteins face predominantly towards the hydrophobic lipid environment of the membrane. Prediction programs exploit these properties by plotting the hydropathy along the polypeptide chain length, indicating patches of high hydrophobicity of helical membrane proteins together with secondary structure prediction, or by use of specially trained programs computing the probability of residues being located within or outside membranes (Figure 3.4a).

The basic rule, by which membrane proteins "decide" whether a particular loop is in the cytoplasm or in the periplasm, is the so called "positive-inside rule" (von Heijne et al. 1988). According to this rule, all fragments that are located on the non-translocated (inner) side (where translation takes place) bear two to four times as many positively charged residues as those on the translocated (outer) side (Figure 3.4b). Remarkably, it applies universally to all membrane proteins, regardless of whether they are targeted to the bacterial membrane, ER, Golgi, or thylakoid membranes.

Membrane targeting

All secreted and membrane targeted proteins possess an N-terminal signal peptide that is recognized by the signal recognition particle (SRP). The ribonucleoprotein SRP binds to the ribosome, and after recognition of the signal peptide, docks to the membrane-bound SRP receptor (SR). Thereby, the nascent polypeptide chain is targeted from the ribosome to a translocator complex, called Sec (Figure 3.5). In contrast to proteins destined to cross the membrane, membrane-bound proteins possess the above-mentioned positively charged sequence feature. The Sec complex translocates the amino acid chain until it encounters a region with high positive charge, which is kept on the inner side of the membrane. In case of multiple spanning membrane proteins, the Sec complex resumes the translocation after recognition of the next hydrophobic transmembrane part.

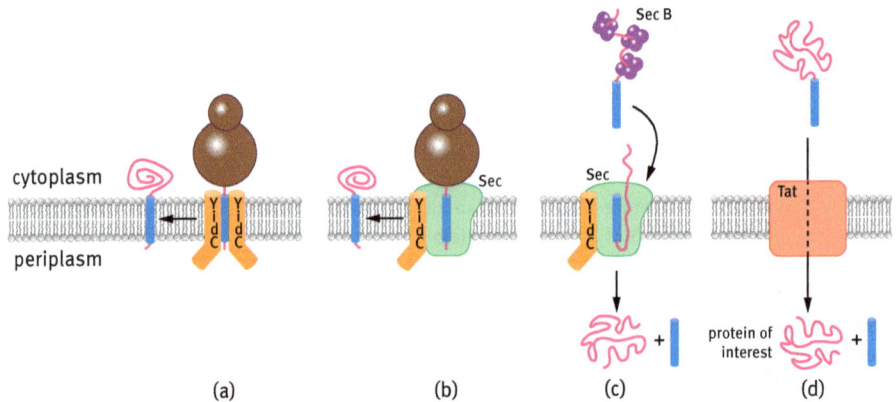

Fig. 3.5: Membrane protein insertion and translocation systems in prokaryotes. (a, b) Two different systems function as membrane protein insertases: The YidC homodimer (a) and the YidC bound Sec complex (b). Both complexes are able to bind the ribosome and transfer polypeptides comprising of transmembrane regions into the membrane. (c, d) Protein translocation across the membrane is also catalysed by two different systems, the Sec complex (c) as well as the Tat (Twin-Arginine Translocation) complex (d). While the Sec complex recognizes positively charged signal sequences and exports unfolded proteins, the Tat complex exclusively translocates folded proteins, if they contain the appropriate signal sequence comprising two arginines.

Production of recombinant membrane proteins

Important consequences have to be considered in terms of recombinant membrane protein production. The intricate membrane targeting and insertion complex acts as a multi-step catalytic system, which can be saturated. At high protein production rates exceeding the catalytic capacity, misfolded proteins could accumulate and either result in (partial) protein degradation or inclusion body formation. For

successful membrane protein production, a few typical parameters should be respected and appropriately adjusted when necessary:
- Low level expression strains should be used if they are available (e.g. the so-called C41/C43 cells in case of an *E. coli* expression host, see below)
- The expression rate should be kept low by attenuating the level of induction (e.g. by a weaker promoter or titratable promoter)
- Keeping the temperature low to increase folding time and reduce protein production)
- A genetic construct should be designed, which provides fused protein tags (e.g. MBP, GFP) generically promoting protein folding (see above).

Despite all recent advances, producing eukaryotic membrane proteins in bacteria still trails behind and is very difficult to achieve, due to profound differences between the eukaryotic ER and the prokaryotic plasma membrane system.

At first, their membrane compositions differ remarkably. Phospholipids are not only present in different ratios, but eukaryotes also contain specific lipids like sphingolipids, have polyunsaturated fatty acid chains, and most importantly, cholesterol. For membrane protein folding, these eukaryotic ingredients are crucial. Also, the Sec translocator complexes show differences in their channel composition: While the ER based channel is positively charged, the bacterial Sec complex lacks these features, and interestingly, the protein translocation rates in eukaryotes are ~4 times lower than the bacterial ones. A fact noted above is the posttranslational modification, which often occurs in eukaryotic proteins, e.g. by acylation or phosphorylation, which can be crucial for folding and activity of the respective protein. Another very important feature of eukaryotic membrane proteins is their glycosylation of the extracellular surface, which can not be performed by prokaryotes. Finally, eukaryotic cells are able to provide cytosolic chaperones, like Hsp70, that may support folding of translocated polypeptides. On the other hand, *E.coli's* periplasm does not have any related chaperones.

The most successful attempts of producing eukaryotic membrane proteins in bacteria are accomplished with β-barrel proteins, which seem to be more robust with respect to their folding process. As yet, the preferred method for eukaryotic α-helical membrane proteins is still to produce them in a eukaryotic system such as in yeast, insect, or mammalian cells.

3.2 *Escherichia coli* – the most popular host for protein production

The Gram-negative γ-proteobacterium *Escherichia coli* is the best-studied prokaryote. Although a few pathogenic variants such as enterohemorrhagic *E. coli* (EHEC) are known, many of the strains (e.g. the K12 derived ones) are considered safe for

total count (not null): 90952

| | 0 | 70039 |

Escherichia coli	
Escherichia coli BL21(DE3)	
Spodoptera frugiperda	
Escherichia coli BL21	
Homo sapiens	
Trichoplusia ni	
Pichia pastoris	
Saccharomyces cerevisiae	
Cricetulus griseus	
Drosophila melanogaster	
Mus musculus	
cell free synthesis	
Rhodobacter Sphaeroides	
Bacillus subtilis	
unidentified baculovirus	
Thermus thermophilus	
Komagataella pastoris	
Escherichia coli K12	
cell-free synthesis	
Aspergillus oryzae	
Streptomyces lividans	
Drosophila	
Mesocricetus auratus	
Halobacterium salinarum	
Dictyostelium discoideum	
Cricetinae	
cell-free protein synthesis	

last updated: Tuesday Apr 29, 2014 at 5 PM PDT

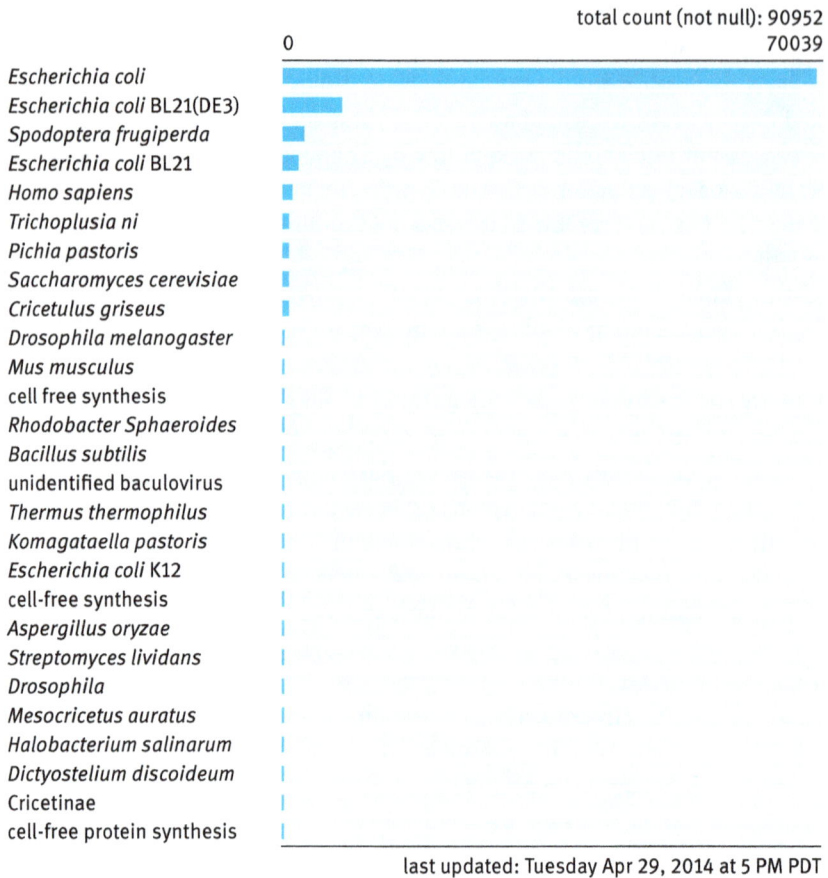

Fig. 3.6: Expression host of proteins listed in the PDB data base (protein data bank). The Protein Data Base (PDB) has currently about 100,000 entries and mainly provides information on high-resolution structures of proteins and nucleic acids. It also provides many additional hints on the production and the biochemical state of the protein. The predominant role of the prokaryote *Escherichia coli* as production plant of the majority of structurally investigated proteins becomes obvious by comparison with other expression hosts.

routine laboratory use. The organism grows rapidly (20 min doubling time under optimal conditions) on relatively cheap media. Its genome, sequenced in 1997, has a size of 4.6 MBp and consists of 4288 genes accounting for the considerable metabolic flexibility. During the last century, an enormous wealth of molecular biological knowledge has been accumulated, which has made the bacterium the easy-to-use "workhorse" in Molecular Biotechnology. Many lab strains with highly specialized features have been developed for innumerable needs, and a large genetic toolbox is available. These facts explain why *E. coli* is the prime choice host for homologous and heterologous expression of proteins in research and industrial applications (Table 3.4). Figure 3.6 indicates that more than 90 % of the ~100,000 proteins

Table 3.4: Pharmacologically relevant proteins produced in *E. coli*.

Active compound	Function	Medical indication	Year	Company
Anakinra	Interleukin-1-receptor antagonist	Rheumatoid arthritis	2002	Swedish Orphan Biovitrum
Certolizumab pegol	Fab-fragment, monoclonal antibody	Rheumatoid arthritis	2009	UCB
Filgrastim	Granulocyte Colony-Stimulating Factor (GCSF)	Neutropenia in chemotherapy and HIV infection	1991	Amgen
Insulin	Insulin-receptor agonist	Diabetes	first: 1996	Lilly/Sanofi-Aventis
Interferon α-2b	Interferon	Hepatitis B/C, Cancer	2000	MSD Sharp & Dohme
Interferon β-1b	Interferon	Multiple Sclerosis	first: 1995	Bayer, Novartis
Lipegfilgrastim	Granulocyte Colony-Stimulating Factor (GCSF)	Neutropenia in chemotherapy	2013	Sicor Biotech
Mecasermin	Insulin-like Growth Factor 1 (IGF-1)	Primary IGF-1 deficiency	2007	Ipsen Pharma
Palifermin	Keratinocyte Growth Factor (KGF)	Oral mucositis	2005	Swedish Orphan Biovitrum
Pegfilgrastim	Granulocyte Colony-Stimulating Factor (GCSF)	Tumour therapy	2002	Amgen
Peginterferon α-2a	Interferon	Hepatitis B/C	2002	Roche
Peginterferon α-2b	Interferon	Hepatitis C	2000	MSD Sharp & Dohme
Pegloticase	Uricase	Gout	2013	Savient Pharma
Pegvisomant	Growth Hormone – antagonist	Acromegaly	2002	Pfizer
Ranibizumab	Fab-Fragment, monoclonal antibody	Macular degeneration	2007	Novartis
Reteplase	Serine protease	Thrombolytic treatment	1996	Actavis
Romiplostim	Thrombopoetin-receptor agonist	Idiopathic thrombocytopenic purpura	2009	Amgen
Somatropin	Growth Hormone (GH)	Growth deficiancies	2001	Ipsen Pharma
Tasonermin	Tumor Necrosis Factor (TNF)	Tumour therapy	1999	Boehringer Ingelheim
Teduglutid	Glucagon-like peptide-2 (GLP-2)	Short bowel syndrome	2012	NPS Pharma Holdings
Teriparatid	Parathyroid hormone (PTH)	Osteoporosis	2003	Lilly

(taken and modified from http://www.vfa.de/de/arzneimittel-forschung/datenbanken-zu-arzneimitteln/amzulassungen-gentec.html) listed in the Protein Data Bank (PDB) (http://www.rcsb.org/pdb/) have been expressed in *Escherichia coli*.

Box 3.3: The gold standard: Molecular genetic techniques for *Escherichia coli*.

Pioneering work of numerous researchers, has extensively characterized the molecular biological system of *Escherichia coli*, and laid the ground work for successful gene expression in this prokaryote. Many tools are available nowadays for gene disruption, allelic replacement, and for mutagenesis, e.g. miscellaneous bacteriophages and transposons that can be delivered by various methods. One can select from a large set of customized plasmid vectors, equipped with different origin of replication, selective markers, promoters, multiple cloning sites, and a lot of extra features, to achieve distinct molecular biotechnological tasks, such as cloning and expression of homo- and heterologous genes. Standard procedures for plasmid vector uptake are transformation of chemically or electrically competent *E. coli* cells. Plasmids can also be im- or even exported by a process named bacterial conjugation, which is especially relevant for the gene transfer from *E. coli* to other prokaryotic genera (see below, section 3.3). The techniques in genetic engineering are easy to learn, and a wealth of laboratory protocols can be obtained from the literature.

The presence of its lipopolysaccharides (LPS) is worth noting, as this endotoxin, located in the outer membrane, is responsible for the potential dangerous characteristics of this organism. Although most of the strains used in everyday laboratory work are innocuous, safe biotechnological production, especially for drug and food applications, is, therefore, more elaborate compared to other hosts. The production of proteins in *E. coli* is often termed as "protein (over)expression", which is rather lab jargon. Correctly spoken, "over" applies only for endogenous genes being expressed to higher than physiological levels, and protein production always follows "gene expression". The choice of the right expression system (Figure 3.1) is crucial for successful protein production.

Most gene expression systems are plasmid-based (Figure 3.7a), due to the simplicity of transforming bacteria with the autonomously propagated plasmids rather than by introducing gene expression cassettes into their chromosomes (see Box 3.3). This immediately bears another level of regulation, as cellular plasmids are usually present in multiple copies within a single bacterium. Typical low copy number plasmids, such as derivatives of pBR322 bearing a *colE*1-type origin of replication (*ori*) are present at about 10–20 copies per cell, (e.g. pBAD and pET vectors). High copy number plasmids like the pUC vectors can show up to a few hundred copies per cell, which is due to an altered *ori*. Obviously, the choice of promoter strongly influences the level of gene expression that can be achieved, so called gene dose rate. When co-expressing two different genes from two plasmids, the choice of *ori* becomes particularly important. Not only is the *ori* combination important for comparable levels of gene expression, it is further necessary to combine different origins of replication as this ensures unbiased copying of each plasmid.

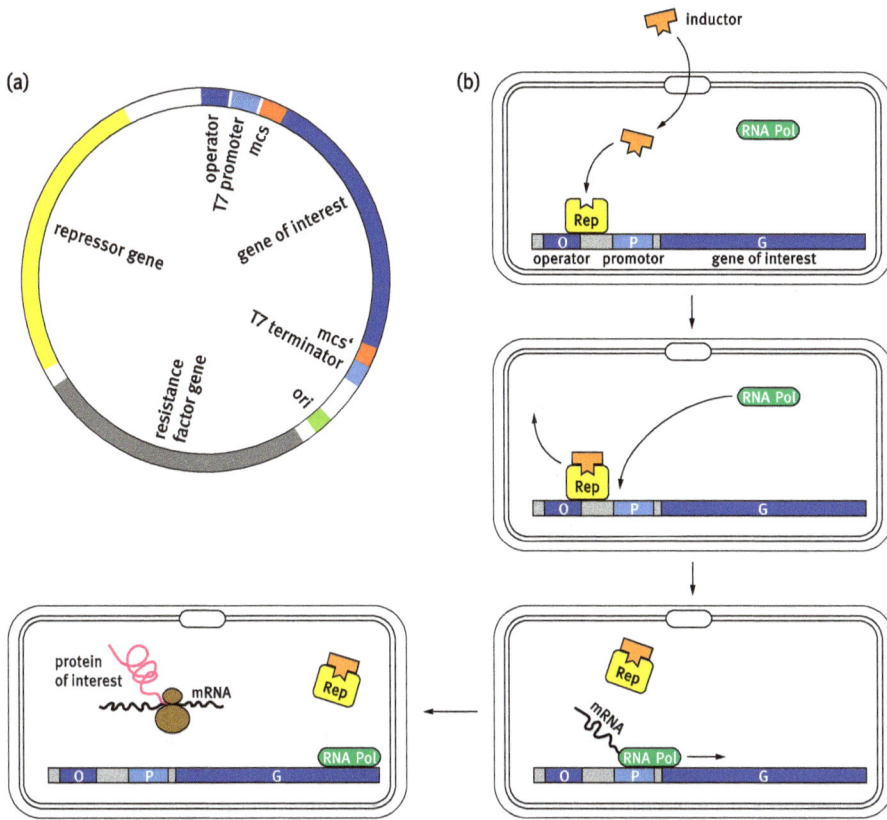

Fig. 3.7: Induced gene expression. (a) Example of an expression vector. A standard expression vector contains an origin of replication (*ori*), a gene for resistance selection, a gene that codes for the transcription repressor and a multiple cloning site that harbours the gene of interest. (b) Example of gene expression mediated by an inductor. In the uninduced system the repressor (Rep, yellow) binds to the promoter site and inhibits gene expression. Once inductor, e.g. IPTG (orange), is added, it passes the outer membrane (thin line) via nonspecific pores and enters the cytoplasm by a specific inner membrane-associated transporter (white oval). The inductor binds to the repressor with high affinity, causing a conformational change, which facilitates the dissociation from the promoter site. This enables the RNA polymerase (Pol, green) to bind at the promoter region and to start transcription. The formed mRNA is directly taken up by ribosomes leading to immediate translation of the desired gene.

If antibiotics are to be used as selection markers in molecular biology, plasmids need to exhibit a specific gene that codes for proteins conveying resistance to antibiotics. For example, β-lactam (beta-lactam) antibiotics are destroyed by a protein called beta-lactamase that cleaves the functional beta-lactam ring of the molecule. Aminoglycoside antibiotics cannot be hydrolyzed; instead they are phosphorylated or adenylated by transferases, thereby inhibiting their function. Tetracyclin resistance is achieved by expressing a specific and efficient transporter, which pumps the antibiotic molecules outside the cell.

Table 3.5: Commonly used antibiotics.

Name	Type	Function	Resistance	Std. Concentration
Ampicillin	Beta-lactam	Inhibitor of transpep-tidase in cell wall synthesis	Beta-lactamase	100 µg/ml
Kanamycin	Aminoglyco-side	Binder of 30S riboso-mal subunit to block initiation complex	Kanamycin phso-photransferase	50 µg/ml
Chloramphenicol	Chlor-amphenicol	Binder of 50S riboso-mal subunit to block peptide transfer	Chloramphenicol acyetyltransferase	30 µg/ml
Tetracycline	Tetracycline	Prevents binding of aminoacyl-tRNA to 30S ribosomal subunit	Tetracycline Exporter	50 µg/ml
Carbenicillin	Beta-lactam	Inhibitor of trans-peptidase in cell wall synthesis	Beta-lactamase	100 µg/ml
Streptomycin	Aminoglyco-side	Binder of 16S riboso-mal subunit to block initiation complex	Streptomycin Adenyltransferase	50 µg/ml

The intracellular maintenance of this extra DNA requires metabolic resources, resulting in a considerable growth handicap of plasmid carrying bacteria. Plasmids in transformed cells can be lost at a certain frequency, resulting in low levels or even stalled gene expression. Hence, plasmids contain antibiotics resistance factor genes that code for proteins that ensure the cell's survival in media containing antibiotics (Box 3.4, Tab 3.5).

Box 3.4: Antibiotics and their mode of action.

Antibiotic selection is important for maintaining a constant replication of transformed plasmids. Most antibiotics used in laboratories target two different cellular processes (Table 3.5), either interfering with cell wall synthesis or translation.

Beta-lactam antibiotics (such as ampicillin) covalently bind to penicillin-binding-proteins (PBPs) that sit in bacterial cell membranes, thereby inhibiting PBPs to catalyze cell wall formation.

Translation can be inhibited by antibiotics in different ways. Aminoglycosides (such as kanamycin or streptomycin) bind to ribosomal subunits and block different stages of protein production, e.g. the formation of the initiation complex.

3.2.1 Induced recombinant gene expression

As opposed to expressing a gene in its natural host at physiological level, recombinant expression always causes artificial cellular states. Hence, *Escherichia coli* will

Table 3.6: Inducible expression systems and promoter types.

	Operator	Promoter	Induction	Properties
T7 system	*lacO*	T7	IPTG binding to LacI	IPTG induces expression of T7 polymerase (e.g. DE3 strain) and target gene Very efficient, high expression
Arabinose system	*araO*	*araI*	Arabinose binding to AraC	Medium to high expression Independent of T7 system Combination with BL21(AI) cells reduces leaky expression
Trp system	*trpO*	*trpI*	3-β-indole-acrylic acid binds to TrpR and replaces tryptophan	Medium expression Leaky expression can be very high
Tac system	*lacO*	(−35 region) of *lac* (−10 region) of *trp*	IPTG binding to LacI	Combination of *lac* promoter and *trp* promoter

suffer from extensive stress, and needs to be constrained to express the foreign protein at high rate and high quality. Two major factors are generally considered, the choice of operon and the promoter type (Figure 3.7a, b). The operon regulates the production of mRNA by precisely controlling the time point of transcriptional onset, and the promoter type determines by its strength, how much of mRNA is produced (Table 3.6). The majority of researchers focus on only a few different types with variable properties, frequently using a repressor-based gene regulation system. DNA transcription repressors are often regulated by small molecules, typically the ones that are produced or needed by those proteins whose genes are controlled by positive or negative feedback loops.

The *lac* operon
The first biotechnological attempts to control gene expression have used the *lac* promoter, which is copied from the well-known *lac* operon (coding for genes needed for bacterial lactose catabolism). In fact, instead of the *lac* promoter, most expression vectors use the so-called *lac*UV5 promoter variant that is altered at two base positions and enables a higher transcriptional initiation rate. The promoter is located immediately upstream of the DNA operator region (called *lacO*), to which the inhibitory *lac* repressor LacI binds. The *lacI* gene is often cloned into the expression plasmid and is constitutively transcribed. After transformation by the vector, the LacI repressor protein immediately docks to its binding site on the plasmid and inhibits gene expression.

Isopropyl-β-D-thio-galactoside (IPTG), a molecular mimic of allolactose, is used to induce expression. IPTG binds with a dissociation constant of about 6 µM to the repressor. Binding induces a conformational change accompanied with drastically reduced affinity of the LacI repressor to the DNA binding site which initiates the efficient synthesis of mRNA (Figure 3.7).

The *lac/T7* system

This system was already developed in the eighties of the last century (Studier et al. 1986) and is now the most widely used gene expression system (e.g. the so-called pET or pGEX vectors). It relies on the strong T7 promoter of the bacteriophage T7 placed before the target gene. As the T7 promoter is unaffected by repressors, the intracellular level of bacteriophage T7 polymerase regulates its activity, because the endogenous *E. coli* RNA polymerase does not recognize the T7 promoter. To avoid transcriptional read-through, vectors of the T7 promoter type must also accomodate a T7 terminator motif after the gene of interest in order to stop the progression of RNA polymerase and to initiate another turn of transcription at the promoter site.

The currently most popular strategy involves bacterial strains (e.g. "DE3" strains, see below) that harbour a gene for T7 RNA polymerase under control of the *lac* promoter/*lac* operator on their chromosome. These DE3 strains are transformed by a vector containing the T7 promoter preceding the target gene of interest. In this set-up, the inducer IPTG releases the LacI repressor from the *lacO* operator of the chromosomally encoded T7 polymerase gene, and the T7 polymerase transcribes the vector encoded gene following the T7 promoter to enormously high amounts of mRNA. The system is extremely efficient: Almost all translationally active ribosomes will produce only the one specific protein, which can make up to 50 % of all bacterial proteins within 3 hours after induction.

Due to the high affinity between ITPG and its receptor, the inducer concentration cannot be adjusted for precise control of gene expression rates. Commonly used standard concentrations higher than 100 µM IPTG immediately saturate all repressor proteins and lead to an all-on answer. As T7-based expression is so powerful, it is difficult to stringently control the *lac* operator. Very often, basal (uninduced) expression levels (also named "leaky expression") are observed. This effect may be negligible for simple standard proteins, but it can be of crucial relevance for "difficult-to-express proteins" which exhibit strong cell toxicity. In extreme cases, bacterial growth may be dramatically diminished or even stalled before the inducer has come into play.

The arabinose (*ara*) system

The arabinose expression system makes use of the well-studied *araBAD* operon, whose gene expression is controlled by the activator arabinose (Guzman et al. 1995).

Herein, *araC* codes for the AraC protein that functions as transcriptional repressor and as activator at the same time. When the cells are depleted of arabinose, AraC binds to the *araI* ("I" stands for "induction") and the *araO1/araO2* operator DNA regions and inhibits transcription. After addition, arabinose binds to AraC, which stays at the *araI* and *araO1* sites. In turn, the AraC-arabinose complex triggers transcriptional activation, which in contrast to the previously described T7 system fully relies on *E. coli's* endogenous RNA polymerase and the *araI* region (Guzman et al. 1995). The resulting switch from repressor to activator function strongly amplifies the expression signal. Another benefit is that the most commonly used *araBAD*-based vector, pBAD, contains a low copy number *ori* and generally produces mRNA at smaller level. The low expression rate in turn, very often determines the success of protein production, especially in cases of toxicity or low solubility of the protein of interest. Furthermore, the *araBAD* operon can be generally inhibited by "catabolite repression" in presence of high intracellular glucose concentration. Hence, researchers frequently use growth media supplemented with glucose in order to suppress leaky, i.e. uninduced gene expression. To get rid of the repressor agent glucose, it is necessary to extensively wash the bacteria prior to induction. At best, the precultures are spun down and suspended in fresh glucose-free expression media containing arabinose.

The *trp* system

Before the advent of the *lac* system, recombinant gene expression has been largely controlled by use of the tryptophan operon. Herein, the repressor is active as long as tryptophan is available, but leaves its DNA binding site upon depletion of tryptophan. Accordingly, cultures had to be diluted into minimal media at first, and were secondly supplied with indole-3-acrylic acid in order to induce gene expression. The promoter is highly active, and results in protein production levels comparable to the T7-system. However, handling problems and high basal gene expression due to low *trp* repressor levels soon led to disuse of the original method, and prompted its further advancement to the "*tac*-system" (see below).

The *tac* system

Different research groups attempted to further improve the highly productive but difficult *trp*-system, and came to congruent conclusions and published similar ideas. They merged the *lac* and the *trp* promoter regions, and termed the resulting hybrid as "*tac* promoter" (de Boer et al. 1983). The DNA sequence upstream of position –20 bp (transcriptional initiation) was taken from the *trp* promoter, and the *lacUV5* promoter region, including the binding site for the lac repressor, was inserted downstream of position –20 bp (bearing the Pribnow box). The resulting artificial promoter construct combined the advantages of the high level gene expression inherent to the *trp* system with the facile control by IPTG induction through the *lac* repressor.

This novel system significantly boosted the field of recombinant protein production, eventually leading to the major breakthroughs in protein biochemistry of the late eighties and nineties of last century.

The autoinduction system

The autoinduction system is very similar to the lactose-induced T7 system (Studier 2005). However, as the name already suggests, this method does not rely on the induction by IPTG, but uses lactose from the media for induction. The media is based on standard constituents (see Table 3.3) supplemented with glucose, lactose, and glycerol. Glucose will initially suppress gene expression and lead to very high cell densities. Once the glucose level drops, the cell starts importing lactose and glycerol. Glycerol serves as energy source, when glucose is unavailable, the imported lactose will induce the T7 system based gene expression. Autoinduction provides one main advantage: Protein production starts at very high cell mass, often resulting in overall increase of total protein. However, it is more difficult to define the induction point and several ratios of glucose to lactose/glycerol need to be tested.

3.2.2 Expression conditions

Bacterial cell division typically obeys the exponential growth kinetics (Figure 3.3). If the cells are undisturbed, rates depend on the strain used, temperature, and availability of nutrients. The growth rate of a bacterial cell should be unaffected, if the recombinant protein of interest is produced in an ideal biotechnological process environment. This assumption is, however, rarely correct, because most of the cell's full translation machinery will be used to overproduce foreign proteins, which elicits obvious consequences: At first, the capacity for synthesis of endogenous proteins needed for fast growth and reproduction of the bacterial cell is drastically lowered. Another factor is based on the stress conditions generated within the host cell, if protein folding kinetics are concerned. Intracellular stress evokes the "Unfolded Protein Response" (UPR, see below) which may ultimately cause either stalled growth or even cell death.

Hence, optimal growth conditions are needed in order to maintain the cell's own ability to continue dividing exponentially and producing cell mass. Fortunately, biotechnological processes are easily governed by adjusting the following accessible parameters, i.e. the growth temperature before and after induction, cell density at the moment of induction, concentration of inducer, and duration of expression.

At the temperature of 37 °C, *E. coli* displays optimal growth and maximum speed of reproduction. In other words, the metabolism is programmed and fully focussed on cell division. The entire translation machinery is at the limit for protein synthesis required for cell growth. Once recombinant gene expression is induced, translation switches over to extraneous protein production, which weakens the cell infrastructure for biosynthetic protein supply needed to maintain the fast 37 °C cell growth. This often leads to a decrease of growth speed at the least, sometimes to stalled division. Hence, reduction of the growth temperature at the time of induction can have a huge impact on successful protein production. Of course after induction, the growth temperature influences the very same parameters: Reduced temperature can often grant the cell more time to firstly translate the protein (e.g. in difficult cases, where codon usage is not fully optimized), and secondly, to correctly fold the protein. The chosen temperature also affects the stability of the protein as many proteins tend to be more stable at lower temperature.

It is understandable from the kinetic growth behaviour of bacteria that the time point of induction critically influences the yield of expressed protein. The bacterial growth curve usually shows the number of cells (measured as media turbidity at a specified wavelength, named optical density (OD)) being generated by division over time (Figure 3.3). Three different concepts exist to meet this purpose and should be tested experimentally:
(i)   Induction at low optical density results in low cell mass, but experiments have shown that this can be optimal for the production of "difficult" proteins.
(ii) Induction at medium OD_{600} of about 0.7 represents the most commonly used condition. Here, the bacterial population is at exponential growth in standard media and is only slightly disturbed by recombinant protein production.
(iii) Induction at high OD_{600} (e.g. about 1.0) occurs, when the cell population is in transition towards the lag phase. Hence, a high cell mass already exists, but further growth continuation should not be expected after induction. This method is often accompanied with very short expression times.

The duration of expression time is typically tested in initial test experiments at small scale, together with the temperature after induction. Many proteins are (partially) degraded over time in *E. coli*, either due to incorrect folding or digestion by endogenous proteases.

The appropriate choice of media determines the supply with nutrients. Table 3.3 summarizes the most commonly used media types. However, the optimal composition is hard to predict and has to be found for each expression task individually.

Typically, different ratios of peptone, a pepsin-hydrolysed mixture of peptides and amino acids, and yeast extract are chosen in order to adjust the "richness" of media. Rich media allow for very high cell densities as the culture is not nutrient-deprived, or in other words, they are not "limited" by one or more non-specified factors. They also tend to produce more recombinant protein, but are usually more expensive. Another type of media is the rarely applied "minimal media" (e.g. M9-based media, see Table 3.3), which offers only scarce supplies. There are reports

Box 3.5: Increased folding efficiency of thermophilic proteins at standard growth temperature.

If mesophilic proteins are difficult to express due to instability, it may help to try thermophilic homologs. For these, the prokaryotic host cell may provide a favorable folding environment at the standard growth temperature of 37 °C, because the actual fermentation temperature is much below that of normal living conditions of the thermophilic genetic source organism. Consequently, the thermophilic protein should exhibit reduced internal mobility and "freeze" the protein conformation, which allows higher stability and much better folding. Structural biochemists agree that the same reason explains the better suitability of thermophilic orthologs in protein crystallization, where reduction of molecule movement is indispensable for contact formation.

No orthologs exist for certain mammalian proteins, such as the G protein coupled receptors (abbreviated GPCR). They exhibit remarkable inherent dynamics related to their biological function, i.e. to readily alter their conformations after ligand binding. For the purpose of crystallization, the protein flexibility could be drastically constrained by means of mutagenesis, in a process named conformational thermostabilization (Serrano-Vega et al. 2008). The site-specifically altered variants, expressed in *Escherichia coli*, were resistant to thermal unfolding by more than 20 °C compared to the wild type GPCR.

It should be noted that the relation of low temperature and folding efficiency is not a rule without exceptions: Many proteins even depend on elevated temperatures to gain a correct fold.

that some membrane proteins show higher yields in M9 compared to nutrient rich media. Most likely, this fact could be explained by the strong reduction of growth rates in this media owing to the low food offer, thus providing the cells with more time for protein production. An additional advantage is that the composition of minimal media can be easily changed, which is required for *in vivo* labelling with isotopically enriched compounds, e.g. $^{15}NH_4Cl$ and/or ^{13}C-glucose for NMR structure determination, or with unnatural amino acids, such as seleno-methionin instead of methionin, which is often used for protein crystallography.

3.2.3 *E. coli* strains used for high-level protein expression

Finding the right combination of a suitable bacterial strain with appropriate media can improve protein production up to more than 100-fold. Many of the existing genetic variants of *E. coli* can be obtained from culture collections (see above). An extensive overview on the genetic backgrounds of these genotypes, i.e. listings of the individual mutations annotated in the well-known four-letter gene code (see Table 3.2 for examples), can be found at http://openwetware.org/wiki/E._coli_genotypes.

The two most commonly used strains *E. coli* K-12 and *E. coli* B have been extensively modified to create genetic derivatives (Studier et al. 2009). Their absence of the F-plasmid precludes horizontal DNA transfer to other strains, which is an important prerequisite for biological safety reasons as the plasmid-transformed bacteria should not act as uncontrolled DNA donor to other prokaryotes.

The K-12 strain is the genetic ancestor of *E. coli* DH5α cells, which have been used for protein expression until about 2005, but are nowadays overrun by strains derived from *E.coli* B. These B strains lack the multi-specific proteases Lon and OmpT, proven to be crucial for generating high levels of recombinant protein.

Today, the most popular hosts for protein production are BL21 and its relatives (see below), which are derived from *E. coli* B, actually originally being selected for bacteriophage λ production. Hence, the BL21 strain bearing a λ prophage and a few additional sequence modifications, became a template used to create many different commercial strains with explicit additional functions (Table 3.6). BL21 (DE3) is the most frequently used expression host, in which the prophage-encoded T7 RNA polymerase gene is put under control of the *lac*UV5 promoter. Use of this strain is mandatory for gene expression from plasmids based on the T7 promoter. In the strain BL21 (DE3) pLys, the BL21 is transformed by chloramphenicol resist-ance-mediating pLys plasmids, either pLysE or pLysS, introduced in order to allow for tighter control of expression. pLys codes for T7 phage lysozyme (Zhang et al. 1997). In addition to its function of cleaving glycosidic bonds of bacterial cell walls, the T7 lysozyme has the unusual ability to bind to the T7 RNA-polymerase enzyme, resulting in its catalytic inhibition. By this means, the lysozyme strongly suppress-es T7 polymerase activity, which could otherwise lead to unwanted leaky expres-sion prior to induction: This tightly controlled setup initiates the transcription of the T7 polymerase gene only after IPTG addition, and ensures that potentially toxic target genes are unable to corrupt protein production by prematurely killing the host cell. BL21 (AI) is similar to (DE3) strains. Their difference is that (AI) strains have a T7 polymerase gene under control of the *ara* promoter and are induced by arabinose instead of IPTG. Hence, a precise control of T7 RNA-Polymerase levels can be achieved, often improving protein production of difficult proteins like mem-brane proteins.

A typical strain that harbours a plasmid that codes for rare tRNAs is named BL21 (DE3) Rosetta (another example is Codon+(RIL), see Table 3.7). These strains can successfully be used, if the expressed gene contains codons with low codon usage in *E. coli*, e.g. eukaryotic genes.

The strains BL21 (DE3) C41 and BL21 (DE3) C43 are expression hosts highly suited for the production of integral membrane proteins that have been discovered by Bruno Miroux and the nobel laureate John E. Walker (1996). As membrane pro-teins are known to be often toxic to standard *E. coli* cells upon high-level produc-tion, leaky expression may already lead to host cell death. These so-called "Walker-strains", have been isolated from colonies of mutant bacteria that survived toxic membrane protein expressions. This was a landmark discovery for membrane pro-tein research as it was then possible to overexpress genes that were hitherto detri-mental for cell survival. Both strains exhibit a mutation in the *lac*UV5 promoter for the T7 RNA polymerase, causing reduced levels of gene transcription, which results in proper folding and marked reduction of toxic effects of the target protein. A commercialization of this concept is used in Lemo21 (DE3) cells (Wagner et al. 2008), where the *lac*UV5 promoter is replaced by an accurately titratable rham-nose-inducible promoter, which acts analogous to BL21 (AI) cells.

Table 3.7: Commonly used *E.coli* expression strains.

Name	Genotype	Properties
BL21	F⁻ *dcm ompT hsdS*(r_B^- m_B^-) *gal* [*malB*⁺]$_{K-12}$($λ^S$)	Standard *E. coli* B expression strain, which lacks Lon and OmpT protease
BL21 (DE3)	F⁻ *ompT gal dcm lon hsdS_B*(r_B^- m_B^-) λ(DE3 [*lacI lacUV5-T7 gene 1 ind1 sam7 nin5*])	Strain with λ prophage, carrying T7 RNA polymerase under IPTG regulation in lacUV5 promoter
BL21 (AI)	F⁻ *ompT gal dcm lon hsdS_B*(r_B^- m_B^-) *araB*::T7RNAP-*tetA*	Strain carrying T7 RNA polymerase under arabinose regulation in *araBAD* operon
BL21 (DE3) pLysS	F⁻ *ompT gal dcm lon hsdS_B*(r_B^- m_B^-) λ(DE3) pLysS(cm^R)	DE3 strain, carrying T7 phage lysozyme which suppresses basal expression of recombinant proteins
BL21 (DE3) Rosetta	F⁻ *ompT hsdS_B*(r_B^- m_B^-) *gal dcm* λ(DE3 [*lacI lacUV5-T7 gene 1 ind1 sam7 nin5*]) (Cam^R)	DE3 strain, carrying a vector for expression of rare tRNAs
BL21 (DE3) C41	F⁻ *ompT gal dcm hsdS_B*(r_B^- m_B^-)(DE3)	DE3 strain for expression of toxic/membrane proteins. Carries uncharacterized mutation
BL21 (DE3) C43	F⁻ *ompT gal dcm hsdS_B*(r_B^- m_B^-)(DE3)	C41 strain for expression of toxic/membrane proteins. Carries additional mutations compared to C41
C600	F⁻ *tonA21 thi-1 thr-1 leuB6 lacY1 glnV44 rfbC1 fhuA1* λ⁻	Strain used for *tac* promoter expression

3.2.4 Peptide and protein tags used for protein localization or affinity purification in *E. coli*

The choice of a tag is crucial for successful purification fused to the protein, and also strongly influences protein production in bacteria. In this chapter we will focus on the consequences of protein tag types on protein production. Generally, the biological nature of a tag as well as the position of the tag (N- or C-terminally fused to the protein of interest) need to be considered. The two most commonly used tag types in molecular biology are short peptides and protein tags.

Frequently used peptide tags are the Flag-, Strep-, c-Myc- and S-tag, and the famous oligohistidine tag, which is most abundantly employed in molecular biology (Table 3.8). At large, the tag-modified derivatives have similar properties to the naturally arising protein, e.g. yield and folding are typically at the same level compared to untagged proteins. However, possible effects of tag positioning on either N- or C-terminus of the target protein should be tested. Tagged variants may indeed show less *in vivo* degradation than native proteins and hence appear to be more stable. Peptide tags and drawbacks for protein purification, are reviewed in (Terpe 2003).

Table 3.8: Peptide and protein tags.

Name	Size/kDa	Sequence	Binder/matrix	Elution
His (6–10)	0.84 (1.4)	HHHHHH(HHHH)	Ni/Co/Cu – matrix (e.g. Ni-NTA)	Imidazole/low pH
GST	26	Uniprot: P08515	GSH	GSH
MBP	40	Uniprot: P0AEX9 without signal sequence	Amylose	Maltose
GFP/YFP/RFP	~27	different	–	–
Intein	5.59	NPGVSAWQVNTAYTAGQLV-TYNGKTYKCLQPHTSLAGW-EPSNVPALWQLQ	Chitin	DTT
FLAG	1.01	DYKDD	Anti-flag antibody	low pH or EDTA
c-Myc	1.20	EQKLISEED		low pH
S-tag	1.75	KETAAAKFERQHMDS	S-Fragment RNAseA	Guanidinium thiocyanate, citrate, MgCl$_2$
Strep-tag	1.06	WSHPQFEK	Strep-Tactin	Desthiobiotin

Nevertheless, protein tags have huge impact on recombinant protein production. Tag encoding sequences are advantageously cloned upstream of the gene of interest: The protein tags in turn often represent well-expressed genes. As their domains are robust and readily folding upon translation, they drag along the formation of the full-length fusion protein. However, the latter aspect may also be disadvantageous, if the asymmetry between tag and target folding behaviour becomes too strong, as occasionally being observed in case of the GST-tag (see below): The ribosome may interrupt translation, resulting in a polypeptide comprising the tag (and a few more amino acids) only. In extreme cases two protein populations would be produced in parallel, the protein tag alone plus the fusion. Furthermore, the high solubility of protein tags may critically help to keep the target protein in solution. This can become obvious, when the target protein precipitates after chemical cleavage of the fusion product.

Glutathion-S-transferase (GST)

The 26 kDa glutathion binding protein is one of the two most commonly used protein tags. Originally, it helps to maintain the reducing environment within a bacterial cell. The protein exists as dimer, dependent on the buffer salt concentration. It is produced to very high yields and is very well folded. Its strong affinity to immobilized glutathion is used in affinity purification.

Maltose binding protein (MBP)

Maltose binding protein is a bacterial periplasmic 43 kDa protein that delivers maltose to the ABC transporter MalFG for maltose uptake of the cytoplasm. Hence, for use as a fusion tag for expression of cytoplasmic proteins, MBP is N-terminally shortened in order to leave out the signal sequence that directs MBP for export into the periplasm. Just like GST, it is very well expressed and folded, and has been successfully used for boosting protein production.

Green fluorescent protein (GFP) (see also chapter 5)

As its name suggests, the green fluorescent protein bears the remarkable feature of being fluorescent on its own, without the need of an external cofactor. In fact, the protein-bound chromophore, absorbing light at 395 nm and emitting at 509 nm, is intrinsically generated by autocatalytic rearrangement of the by the amino acid side chains Ser65-Tyr66-Gly67, which is strictly dependent on correct folding of the protein backbone. By means of a strategy named "spectral tuning", a number of variants exhibiting a wide array of colours have been constructed by extensive genetic manipulation.

Though not being required for its fluorescence, the highly stable GFP domain tends to form dimeric complexes. To alleviate this unwanted effect, the site-specific variant Ala206Lys being unable to dimerize, is often used in biochemical and physiological studies.

In terms of protein production, GFP fusion proteins are widely used as markers for stability of soluble and membrane bound proteins. GFP only exhibits fluorescence if the protein domain is correctly folded. Hence, if the majority of produced protein accumulates in inclusion bodies in unfolded form, no fluorescence is detected. On the contrary, if a protein is soluble and in correct structure, GFP is highly fluorescent and can be observed microscopically either *in vivo*, or – after centrifugation – in bacterial cell pellets. A prerequisite for the use of GFP as an indicator of folded protein is a fusion of GFP to the C-terminus of the protein of interest. This ensures that unfolded fusion protein results in unfolded GFP backbone and deficient chromophore formation.

GFP has been successfully used to study the correct incorporation of recombinantly produced integral membrane proteins into the plasma membrane. Only the

Box 3.6: Engineered protein tags may support the folding of recombinant proteins.

Tag proteins, coupled to either N- or C-terminus of the target, can often act as intrinsic folding helpers. This has been especially reported for fusions of GST, known to be a readily folding protein which even aids to prevent the formation of inclusion bodies. Moreover, the N-terminal addition of the MBP serves in guiding membrane proteins to the proper membrane orientation, as its signal peptide drives the first loop of the fusion protein to the periplasm of *E. coli*. To meet various purposes, distinct tags can be fused to both termini of the target protein.

intracellular, correctly folded GFP is fluorescent, whereas no fluorescence is generated if the non-folded GFP portion faces to the periplasmic side. This leads to a simple assay, where the positioning of the N- or C-terminus of a membrane protein can be directly detected by observing the fluorescence of bacteria after production of GFP-fusion membrane proteins.

3.3 Other prokaryotic organisms for protein production

If protein production in *E.coli* fails, other organisms can be more appropriate for high-level protein expression. Particularly membrane protein expression is often tested in a diverse set of organisms. Apart from switching to eukaryotic expression hosts, several prokaryotic organisms are used. This is due to the fact that membrane lipid compositions can vary extensively between different systems, variable chaperone systems, and generally different metabolism (e.g. growth rate, protein composition).

Other Gram-negative eubacteria such as *Rhodobacter capsulatus* and *sphaeroides*, or archaeal organisms like *Halobacterium salinarum*, are further expression hosts that are in particular suited for special tasks such as for the production of membrane proteins. Species of the *Bacillus* genre or the *Lactococcus* family are Gram-positive, they lack the outer membrane and have a solid cell wall instead. This architecture enables these prokaryotes to efficiently produce proteins which are secreted into the media. Many other prokaryotes have been explored or are being investigated for their potential to synthesize proteins with special needs or under various circumstances (temperature, salt concentration, redox potential, and co-factor production). The following focusses only on a few of them, more examples have been discussed in (Recombinant Gene Expression: Reviews and Protocols, 2011).

3.3.1 *Rhodobacter sphaeroides* and *capsulatus*

Purple bacteria, such as *Rhodobacter species* (Figure 3.8), are Gram-negatives that are highly attractive for the investigation of bacterial physiology. They enable studies of photosynthesis, nitrogen fixation, respiration, and signal transduction in one and the same organism. Although *Rhodobacter* is more difficult to access than *E. coli* by molecular genetic techniques, tools for targeted gene disruption, complementation of mutants, and gene expression are sufficiently available. Some expression plasmids employ light- or oxygen-induced promoters, which strongly influence their practical significance in functional studies, as about 35 % of the endogenous genes are controlled by these effectors.

Rhodobacter strains allow the synthesis of many different co-factors and prosthetic groups, which are not produced by *E. coli*. Moreover, they provide the natu-

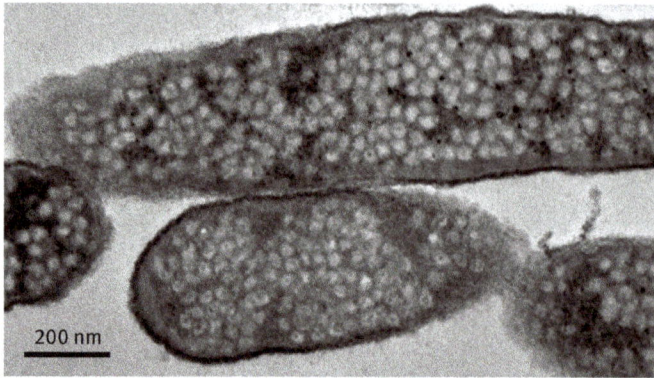

Fig. 3.8: Ultrastructure of *Rhodobacter sphaeroides* filled with ICM vesicles. Cells were grown at low light intensity, thin-sectioned, stained, and viewed by transmission electron microscopy. The chromatophores consist of complex inner membrane invaginations, whose tubular structures are filled with unstained periplasm. In cross-sections, they appear as membrane-surrounded circles. Reproduced from Adams and Hunter, 2012, Copyright 2012 Elsevier.

ral arsenal of catalysts needed for proper assembly. These special features have been successfully applied in the overproduction of the cytochrome oxidases aa_3 in *Rhodobacter sphaeroides* (total genome size 4.6 Mbp, consisting of two chromosomes and some large plasmids) and of cbb_3 in *Rhodobacter capsulatus* (total genome size of 3.7 Mbp, one chromosome and one plasmid). The production of these respiratory multisubunit proteins requests the functional incorporation of several different co-factors, the metal centers Cu_A and Cu_B as well as the synthesis and integration of various iron-porphyrin complexes. *Rhodobacter* strains are also capable to provide another large array of exceptional co-factors, namely the pigments of bacterial photosynthesis, and of nitrogen fixation. As a unique prokaryotic feature, they produce so-called intracytoplasmic membranes (abbreviated ICM) or chromatophores (Figure 3.8), which are strongly activated by low oxygen pressure and low light intensity. ICMs are extensive invaginations of the cytoplasmic (inner) membrane, providing ample space for the accomodation of the photosynthetic apparatus. This property is suggested to be particularly advantageous for the expression of large amounts of membrane proteins. The platform vectors for protein expression are "broad host range" plasmids (named by their exchangeability between many Gram-negative hosts) transmitted via conjugation, because *Rhodobacter* strains lack natural competence for external DNA uptake.

Box 3.7: DNA transfer to α-proteobacteria such as *Rhodobacter* for gene expression.

Rhodobacter and other Gram-negative bacteria receive broad host range expression plasmids by a process named conjugation, which takes place as tri- or bi-parental mating between plasmid donors and recipient strains. During bi-parental mating, the recipient strain (e.g. *Rhodobac-*

ter) comes into close physical contact with a donor strain (in most cases *E. coli* bearing the mobilizable broad host range expression vector). The donor also provides transfer genes, which amongst others mediate cell-cell contacts (e.g. by formation of sex pili). The broad host range vector contains the gene of interest and must carry a transfer origin (*oriT*).

The collection of molecular cloning tools for *R. capsulatus* has been expanded by introduction of the potent well-known T7 polymerase-expression system (see above). In this concept, the gene of interest under control of the T7 promoter is placed on a broad host range vector. Transcription of the chromosomally integrated RNA polymerase gene is controlled by the promoter P_{fru} (promoter of the *fruBKA* operon encoding genes involved in fructose metabolism). This setup has been reported to yield up to 80 mg per liter culture volume of the model protein YFP (yellow fluorescent protein) (Katzke et al. 2010).

3.3.2 *Bacillus subtilis* and other *Bacilli*

Bacillus subtilis expression systems are available since the late 1980s (Zweers et al. 2008). Although it has not been able to displace *E. coli* as standard expression host in basic research, it is widely used for expression of medically relevant substances. This is due to two important properties. Firstly, its lack of the outer membrane and thus absence of lipopolysaccharides (LPS). LPS are strong contaminants in protein purification and induce endotoxic reactions in humans, hence, need to be avoided in production of medically used proteins. Secondly, industrial production of proteins normally employs the secretion of the protein into the cultivating media, simplifying large scale protein purification. *Bacilli* possess a very efficient secretory system, superior to the one found in *E. coli*.

Many representatives of the genus *Bacillus* are capable to secrete large quantities of synthesized protein into the media. Examples are the highly pathogenic *B. anthracis* which releases the anthrax exotoxin, a protein mixture consisting of the protective antigen, the so-called edema factor, and lethal factor. *Bacillus thuringiensis* secretes crystals of the so-called Bt toxin acting as a biological insecticide. Animal-eaten Bt crystals are solubilized in the insect gut, from which the poison is proteolytically liberated. The cyclic antibiotic peptide bacitracin is synthesized nonribosomally by *Bacillus licheniformis* and is released to the growth media. It is industrially produced in the order of two hundred tons per year and is being medically used for wound healing and agronomically applied in animal farming.

Bacillus subtilis is probably the best studied microorganism next to *E. coli*. The Gram-negative soil bacterium has a genome size of 4.2 MBp and contains 4175 genes. It is especially adapted to external stress conditions: In addition to its fitness to survive as endospore, the organism is capable to vegetatively subsist on various organic sources. Its plasmid arsenal is less well developed, which is in part due to

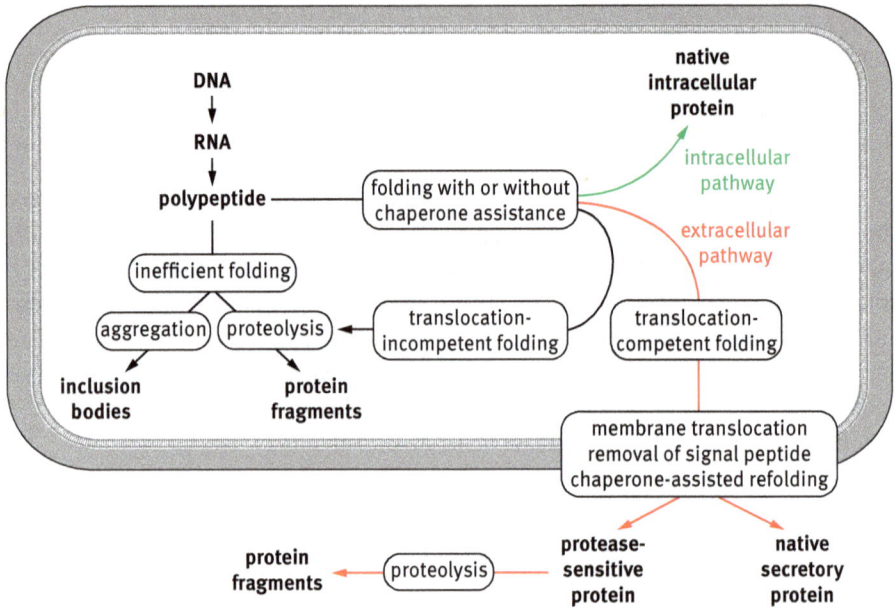

Fig. 3.9: Protein expression modes in *Bacillus subtilis*. The factors and events affecting the expression and secretion of proteins of the Gram-negative bacterium are shown. Expressed proteins are accumulated after correct folding in the cytoplasm; if not folded, they are proteolytically degraded (intracellular pathway). If proteins are destined by means of N-terminal signal sequences to enter the external media via the cytoplasmic membrane in translocation-competent state, the proteins are either stably kept in solution or cleaved by exoproteases (extracellular pathway). Naturally, *B. subtilis* secretes proteases and various hydrolytic enzymes degrading other macromolecules to exploit external material as carbon sources for growth. For efficient production of stable secreted proteins, these proteolytic capacities have been deleted by extensive genetic engineering (Westers et al. 2004).

the low stability in *B. subtilis*. However, many vectors have been developed for heterologous expression of proteins. On the other hand, manipulations at the genome level are greatly facilitated, because the microorganism has natural competence, it takes up DNA very efficiently and readily undergoes homologous recombination. *B. subtilis* has become popular for protein production, because it is considered as a safe organism ("GRAS = generally regarded as safe"). The potency to secrete proteins into the media is a great benefit for using *B. subtilis* in comparison to *E. coli*, where secretion ends up in the periplasmic space. In addition, secreted proteins could be conveniently purified from the *Bacillus* growth media without the need to break cells (Figure 3.9).

Naturally *B. subtilis* secretes hydrolases, in particular proteases and amylases, into the media in order to exploit external carbon sources. In fact, these enzymes are important biotechnological products. On the other hand, the biosynthesis of external proteases is critical if heterologous proteins were to be expressed, and to this end strains carrying deletions of up to eight distinct protease encoding genes have been constructed (Westers et al. 2004). Secretion of protein in translocation-competent conformation is performed by the membrane-associated Tat and Sec transport systems (Figure 3.5c, d), of which the latter requests an N-terminal signal peptide being responsible for proper membrane targeting, and is cleaved during passage through the translocase. As proper folding is essential for biological function of proteins, chaperones of the class GroEL, DnaK, DnaJ, and GrpE, known to occur in *E. coli*, are present in the cytoplasm. As yet, the only known external folding helper acting in the maturation of translocated proteins is the lipoprotein PrsA, which is anchored to the outer membrane leaflet by a cystein-linked diacyl-glycerol molecule. Optimal promoter strength, chaperone folding, suitable signal peptide structure, and restraining of protease background have been recognized as critical factors strongly affecting the biotechnological power of protein synthesis and the excretion system of *B. subtilis*.

Bacilli are fermented as easily and productively as *E. coli*, plasmid systems can be used, growth media are similar to those used for *E. coli* cultures. Many different inducer systems have been used in *Bacilli*, including the *lac* system by IPTG induction, sucrose or xylose induction, and many more. The most powerful system until now is the subtilin-regulated gene expression (SURE) system with expression levels up to 350fold, however, strong leaky expression is also observed. Subtilin (not to be confused with the protease subtilisin) is a cyclic peptide antibiotic that induces bacterial signal transduction from a sensor histidine kinase (SpaK) to its response regulator (SpaR).

As the secretory pathway, using the Sec complex is very efficient, it is believed that *Bacilli* might have advantages in membrane protein production as well, since their synthesis also relies on the Sec system. This is currently been tested extensively.

3.3.3 Lactococcus lactis

Unlike *Bacilli*, *Lactococcus lactis* has been tested extensively for production of membrane proteins (Kunji et al. 2003). *L. lactis* is a Gram-positive bacterium, mostly known for its application in milk fermentation such as cheese production. Its use in production of proteins relies on similar features as those of *Bacilli* – the lack of an outer membrane, an efficient Sec translocator complex, and the absence of inclusion bodies. However, growth rate of *L. lactis* is slower (~1 h doubling time) compared to *Bacilli* or *E. coli*. *L. lactis* shows only mild proteolytic activity and can be grown without aeration to high cell density.

A similar induction mechanism as in the *Bacillus* SURE system is applied, using nisin as inducer. *L. lactis* has been used for production of membrane proteins, particularly for *in vivo* tests of these proteins, which are stably incorporated in the *Lactococcus* membrane.

3.3.4 *Halobacterium salinarum*

The strongly pigmented prokaryote *Halobacterium salinarum* (Figure 3.10a) belongs to the extremophilic archaea thriving in hypersaline (up to 5 M NaCl) environment. In part, the high osmotic pressure is balanced by intracellular accumulation of the so-called "compatible solute" ectoine, which is commercialized for skin care and UV protection, or used as a stabilizer of proteins and nucleic acids. K^+ is the predominant intracellular cation present in molar concentrations. Many proteins of *H. salinarum* denature at low salt conditions, and are strictly adapted to this high potassium concentration, as their low proportion of hydrophobic amino acids and their unusually high content of acidic amino acid side chains prevents the aggregation of protein. The best strategy to produce halobacterial proteins is by homologous overexpression, because only the genuine halobacterial host provides the hypersaline cytoplasm that is vital for native folding of the desired protein.

Although growing very slowly, *H. salinarum* exhibits remarkable metabolic versatility by undergoing fermentation or oxidative phosphorylation, or even generating the proton motive force *via* photon-driven energy conversion. *H. salinarum* is capable to move into regions of optimal light quality by means of flagellae that are located to either one or both cell poles. Attractive and repulsive phototaxis is controlled by sensory rhodopsins I and II (sRI and sRII) which are related membrane-integral retinal proteins as well as the chloride pump halorhodopsin, and the light-driven proton pump bacteriorhodopsin (bR, encoded by the *bop* gene) (Figure 3.10a). Like in all microbial rhodopsins, light activation of bR triggers the conformational change of the covalently attached retinal cofactor and of the apoprotein, which executes transmembrane proton translocation and, after passing through a series of photointermediates, relaxes back to the ground state (Figure 3.10c). Physiologically, this process leads to the build-up of a proton concentration gradient and a membrane potential.

At present, up to 4000 mostly eubacterial proteorhodopsin genes homologous to *bop* are known, predominantly from metagenomic analyses, but only bR has the unique feature to form "purple membranes" being formed at low oxygen pressure. After lysis of halobacteria, the purple membranes can be readily isolated by centrifugation. They form 2D-crystalline, hexagonal lattices that exclusively consist to ¼ part of lipid molecules and to ¾ part of trimer-associated bacteriorhodopsin.

Due to its enormous compactness, the purple membrane lipid-protein complex is stable in water and largely withstands thermal, chemical, or photochemical degradation. These features, in addition to the unique and robust photobiochemical properties, enable to exploit the halobacterial protein complex as a photonic device in biomolecular electronics, recently reviewed (Jin et al., 2008; Trivedi et al., 2011). Among others, the most exciting bioelectronic inventions are the potential applications as photochemical switch, artificial retina, or as data storage material. To meet these purposes, the photochemical and kinetic properties of bR could be specifically modulated by targeted amino acid changes, without affecting purple membrane architecture.

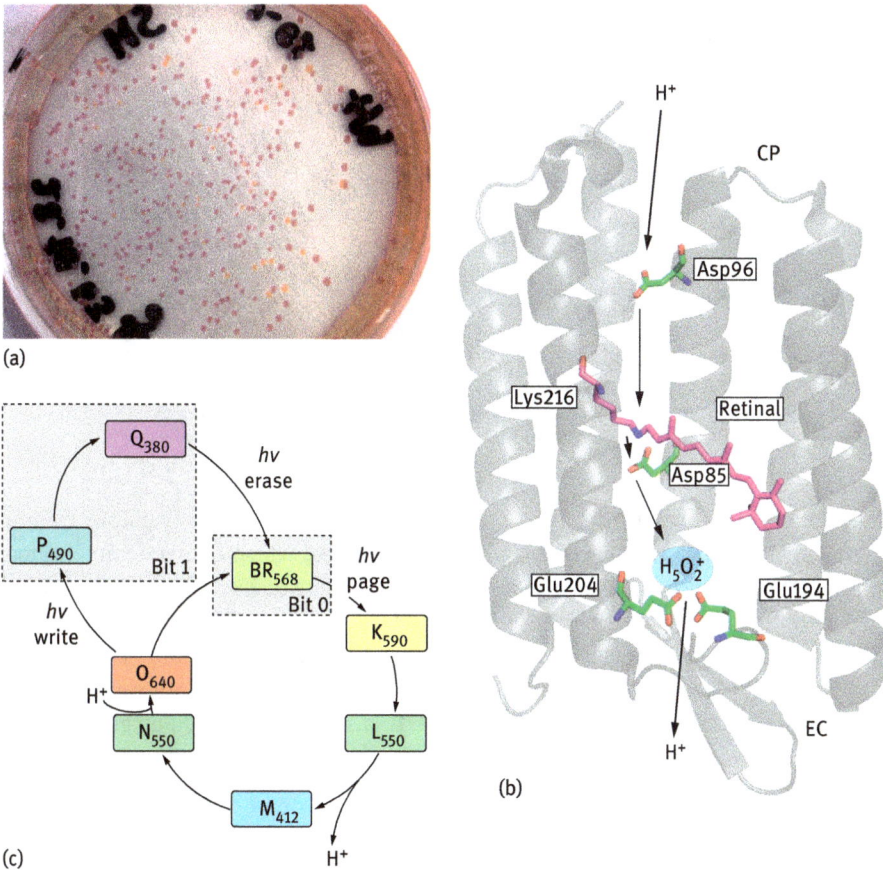

Fig. 3.10: Use of *Halobacterium salinarum*. (a) Colonies of *Halobacterium salinarum*. The archaea were grown microaerobically in rich media under illumination to the stationary phase and plated on agar. The purple colour indicates a high level of bacteriorhodopsin occurring in purple membranes. Different coloration of colonies indicates alterations on pigment expressing genes by spontaneous mutation. (b) 3D structure of bacteriorhodopsin (bR): The integral membrane protein consists of 7 transmembrane α helices. Its co-factor retinal is embedded to the hydrophobic interior and is covalently attached to a lysine side chain amino group in a Schiff base linkage, which causes the absorbance maximum at 568 nm in the so-called ground state. Together with the aspartic acids 85 and 96, and a number of other residues as well as protein bound water molecules, it forms the transmembrane proton channel (outlined by black arrows) connecting the cytoplasmic (CP) and external (EC) spaces. Retinal acts as a conformation-dependent switch controlling the light-driven proton pump activity. The prosthetic retinal group, the amino acid side chains, and the internal water cluster involved in proton translocation are highlighted. (c) Photocycle of bR. A number of photo-intermediates K, L, M, N, O, characterized by their specific absorbance maxima, are generated after absorption of a single photon (568 nm) in the ground state named BR (capital letters).

Box 3.8: Use of bacteriorhodopsin as a potential optical storage media in bioelectronics.

A very sophisticated technical example is the use of purple membranes as photochemically encoded data storage material. Due to its photochromic features, bR can adopt two stable states, (i) the ground state (named BR) having the "Bit 0" form and (ii) the unnatural Q state, characterized by "Bit 1". The cyclic reaction series of bR (Figure 3.10c) following photoexcitation of the BR ground state, going naturally through the intermediates K-L-M-N-O, can be expanded by absorption of a second photon at 640 nm from the O state: This step converts the all-*trans* retinal into the 9-*cis* conformation and forms the P490 state. Because the 9-*cis* conformation is instable, the covalent Schiff-base bond is hydrolyzed and leaves the retinal molecule being trapped in the binding site, resulting in the Q380 state. In the absence of blue light or high temperatures, this state is unchanged for decades. However, irradiation at 380 nm reconstitutes the covalent Schiff-base bond of retinal in all-*trans* conformation, and converts the Q intermediate back to the BR ground state.

Thus, data can be written on bR after initialization of the photocycle (hv paging), by irradiation at 640 nm (hv write), and erased by irradiation with 380 nm (hv erase).

Many techniques to alter gene expression of *H. salinarum* are available. Several genus-specific plasmid vectors have been constructed, which carry unconventional antibiotic resistance factor genes (providing resistance to Novobiocin and Mevinolin), owing to its archaeal nature. However, *halobacteria* suffer from genetic instability due to the presence of many insertional (IS) elements. Recombinants are significantly stabilized by insertion of the vector-bound gene of interest via allelic replacement into the halobacterial genome. This strategy has been elegantly facilitated by means of the negative selection on the *ura3* gene using 5-fluoro-orotic acid (Peck et al. 2000).

Key-terms

Affinity purification, antibiotic selection marker, archaea, bacterial transformation, bacteriorhodopsin, bicistronic gene expression, broad host range vector, chaperone, chromatophores, cloning, codon usage, co-factor biosynthesis, compatibility, competence, conjugation, copy number, enzyme production, eubacteria, *Escherichia coli*, Gram-positive and -negative bacteria, growth curve, growth media, high-level protein expression, homologous and heterologous expression, inclusion bodies, inducible promoter, insulin, lipopolysaccharide, peptide and protein tags, membrane protein insertion, proteobacteria, minimal media, plasmid vector, protease, protein folding, protein toxicity, rare codon, refolding, replicon, secretion, structure determination, T7 polymerase, photocycle, vaccination

Questions

– For which purposes is high-level expression of proteins needed?
– What are the advantages of prokaryotic protein expression?

- Which methods of DNA transfer into prokaryotes are known?
- What does the term "bacterial competence" mean?
- What are "inclusion bodies" and how are they generated?
- Which affinity tags are used for efficient purification of proteins?
- How can leaky expression of potentially toxic proteins be suppressed?
- Which measures can be taken to prevent inclusion body formation?
- How do chaperones aid in cellular protein folding?
- What are selective markers and when are they used?
- By which mechanisms do antibiotic resistance factors act?
- How does T7 polymerase control the high-level expression of proteins?
- What can be done if the protein of interest requests posttranslational modifications?
- For what reasons are purple bacteria suitable for the heterologous expression of proteins?
- What is the mechanism of conjugative DNA transfer to a purple bacterium?
- What is the function of chromatophores?
- How are halobacterial proteins adapted to high salt conditions?
- What makes the association of bacteriorhodopsin in purple membranes favorable?
- How can recombinant genomic changes of *Halobacterium salinarum* be stabilized?

Further readings

Higgins, S. J. & Hames, B. D. 1999. *Protein Expression: A Practical Approach*. 1st ed.

Robinson, A. S. 2011. *Production of Membrane Proteins: Strategies for Expression and Isolation*. 1st ed.

Freigassner, M., et al. 2009. Tuning microbial hosts for membrane protein production. *Microb Cell Fact* 8: 69.

Hwang, P. M., et al. 2014. Targeted expression, purification, and cleavage of fusion proteins from inclusion bodies in *Escherichia coli*. *FEBS Lett* 588: 247–252.

Zweers, J. C., et al. 2008. Towards the development of *Bacillus subtilis* as a cell factory for membrane proteins and protein complexes. *Microb Cell Fact* 7: 10.

Rosano, G. L., et al. 2014. Recombinant protein expression in *Escherichia coli*: advances and challenges. *Front Microbiol* 5: 172.

Terpe, K. 2006. Overview of bacterial expression systems for heterologous protein production: from molecular and biochemical fundamentals to commercial systems. *Appl Microbiol Biotechnol* 72: 211–122.

Sørensen, H. P. & Mortensen, K. K. 2005. Advanced genetic strategies for recombinant protein expression in *Escherichia coli*. *Biotechnol* 115: 113–128.

Hochkoeppler, A. 2013. Expanding the landscape of recombinant protein production in *Escherichia coli*. *Biotechnol Lett* 35: 1971–1981.

Royle, K. & Kontoravdi, C. A. 2013. Systems biology approach to optimising hosts for industrial protein production. *Biotechnol Lett* 35: 1961–1969.

References

Adams, P. G. & Hunter, C. N. 2012. Adaptation of intracytoplasmic membranes to altered light intensity in *Rhodobacter sphaeroides*. *Biochim Biophys Acta* 1817: 1616–1627.

Almen, M. S., Nordström, K. J., Fredriksson, R., et al. 2009. Mapping the human membrane proteome: a majority of the human membrane proteins can be classified according to function and evolutionary origin. *BMC Biol* 7: 50.

Baneyx, F. & Mujacic, M. 2004. Recombinant protein folding and misfolding in *Escherichia coli*. *Nat Biotechnol* 22: 1399–1408.

Burgess, R. R. 2009. Refolding solubilized inclusion body proteins. *Methods Enzymol* 463: 259–282.

de Boer, H. A., Comstock, L. J. & Vasser, M. 1983. The *tac* promoter: a functional hybrid derived from the *trp* and *lac* promoters. *Proc Natl Acad Sci USA* 80: 21–25.

Fu, H. Y., Lin, Y., Chang, Y., et al. 2010. A novel six-rhodopsin system in a single archaeon. *J Bacteriol* 192: 5866–5873.

Georgiou, G. & Valax, P. 1996. Expression of correctly folded proteins in *Escherichia coli*. *Curr Opin Biotechnol* 7: 190–197.

Goodman, D. B., Church, G. M. & Kosuri, S. 2013. Causes and effects of N-terminal codon bias in bacterial genes. *Science* 342: 475–479.

Guzman, L. M., Belin, D., Carson, M. J., et al. 1995. Tight regulation, modulation, and high-level expression by vectors containing the arabinose P_{bad} promoter. *J Bacteriol* 177: 4121–4130.

Henderson, R. 1977. The purple membrane from *Halobacterium halobium*. *Annu Rev Biophys Bioeng* 6: 87–109.

Jin, Y., Honig, T., Ron, I., et al. 2008. Bacteriorhodopsin as an electronic conduction medium for biomolecular electronics. *Chem Soc Rev* 37: 2422–2432.

Katzke, N., Arvani, S., Bergmann, R., et al. 2010. A novel T7 RNA polymerase dependent expression system for high-level protein production in the phototrophic bacterium *Rhodobacter capsulatus*. *Protein Expr Purif* 69: 137–146.

Kunji, E. R., Slotboom, D. J. & Poolman, B. 2003. *Lactococcus lactis* as host for overproduction of functional membrane proteins. *Biochim Biophys Acta* 1610: 97–108.

Mayer, M. & Buchner, J. 2004. Refolding of Inclusion Body Proteins, in: Molecular Diagnosis of Infectious Diseases, Decler, J. & Reischl, U., Editors, *Humana Press*, 239–254.

Miroux, B. & Walker, J. E. 1996. Over-production of proteins in *Escherichia coli*: mutant hosts that allow synthesis of some membrane proteins and globular proteins at high levels. *J Mol Biol* 260: 289–298.

Peck, R. F., DasSarma, S. & Krebs, M. P. 2000. Homologous gene knockout in the archaeon *Halobacterium salinarum* with *ura3* as a counterselectable marker. *Mol Microbiol* 35: 667–676.

Ramon, A., Senorale-Pose, M. & Marin, M. 2014. Inclusion bodies: not that bad. *Front Microbiol* 5: 56.

Recombinant Gene Expression: Reviews and Protocols, Third Edition. *Methods in Molecular Biology*. Vol. 824. 2011. Springer Science and Business Media.

Serrano-Vega, M. J., Magnani, F., Shibata, Y., et al. 2008. Conformational thermostabilization of the beta1-adrenergic receptor in a detergent-resistant form. *Proc Natl Acad Sci USA* 105: 877–882.

Structural Genomics Consortium, et al. 2008. Protein production and purification. *Nat Methods* 5: 135–146.

Studier, F. W. & Moffatt, B. A. 1986. Use of bacteriophage T7 RNA polymerase to direct selective high-level expression of cloned genes. *J Mol Biol* 189: 113–130.

Studier, F. W. 2005. Protein production by auto-induction in high density shaking cultures. *Protein Expr Purif* 41: 207–234.

Studier, F. W., Daegelen, P., Lenski, R. E., et al. 2009. Understanding the differences between genome sequences of *Escherichia coli* B strains REL606 and BL21(DE3) and comparison of the *E. coli* B and K-12 genomes. *J Mol Biol* 394: 653–680.

Thompson, A. K., Gray, J., Liu, A., et al. 2012. The roles of *Rhodobacter sphaeroides* copper chaperones PCu$_A$C and Sco (PrrC) in the assembly of the copper centers of the aa_3-type and the cbb_3-type cytochrome *c* oxidases. *Biochim Biophys Acta* 1817: 955–964.

Westers, L., Westers, H. & Quax, W. J. 2004. *Bacillus subtilis* as cell factory for pharmaceutical proteins: a biotechnological approach to optimize the host organism. *Biochim Biophys Acta* 1694: 299–310.

Terpe, K. 2003. Overview of tag protein fusions: from molecular and biochemical fundamentals to commercial systems. *Appl Microbiol Biotechnol* 60: 523–533.

Trivedi, S., Choudhary, O. P. & Gharu, J. 2011. Different proposed applications of bacteriorhodopsin. *Recent Pat DNA Gene Seq* 5: 35–40.

von Heijne, G. & Gavel, Y. 1988. Topogenic signals in integral membrane proteins. *Eur J Biochem* 174: 671–678.

Wagner, S., Klepsch, M. M., Schlegel, S., et al. 2008. Tuning *Escherichia coli* for membrane protein overexpression. *Proc Natl Acad Sci USA* 105: 14371–14376.

Yoshida, M. 2014. personal communication to be published in Nature Biotechnology.

Zhang, X. & Studier, F. W. 1997. Mechanism of inhibition of bacteriophage T7 RNA polymerase by T7 lysozyme. *J Mol Biol* 269: 10–27.

Zweers, J. C., Barák, I., Becher, D., et al. 2008. Towards the development of *Bacillus subtilis* as a cell factory for membrane proteins and protein complexes. *Microb Cell Fact* 7: 10.

Ulrich Kück and Ines Teichert

4 Fungal biotechnology

Fungi are distributed worldwide and have a huge impact on ecology, economy, and human society. Today, about 100,000 fungal species are characterized, but even conservative estimates believe that between 1.5 and 6 million species exist on earth. In biotechnology, fungi are important for the industrial production of a variety of products. For example, fungi are used in the food production industry (bread making), for producing alcoholic drinks and citric acid (e.g. the soft drinks

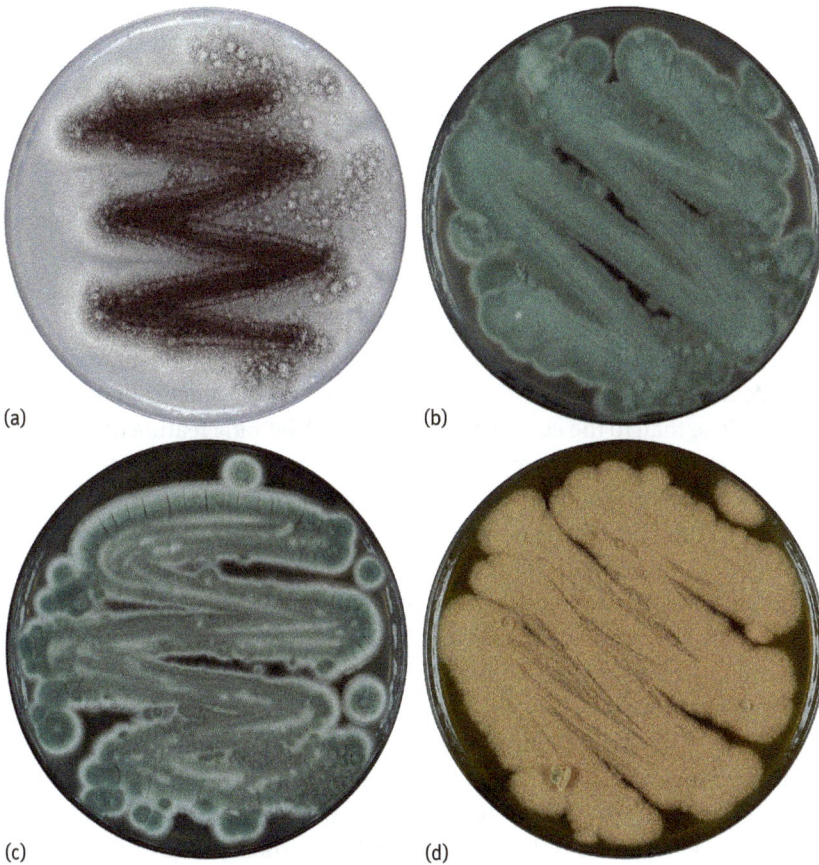

Fig. 4.1: Petri dishes with filamentous fungi of biotechnological relevance. (a) *Aspergillus niger* (Citric acid). (b) *Penicillium citrinum* (Compactin). (c) *Penicillium chrysogenum* (Penicillin). (d) *Aspergillus terreus* (Lovastatin). For comparison of morphology, all strains were grown on the same complete medium.

industry), and fermented food (soy sauce), as well as the ripening and flavoring of cheese. In the pharmaceutical industry, a wide range of applications use filamentous fungi to produce antibiotics (e.g. penicillin, cephalosporin C, griseofulvin), alkaloids (ergotamine), immunosuppressants (cyclosporine A), steroids (progesterone), or cholesterol-lowering drugs (statins).

Prominent examples of filamentous fungi with biotechnological relevance are displayed on solid media in Figure 4.1. To optimize strains for production processes, strain development programs use conventional random mutagenesis, and more recently, recombinant technologies to generate microbial production strains with novel and advantageous properties.

This chapter provides an overview of fungal biology, followed by a description of how fungi are used for food production. Subsequently, it focuses on fungal products derived from primary and secondary metabolism, and finally the application of recombinant technologies used to manipulate fungi of biotechnical relevance.

4.1 Fungal biology

This section briefly reviews the taxonomy of fungi, and then describes general aspects of fungal morphology and reproduction.

Different textbooks classify fungi rather diversely. In general, fungi can be divided into two types, the oomycota (oomycetes) and eumycota (eumycetes). The cell wall of the oomycota contains cellulose as a major component, as in higher plants. Members of the oomycota are believed to be phylogenetically more closely related to algae than to the eumycota. In contrast, a defining feature of the eumycota is their cell wall, which comprises three major layers of different compounds. One layer contains chitin (N-acetyl-D glucosamine), a second layer contains β-1,3-glucan, and a third outer layer contains glycoproteins heavily glycosylated by polysaccharides. In yeasts, the chitin content of the cell wall is much lower than in filamentous fungi. The eumycota are sub-classified into the following classes: chytridiomycetes, zygomycetes, ascomycetes, and basidiomycetes. However, this chapter focuses on species belong to the three latter classes, with ascomycetes as the most dominant group.

Members of the fungal kingdom display a great variety of phenotypes, but in general two major morphological groups can be distinguished: yeasts and filamentous fungi. Yeasts are unicellular fungi, usually with a single nucleus per cell (Figure 4.2a). However, under certain growth conditions some yeasts can generate filaments. Often, the term "yeast" is applied to species of the genus Saccharomyces, and therefore the term "non-conventional yeasts" has been introduced for species of other genera. Depending on the mode of cell division, budding yeasts (e.g. *Saccharomyces cerevisiae* [Box 4.1]) can be distinguished from fission yeasts (e.g. *Schizosaccharomyces pombe*). Yeasts are further classified by their enzymatic reper-

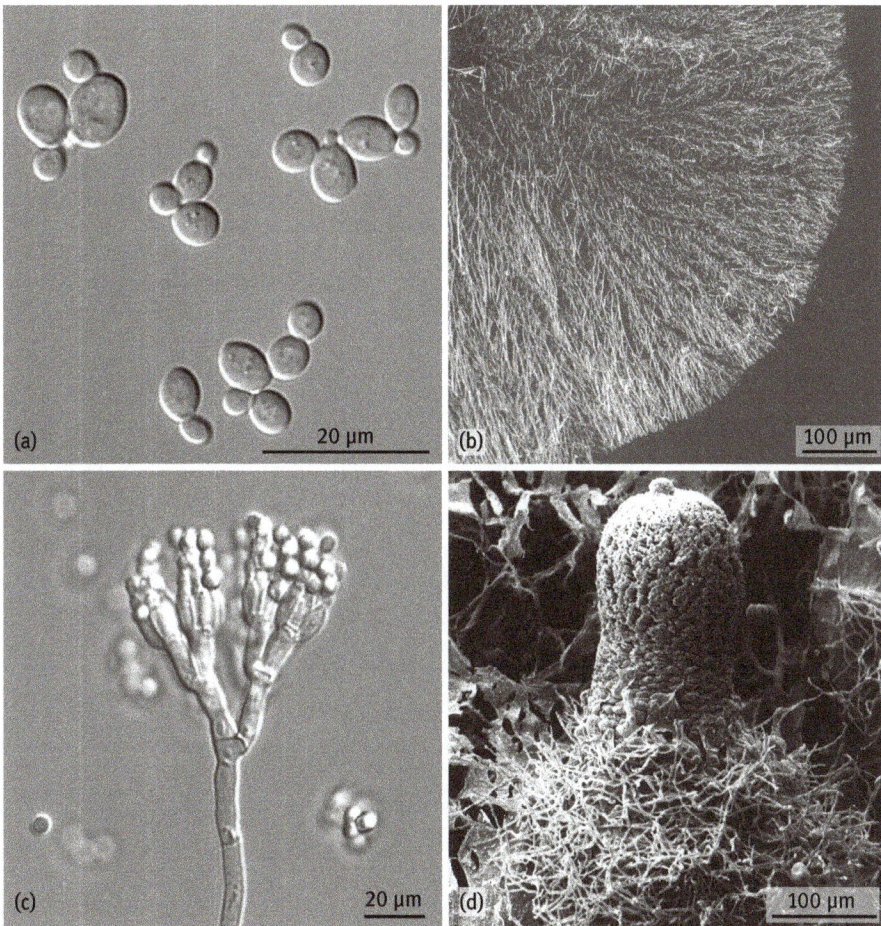

Fig. 4.2: Morphology of yeasts and filamentous fungi. (a) Light microscopy of budding cells from the yeast *Saccharomyces cerevisiae*. (b) Light microscopy of branching hyphae from a filamentous fungus. (c) Light microscopy of a conidiophore from the filamentous fungus *Penicillium chrysogenum*. (d) Scanning electron microscopy of a fruiting body from the filamentous fungus *Sordaria macrospora*.

toire, using different sugars and/or alcohols as carbon and energy sources (Table 4.1).

In biotechnological applications, two "non-conventional" methylotrophic yeasts are of major importance: *Hansenula polymorpha* and *Pichia pastoris*. *P. pastoris* has been renamed *Komagataella pastoris* recently, but is mostly still known under its original name. Both yeasts can use methanol as a sole carbon source for energy consumption, and both yeasts are used as expression platforms to generate heterologous proteins (see section 4.5.2). All yeasts mentioned above belong to the ascomycetes, but it should be emphasized that even within the basidiomycetes several genera grow mostly in a yeast-like stage (e.g. *Ustilago spec.*).

Box 4.1: The model yeast *Saccharomyces cerevisiae*.

Due to its major contribution to basic and applied science, the yeast *Saccharomyces cerevisiae* (see also chapter 1), also known as baker's or brewer's yeast, is considered in more detail. *S. cerevisiae* can grow as a diploid or haploid cell. Both forms propagate vegetatively by budding, and thus are clearly distinguished from fission yeast (*Schizosaccharomyces pombe*). During budding, a small daughter cell is formed at the side of the parent cell, and the size of the daughter cell increases, followed by mitotic division of the nucleus. One daughter nucleus then migrates into the daughter cell and a cross wall forms to bud off the daughter from the parent cell. This type of cell division can occur indefinitely in culture, and often parent and daughter cells stay attached to form a budding mycelium (Figure 4.2a). During this vegetative propagation, all cells are genetically identical.

Budding yeasts can also propagate sexually. In *S. cerevisiae* two genders exist that can be distinguished by their mating-type loci, comparable to sex chromosomes in animals. These loci are called either a or α. During sexual propagation, two haploid cells carrying opposite mating-type loci can mate and undergo karyogamy to form a diploid nucleus. This is followed by meiotic division, which leads to the formation of four haploid cells. These cells are called ascospores (products of meiosis) and are contained in a meiosporangium (ascus). At this stage, haploid ascospores can be isolated and used for further cultivation. Thus, *S. cerevisiae* can be grown as either haploid or diploid cells, making it a useful model for molecular and genetic studies.

Table 4.1: Substrates and products of yeast fermentation.

Substrate	Enzyme/Pathway	Product	Organism
Cellobiose	β-Glucosidase	2 Glucose	*Brettanomyces spec.*
Ethanol	Alcohol dehydrogenase	Glucose	*Saccharomyces cerevisiae*
Glycerol	Gluconeogenesis	Glucose	*Saccharomyces cerevisiae*
Lactate	Gluconeogenesis; Lactate Dehydrogenase	Glucose	*Saccharomyces cerevisiae*
Lactose	β-Galactosidase	Glucose + Galactose	*Kluyveromyces spec.*
Maltose	Maltase	2 Glucose	*Saccharomyces cerevisiae*
Melibiose	α-Galactosidase/Melibiase	Glucose + Galactose	*Saccharomyces cerevisiae*
Methanol	Alcohol oxidase	Hydrogen peroxide + Formaldehyde	*Hansenula polymorpha, Pichia pastoris*[a]
Sucrose	β-D-fructofuranoside fructohydrolase/Invertase	Glucose + Fructose	*Saccharomyces cerevisiae*

[a] *Pichia pastoris* has been renamed *Komagataella pastoris* recently, but is mostly still known under its original name.

The term filamentous fungi refers to a morphologically different type of fungi that form hyphae. These cells have a tubular structure and often show extensive branching (Figure 4.2b). The sum of all hyphae is called the mycelium, and between individual hyphae so-called "anastomosis" occurs that allows cytoplasmic flow between different compartments.

In zygomycetes, the hyphae are not segmented, while in ascomycetes and basidiomycetes the mycelium is divided by septae. In ascomycetes the septae are characterized by a porus that allows the movement of cytoplasm and organelles from one compartment to another. Depending on the physiological growth conditions, the number of nuclei per compartment is rather variable. In basidiomycetes, the septum shows a more complex structure, seen at the ultrastructure level as a doliporus septum. This type of septum guarantees that in most basidiomycetes only two genetically different nuclei occur per compartment. Due to their hyphal growth, filamentous fungi are able to form complex structures with pseudo-tissues for sexual or asexual propagation (Box 4.2).

Box 4.2: Sexual and asexual propagation of filamentous fungi.

In filamentous fungi, asexual as well as sexual spores can be generated, both showing highly diverse morphology. Vegetative propagation is often accompanied by the formation of a huge number of spores (conidiospores). Depending on the mode of spore formation, one may distinguish sporangia from conidiophores. While sporangia are enclosures where millions of spores are formed, conidiophores constrict off a chain of conidiospores. Fungi producing millions of spores are often referred to as molds. The morphology of sporangia or even conidiophores is often used to identify filamentous fungi (Figure 4.2c).

In contrast to yeasts, most filamentous fungi produce their sexual spores within multicellular fruiting bodies that protect the sexual spores from environmental factors (Figure 4.2d). The shape and size (0.1 to several centimeters) of fruiting bodies are very diverse. Most industrially relevant fungi were believed to propagate exclusively asexually. However, recent investigations have shown that even highly derived industrial strains have retained the capacity to undergo a sexual life cycle.

4.2 Food biotechnology

Fungi have a long tradition as foodstuffs. For a start, fruiting bodies of basidiomyceteous fungi (e.g. *Cantharellus cibarius* [chanterelle], *Agaricus bisporus* [button mushroom]) are consumed in salads or cooked dishes, and the ascomycete *Fusarium venenatum* is used to produce Quorn™, a meat substitute. Moreover, fungi are used to produce food additives, such as organic acids (see section 4.3.1), flavorings, and colorants, or enzymes used for industrial food production (see section 4.3.2). This section will focus on the use of fungi during food fermentation. An overview of food generated by fungal fermentation is given in Table 4.2.

Fermentation is the degradation of organic materials by microorganisms. The production of food by fermentation is rather expensive, but there are several advantages, namely preservation, increased digestibility, accessibility of nutrients, detoxification, flavor enhancement, and enrichment with certain essential molecules, such as vitamins or essential amino acids. Importantly, the microorganisms used for fermentation must be approved for food production. Many fungi are "gen-

Table 4.2: Fermented food produced with fungi. Modified from Reiß 1987; Kück 2009.

Designation	Substrate	Consistence and Application	Fungus
Ang-kak (red rice)	Rice	Powder, food colorant, seasoning, dietary supplement	*Monascus pilosius, Monascus purpureus, Monascus ruber*
Beer	Barley, wheat	Beverage	*Saccharomyces cerevisiae*
Bread	Flour	Solid, staple food	*Saccharomyces cerevisiae*
Hamanatto	Soybeans, wheat flour	Soft beans, flavor enhancer	*Aspergillus oryzae*
Miso (soybean paste)	Soybeans, rice / barley	Paste, high protein content, seasoning	*Aspergillus oryzae, Aspergillus sojae*
Mold cheese	Milk	Semi-solid, high protein content	*Penicillium camemberti, Penicillium candidum, Penicillium roqueforti*
Ontjom	Peanut press-cake	Solid cake, roasted or deep fried as meat substitute	*Neurospora intermedia, Neurospora sitophila*
Peuyeum	Maniok	Semi-solid, snack	*Amylomyces rouxii*
Sake (rice wine)	Rice	Beverage	*Aspergillus oryzae*
Salami	Sausage	Solid, high protein content	*Penicillium spec.,* especially *Penicillium chrysogenum, Penicillium nalgiovense*
Soy sauce	Soybeans, wheat (in Japanese soy sauce), salt	Salty, dark fluid, seasoning	*Aspergillus oryzae, Aspergillus sojae*
Sufu	Tofu (soybean cheese)	Salty cubes, main meal	*Actinomucor elegans*
Tempeh	Mostly soybeans, also cereals, peanut and coconut press-cake	Semi-solid cake, roasted or deep fried or as meat substitute in soups	*Mucor indicus, Rhizopus chinensis, Rhizopus oligosporus, Rhizopus oryzae*
Wine	Grapes	Beverage	*Saccharomyces cerevisiae*

erally recognized as safe" (GRAS, see Box 4.3) for human health and the environment. Moreover, fungi used in food production must fulfill some prerequisites. The of use pathogens or mycotoxin producers is not allowed. Instead, the fungi used should be able to prevail against other microorganism and produce the required enzymes.

Box 4.3: GRAS status.

The GRAS status is an American Food and Drug Administration (FDA) designation to confirm that an organism has been shown to be "generally recognized as safe" for human health and the environment under the conditions of its intended use.

It is estimated that fermentation has been used for food preservation for over 6000 years. For example, the Egyptians used yeast to make bread, wine, and beer (Chapter 1). In those days, food fermentation occurred naturally and the microorganisms were derived from the environment. Today, industry uses defined starter cultures, and is interested in constantly improving microorganism to increase yield and food quality. Besides products generated by yeast fermentation, only a few fungi-fermented foods are well known in Western countries; however, they are very common in Asia (Table 4.2). The production of mold cheese, soy sauce and tempeh will be outlined in more detail below.

Mold cheese

In Europe, the best-known food produced by filamentous fungal fermentation is mold cheese. Two varieties can be distinguished in general, namely white mold cheese, produced with *Penicillium camemberti*, and blue mold cheese, produced with *P. roqueforti*. Interactions between bacteria, yeasts and the mold are important for successful fermentation. The use of milk from different bovine species further enhances the variety of mold cheeses. Examples for white mold cheese are Brie and Camembert, and examples for blue mold cheese are Gorgonzola, Roquefort, and Stilton.

The production of mold cheese starts with producing the cheese basis (see section 1.5.3). Milk is mixed with lactic bacteria and rennet, an enzyme mixture containing the proteases chymosin and pepsin (see section 4.3.2), causing milk casein to coagulate. By pressing, the whey is separated from the moist mass, which is further cut into curd. The curd can be pressed into forms, is then salted, e.g. by adding brine and left to ripen. For mold cheese, *P. camemberti* and *P. roqueforti* starter cultures are added to the milk directly with the rennet and lactic bacteria, or (for white mold cheese) spore suspensions are sprayed on the curd surface. For blue mold cheese, the salted curd has to be piqued to generate air channels for *P. roqueforti* to grow. The fungi cause extensive lipolysis and proteolysis by secreting enzymes. The main aroma compounds of mold cheese are generated by lipolysis of milk fat to methyl ketones and secondary alcohols. Small peptides and amino acids, generated by proteolysis, and derived carbonic acids and amines further contribute to texture, taste, and aroma.

Soy sauce

Soybeans are the basis of many fermented foods (Table 4.2). In the Western world, the best-known fermented food based on soybeans is soy sauce. Chinese soy sauce is made solely from soybeans, while Japanese soy sauce is made from soybeans and wheat. Within these groups, many different varieties of soy sauce exist, and

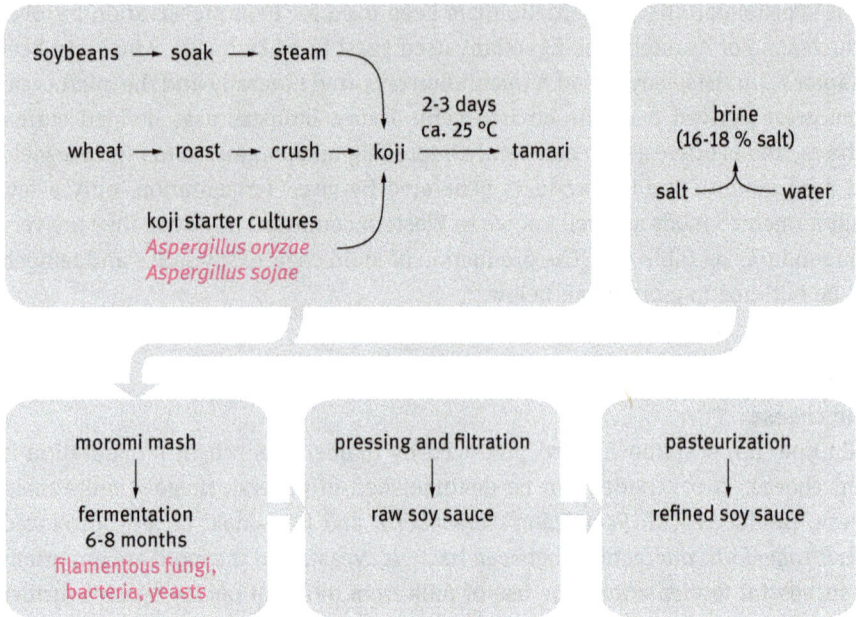

Fig. 4.3: Flow chart of the industrial production of Japanese soy sauce using the filamentous fungi *Aspergillus oryzae* and *Aspergillus sojae*, and the yeast *Zygosaccharomyces rouxii*.

the Japanese koikuchi-shoyu type is commercially available all over the world. An overview of its production is shown in Figure 4.3.

Essentially, wheat and soybeans are mixed with fungal spores (from *Aspergillus oryzae* or *Aspergillus sojae*), generating the Koji. The fungus grows and secretes enzymes (cellulases, proteases, amylases) that digest the plant material, leading to Tamari. Tamari is mixed with brine to inhibit decay and fermented for 6–8 months in stainless steel tanks. The fungal enzymes remain active during fermentation and further degrade the plant polymers to sugars and amino acids, which are then catabolized by lactobacteria to organic acids. Lactic acid fermentation lowers the pH, enabling the yeast *Zygosaccharomyces rouxii* to grow and start alcoholic fermentation. Later during fermentation, the yeast *Candida versatilis* takes over and produces phenolic flavoring compounds. Finally, the fermented Moromi is pressed, filtered and pasteurized to yield refined soy sauce.

This rather time-consuming fermentation process has several advantages. First, large indigestible polymers are degraded to easily digestible monomers such as monosaccharides and amino acids. Second, several substances causing indigestion in humans are degraded by fungal enzymes, such as an inhibitor of the intestinal enzyme trypsin, or phytic acid, which binds essential metal ions in the human intestinal tract. Third, fermentation leads to an enrichment of flavors and tastes due to small molecules produced by the yeasts and bacteria.

Tempeh

Tempeh is the generic term for fungal-fermented beans or cereals and is made using zygomycetes of the genera *Mucor* or *Rhizopus*. Tempeh originates from Indonesia and nowadays is produced industrially as a highly nutritious meat replacement. It is often made solely from soybeans and is then called tempeh kedele. In short, lactic acid-fermented beans are cooked, cooled down and inoculated with fungal spores. The fungus grows into a mycelium, which covers the beans and holds them together. After 1–2 days of fungal fermentation, the raw tempeh is ready and must be cooked or fried before consumption.

During fermentation, the filamentous fungi, together with bacteria and yeasts, degrade proteins and lipids, causing high amino acid and fatty acid contents and better digestibility. Furthermore, fungal fermentation leads to an enrichment of B vitamins (biotin, folate, nicotinic acid, pantothenic acid, pyridoxine, riboflavin). Some of these vitamins, however, are only produced when certain microorganisms are present during the fermentation process, underlining the importance of microbial interactions for successful fermentation.

4.3 Primary metabolism

Primary metabolites are necessary for the survival and growth of organisms. Typical products of primary metabolism are organic acids, which have long been known as major biotechnologically relevant fungal products (see section 4.3.1). Further, fungi use a wide variety of enzymes for metabolic reactions, and these enzymes can be isolated and used for many industrial applications (see section 4.3.2). Based on their enzymatic content and the reaction specificity of these enzymes, fungi are further used in a process termed biotransformation (see section 4.3.3).

4.3.1 Organic acids

Organic acids are typical products resulting from microbial fermentations (see Box 4.4). In particular, the development of large-scale fermentation technologies has reached an industrial level that is superior to pure chemical synthesis with its harsh production conditions. Further advantages of microbial fermentations are their cost effectiveness and rather simple methodology. Organic acids play a major role in our everyday life, and they are primarily important for the production of food, as well as care and household products (Table 4.3). Besides bacteria, several filamentous fungi are the main producers of a diverse number of organic acids. The most important one is citric acid, with a yearly world production of about 1 million tons and a yearly market value of about 0.9 billion Euros.

Box 4.4: Fermentation processes.

Solid-state fermentation is defined as a process where microbial growth and product formation occur on the surfaces of solid substrates, whereas liquid-state fermentation occurs in liquid media. Fed-batch cultures are characterized by feeding one or more nutrients to the microbial culture.

Table 4.3: Organic acids produced by fungi. Modified from Archer et al. 2008; Kubicek et al. 2010.

Acid	Application	Producer
Arachidonic acid	Baby milk supplement	*Mortierella alpina*
Ascorbic acid	Preservative	*Aspergillus spp., Penicillium spp., Torula spp.*
Citric acid	pH regulator, acidulant, preservative, antioxidant, anticoagulant (used in food, beverage, technical, pharmaceutical, and cosmetic industry)	*A. niger*
Epoxisucchynic acid	Precursor for tartaric acid	*A. fumigatus*
Fumaric acid	Acidulant (strong acid flavor), coagulant (food industry)	*Rhizopus spp.*
Gallic acid	Dyeing (cosmetic industry)	*Aspergillus spp.*
Gluconic acid	slow-acting acidulant, preservative, chelating agent (food and technical industry)	*A. niger, Penicillium spp.*
Itaconic acid	Detergent	*A. itaconicus, A. terreus*
Kojic acid	Antioxidant, preservative, prevents oxidative browning, skin whitening (food and cosmetic industry)	*A. oryzae, A. sojae, Penicillium spp.*
Lactic acid	Acidulant, preservative (food industry)	*Rhizopus oryzae*
Linolenic acid	Dietary supplement	*Mucor circinelloides, Mortierella isabellina*
Malic acid	Acidulant (smooth acid flavor), preservative (food industry)	*Paecilomyces varioti, Schizophyllum commune,* also produced by biotransformation of fumaric acid (section 4.3.3)
Succinic acid	Flavoring agent	*Rhizopus spp.*
Tartaric acid	Flavoring agent	*Aspergillus spp., P. notatum*

bold: mostly chemical synthesis

Citric acid

Citric acid, a tricarboxylic acid, is a widely applied organic acid that was initially extracted from citrus fruits. Citric acid is applied for a huge and broad variety of processes. Typical commercial applications are its use in food, soft drinks, pharma-

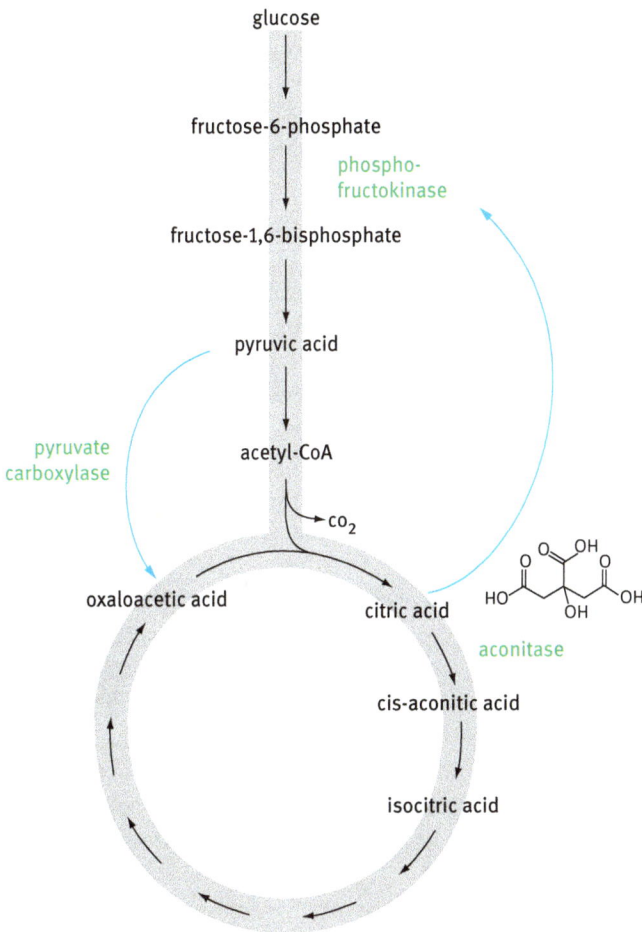

Fig. 4.4: Metabolic pathways during citric acid production in *Aspergillus niger*. Enzymes that are critical for citric acid production are highlighted in gray.

ceutical, leather tanning, and electroplating industries. Furthermore, citric acid is used in the detergent industry as a substitute for phosphate, and in the construction industry to slow down the hardening of cement.

Citric acid is an intermediate of the tricarboxylic acid cycle (TCA or Krebs cycle), a series of reactions found in almost all aerobically growing organisms. The TCA cycle generates energy through the oxidation of acetate derived from, for example, carbohydrates. Products of the TCA cycle are carbon dioxide and adenosine triphosphate (ATP) as a form of chemical energy. In fungi, as in all eukaryotes, the TCA cycle takes place in the mitochondria.

In the scheme presented in Figure 4.4, the main metabolic reactions leading to the formation of citric acid are shown. Glucose as a carbon source is degraded by glycolysis to produce pyruvic acid. In a further step, pyruvate dehydrogenase generates acetyl-CoA. This molecule is then the starting point for the TCA cycle, with citrate as a first intermediate. Subsequent oxidative steps and chemical transformation lead to the loss of two carboxyl groups from citrate. During the TCA cycle, several energy-rich electrons are transferred to NAD^+, leading to the formation of NADH. At the end of this cycle, carbon oxaloacetate has been regenerated, and the cycle can start again.

Already in 1893, the German chemist C. Wehmer discovered that Penicillium molds are able to produce citric acids from sugar. In 1917, the American food chemist James Currie discovered that certain strains of the mold *Aspergillus niger* can efficiently produce citric acid. In 1929, the company Pfizer succeeded in producing citric acid at an industrial level by feeding *A. niger* on sucrose or glucose-containing media. The source of sugar was usually corn steep, liquor, molasses, hydrolyzed cornstarch, or other inexpensive sugar solutions. Initially, the fungus was grown in batch cultures, where citric acid is secreted into the supernatant. The acid was converted into calcium citrate by precipitation with lime (calcium hydroxide), and in a subsequent step, citric acid was regenerated by treatment with sulfuric acid.

High accumulation of citric acid is believed to require the deactivation of enzymes of the TCA cycle. Two candidate enzymes are aconitase and isocitrate dehydrogenase, which might be degraded or deactivated in high-producing *A. niger* strains. An alternative explanation for citric acid accumulation is a highly active tricarboxylic acid transporter that competes with aconitase for citric acid. If the transporter has greater affinity to citric acid than aconitase, citric acid is transported out of the mitochondria without inhibiting the enzymes of the TCA cycle. Finally, it was also suggested that citric acid accumulation is more or less a result of enhanced metabolism in general.

Nowadays, *A. niger* is grown in submerged aerobic cultures, where the composition of the media, e.g. keeping trace element deficiency, is critical to obtain optimal citric acid concentrations. The best conditions to produce citric acid with *A. niger* are based on the following criteria. Sucrose is preferred as the carbon source, since *A. niger* has an extracellular invertase (Table 4.4), converting sucrose into glucose and fructose, that is particularly active at low pH. Usually, the culture medium has a pH of about 5 when fermentation starts with germinating conidiospores. The germinating spores absorb ammonia in the medium to further release protons that lower the pH, and thus improve the production of citric acid. A pH of about 2 also reduces the production of unwanted byproducts such as gluconic and oxalic acid, and substantially reduces the risk of contamination by other microorganisms. Finally, a combination of high concentrations of glucose and ammonium is preferable, since these two compounds repress the synthesis of α-ketoglutarate dehydrogenase, which itself catabolizes citric acid. Thus, this combination favors the overproduction of citric acid. Another important aspect in the fermentation with fungal species is the morphology of the producer strain. Although the underlying reasons have not been finally clarified, optimal fungal morphology during citric acid production is characterized by short, swollen hyphal branches, leading to the formation of pellets. It has long been observed that pellets of fungal mycelia are preferred over dispersed growth during fermentation.

Gluconic acid

This acid and its derivatives are produced with a yield of about 100,000 tons per year. Gluconic acid is used in the food industry, and some sodium and calcium salts of gluconic acid are used as cleaning products, and as metal complexing agents in metal, pharmaceutical, and textile industries. Besides bacteria, mainly filamentous fungi of the genera Aspergillus and Penicillium are used as producers. Gluconic acid is synthesized at high glucose concentrations. The responsible enzyme is a FAD-dependent glucose oxidase that converts glucose to glucono-Δ-lactone, which hydrolyzes spontaneously in water to gluconic acid.

In the early twentieth century, surface cultures of *Penicillium luteum-purpurregenum* were used to produce gluconic acid. This production process is superior to bacterial fermentation, which showed a low tolerance to high glucose concentrations. Today, *Aspergillus niger* is mainly used in submerged processes to produce gluconic acid. In optimal fermentation processes, 99 % of the substrate glucose is converted into gluconic acid.

Itaconic acid

Itaconic acid or methylenesuccinic acid is an organic acid used in diverse applications, for example, in paper and architectural coating industries. It is further used for manufacturing glass fibers and synthetic resins. Itaconic acid is a by-product of the TCA cycle and is produced industrially by diverse strains of *Aspergillus terreus*. Alternatively, itaconic acid can be obtained during the distillation of citric acid.

Lactic acid

This organic acid is used to conserve food and drinks, but it is also applied in pharmaceutical and textile industries for its preserving properties. Although lactic acid is mostly produced by bacteria of the genus Lactobacillus, some filamentous fungi are economically important producers due to their capacity to use starch as a carbon source. Examples of fungal producers are *Aspergillus niger* and *Rhizopus oryzae*. It can be foreseen that metabolic engineering will enable industrial microbiologists to also use baker's yeast as a producer of lactic acid.

4.3.2 Technical enzymes

Fungi are a rich source of a great variety of technical enzymes, which have tremendous potential, e.g. in pulp, paper, food, and cosmetic industries. Most important-

Table 4.4: Technical enzymes from filamentous fungi. The given producers represent typical examples and do not reflect the broad range of different species that have been used for diverse industrial applications. Modified from Kück et al. 2009; Archer et al. 2008.

Enzyme	Function/Activity	Application	Producer
α-Amylase	Reduction of starch, 1,4-α-D-Glucan-Glucanhydrolase	Food/beverage production, Production of high-maltose syrup, starch processing industry, alcohol production	*Aspergillus niger, Aspergillus oryzae, Thermomonspora spec.*
Arabinanase	α-L-Arabinanase, Arabinan endo-1,5-alpha-L-arabin-osidase, Degradation of Lignocelluloses	Bioethanol production, vegetable processing	*Aspergillus spec.*
Asparaginase	L-Asparaginamidohydrolase; Hydrolysis of asparagine to aspartic acid	Baking industry, reduction of acrylamide production	*Aspergillus spec.*
Cellulase	endo-1,4-beta-D-glucanase, degradation of lignocellu-loses, cellulose and hemi-cellulose modification	Bioethanol production, vegetable processing, malting and brewing industry, detergents	*Trichoderma longibrachiatum*
Glucoamylase	Glucan-1,4-alpha-glucosidase	Soluble starch processing	*Aspergillus niger*
β-Glucosidase	Degradation of Cellulose β-1,4-D-Glucosid-Glucohydrolase	Flavor enhancers in wine	*Aspergillus spec., Penicillium spec.*
Invertase	β-D-fructofuranoside fructohydrolase, Degradation of sucrose to glucose and fructose	Confectionery	*Saccharomyces cerevisiae*
Laccase	Oxidation of phenolic substances combined with oxygen reduction, degradation of lignocellulose	Beverage (breweries), personal care, textile industry	*Myceliopthora spec. Polyporus spec. Thielavia spec. Trametes spec.*
Lactase	Hydrolysis of lactose to D-galactose and D-glucose	Dairy	*Kluyveromyces spec.*
Lipase	Generic term for enzymes that catalyze the hydrolysis of lipids	Various applications; e.g. vegetable oil processing, cheese manufacture, detergents, biocatalysis	*Rhizopus oryzae*
Pectinase	Polygalacturonase, degradation/modification of pectins	Soft drinks industry	*Aspergillus niger*
Phytase	Phosphorus release from phytic acid	Legume processing, bread making, animal feed	*Aspergillus niger*
Protease	Generic term for enzymes that catalyze the hydrolysis of peptide bonds	Various applications; e.g. cheese manufacture, baking, leather, dairy industry	*Aspergillus oryzae, Rhizopus oligosporus*

Table 4.4 (continued)

Enzyme	Function/Activity	Application	Producer
Raffinase	hydrolyzes raffinose, yielding fructose in the reaction	Food processing, Removal of raffinose from molasses, corn and soybean milk	*Aspergillus niger, Saccharomyces cerevisiae*
Rennet	Enzyme mixture, naturally from calf abomasum, contains proteases (chymosin) and lipases	Cheese manufacture	*Mucor miehei*
Xylanase	Endo-1,4-beta-xylanase, Endo-1,3-beta-xylanase, Xylan degradation	Brewing industry	*Trichoderma reesei*

ly, such enzymes are biocatalysts that can replace cost-intensive chemical catalysts with their many side reactions. Technical enzymes perform specific modifications of substrates without forming unwanted byproducts. Several enzymes from diverse filamentous fungi are listed in Table 4.4, and a few cases are described in more detail above.

α-Amylase

Amylases are enzymes that catalyze the hydrolysis of starch into sugars. α-amylase is a 1,4-α-D-glucan-glucanhydrolase and, together with β-amylase and glucoamylase, is among the most important enzymes in biotechnology. The first α-amylase enzyme produced industrially was isolated from fungal sources as early as 1894, and since then, many applications have been demonstrated. For example, industries such as food, fermentation, textile, and paper industries use amylases. α-amylase can be derived from bacteria, yeasts, and filamentous fungi. Important fungal producers are *Aspergillus niger* and *Aspergillus oryzae*. Currently, amylases are used to convert cheap carbon sources, e.g. raw starch, into fermentable sugars. Thus, these amylases are used for cost-effective starch hydrolysis, which is of major interest for the ethanol industry. Another challenge is to isolate amylases that are thermostable and, thus, suitable for hydrolyzing starch in solid-state fermentations.

Cellulases

Cellulases are a group of enzymes that catalyze the hydrolysis of cellulose. Besides bacteria and protozoa, many fungi show cellulase activity. Several types of cellulases can be distinguished based on their type of reaction, for example endocellulases, exocellulases, cellobiases, and oxidative cellulases. For commercial production of cellulases, *Trichoderma longibrachiatum* (formerly *reesei*) is used in fermentation processes where the enzymes are secreted. A common application is the use of cellulases to prepare wine, beer, and fruit juices. Cellulases break down insoluble compounds of the plant cell wall, and this releases soluble sugars. As a side effect, cellulase treatment results in a clearer product.

Another application is the incorporation of cellulases into detergents for laundry purposes. During washing, cellulases remove cellulose microfibres from cotton-based clothes. In this case, they are preferred as additives to detergents, since cellulases are active at low temperatures and under alkaline conditions.

Similar to amylases, cellulases have emerged in recent years as important biotechnical tools for ethanol production from renewable resources. Converting lignocellulotic biomass to fermentable sugars using cellulases is a major challenge for current biofuel production.

Laccases

Laccases are enzymes that catalyze the oxidation of phenolic substances by reducing oxygen. Usually they are copper-containing enzymes that are found in bacteria, plants, and fungi. Many wood-decaying fungi, which are known as brown rot, soft rot, and white rot, are rich sources of laccases. Industrial fungal producers of diverse laccases are mostly basidiomycetes of the genera Polyporus or Trametes.

Since laccases do not release toxic peroxide intermediates after oxidation, they are used in diverse applications to detoxify phenolic compounds. Examples are their application for teeth whitening and bioremediation processes. In the food industry, laccases are applied to remove polyphenols, which may remain in beer after the brewing process. Laccases can form polyphenol complexes that can be separated from the substrates by filtration. Similarly, fruit juices are treated with laccases to remove phenolic compounds that have negative effects on taste or color. Finally, the application of laccases in the pulp and paper industry should be mentioned, for example, to improve the properties of pulp, or to bioremediate the mill effluence during paper production.

Lipases

Lipases are a class of enzymes that catalyze the hydrolysis of long-chain triacylglycerols to glycerol and free fatty acids. Lipases are only activated when adsorbed to an oil–water interface. They have broad industrial applications in pulp, paper, food, detergent, and textile industries. Specific applications are organic synthesis, hydrolysis of fats and oils, modification of fats, flavor enhancement in food processing, resolution of racemic mixtures, and chemical analyses. Furthermore, they are used as anti-thrombosis and anti-inflammatory agents in diverse medical therapies, or as diagnostic tool to determine blood glucose, urea or cholesterol levels. Significant fungal producers are from the genera Aspergillus, Candida, Penicillium, and Rhizopus. Currently, batch, fed-batch, and continuous mode fermentations in submerged and solid-state fermentations (see Box 4.4) are used for production.

Pectinases

Pectins are polysaccharides of the cell walls of higher plants. Pectinases break down the glycosidic bonds of the long-chain galacturonic acid residues of pectin, and generate simpler molecules such as galacturonic acid. Pectinases can be distinguished further into pectin methylesterases, which de-esterify pectins to low methoxyl pectins or pectic acid, and pectin depolymerases, which split the glycosidic linkages between galacturonosyl residues.

In technical applications, pectinases are favorable for fruit and textile industries. For example, pectinases are used to reduce the bitterness and cloudiness of fruit juices. In the textile industry, these enzymes are applied for the retting and degumming of fiber crops, or for the production of good quality paper. Pectinases are further used to degrade agro-industrial by-products such as orange and sugar cane bagasse, or food processing waste.

Fungi are ideal decomposers of plant material, because they are high-yield producers of cellulases and pectinases. Since these enzymes are secreted into the culture medium, they can easily be harvested for further purification. The most potent fungal producers of pectinases (and xylanases) have been found within the genera Aspergillus, Penicillium, Sporotrichum, and Trichoderma.

4.3.3 Biotransformation

Biotransformation is an important biotechnical tool to modify structurally diverse organic compounds using living or fixed cells or isolated enzymes. A precursor is used as substrate for the cells or enzymes, and the product is then isolated after biotransformation. In the pharmaceutical and food industry, fungi, bacteria, plant or animal cell cultures, as well as organelles and isolated enzymes, have been used as biocatalysts to generate novel compounds. With biotransformation, it is possible to perform reactions that are chemically laborious, time-consuming, or even impossible. This includes site- and stereo-specific reactions at positions that are hardly available for chemical agents. Biotransformation results in either biologically active compounds or key intermediates that have to be further modified. Thus, industry commonly applies a combination of chemical and biotransformation processes to generate pharmaceutically active products.

Diverse mold species are applied as catalysts for biotransformation procedures. The strains used are characterized by certain properties, e.g., the ability to perform specific reactions such as hydrolysis, condensation, oxidation, reduction, or isomerization. These reactions are possible because of the wide variety of enzymes that are produced by different fungi (compare section 4.3.2). The fungal enzymes are often able to specifically generate one of two possible enantiomers (see Box 4.5, compare section 2.1.3). Thus, biotransformation is superior to chemical catalysis, which usually generates both enantiomers of a chemical reaction.

Box 4.5: Enantiomers.

Enantiomers are a pair of stereoisomers. Two enantiomers are mirror images of each other, like the left and right hand. Because of this asymmetry, enantiomers often behave chemically different, which is a critical point in drug production. By biotransformation, it is possible to generate enantiopure products, i.e., the product consists only of one of the two enantiomers.

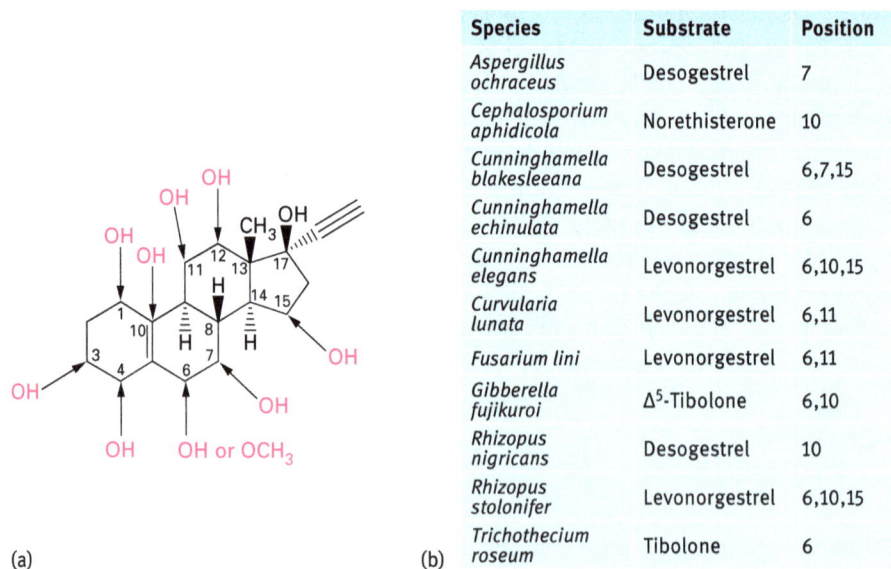

Species	Substrate	Position
Aspergillus ochraceus	Desogestrel	7
Cephalosporium aphidicola	Norethisterone	10
Cunninghamella blakesleeana	Desogestrel	6,7,15
Cunninghamella echinulata	Desogestrel	6
Cunninghamella elegans	Levonorgestrel	6,10,15
Curvularia lunata	Levonorgestrel	6,11
Fusarium lini	Levonorgestrel	6,11
Gibberella fujikuroi	Δ^5-Tibolone	6,10
Rhizopus nigricans	Desogestrel	10
Rhizopus stolonifer	Levonorgestrel	6,10,15
Trichothecium roseum	Tibolone	6

(a)

(b)

Fig. 4.5: Biotransformation of steroids by filamentous fungi. (a) Structure of a 17α-ethynyl substituted steroid. Numbers define C atoms used as target positions for hydroxylation or methoxylation. (b) Target positions for hydroxylation of 17α-ethynyl substituted steroids by different fungal species. If multiple target positions are given, choice is dependent on physiological growth conditions of the fungal culture. Modified from (Shah et al. 2013).

Steroids are among the most prominent compounds that are produced by bio-transformation. These hormones are applied in a wide range of therapeutic uses, e.g., as anabolic, anti-inflammatory, anti-rheumatic, contraceptive, and immuno-suppressive agents. Approximately 1,000,000 tons of steroids are produced per year, with an estimated market value of US$ 10 billion.

In the 1950th, the diosgenin-rich yam species *Dioscorea mexicana* was the major source for the conversion of diosgenin to the steroid cortisol. This was done by combination of chemical reactions and microbial biotransformation. The first patented biotransformation process was the 11α-hydroxylation of progesterone by *Rhizopus arrhizus* in 1952. Since then, a huge number of biotransformation processes were patented, and fungal species belonging to the zygomycetes, ascomycetes, and basidiomycetes have been used as biocatalysts.

Hydroxylation is the most wide-spread type of steroid bioconversion. As with many biotransformation products, the obtained steroids are mostly used as intermediates for further chemical synthesis. Different target positions for hydroxylation are used by different microorganisms; and this variety results in modified drugs that have therapeutic advantages such as high potency, longer half-lives in blood, and reduced side effects. As an example, the regio- and stereoselective hydroxylation of a 17α-ethynyl steroidal drug is shown in Figure 4.5a, and fungal species that substitute different target positions of the steroid are given in Figure 4.5b. This example illustrates that biotransformation processes are highly dependent on the availability of natural microbial resources for finding appropriate strains for biotransformation.

4.4 Secondary metabolism

Secondary metabolites are not absolutely required for the survival of an organism, possess great structural diversity in different fungi, and are clearly distinguished from essential primary metabolites. Whereas yeasts have either no or a rather restricted secondary metabolism, filamentous fungi possess a cornucopia of secondary metabolites, which are produced only under distinct physiological conditions. In most cases, secondary metabolites are small molecules produced from precursor molecules of the primary metabolism by certain enzymes (Table 4.5). Besides the physiological conditions, a number of factors such as substrate composition, atmospheric pressure, temperature, pH, or water activity are important parameters for the synthesis of secondary metabolites. Frequently, their synthesis is correlated with certain stages of the fungal life cycle.

Table 4.5: Examples of secondary metabolites, derived from different primary metabolite precursors.

Primary metabolite	Derived primary metabolites	Derived secondary metabolites (compound class)	Secondary metabolites (examples)
Glucose	Aromatic amino acids	Alkaloids	Fumitremorgen C, Lysergic acid
Pyruvate	Aliphatic amino acids	Non-ribosomal peptides	Cephalosporins, Penicillins
Acetyl-CoA	Isopentenyl diphosphate (IPP)	Terpenes	Geosmin, Gibberellins
		Steroids	Cortisol
	Malonyl-CoA	Polyketides	Aflatoxins, Lovastatin
		Fatty acid derivatives	Griseofulvin

Table 4.6: World market of selected pharmaceuticals produced with fungi. For comparison, the value of a bulk fungal product, citric acid, is given.

Product	Yearly production t/a	Price EUR/kg	Market value Mio EUR
Citric acid	1,000,000	0.80	900
Cephalosporins	30,000	283	8,500
Cyclosporin A	3	5,200	15.6
Penicillins & derivatives	45,000	300	13,500
Statins	NA	NA	18,000
Tetracyclins	5,000	50	250

NA, not available

Secondary metabolites usually have a low molecular weight, but show a high complexity that is characteristic for a certain strain of a given species. Typical fungal secondary metabolites are mycotoxins, such as aflatoxin, ergot alkaloids, and trichothecene. Other secondary metabolites of pharmaceutical importance are cephalosporins, cyclosporins, penicillins, and statins. Although the latter products have high pharmaceutical relevance, their biological function for the fungi themselves is almost unknown. Sometimes secondary metabolites are able to provide growth advantages over other microorganisms living in the same environment. Besides that, it is assumed that secondary metabolites act as signaling molecules to govern the biological function of a fungus.

To date, several thousand secondary metabolites have been characterized from fungi, some of them having antibacterial, antiviral, or cytostatic activity. Indeed, fungal secondary metabolites are among those pharmaceuticals with the highest yearly world market value (Table 4.6). Despite their enormous chemical complexity and diversity, secondary metabolites from fungi can be grouped into a few classes. This classification is based on the precursor molecules derived from primary metabolism and the enzymes used for the metabolite's biosynthesis. The four major chemical groups are polyketides and fatty acid derivatives, non-ribosomal peptides, isoprenoids, and alkaloids. The following sections describe the biosynthesis of typical secondary metabolites and give examples of their biotechnical relevance.

Polyketides and fatty acid derivatives
Polyketides belong to the major group of fungal secondary metabolites and display great heterogeneity in terms of their chemical structure and pharmacological properties. Typical polyketides are mycotoxins such as aflatoxin B1, fumonisin, fusarin C, and ochratoxin. However, there are several commercially available polyketide pharmaceuticals such as the cholesterol-lowering drug Lovastatin.

Acetyl-CoA or malonyl-CoA is the precursor molecule for the biosynthesis of all polyketides (Figure 4.6a). In filamentous fungi, polyketides are synthesized by type I polyketide synthases (PKSs), which represent large, multi-domain proteins used exclusively to synthesize secondary metabolites. These fungal PKSs possess different domains for iterative biosynthetic reactions, which is different from bacterial PKSs or fungal NRPSs (see below) that carry several modules, each for the addition of one starter molecule. Figure 4.6a depicts a typical type I PKS with essential and optional domains.

Synthesis of a polyketide is initiated when the acetyl transferase (AT) domain of a PKS recognizes acetyl-CoA and malonyl-CoA as starter molecules, which subsequently bind to the acyl carrier protein (ACP) domain of the PKS. In the following step, another acetyl-CoA or malonyl-CoA molecule binds to the ketoacyl synthase (KS) domain, which catalyzes the condensation of both molecules. A diketide product bound to the ACP domain then functions as a substrate for the next chain elongation step, as soon as a further unbound KS domain binds acetyl-CoA or malonyl-CoA again. These biosynthetic steps are repeated until the final length of the chain is obtained. Finally, the cleavage of the linear polyketide from the PKS occurs by hydrolysis or cyclization. These termination reactions are catalyzed by the thioesterase (TE) or cyclization (CYC) domain, which are located at the C-terminus of the PKS (Figure 4.6a).

The number of domains of a PKS is not predictive for the actual structure of the polyketide itself (i.e. the polyketide's structure does not directly depend on the number of domains contained in the PKS). Modifications of the polyketide's basic structure, such as methylation by methyl transferase domains (MT), lead to the structural diversity of these secondary metabolites. Further, different combinations of domains (e.g. dehydrogenase domains (DH), enoyl reductase domains (ER), and ketoacyl reductase domains (KR), Figure 4.6a) lead to a variable number of reductions in the polyketide molecule. Thus, the diversity among polyketides arises from (i) the length of the chains, i.e. the number of iterations, (ii) the number of reduction reactions, (iii) a facultative cyclization and other modifications by the PKS, and (iv) post-polyketide synthesis steps.

PKSs are structurally related to fatty acid synthases. The main difference between polyketides and fatty acids is the full reduction of the beta-carbon in fatty acids, which is only an optional event during polyketide biosynthesis (Figure 4.6a. Furthermore, fatty acid synthases usually possess all the essential domains necessary for fatty acid biosynthesis, while PKSs often produce molecules that are further modified by additional enzymes.

Fatty acid derivatives are an important group of secondary metabolites that also have medical significance. For éxample, griseofulvin, a fatty acid synthesized by *Penicillium griseofulvum*, is characterized by fungistatic activity against various dermatophytic species of Epidermophyton, Microsporum, and Trichophyton. So far, no effect on yeasts or other filamentous fungi has been observed. Griseofulvin has been shown to bind with great affinity to keratin in diseased tissues colonized by the dermatophytes. The keratin-griseofulvin complex also binds to fungal micro-

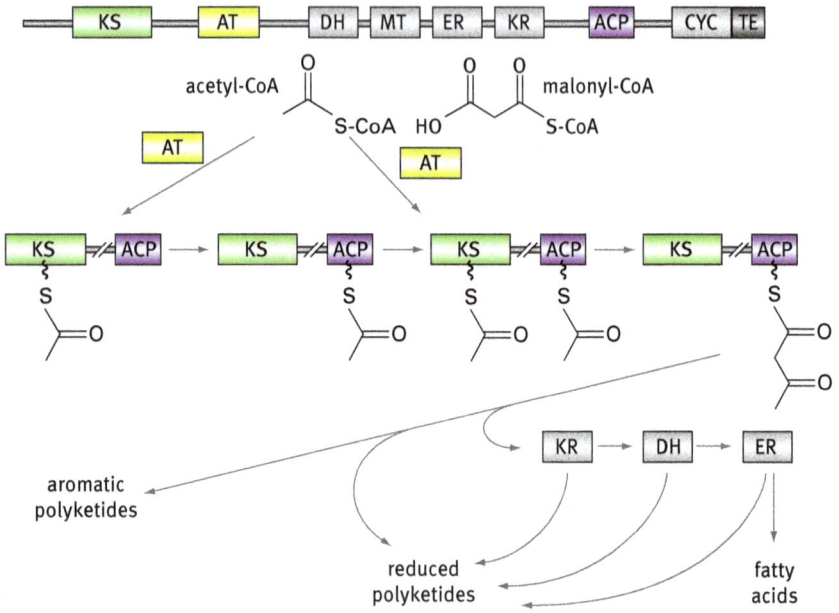

(a)

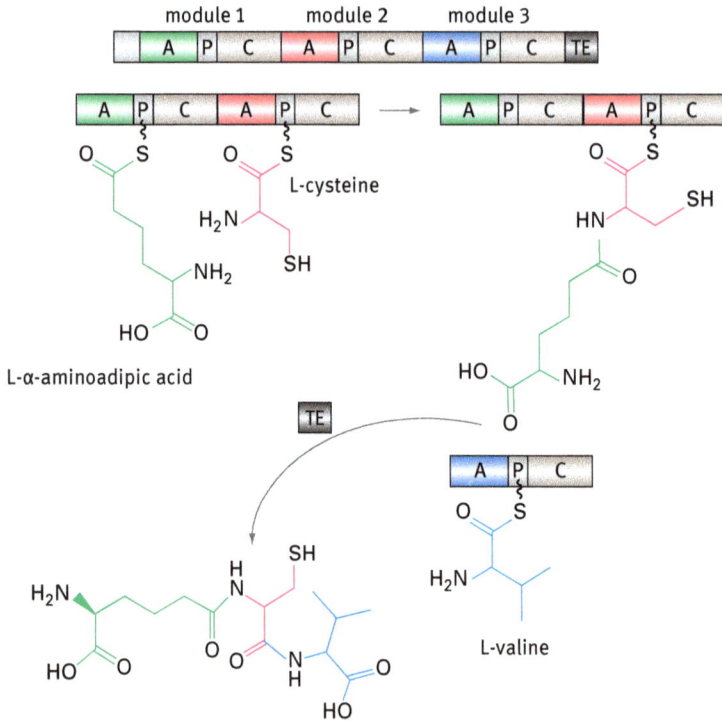

(b)   δ-(L-α-aminoadipyl)-L-cysteinyl-D-valine

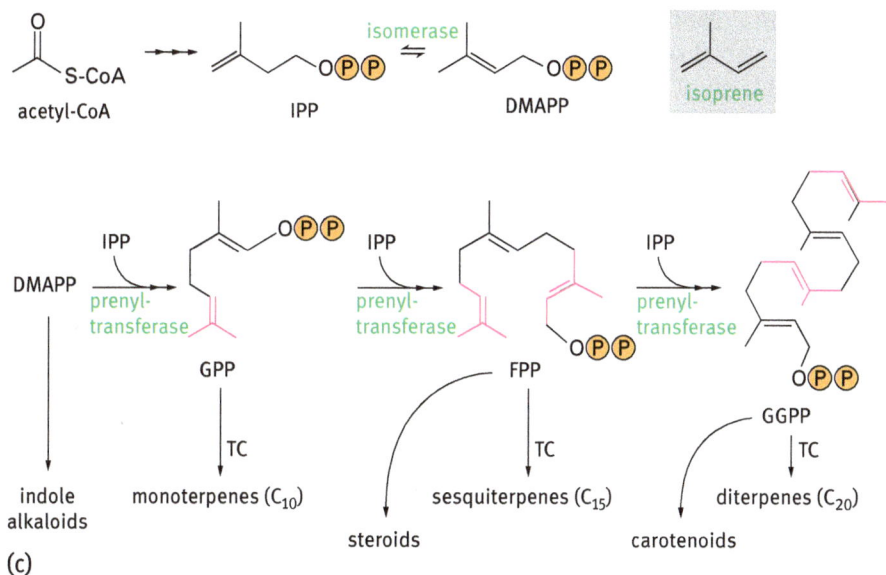

Fig. 4.6: Biosynthesis of polyketides, non-ribosomal peptides, and isoprenoids. Details are described in the text. (a) Polyketides are synthesized by modular enzymes, the polyketide synthases. (b) Non-ribosomal peptides (NRPs) are synthesized by non-ribosomal peptide synthetases (NRPSs). As substrates, proteinogenic and non-proteinogenic amino acids are used. Shown is the first step of penicillin and cephalosporin C biosynthesis. (c) Biosynthesis of isoprenoids. Several enzymes use the primary metabolite acetyl-CoA for biosynthesis of isopentenyl diphosphate (IPP). This can be converted by an isomerase into dimethylallyl diphosphate (DMAPP). IPP and DMAPP are the starter molecules for carotenoid, indole alkaloid, steroid, and terpene biosynthesis. A, adenylation domain; ACP, acyl carrier protein domain; AT, acetyl transferase domain; C, condensation domain; CYC, cyclase domain; DH, dehydratase domain; ER, enoyl reductase domain; FPP, farnesyl diphosphate; GGPP, geranylgeranyl diphosphate; GPP, geranyl diphosphate; KR, ketoacyl reductase domain; KS, ketoacyl synthase domain; MT, methyl transferase domain; P, peptidyl carrier protein; TC, terpene cyclase; TE, thioesterase domain.

tubules (tubulin), and thus inhibits fungal mitosis and nucleic acid biosynthesis. However, the exact mechanism of the fungistatic action of griseofulvin is not yet fully understood.

In recent years, genome analysis has shown that filamentous fungi possess a huge set of genes for PKSs and fatty acid synthases. For example, the filamentous fungi *Aspergillus oryzae*, *Aspergillus nidulans*, and *Aspergillus fumigatus* have 30, 27, and 14 genes, respectively, for diverse PKSs. The number of fatty acid synthases is rather low; so far, one to six genes for these enzymes have been found within fungal genomes. These examples demonstrate that even the number, not least the structure of secondary metabolite enzymes, is highly variable between species of the same genera, indeed illustrating that every fungus produces a particular set of secondary metabolites.

Non-ribosomal peptides

All non-ribosomal peptides (NRPs) are synthesized by non-ribosomal peptide synthetases (NRPSs). Similar to polyketide and fatty acid synthases, these multifunctional fungal enzymes are multidomain, multimodular proteins. For the biosynthesis of secondary metabolites, NRPSs use both proteinogenic and non-proteinogenic amino acids. Although the number of proteinogenic amino acids is rather constant (20–22), the number of non-proteinogenic amino acids can be more than 300, thus guaranteeing high structural diversity of NRPs. In contrast to polyketides, the amino acid sequence and the length of a final NRP product depends on the number and order of modules within the NRPS. Obviously, this type of peptide biosynthesis is different from RNA-dependent ribosomal protein biosynthesis, where the amino-acid sequence is encoded by genes.

Every module of an NRPS comprises three domains that are essential for peptide synthesis (Figure 4.6b). These domains are an adenylation (A) domain, a peptidyl carrier protein (P) domain, and a condensation (C) domain. Every module specifically binds a certain amino acid. In the first step, the A domain of every module recognizes its specific amino acid and transfers it to the P domain. Next, the C domain catalyzes the formation of a peptide bond between two amino acids bound by a P domain. This process is repeated until the polypeptide reaches its complete length at the final module. During growth, the amino acid chain is covalently linked to a phosphopantetheine cofactor which is attached to a conserved serine. In the final step, a thioesterase domain (TE) cleaves off the product from the NRPS.

Similar to polyketides, NRPs are highly diverse due to their variable length and high number of potential substrates. Moreover, amino acid modifications are possible through specific domains that exist in addition to the above-mentioned essential module domains. A typical modification domain is the cyclization domain, which catalyzes the formation of cyclic peptides.

NRPs that are of major importance as anti-infectives are the β-lactam antibiotics penicillin and cephalosporin. Both antibiotics are characterized by the four-membered amid ring system. β-lactam antibiotics used to treat humans are exclusively produced by *Acremonium chrysogenum* (cephalosporin C) and *Penicillium chrysogenum* (penicillin). β-lactam antibiotics have a bactericidal activity that prevents the propagation of dividing bacterial cells. The antibiotics are recognized by the bacterial D-adenine transpeptidase and are used as substrates during bacterial cell wall biosynthesis. Incorporation of the antibiotics prevents crosslinking between bacterial cell wall components, and thus, propagation of the bacterial cell itself.

A significant aspect of penicillin and cephalosporin biosynthesis is that the first two steps are identical. Isopenicillin N, an intermediate in both pathways, already possesses antibiotic activity. After this intermediate, the biosynthetic pathways differ. Only a single biosynthetic step is necessary to convert isopenicillin N into penicillin G, but six further steps are needed to generate cephalosporin C.

The first step of penicillin biosynthesis in *A. chrysogenum* and *P. chrysogenum*, the generation of a tripeptide by the NRPS ACV synthase, is shown in Figure 4.6b.

P. chrysogenum can produce two different forms of penicillin. Penicillin G must be injected while penicillin V is acid-stable, meaning that patients can be treated with tablets. Penicillins are usually most effective against gram-positive bacteria. However, ampicillin, a penicillin derivative generally used for recombinant DNA technology, has activity against both gram-positive and gram-negative bacteria.

The current world market of penicillin and its derivatives is about 13.5 billion Euros and thus corresponds to about one third of the total anti-infectives market. Besides penicillins and cephalosporin C, several fungal NRPs such as cyclosporin A are of major pharmaceutical significance. Another NRP with applied relevance is the mycotoxin gliotoxin, which was shown recently to accelerate apoptosis. Currently, research is focusing on the use of gliotoxins in cancer therapy.

Isoprenoids

Isoprenoids or terpenoids are carbohydrates that are assembled from isoprene units containing five C atoms (Figure 4.6c, gray box). The isoprenoid biosynthesis pathway is depicted in Figure 4.6c. Starting from acetyl-CoA, mevalonic acid is synthesized, which is then catalyzed to isopentenyl diphosphate (IPP), which in turn is isomerized to dimethylallyl diphosphate (DMAPP). IPP as well as DMAPP are activated isoprene units generated within the fungal primary metabolism (Table 4.6). Both molecules are used as substrates by prenyltransferases to synthesize the linear polyprenyl diphosphates geranyl diphosphate (GPP), farnesyl diphosphate (FPP), and geranylgeranyl diphosphate (GGPP). Linear or cyclic isoprenoids that are unmodified multiples of linked isoprene units are termed terpenes and are classified according to the number of C atoms into e.g. monoterpenes (composed of two isoprene units), sesquiterpenes (three isoprene units), or diterpenes (four isoprene units) (Figure 4.6c). Terpenes can be further modified, and depending on their chemical structure are then termed carotenoids, steroids, or indole alkaloids.

Isoprenoids are secondary metabolites produced by many different filamentous fungi. A prominent example of terpenes is trichothecenes, which belong to the group of mycotoxins. Typical trichothecene producers are species of the genera Fusarium, Stachybotrys, and Trichoderma. Another example of a terpene-producing fungus is *Gibberella fujikuroi*, a plant pathogen that produces gibberelins. These secondary metabolites are diterpenoid acids that act as phytohormones and are used commercially in agriculture and horticulture.

Alkaloids

Alkaloids are secondary metabolites containing at least a single nitrogen atom as part of a heterocyclic molecule. Amino acid primary metabolism is usually the starting point for the biosynthesis of alkaloids. However, the subsequent biosynthetic steps are innumerous and diverse, and cannot easily be classified within this context.

The most commonly known fungal alkaloid is the ergot alkaloid produced by the plant pathogenic fungus *Claviceps purpurea*. This fungus infects various cereal plants and forms compact sclerotia, also termed ergot kernels, that replace the grains of the infected plants. In the middle ages, consumption of grains infected by the ergot fungus was responsible for a famine phenomenon called ergotism, also known as "St. Antony's fire". Ergot alkaloids are structurally similar to various neurotransmitters and cause severe symptoms, such as dysfunctions of the nervous system (psychoses, dementia), or gangrene leading to loss of extremities.

Today, low-dosed ergot alkaloids are used pharmaceutically; e.g. ergotamine is used to treat headaches and migraines. This is due to the ability of ergotamine to act as an α-adrenergic blocker that directly stimulates the smooth muscle of peripheral and cranial blood vessels. Ergotamine is also an antagonist of serotonin. Another ergot alkaloid derivative is ergonovine, which is used to control postpartum hemorrhage.

4.5 Recombinant technologies

Many yeasts and filamentous fungi possess the GRAS status (see Box 4.3), thus making them ideal organisms to produce recombinant proteins for use in human food production and medical treatment. In traditional strain improvement programs, random mutagenesis was the method of choice to generate improved fungal producers. Since the advent of DNA-mediated transformation procedures in the late 1970s (see below), targeted strategies have been set up to generate recombinant fungal strains. Shortly after the first successful transformations, recombinant strains were constructed for commercially valuable fungi, a process called "recombinant fungal biotechnology".

To date, more than 100 completely sequenced fungal genomes are published, including biotechnologically relevant species such as *Aspergillus niger*, *Penicillium chrysogenum*, *Pichia pastoris*, *Saccharomyces cerevisiae*, and *Trichoderma viridae*. The availability of genome sequences facilitates functional genomics approaches, e.g. the analysis of candidate genes potentially involved in the regulation or biosynthesis of primary or secondary metabolites.

4.5.1 Generation of recombinant fungal strains

As mentioned above, the establishment of DNA-mediated transformation was the first step for the generation of recombinant fungal strains. The first successful DNA-mediated transformation of yeast was in 1978, when a stable leucin-auxotrophic

(a)

Mycelium → A → Protoplasts → B → Plasmid DNA + protoplasts → C → Selection on media containing antibiotics → D → Analysis of selected strains

(b)

Selection Marker$_{Bac}$
T$_{Fun}$
Selection Marker$_{Fun}$
P$_{Fun}$
ars
ori

(c)

P$_{Fun}$ T$_{Fun}$ dsRNA

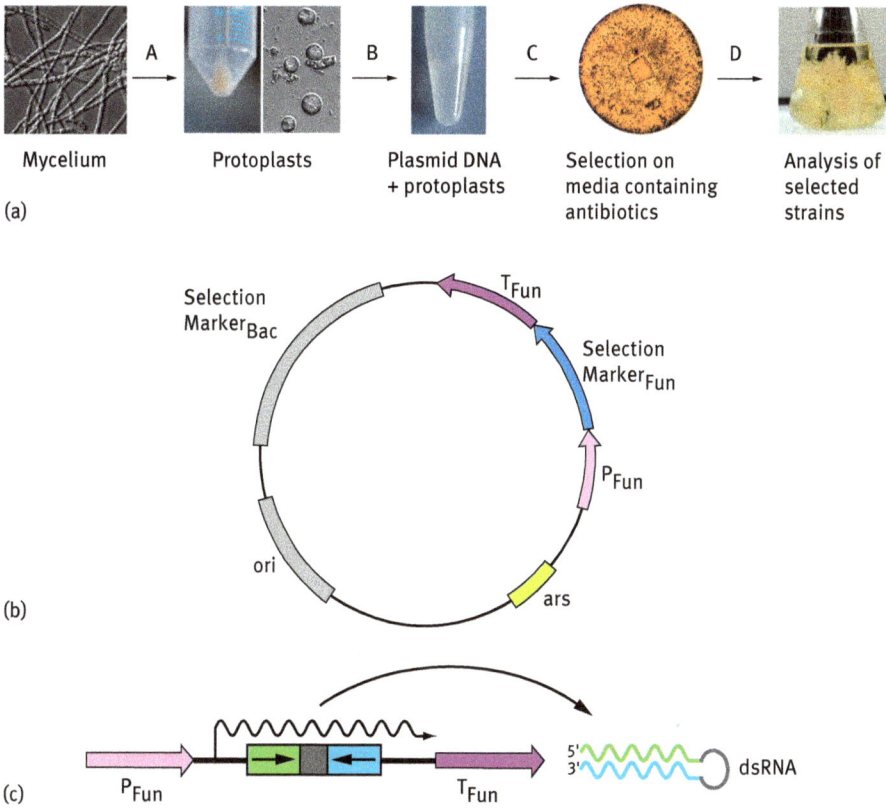

Fig. 4.7: Fungal DNA-mediated transformation and vector systems. (a) Procedure of protoplast transformation in filamentous fungi. A, cell wall degradation by chitinases; B, mixing of protoplasts with plasmid DNA; C, plating of transformed protoplasts on solid media containing antibiotics for selection of transformed cells; D, large scale fermentation of selected strains. (b) Map of a standard fungal transformation vector. Note that fungal transcription promoter and terminator sequences are necessary to express bacterial selection marker genes. (c) Schematics of a vector-based RNAi construct used for RNAi in filamentous fungi. ars, autonomously replicating sequence for autonomous plasmid propagation in the fungal host; Bac, bacterial; Fun, fungal; ori, origin of replication for propagation in *E. coli*, P$_{Fun}$, fungal promoter; selection marker, auxotrophic or resistance gene; T$_{Fun}$, fungal terminator.

Saccharomyces cerevisiae leu2 strain was transformed with the *LEU+* gene. In this case, the transformed plasmid DNA integrated stably into the yeast chromosome.

The first efficient transformation system for the filamentous fungus *Neurospora crassa* was reported in 1979. Since then, many different methods have been developed to generate recombinant strains in yeasts and filamentous fungi. Most of these methods start with digesting the cell wall, which is always a barrier for the uptake of DNA. Resulting protoplasts (or spheroplasts, still having a fragmented cell wall) are used for the DNA-mediated transformation procedure. Figure 4.7a gives an overview of the protoplast transformation procedure.

Table 4.7: Selection markers for DNA-mediated transformation of fungi.

Gene	Gene product	Selection	Organism
ble^R	Bleomycin resistance gene from *Streptoalloteichus hindustanus*	Phleomycin resistance	*N. crassa*
his3	multifunctional enzyme of the histidine biosynthesis pathway	Histidine prototrophy	*N. crassa*
hph	Hygromycin B phosphotransferase from *E. coli*	Hygromycin B resistance	*A. chrysogenum, N. crassa*
nat1	Nourseothricin N-actetyltransferase from *Streptomyces noursei*	Nourseothricin resistance	*A. chrysogenum, P. chrysogenum, S. macrospora*
ptrA	Expression of genes of the thiamine metabolism riboswitch	Pyrithiamine resistance	*P. chrysogenum*
pyr4/ura3	Orotidine-5′-phosphate decarboxylase	Uracil prototrophy	*N. crassa, S. macrospora*
$tubB/Ben^R$	mutated β-tubulin	Benomyl resistance	*A. chrysogenum, N. crassa*

An alternative to protoplast transformation is the electroporation of (conidio) spores that has been described for e.g. *N. crassa*. Recently, the soil bacterium *Agrobacterium tumefaciens* was used to transfer foreign DNA into fungal cells. This system is similar to already developed plant techniques, where the T-DNA from *A. tumefaciens* is used to generate recombinant plant cells under laboratory conditions (compare section 9.3.2).

To select for recombinant fungal cells, two alternative marker systems are usually used, either auxotrophic or resistance markers. Auxotrophic markers require auxotrophic recipient strains unable to synthesize a particular organic compound necessary for their growth. Successful transformation can be observed using the wild-type genes as markers, since the auxotrophic strains are then able to grow on minimal medium lacking the required organic compounds: the successful transformants are prototrophic. Resistance markers confer resistance against antibiotics, resulting in growth on media containing the corresponding antibiotics. The resistance marker genes are often obtained from bacterial genomes and cloned into fungal transformation vectors. Both marker systems have been used successfully in yeasts as well as in filamentous fungi. A number of selection markers frequently used for fungal DNA-mediated transformation are listed in Table 4.7.

A standard fungal transformation vector is shown in Figure 4.7b. This vector carries a selection marker gene surrounded by fungal transcription promoter (P_{Fun}) and terminator (T_{Fun}) sequences. Promoters used for expression of selection marker genes have to be strong and constitutive, i.e. they have to be active throughout the lifecycle of the fungus and the fermentation process. A bacterial selection marker gene and an origin of replication (ori) are necessary for propagation in bacteria, a

Table 4.8: Comparison of knock-out/knock-in and knock-down approaches to inactivate gene expression for a functional analysis of fungal genes.

Feature	Knock-out / knock-In	Knock-Down
Feasibility	Technically laborious	Easy technical application
Throughput	Only a few genes can be analyzed	A high number of genes can be analyzed
Essential genes	Not applicable when mutation is lethal	Application is possible when mutation is lethal
Ambiguity	Clear and unambiguous results	Non-specific effects
Interpretation	No revertants	Incomplete and reversible mutations

necessary step to produce sufficient vector DNA amounts for fungal transformation. The ars (autonomously replicating sequence) is needed for autonomous propagation of plasmids in a fungal host. These sequences are mostly found in yeast vectors, where freely replicating plasmids are more common (e.g. the *S. cerevisiae* 2 μm plasmid), while in filamentous fungi foreign DNA integrates into genomic DNA due to the lack of ars sequences. However, some filamentous fungi are known to support free replication of transformed vectors with artificial ars sequences.

Transformation vectors are important tools for manipulating fungal genomes. For example, in cases where a certain gene is known to be a regulator of a biosynthetic pathway or a gene is supposed to act in a biosynthetic pathway, it is of interest to completely delete the gene or to lower its expression level. A frequently applied strategy is disrupting target gene expression by inserting a marker gene (Table 4.7) into the open reading frame of the target gene ("knock-in") or substituting the target gene sequence by a marker gene ("knock-out") using homologous recombination (see also chapter 8). Although technically laborious, these methods provide clear and unambiguous results with no revertants (Table 4.8). In the yeast *S. cerevisiae*, 25 bp of overlapping homologous sequences are sufficient to promote homologous recombination. However, in filamentous fungi, homologous recombination is a rather rare process, a property they share with plants and animals. Thus, the rate-limiting step for functional molecular analysis of genes in filamentous fungi is the low gene targeting frequency due to a highly efficient "non-homologous end joining" (NHEJ) system (Box 4.6). Interestingly, the efficiency of homolo-

Box 4.6: Non-homologous end joining (NHEJ).

The NHEJ process is a DNA double strand break repair mechanism mediated by a multi-protein complex containing several subunits. Disruption of a gene for one of these subunits results in fungal recipient strains where random DNA integration is blocked, thus leading to a significant relative increase in homologous recombination frequencies. NHEJ-deficient recipients are used as tools for functional genomic approaches to construct recombinant knock-out or knock-in strains.

gous recombination in filamentous fungi was significantly improved by constructing recipient strains lacking genes necessary for the NHEJ process.

An alternative to knock-out or knock-in approaches for functional gene analysis is the application of RNA interference (RNAi), a process also referred to as "knock-down" (Box 4.7). Although knock-down approaches can have non-specific effects that distort the results, they are easy to perform, and a large number of target genes can be tested without major efforts (Table 4.8). Another important advantage of RNAi compared to knock-out or knock-in strategies is the fact that essential genes whose knock-out would be lethal can be tested.

Most fungal RNAi systems use vector systems that contain intron or spacer sequences between two inversely orientated target fragments to express double-stranded RNA with a hairpin structure (Figure 4.7c). These stem-loop systems provide stable and high RNA silencing frequencies and, in some cases, even lead to completely silenced phenotypes. A disadvantage, however, is the time-consuming construction of stem-loop vectors containing two repeated sequences with an opposite orientation.

Box 4.7: RNA interference.

RNA interference (RNAi), or RNA silencing, is a biological process that occurs in cells of almost all eukaryotes. During the RNAi process, gene expression is inhibited by RNA molecules. In 2006, A. Z. Fire and C. C. Mello were awarded the Nobel Prize for the discovery of the RNAi system. The process of RNAi can be described as follows: In the nucleus, mRNA is transcribed into a double-stranded (ds) RNA by an RNA polymerase. The dsRNA molecule is transported through nuclear pores into the cytoplasm and subsequently cut into small interfering (si) RNAs by a dicer protein. The siRNAs have a size of 20 to 25 nt and are distinctive for the RNAi signaling pathway. siRNAs are recognized by the RNA-induced silencing complex (RISC), which guides the degradation of the corresponding mRNA. As a result, the mRNA can no longer be translated, leading to a knock-down phenotype. While the RNAi system has been described in many filamentous fungi, it is lacking in the yeast *S. cerevisiae*.

4.5.2 Recombinant fungal products

Fungi are frequently used as an alternative to bacterial expression systems (see chapter 3) to produce recombinant proteins. Major advantages over prokaryotic expression platforms are the effectiveness of protein folding, the generation of disulfide bonds, and the glycosylation of eukaryotic proteins, all prerequisites for functional recombinant products. In particular, baker's yeast has a long tradition in biotechnical fermentation processes. Yeast strains are genetically well characterized and provide the opportunity to work with stable production strains that generate a high yield of recombinant proteins. Furthermore, this unicellular fungus can

Table 4.9: Recombinant products made in yeasts and filamentous fungi. Modified from Gerngross 2004; Ward 2012.

Protein	Source	Expression system
Angiostatin	Human	*Pichia pastoris*[a]
Aspartic proteinase	*Rhizomucor miehei*	*Aspergillus oryzae*
Chymosin	Bovine, prochymosin B	*Aspergillus awamori*, *Trichoderma reesei*
Endostatin	Human	*Pichia pastoris*[a]
Epidermal growth factor analog	Human	*Pichia pastoris*[a]
Glucagon	Human	*Saccharomyces cerevisiae*
Glucose oxidase	*Aspergillus niger*	*Aspergillus oryzae*
Granulocyte-macrophage colony stimulating factor	Human	*Saccharomyces cerevisiae*
Hepatitis B surface antigen	Hepatitis B virus	*Saccharomyces cerevisiae*
Hirudin/lepirudin	*Hirudo medicinalis*	*Saccharomyces cerevisiae*
Insulin	Human	*Saccharomyces cerevisiae*
Insulin-like growth factor-1	Human	*Pichia pastoris*[a]
Lactoferrin	Human	*Aspergillus awamori*
Leucin aminopeptidase	*Aspergillus sojae*	*Aspergillus oryzae*
Lipase	partly *Thermomyces lanuginosus* (modified) and *Fusarium oxysporum*	*Aspergillus oryzae*
Lipase	*Candida antarctica*	*Aspergillus niger*
Platelet-derived growth factor	Human	*Saccharomyces cerevisiae*
Serum albumin	Human	*Pichia pastoris*[a]
Urate oxidase	*Aspergillus flavus*	*Saccharomyces cerevisiae*

[a] *Pichia pastoris* has been renamed *Komagataella pastoris* recently, but is mostly still known under its original name.

be grown to high densities on cheap and defined media, providing a cost-effective production process. Yeast products available as pharmaceuticals include granulo-cyte macrophage colony stimulating factor, hepatitis B surface antigen, and hirud-in. Further examples are found in Table 4.9.

The methylotrophic yeasts *Hansenula polymorpha* and *Pichia pastoris* are alter-native and preferred expression systems for producing single cell proteins. These yeasts are attractive for producing recombinant proteins for several reasons. First, their molecular genetic manipulation is rather simple, as in *S. cerevisiae*, and a large number of protocols are in use. Second, an efficient and tightly regulatable promoter for the alcohol oxidase gene (*aox1*) is available, allowing controlled ex-pression of foreign genes (compare section 4.1). Third, these yeasts can be grown to high cell densities in bioreactors; and fourth, they efficiently secrete foreign proteins. Finally, the yeasts can be grown on simple mineral media that allows easy recovery and purification of the desired product.

The *P. pastoris* expression system is a highly valuable alternative to bacterial expression systems, in particular when glycosylation of the product is necessary for its function. Another advantage compared to *S. cerevisiae* is that there is no tendency for hyperglycosylation of recombinant products. Most importantly, the productivity of *P. pastoris* and *H. polymorpha* is high. For example, a yield of 1500 mg/l of hirudin has been reported for these yeasts. Published examples of the biotechnological production of recombinant proteins in methylotrophic yeasts are insulin precursor, tumor necrosis factor, and human chitinase. Further examples are given in Table 4.9.

Similar to yeasts, several filamentous fungi have successfully been used as expression platforms for foreign genes, in particular because they have an effective secretion system and recombinant proteins show posttranslational modifications similar to those in animal cells. Compared to yeasts, the yields of secreted recombinant proteins are often rather low. These difficulties have been circumvented by the use of strong promoters and multi-copy strains, i.e. strains where multiple copies of the transformation vector have integrated into the genome, to increase the transcription level of the heterologous gene, or gene fusions, where a heterologous gene is fused to a fungal gene, leading to the generation of a fusion protein. A preferred gene fusion contains a truncated version of the fungal glucoamylase (compare Table 4.4) gene in frame with the heterologous gene. These fusion proteins are often more effectively secreted into the extracellular medium than the heterologous protein alone.

Another advantage of filamentous fungi is the availability of protease-deficient strains to promote stability of the recombinant proteins (some fungal species have been found to have up to 80 different proteases). Attractive hosts for the heterologous expression of foreign genes are several Aspergillus species, in particular *Aspergillus niger* and *Aspergillus oryzae*. In *Aspergillus awamori*, a genetically engineered bovine chymosin was produced with a yield of about 1.1 g/l. Another example is the production of human lactoferrin in the same host with a yield of 2 g/l. However, the most successful examples of expressing heterologous genes in fungi are the production of technical enzymes (compare section 4.3) in established fungal expression platforms. Several examples of filamentous fungi used commercially to produce heterologous enzymes are listed in Table 4.9.

Key-terms

Biotransformation, cellulases, cheese, citric acid, conidiospores, fatty acid synthase, fermentation, filamentous fungi, fungal transformation, fruiting bodies, fungal expression systems, isoprenoids, non-ribosomal peptide (NRP), non-ribosomal peptide synthetase (NRPS), organic acids, pectinases, polyketide synthase, polyketides, proteases, recombinant proteins, soy sauce, spores, technical enzymes, Tempeh, vegetative and sexual development, yeast

Questions

- What are the starting molecules for isoprenoid biosynthesis?
- What are typical organic acids? Name fungal producers and industrial applications!
- What are the differences between yeasts and filamentous fungi?
- What are the advantages of producing food by fungal fermentation?
- What sugars can be fermented as carbon sources by yeasts?
- What are the main steps in soy sauce production?
- What are the differences between blue and white mold cheese?
- What are the prerequisites for fungal strains to be used for food fermentation?
- Why is *Aspergillus niger* used for biotechnological citric acid production?
- Give examples for pharmaceutically relevant fungal polyketides and non-ribosomal peptides!
- Give examples of fungal technical enzymes and their industrial applications!
- Which alkaloid is produced by *Claviceps purpurea*?
- Why are non-conventional yeasts important for the production of recombinant proteins?
- What are advantages and disadvantages when recombinant proteins are produced with fungi instead of with bacteria?
- What are appropriate genetic markers for the generation of recombinant fungal strains?
- What is the difference between knock-down and knock-out approaches to generate recombinant fungal strains?
- Give examples of fungal recombinant proteins and their pharmaceutical application!
- What is the advantage of using fusion proteins when human proteins are synthesized in a fungal host?

Further readings

Archer, D. B, Connerton, I. F. & MacKenzie, D. A. 2008. Filamentous fungi for production of food additives and processing aids. *Adv Biochem Eng Biotechnol*, 111, 99–147.
Hofrichter, M. 2010. *The Mycota X* (2nd ed.) Springer Verlag, Heidelberg, New York, Tokyo.
Keller, N. P., Turner, G., Bennett, J. W. 2005. Fungal secondary metabolism – from biochemistry to genomics. *Nat Rev Microbiol.* 3, 937–947.

References

Archer, D. B., Connerton, I. F. & MacKenzie, D. A. 2008. Filamentous fungi for production of food additives and processing aids. *Adv Biochem Eng Biotechnol* 111: 99–147.
Gerngross, T. U. 2004. Advances in the production of human therapeutic proteins in yeasts and filamentous fungi. *Nat Biotechnol* 22: 1409–1414.

Kubicek, C. P., Punt, P. & Visser, J. 2010. Production of organic acids by filamentous fungi. In: Hofrichter M., ed. *The Mycota* X. 2nd ed.: 215–234, Heidelberg, New York, Tokyo, Springer Verlag.

Kück, U., Nowrousian, M., Hoff, B., et al. 2009. *Schimmelpilze*. Springer-Verlag, Heidelberg.

Reiß, J. 1987. Herstellung von Lebensmitteln durch den Einsatz von Schimmelpilzen. *Biologie in unserer Zeit* 17: 55–63.

Shah, S. A., Sultan, S. & Hassan, N. B., et al. 2013. Biotransformation of 17α-ethynyl substituted steroidal drugs with microbial and plant cell cultures: a review. *Steroids* 78: 1312–1324.

Ward, O. P. 2012. Production of recombinant proteins by filamentous fungi. *Biotechnol Adv* 30: 1119–1139.

Franz Narberhaus

5 Blue biotechnology

Water covers almost 75 % if the earth's surface. Due to the depth of the water column of up to 10,000 meters, the marine environment makes up more than 300 times the inhabitable volume of terrestrial habitats. This enormous ecosystem is home to numerous organisms, most of which remain to be discovered. It has been estimated that more than 80 % of all organisms on earth live in aquatic habitats. Their highly adapted lifestyle promises solutions to a wide range of global problems.

Blue biotechnology, also referred to as aquatic or marine biotechnology, exploits the rich potential of water organisms. It involves the use of marine and freshwater organisms to increase the food supply and to develop compounds of commercial value. Areas of blue biotechnology include the growth of seafood in aquaculture and the use of drugs, enzymes and other products from marine organisms. Some aspects of blue biotechnology overlap with other disciplines, for example green biotechnology (see chapter 7 of this book), when photosynthetic activities are concerned. To prevent redundancies, applications such as the potential of algae to produce biofuels will not be covered in this chapter.

5.1 The challenges of marine life

Most aquatic organisms, in particular marine species, are confronted with conditions that pose a serious threat to life. In the open ocean, the nutrient supply is extremely low. This does not only apply to carbon sources but also to phosphate, nitrogen, iron and other micronutrients. Most water is cold with a temperature around 4 °C but temperatures can range from below 0 °C in polar regions to several hundred degrees close to hydrothermal vents at the ocean floor. The pressure increases by about 1 atmosphere for every 10 m of depth. The salt concentration of sea water is about 35 g per liter. Light does not penetrate far into the water column, and photosynthesis is only possible in surface areas.

Life under such hostile and sometimes extreme conditions requires adaptation strategies based on cellular enzymes with biotechnologically attractive properties, like cold or heat resistance. Of particular interest are enzymes that produce novel bioactive compounds of medical or industrial interest.

5.2 The census of marine life and metagenome projects

A prerequisite for harnessing its biotechnological potential is a comprehensive picture of marine life. Aquatic ecosystems are inhabited by incredibly diverse organisms. Life forms range from submicroscopic viruses and phages to bacteria and archaea in micrometer dimensions up to huge plants and animals. Vast areas of the oceans are still unexplored and it is believed that more than 99 % of all microorganisms are not culturable under standard laboratory conditions. Various exploratory approaches have collected a wealth of information on the inventory of our oceans. Metagenomics (Box 5.1) approaches provided insights into the biodiversity by culture-independent molecular methods.

Box 5.1: Metagenomics.

This culture-independent method, also called environmental genomics, ecogenomics, population genomics, or community genomics, is based on sequencing of genetic material (DNA or reversely transcribed RNA) from environmental samples. In traditional metagenomics, phylogenetically informative genes, like the 16S rDNA, are amplified and sequenced to obtain a community profile. Next-generation, high-throughput sequencing approaches that do not require DNA cloning prior to sequencing now provide unbiased insights into the genetic diversity of an ecosystem. Provided that sufficient sequence information is generated, entire microbial chromosomes and plasmids can be re-constructed.

The motivation of functional metagenomics is to mine the genomic diversity for the development of novel biotechnological and pharmaceutical products (Figure 5.1). In a sequence-based approach, a metagenomic dataset is searched for sequences coding for proteins with potential value in medicine, agriculture, or industry. Synthetic genes of interesting candidates will be cloned for recombinant protein production and further analysis. In a more direct function-driven strategy, metagenomic libraries are constructed in hosts like *Escherichia coli* and screened for specific traits, like enzyme activity (e.g. cellulase, chitinase, lipase) or production of bioactive compounds (e.g. antimicrobial activity). DNA from clones of interest is then sequenced and used for recombinant production, biochemical characterization, and optimization of the gene products.

The census of marine life, a decade-long (from 2000 to 2010) international enterprise involved 2,700 scientists from more than 80 nations. They went on 540 expeditions to assess the diversity, distribution, and abundance of marine life around the globe (http://www.coml.org/). Aircrafts, boats, manned and unmanned

Box 5.2: DNA barcoding.

Morphological features are often not sufficient to distinguish closely related organisms. DNA Barcoding or phylotyping is a genetic technique to identify species. It requires only small amounts of material and uses a sequence from a standard region of the genome. DNA extraction is followed by amplification, sequencing, and computational clustering of an informative gene locus. The sequence must be present in all species of interest and should be short enough to be easily sequenced. Flanking regions must be identical for PCR amplification, and the internal region must be sufficiently diverse to allow distinction on the species level. A commonly used region for animals contains approx. 650 basepairs from the mitochondrial cytochrome oxidase gene. The DNA code is displayed in black stripes resembling the label on supermarket products.

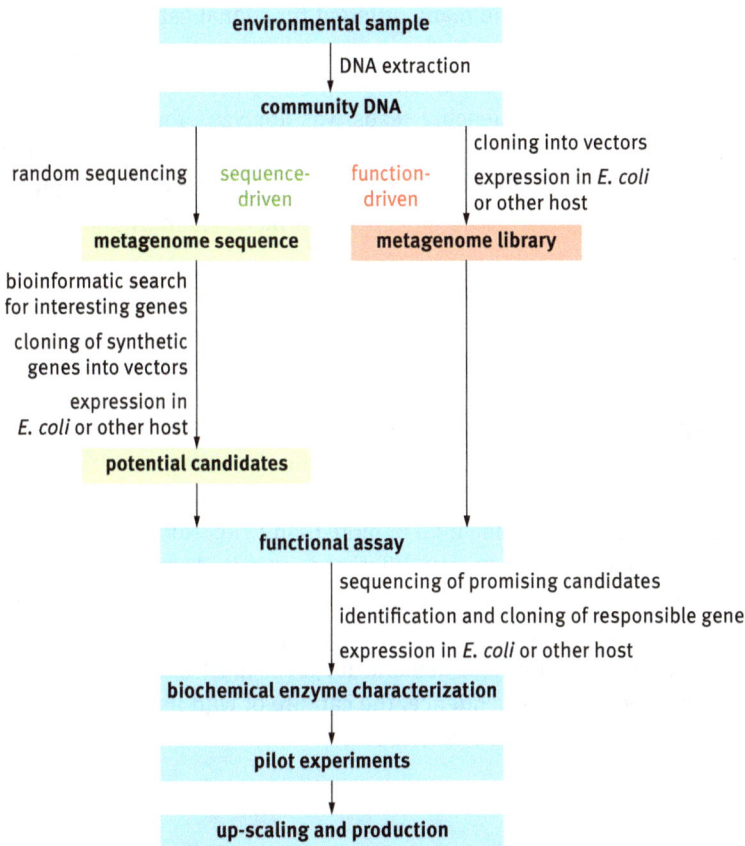

environmental sample

DNA extraction

community DNA

random sequencing | sequence-driven | function-driven | cloning into vectors
expression in *E. coli* or other host

metagenome sequence

metagenome library

bioinformatic search for interesting genes

cloning of synthetic genes into vectors

expression in *E. coli* or other host

potential candidates

functional assay

sequencing of promising candidates
identification and cloning of responsible gene
expression in *E. coli* or other host

biochemical enzyme characterization

pilot experiments

up-scaling and production

Fig. 5.1: Functional metagenomics. Sequence-driven (left) and function-driven (right) approaches to industrially relevant enzymes from environmental communities.

submarines, tracking devices, sonar systems, videotaping, and other innovative technology has been used to profile the abundance and diversity of marine life. The scientists discovered thousands of new species from various taxonomic groups and set up a global marine life database. DNA barcoding (Box 5.2) was used to distinguish species by their DNA sequence.

An interesting observation of the Census project is that microbes dominate the oceans. They outnumber larger species between 100 and 1000 times. By weight up to 90 percent of marine life is microbial. 18 million DNA sequences from 1,200 sites worldwide revealed an unexpected biodiversity. A liter of water can contain almost 40,000 kinds of bacteria and the Census scientists estimated that up to a billion different microbial species may exist.

Various other metagenomic studies have been devoted to the microbial diversity in the ocean. Sequencing of surface water samples from the Sargasso sea by Craig Venter, and colleagues in 2004, revealed more than 1.2 million previously unknown genes that were assigned to many different functional categories (Venter et al. 2004). Probably best suited for functional metagenomics approaches is a follow-up study from 2007, the Global Ocean Sampling (GOS) project (Yooseph et al. 2007). A dataset of 7.7 million sequencing reads was analyzed for novel protein families. More than 6 million predicted proteins were clustered into almost 4000 protein families. The dataset is publically available and opens the door for numerous biotechnological applications based on functional metagenomics approaches (Box 5.1).

5.3 Food from the sea

The most obvious products from the sea are fish and other sea food. Oceans, lakes, and rivers have provided food to the human population for many centuries. Fish provides a good source of high-quality proteins, and the average annual consumption per person has reached an all-time high of more than 17 kg. Notably, consumption per capita varies substantially between different countries. According to recent surveys, it ranges from less than one pound in Mongolia to 15 kg and 24 kg in Germany and the USA, respectively, to 60 kg in Portugal and Japan. Due to the growing human population and a high demand for food, most natural lakes and oceans are overharvested. As a consequence, the capture of wild fish has declined and the average size of caught fish has decreased.

5.3.1 Aquaculture

Aquaculture, mariculture or aquafarming is playing an increasing role in satisfying the increasing demand for seafood. It is one of the most rapidly growing segments of industrial agriculture and refers to a variety of land-based or water-based technologies to farm fish, shellfish, and aquatic plants. It involves the breeding, rearing, and harvesting of all types of freshwater and marine organisms. The products of aquaculture are not only used for human consumption, but in part processed into fishmeal and oil to feed other animals including fish, shrimps, pigs, and chicken. Non-nutritional products from aquaculture are ornamental fish and plants for aquariums, and plants for use in pharmaceutical and biotechnological industry.

According to the Food and Agriculture Organization of the United Nations (http://faostat3.fao.org/faostat-gateway/go/to/home/E), wild captures still dominate the food market, but aquacultural products have been catching up from 4 % in 1970 to 40 % in 2010 (Figure 5.2). It is estimated that by the year 2030, more than half of our seafood supply will originate from aquaculture. However, wild catches will remain to play a significant role in the seafood market because not every valuable fish species is amenable to aquaculture. For obvious reasons, large species like tuna or swordfish cannot be cultured in close confinement.

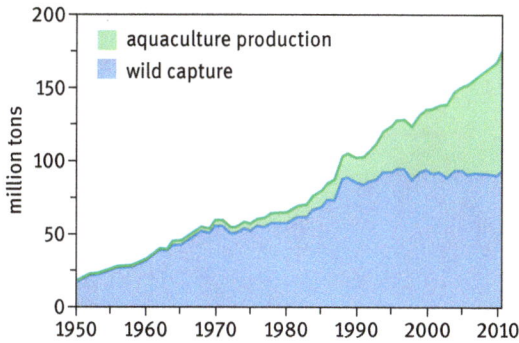

Fig. 5.2: Global importance of aquaculture. Global harvest of aquatic organisms in million tons from 1950 to 2010 (from FAOSTAT, Food and Agriculture Organization of the United Nations, Statistics Division).

Like any other industrial farming process, aquaculture has severe environmental impacts. Large-scale coastal operations can destroy natural habitats. Artificial feeding of large quantities of fish in open rearing pens releases nutrient-rich wastewater into the ecosystem leading to pollution and eutrophication. The same is true if wastewater from land-based operations is discharged into natural watersystems. Wild fish stocks are affected by aquaculture because farmed fish is typically fed with fish because many species are carnivorous depending on a protein-rich, high-energy diet. Even if most bait fish is not sellable for human consumption, removal of these species from the food chain can have serious implications on the ecosystem.

The life of farmed fish in a densely crowded confinement raises a number of problems on its own. The genetic monoculture makes the fish stock more susceptible to infections than a heterogeneous wild-life population. Treatment with antibiotics and food additives, use of disinfectants, pesticides and anti-fouling agents, spread of diseases and parasites into the natural environment, and escape of nonnative or even genetically engineered fish are other issues of concern.

Several different aquaculture systems are in practice. Some are land-based, others open water-based (Figure 5.3a). Various species, including trout and bass, are farmed in closed systems on land, where the fish are contained in tanks, ponds, or raceways diverted from rivers. This method is considered low-risk in terms of pollution, fish escapes, and transfer of diseases into the wild, in particular if the waste water is treated and re-circulated. Semi-closed systems, often used for prawn culture, are a land-based method, in which water is exchanged between the aquaculture and the coastal environment. When salmon has reached a certain size, it is typically farmed in off-shore open-net pens. Floating cages are anchored to the sea floor and stocked with juvenile fish from hatcheries. Both semi-closed and open methods are much riskier than closed confinement systems as they do not prevent the exchange of organic material, chemicals, and pathogens between the aquaculture and the environment. Moreover, escape of some individuals from the system

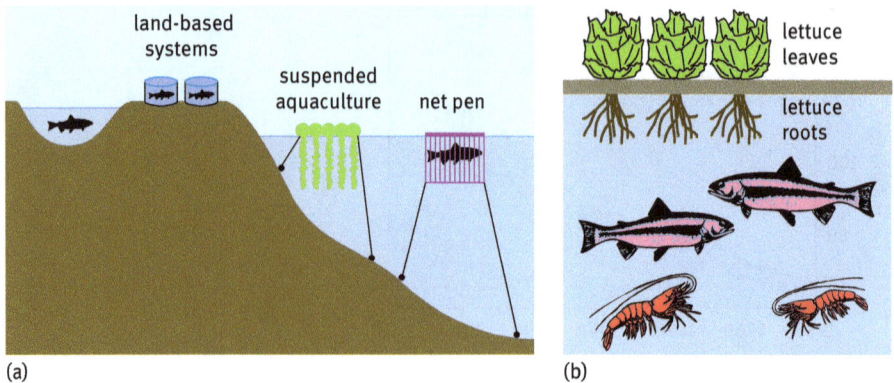

Fig. 5.3: Aquaculture methods. (a) Land-based and open water-based systems. (b) Polyculture.

and breeding with wild populations can hardly be avoided. The organic load introduced into any of these systems is very high because up to 5 kg of bait fish or of
fishmeal are necessary to produce 1 kg of fish. In contrast, no artificial food is
required to farm filter feeders like mussels, oysters, and other shellfish in suspended aquaculture. They are either attached to sticks or ropes or contained in cages
close to the shoreline where they extract nutrients from the water flow.

Polyculture is an interesting emerging concept towards sustainable aquaculture. Compared to conventional aquaculture, it has a number of economic and
ecological advantages. In contrast to monoculture systems, two or more species
with different nutrient requirements are co-cultured in order to produce several
products of economic value and to reduce the contamination and eutrophication
of the environment by metabolic by-products. Such integrated aquaculture practices are based on a mixture of compatible species, for example fish and shellfish or
a combination of top-feeding fish with bottom-feeding fish to prevent spoilage of
sinking foods. In advanced polyculture concepts, the production of seafood products is directly coupled to plant growth. In a simple scenario, lettuce or vegetables
grown in hydroponic racks above an aquaculture are fertilized by the nutrients in
the wastewater (Figure 5.3b). In a more complex water recycling setup, waste water
from the fish tank is pumped into a container and mixed with fresh water. This
nutrient-rich water is used to irrigate plants grown in a gravel-filled bed. The plants
and bacteria in that grow bed reduce the organic and nitrogen load in the water.
The overflow of cleaned water from the plant cultures is then aerated before it is
returned to the fish tank.

5.3.2 Genetically modified (GM) fish

One of the bottle-necks in aquaculture is the time it takes the fish to reach a mar-
ketable size. Depending on the species, this can take up to three years. Reducing

this time period is of course a matter of huge economic interest. Conventional selection and breeding methods have previously been used to enhance the growth
rate, muscle mass, taste, smell, color, and disease resistance of fish. With the event
of recombinant gene technology, new ways of improving these traits have emerged.
Of particular commercial value is farmed salmon that has reached a world-wide
market volume of almost 2 million tons per year.

As in other disciplines of biotechnology, molecular tools have been developed
to introduce genes of interest into the fish genome generating a genetically modified (GM) or genetically engineered (GE) organism. The most prominent example
is the AquAdvantage salmon, a trade name for a genetically modified Atlantic
salmon that was developed by AquaBounty Technologies in Canada already in the
early 1990s, and is still awaiting approval by federal agencies like the Food and
Drug Administration although the application was filed in 1995. No approval has
yet been sought in the EU. The slow legal process reflects the complicated and
controversial risk assessment typically associated with GM food. While at least
90 % of the corn and soybeans grown in the US are genetically modified, the AquAdvantage salmon would be the first transgenic animal approved for human consumption. This makes it a particularly sensitive issue, setting a precedent for future
applications.

What is this debate about, and what distinguishes the GM salmon from conventional salmon? In its genome, the engineered Atlantic salmon contains a single
additional gene coding for a growth hormone from the unrelated Pacific chinook
salmon. This foreign gene is combined with a promoter from the ocean pout that
drives expression of the growth hormone gene, which in turn stimulates the growth
of bones and muscle cells. The growth rate of the genetically modified salmon is
greatly accelerated, in part because it grows year-round, rather than seasonally as
its wild relative. Overall, it won't get much bigger than wild salmon but it reaches
market size twice as fast. It takes only 18 months until it reaches full size instead
of the usual three years (Figure 5.4a). Reports on the effect of the transgenic growth
hormone on fish development, morphology, and fitness are controversial and may
depend on the genetic background and physiology of the engineered fish. Several
studies indicated that the composition of the GM salmon is essentially indistin-
guishable from the conventional relative and that it is safe to eat. Other reports
state that metabolism, skeletal and organ structures, hormone regulation, and disease resistance are altered. There also are concerns about the escape and breeding
of the transgenic fish with wild populations. Farming the GM salmon in enclosed
land-based facilities (physical containment) and growing only sterile females that
cannot reproduce (biological containment) are employed to minimize these risks.
However, even with negligible environmental and health risks, it will always remain to be a personal choice based on ethical decisions whether one wants to eat
genetically modified animals. Several groceries stores in the US have already made
the commitment to not sell transgenic salmon. Time will show whether it will ever
make it into the food market and if yes, if the costumer is going to accept it.

Fig. 5.4: Genetically engineered fish. (a) Size comparison of an AquAdvantage® Salmon (background) vs. a non-transgenic Atlantic salmon sibling (foreground) of the same age. Reproduced with permission from AquaBounty Technologies. (b) GloFish® Fluorescent Fish Display (www.glofish.com).

Now that recombinant technologies have become almost routine practice, the introduction of a growth hormone gene can, in principle, be done in almost any fish species. In fact, various salmon species, rainbow trout, carp, catfish, and tilapia have already been subjected to genetic engineering (Table 5.1). Recent transgenic approaches go a step further and aim at improving other commercially interesting phenotypes. For example, a gene coding for antifreeze protein (see below, chapter 5.6.2) is meant to improve fitness at low temperatures. Various metabolic genes have been introduced to improve nutrient utilization by the fish. Other foreign genes provide better resistance to bacterial or viral diseases.

The given examples of genetically enhanced species are proof-of-concept studies and not yet intended to reach the supermarket. The tedious regulatory requirements, ongoing media protests and the widespread consumer reservations are probably going to discourage companies from the development of GM fish, at least for human consumption.

A GM fish that is commercially available today, at least in some countries, is not meant to be eaten. It is a zebrafish, with the tradename GloFish, that glows when illuminated. Zebrafish are a very popular aquarium fish and a well-studied model organism. Hence, it is not surprising that it was among the first genetically

Table 5.1: Transgenic fish and shellfish species.

Product of transgene	Phenotype of interest	Species
Growth hormone	Fast growth	Atlantic salmon
		Chinook salmon
		Coho salmon
		Rainbow trout
		Tilapia
		Rohu
		Loach
Antifreeze protein	Cold tolerance	Atlantic salmon
Antimicrobial peptide Cecropin	Bacterial resistance	Catfish
		Medaka
Lysozyme	Bacterial resistance	Zebrafish
Human lactoferrin	Virus resistance	Grass carp
Antisense virus-coat protein gene	Virus resistance	Shrimp
Glucose transporter	Carbohydrate uptake and metabolism	Rainbow trout
Hexokinase		
Delta5 or delta6 desaturase	Lipid metabolism (poly-unsaturated fatty acids)	Zebrafish
Phytase	Phosphorus utilization	Zebrafish
Green fluorescent protein	Fluorescence	Zebrafish

engineered animals. It contains the gene coding for green fluorescent protein (GFP), a commonly used tool in molecular biology (see below). The original inventors of the fluorescent zebrafish did not foresee its career as pet animal but intended to create a bioindicator. When the GFP gene is hooked up to a suitable environmentally responsive promoter, the transgenic fish can be used to monitor arsenic, estrogens, or chemicals because the fish will glow when such pollutants are present in the water.

Meanwhile, a variety of fluorescent fish are on the market (Figure 5.4b). The rainbow of colors developed on the basis of yellyfish GFP and fluorescent proteins from various other sources (corals, see anemone) have been exploited to construct red, orange, pink, purple. and blue fish; not only zebrafish but also tetras and tiger barb. While these fluorescent fish species have become quite popular in various countries, their trade and possession in Europe, Canada, and California is strictly prohibited. Genetic manipulation is not restricted to fish. Since GFP and its derivatives are very robust proteins, they have been used to create a zoo of other fluorescent animals, like rabbits, pigs, and monkeys.

5.4 Tools for molecular biology from the ocean

We have already learned that some genetically engineered fluorescent fish produce GFP, a protein that has become one of the most commonly used tools in molecular biology because it bears an intrinsic chromophor build from three neighboring amino acids (see chapter 3). GFP derives from *Aequorea victoria* (Figure 5.5a), a yelly-fish that produces the luminescent protein aequorin, which emits blue light when bound to Ca^{2+}. Along its umbrella, the yellyfish produces another protein, GFP, that is able to absorb the blue light emitted from aequorin and returns it as green light. Understanding this complicated process and developing it into a useful tool started in the 1960s and took several decades. Since gene cloning and recombinant expression was not yet available at that time, more than 800,000 yellyfish were harvested to purify and characterize the responsible proteins. Three key players in this decade-long endeavor, Martin Chalfie, Osamu Shimomura, and Roger Tsien, were awarded the Nobel Prize in Chemistry in 2008 for the discovery and development of this versatile molecular biology tool. In the meantime, one can choose from a whole suite of GFP variants that have been generated by the introduction of site-directed or random mutations into the GFP gene (Figure 5.5b). According to their colors and properties, these variants are then called RFP for Red Fluorescent Protein, YFP for Yellow Fluorescent Protein, eGFP for enhanced GFP, and a brighter green variant or superfolder GFP for a fast folding variant. Fusions to the GFP gene or its derivatives are now widely used to monitor gene expression or to track protein localization in bacteria, archaea, plants, and animals.

In contrast to fluorescence, which requires the absorption of light of a certain wavelength prior to the emission of light of some other wavelength, bioluminescence is a process that does not depend on an external light source. It is defined as the production of light by a living organism. Bioluminescence is very common in marine organisms and is typically generated by bacteria that live in close symbiotic cooperation with their host (Figure 5.5c). Probably the best known examples are deep sea Angler fish and squids. Among the bioluminescent bacteria are Vibrio species, like *Vibrio fisheri* and *Vibrio harveyi*. Some of their so-called *lux* genes encode the subunits of the luciferase. The bacterial enzyme uses the same principle as the one in the firefly. The conversion of a complex organic substrate, the lucifer-in molecule, to oxyluciferin is accompanied by the emission of visible light. This process strictly depends on ATP and oxygen. The substrate requirement is the reason why the luciferase is not as versatile as GFP for applications in cell biology. Still, *lux* reporter gene fusions are often used to measure gene expression in various organisms and cell cultures. Another application is the measurement of ATP in biological samples by applying a cocktail of luciferin and luciferase and reading out the light signal in a luminometer.

Finally, we should not forget that the heat-stable enzymes routinely used in the laboratory for PCR and other purposes originate from thermophilic aquatic microorganisms like *Thermus aquaticus* and *Pyrococcus furiosus*.

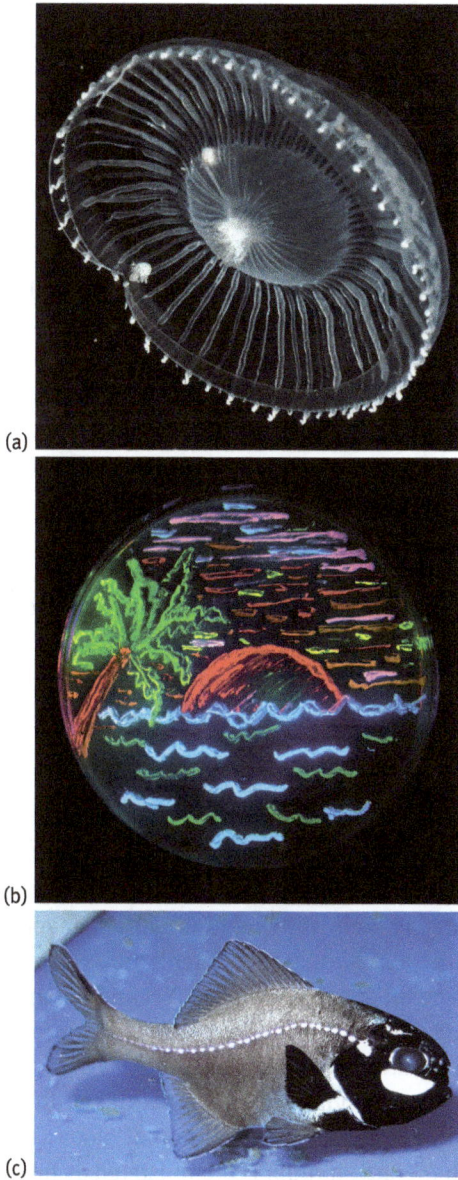

(a)

(b)

(c)

Fig. 5.5: Marine-derived genetic tools. (a) *Aequora victoria*, the source of GFP
(http://commons.wikimedia.org/wiki/Category:Aequorea_victoria).
(b) Agar plate of fluorescent bacteria colonies representing the rainbow of engineered GFP variants (http://www.tsienlab.ucsd.edu/Images.htm) Reproduced with permission from the Tsien lab.
(c) Bioluminescent fish (*Photoblepharon palpebratum, courtesy of Jens Hellinger, Ruhr University
Bochum*).

5.5 Drugs from the sea

The incredible diversity of marine life promises a treasure trove of bioactive compounds, most of them waiting to be discovered. Compared to the large number of natural products from terrestrial (micro)organisms, the portfolio of currently available marine products is rather small. Among these products is salmon calcitonin, a short alpha-helical polypeptide of 32 amino acids. Human calcitonin is a thyroid hormone that stimulates calcium uptake and bone calcification, and it inhibits bone resorption by osteoclasts. Salmon calcitonin resembles human calcitonin but its bioactivity is 20 times higher for reasons not well understood. It is commonly used to treat osteoporosis, the progressive loss of bone mass that creates porous and brittle bones. Recombinant salmon calcitonin is now available as injection or a nasal spray.

An important component of bone, cartilage, and teeth is hydroxyapatite (HA). It is a calcium phosphate mineral also found in corals and sponges. Various porous materials are used in surgery to support bone formation and regeneration. Bone-like HA from corals and sponges are interesting biomaterials for this purpose. They are typically well-accepted by the host tissue and partially resorbed and replaced by natural bone.

Other marine products of pharmaceutical interest have antimicrobial or antiviral activities. Scientists have actively searched for such activities in sharks because they show a natural resistance to bacterial and viral infections despite their primitive immune system. Squalamine was discovered in the dogfish shark. It is an aminosterol with broad antimicrobial activity. The planar cationic molecule interacts with negatively charged bacterial membrane lipids and disturbs the membrane function by a yet unknown mechanism. Squalamine has been reported to have a variety of other interesting properties, among them antiviral and antifungal activity. Its potential to block angiogenesis, the formation and differentiation of blood vessels, promises applications in the treatment of certain cancers. Yet another anti-angiogenic agent is of shark origin. A compound called Neovastat is a shark cartilage extract that exhibits antitumor and antimetastatic effects.

The polypeptide hirudin from the salivary glands of leech is a natural inhibitor of thrombin and has anticoagulant properties preventing formation of blood clots and thrombi. Apart from medicinal leech therapy, hirudin is administered as injections or as cream. Nowadays, recombinant gene technology is used to produce a variety of hirudin-based pharmaceutical products.

Traditionally, soil-dwelling actinomycetes are a rich source of antibiotics. A new antibiotic was recently isolated from a marine *Streptomyces* species in sediments near Santa Barbara in California. The name Anthracimycin acknowledges its potent activity against the anthrax-causing Gram-positive bacterium *Bacillus anthracis*. It effectively kills Gram-positive bacteria by a yet unknown mode of action but is less potent against Gram-negatives. Another antimicrobial compound called

Aplasmomycin is produced by *Streptomyces griseus,* isolated from the sediment in Sagami bay in Japan. It inhibits growth of Gram-positive bacteria.

Marine sponges, molluscs, worms, and algae are also known to produce antimicrobial compounds. In most cases, however, their exact chemical identity and mode of action remains to be explored.

A number of marine animals produce potent neurotoxins. A prominent example is the Fugu or puffer fish, which is considered a delicacy in Japan despite the presence of the extremely dangerous tetrodotoxin (TTX) in its skin, liver, and gonads. Several other fish species, octopus, and even terrestrial animals like frogs and newts produce the same poison, and it is believed that endo-symbiotic bacteria are the actual source of the neurotoxin. TTX is heat stable and one of the deadliest toxins known to date. Numbers in the literature vary but it is clear that TTX is a hundred to several thousand times more lethal than cyanide. Things get worse as there is no known antidote. TTX is a small heterocyclic organic molecule that selectively blocks the voltage-gated sodium channels in nerve cells preventing propagation of action potentials, which can lead to fatal paralysis. As with many other poisons, low doses can be of medicinal value and TTX is a potential therapeutic agent for pain associated with cancer and migraines.

Another effective painkiller has been extracted from the marine cone snail *Conus magus.* The cyclic peptide Ziconotide (trade name Prialt) contains 25 amino acids. Six of them are cysteines that are linked in pairs by three disulphide bonds. The compound blocks N-type voltage-gated calcium channels in the spinal cord and thereby, interrupts pain signaling. Ziconatide is a powerful analgesic drug but has been reported to have profound side effects if delivered orally or intravenously. It must be administered intrathecally (into the spinal fluid) and is used to treat severe chronic pain.

Trabectedin (Yondelis) is an anti-cancer drug. The complex compound was extracted from a sea squirt, a colonial mangrove tunicate. Because a ton of animals is needed to isolate 1 gram of trabectedin, synthetic and semi-synthetic synthesis routes have been established. Although the precise mode of action is not yet known, it was approved in 2007 by the EU for the treatment of soft tissue sarcoma.

The given examples demonstrate that marine compounds can be of great value in the health care sector. Numerous bioactive compounds from marine organisms are in the prospecting phase for their medical potential. Some of them have already entered clinical trials. Traditional approaches starting with the isolation of promising activities from aquatic samples followed by the identification of the responsible compounds are still in practice. A typical problem of this approach is the limited supply of these compounds from natural sources. Recently developed high-throughput metagenomic approaches are probably going to bypass such limitations and steadily provide new candidates into the pipeline of novel compounds. Bioinformatic predictions are able to predict for example synthesis genes for secondary metabolites from global sampling datasets. Recombinant gene technology can then be used to express synthetic genes in suitable host organisms able to produce large quantities of the desired product. However, as many of the novel gene products are unprecedented, their activity cannot necessarily be inferred from sequence comparisons. Therefore, complementary function-driven metagenomic strategies are required to directly screen for genes or gene clusters responsible for antimicrobial or other clinically relevant activities.

5.6 Non-medical marine products

5.6.1 Enzymes

As outlined in the introduction, the marine lifestyle imposes diverse challenges on cellular enzymes. The features rendering marine biocatalysts resistant to cold, heat, salt, or pressure are of prime interest to biotechnologists. Enzymes extracted from fish or marine microorganisms can provide numerous advantages over traditional enzymes. Thermostable enzymes used in molecular biology have already been mentioned. Detergent-resistant amlyases, pullulanases, cellulases, proteases, and lipases are of interest as additives to laundry and dishwasher detergents to degrade natural polymers such as starch, cellulose, proteins, and fat. While previous efforts had primarily been directed at heat-resistant enzymes active at for example 60 °C, the ever increasing energy costs have raised substantial interest in the isolation and genetic engineering of enzymes operating in the cold to efficiently remove dirt stains at low washing temperatures.

Are large spectrum of enzymes from marine organisms have been harnessed for industrial biotransformation processes. Performing biotechnological processes under non-standard conditions with the help of marine microorganisms or biocatalysts derived from these organisms has several advantages. Extreme conditions, in general, reduce the risk of microbial contaminations. High temperatures increase the solubility of substrates and products, decrease viscosity, and result in faster reaction rates. High salt concentrations often enhance the stability of enzymes, and tolerance of a biocatalyst to high or low pH can be useful in various processes.

As we learned in chapter 1, amylases are glycoside hydrolases that act on alpha 1,4-glyosidic bonds to break down complex sugars into simple sugars such as disaccharides and trisaccharides. Efficient amylase enzymes with a range of interesting properties and substrate specificities have been isolated from marine bacteria, archaea, and yeasts. A heat and acid stable alpha-amylase discovered in archaea from hydrothermal vents is able to degrade starch from corn, potatoes, or other plant products into sugars that are further processed into ethanol by microbial fermentation. If algal biomass is used, amylase-producing marine bacteria can be used in saline conditions.

Several other polysaccharide-degrading enzymes have been isolated from marine organisms. Chitinases are frequently found because chitin is a major marine carbon source. Actually, it is the second most abundant polysaccharide in nature after cellulose. Chitin, a linear polymer of beta-1,4-N-acetylglucosamine (GlcNAC), is present in the exoskeleton of insects, crustaceae, fungi, and algae. Chitinases are able to degrade the polysaccharide to low molecular weight chitooligomers, which serve a broad range of industrial, agricultural, and medical functions. Some applications of chitinases are in agriculture to control chitin-containing pathogens, in aquaculture to hydrolyze chitinous waste, in mosquito control, and in the lab to produce protoplasts from fungi and yeast. The potential of the enzyme as an ingredient in anti-fungal creams and lotions is under investigation. Great pharmaceutical interest in chitinase-generated chitooligosaccarides is based on their function as elicitor molecules and their anti-inflammatory and anti-tumor activities.

The most abundant polysaccharide on earth is cellulose, as it is a major structural component of plant cell walls. It consists of a linear chain of up to several thousand beta-1,4-linked D-glucose units. Different types of cellulose enzymes generate different breakdown products, which can be of high economic value. Alkaline cellulases active at high pH are used to produce biofuels from renewable cellulose-rich substrates. Other industrially relevant polysaccharide-degrading enzymes are hemicellulase and xylanases. The latter is used in the paper industry and for clarification of juice and beer. Although filamentous fungi currently are the primary source of these enzymes, temperature or pH adapted hydrolytic enzymes from marine sources have potential for countless applications.

Novel proteases, lipases, esterases, and phosphatases isolated from marine habitats are tolerant to high pressure, salt or organic solvents and promise a wide range of applications in the food and pharmaceutical industry. Alkane hydroxylases from marine microorganisms that utilize alkanes as carbon sources have potential in the bioremediation of oil spills.

5.6.2 Other marine products

An interesting non-enzymatic protein of marine origin is the antifreeze protein (AFP). It has already been mentioned in the context of genetically engineered fish (5.3.2). With the same goal to increase the frost tolerance of the organism, transgenic tomatoes, potatoes, and strawberries producing AFP from an arctic flounder fish have been generated but were never commercialized. AFPs comprise a structurally and functionally heterogeneous group of short polypeptides or proteins with ice-structuring properties. They are not only found in cold-water fish but also in some insects, bacteria, and plants. As a variation to the theme, some fish produce small glycoproteins with antifreeze activity. The best understood AFPs are type I AFPs from arctic fish. They are short polypeptides in the range of 4 kDa in size and consist of a single amphipathic alpha helix. They bind to small ice crystals, affect their morphology, and inhibit their growth. Some AFPs may also protect cell mem-

(a)   (b)

Fig. 5.6: Mussels adhering to rocks (Oregon coast, Franz Narberhaus).

branes from cold damage. The cryoprotectant activity of AFPs promises many inter-esting applications in agriculture and medicine. Crop protection does not require genetic engineering of the plant itself. Recombinantly produced AFPs can be sprayed in green houses or nurseries to protect plants from frostbite on a cold night. The cryogenic protection of blood, human tissues, organ transplants, and oocytes are other areas of interest. In frozen food, AFPs can improve the quality and extend the shelf life by preventing formation of large ice crystals. Several ice cream products contain AFP to inhibit ice crystal formation and improve the tex-ture and taste of the product, in particular of low fat ice creams.

Another food additive of marine origin is carrageenan, a thickening agent present in many preserved foods, toothpaste, shampoo, and cosmetic creams. The sulfated polysaccharide extracted from red seaweeds has been used as an alterna-tive to gelatin for many years. Other seaweed products with gelling activity are the polysaccharides agarose and agar, that we routinely use in the lab for gel electro-phoresis and for solid growth media, respectively. Agarose is a linear polymer con-sisting of alternating D-galactose and 3,6-anhydro-L-galactopyranose units, where-as agar is a mixture of agarose and agropectin. Food applications of agar, for exam-ple, as a thickening agent of deserts have already been established in Asia for several hundred years.

A promising compound that still needs to get ready for the market is superglue from mussels. Mussels and barnacles glue themselves tightly to rocks (Figure 5.6), poles, ropes, and boat surfaces and resist even rough currents. The nature of this strong biological glue has raised considerable interest, and researchers are trying to design biomimetic materials inspired by the biological system. The sticky materi-al of marine mussels is based on cross-linked proteins. Responsible for the crosslinking is the amino acid L-3,4-dihydroxyphenylalanine (DOPA) that is formed by post-translational hydroxylation of tyrosine residues. The sticky fibers are enriched in iron that comes with the sea water, and it is proposed that the transition metal is involved in cross-linking DOPA residues to make the adhesive material strong and flexible. Because the marine glue works in wet environments, it is attractive for applications in surgery and dentistry.

5.7 The prospects of blue biotechnology

Blue biotechnology may still be the least developed branch of biotechnology. It nevertheless represents a highly exciting and dynamic area of research that has already delivered a number of valuable products, some of which are used on a daily basis. Given the enormous and largely unexplored biodiversity in aquatic environments, many more products are to be expected. The development of inno-vative culture-independent discovery routes combined with robotic instrumenta-tion is going to facilitate this process. Often, the compounds and biocatalysts de-rived from screening approaches are suboptimal for medical or industrial applica-tions. Collaborative efforts between environmental and molecular biologists, bioinformaticians, chemists, and engineers are then required to produce designer products with improved properties.

Key-terms

Marine biotechnology, aquaculture, metagenomics, DNA barcoding, enzymes, drugs

Questions

- Why is Blue Biotechnology the least developed branch of biotechnology?
- What properties make aquatic organisms interesting for biotechnology?
- Provide examples of valuable enzymes from the aquatic organisms.
- Describe examples of medical value.
- Consider the advantages and disadvantages of DNA- and RNA-based metage-nomics.
- Discuss different host strains for functional metagenomics approaches.
- What are the problems associated with aquaculture?
- Why is polyculture more sustainable than conventional aquaculture?
- Describe the controversy about genetically engineered fish?
- Why do recombinant techniques out-compete conventional breeding ap-proaches?

Further readings

Schloss, P. D. & Handelsman, J. 2003. Biotechnological prospects from metagenomics. *Curr Op Biotech* 14: 303–310.

Streit, W. R., Daniel, R. & Jaeger, K. E. 2004. Prospecting for biocatalysts and drugs in the genomes of non-culturabel microorganisms. *Curr Op Biotech* 15: 285–290.

DeLong, E. F. 2005. Microbial community genomics in the ocean. *Nat Rev Microbiol* 3: 459–469.

References

http://faostat3.fao.org/faostat-gateway/go/to/home/E

http://www.coml.org/

Venter, C., Remington, K., Heidelberg, J. F., et al. 2004. Environmental genome shotgun sequencing of the Sargasso Sea. *Science* 304: 66–74.

Yooseph, S., Sutton, G., Rusch, D. B., et al. 2007. The Sorcerer II Global Ocean Sampling expedition: Expanding the universe of protein families. *PLoS Biol* 5: e77.

Marc Nowaczyk, Sascha Rexroth and Matthias Rögner

6 Biotechnological potential of cyanobacteria

Cyanobacteria are attractive candidates for an alternative production approach in bioindustrial applications. With approximately 10,000 characterized strains ranging from psychrophilic (i.e. below 0 °C) to thermophilic (i.e. up to about 75 °C), and an even much higher amount of still unidentified species (estimated >100,000), they promise a very high potential as photosynthesis-powered microorganisms – although their exploration is still at an initial stage. Remarkably, they are fast growing prokaryotic microorganisms, which only require some inexpensive components such as sunlight, CO_2, and water (enriched with some inorganic mineral nutrients). In particular, the use of sunlight, which is available nearly everywhere on earth for free, as sole energy source for metabolic processes is one of their major benefits. In contrast to other microorganisms requiring carbohydrate compounds, this photosynthetic capability makes cyanobacteria independent of any carbon-based feedstock. As they also show higher energy conversion efficiency than plants (due to their highly efficient carbon concentration mechanisms), they also require a considerably smaller area for biomass production.

In contrast to eukaryotic microalgae, cyanobacteria – which are gram⁻ bacteria like *E. coli* – contain no organelles and harbor both photosynthetic and respiratory electron transport within one membrane system – the thylakoid membrane (Figure 6.1). While structure-function-regulation of this system has been quite well characterized, energetics of the cell-enclosing cytoplasmic membrane and communication with the thylakoid membrane is still poorly understood, as well as the fragmentation of matrix-localized metabolic processes which – unlike the organelles of eukaryotic systems – are not physically separated. Irrespective of these basics, the short generation time, the ease of genetic transformation, and the possibility to keep most of them in mass cultures make cyanobacteria ideal organisms for biotechnological applications, in particular in combination with the tools of synthetic biology.

This chapter provides first a short introduction in cyanobacteria and gives some principles of genetic manipulation and metabolic engineering of cyanobacteria. The chapter further provides an overview on how to harness cyanobacteria for biotechnological applications with the focus on engineered photosynthesis and the design of specialized photobioreactors.

Box 6.1: Basics of cyanobacterial photosynthesis as power supply for biotechnology.

Oxygenic photosynthesis, i.e. photosynthesis, which is combined with oxygen evolution due to the light-driven splitting of water, was invented by cyanobacteria and is performed with some modifications similarly in green algae and higher plants (see also chapter 7 for a general introduction into eukaryotic photosynthesis). As shown in Figure 6.1, the photosynthetic electron

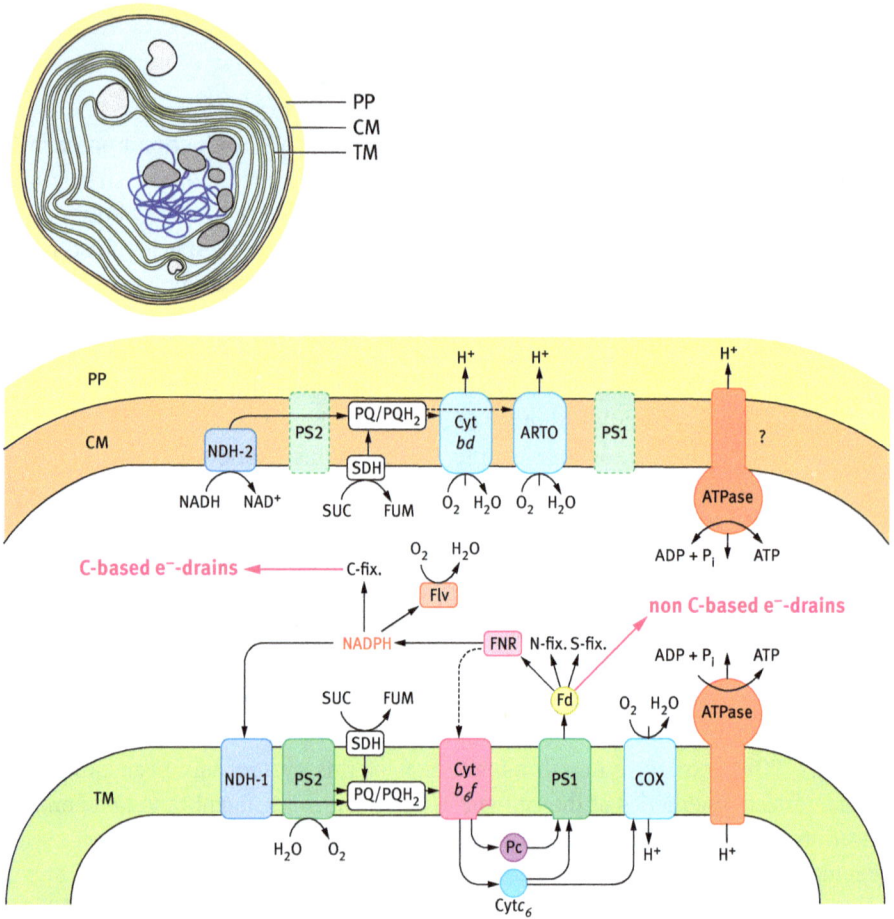

Fig. 6.1: Membrane systems in cyanobacterial cell (top) and major bioenergetic electron flows in cytoplasmic (CM) and thylakoid membrane (TM) connected with their dependent processes in cytoplasm and lumen. Sites for the potential coupling of C-dependent and C-independent biotechnological reactions are also indicated. PP = periplasm.

transport (PET) is localized in the thylakoid membrane of these organisms – however, in case of cyanobacteria, combined with the respiratory electron transport. While most biotechnological processes require energy-rich compounds as feed stock for the involved (micro-)organisms, cyanobacteria are powered by the energy of light and need nothing but fresh- or sea-water with some micronutrients and air. Following the primary steps of photosynthesis and their efficiency on the molecular level – the so called „light reactions" (Figure 6.2) – yields valuable information for the design and the success of photosynthesis-coupled biotechnological processes.

If efficiency is rated as "quantum-efficiency", i.e. efficiency of each light quantum, which hits the thylakoid membrane and triggers a primary reaction in the photosynthetic reaction centers (PS1 and PS2), this efficiency is beyond 99 %, resulting in a measurable charge separation across the thylakoid membrane within a few nanoseconds. However, as natural light trans-

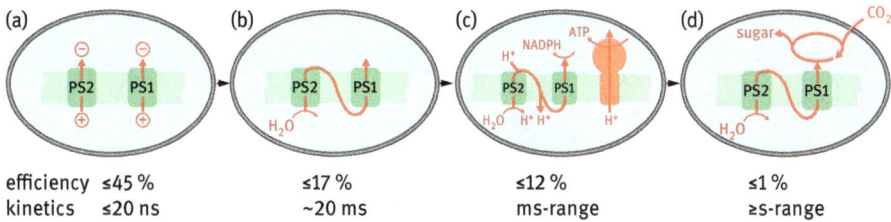

(a)   efficiency ≤45 %
      kinetics ≤20 ns

(b)   ≤17 %
      ~20 ms

(c)   ≤12 %
      ms-range

(d)   ≤1 %
      ≥s-range

Fig. 6.2: (Box 6.1). Efficiency of major steps in consecutive photosynthetic reactions: Starting from light induced charge separation (A), followed by linear electron transport with water splitting at PS2 (B), the „light reactions" are completed with the formation of NADPH (from e-flow) and ATP (from generated proton gradient) (C). In the subsequent „dark reactions" (D) CO_2 is reduced to sugar by the generated reduction equivalents (NADPH) and energy (ATP).

formation efficiency is often compared with technical systems (Blankenship et al. 2011) such as solar cells, and as these systems are based on the full solar spectrum of which the photosynthetically active radiation (PAR) is approximately 45 %, primary photosynthetic events start with an efficiency of max. 45 % (A), which decreases with the subsequently induced PET-chain between PS2 and PS1 to (theor.) ≤17 % (B). Finally, the combined electron and proton transport results in NADPH and ATP as countable products of the light reactions, representing a theoretical efficiency of ≤12 % (C). Practically, however, these efficiencies are lower due to loss of energy in each electron transfer step. ATP and NADPH are then mainly required for the metabolic pathways, resulting in biomass and storage compound formation, which are not directly dependent on light and, therefore, combined as „dark reactions" (D). They occur in seconds or slower. Based on one of the first measurable products, „sugar", the efficiency of most photobiological systems is, however, very low – usually <1 %. This severe drop in efficiency is mainly due to the low turnover frequency (TOF) of the key enzymes involved in CO_2-fixation (e.g. Ribulosebisphosphate carboxylase with TOF = 0.3 s^{-1}), while key steps of the „light reactions" show at least two orders of magnitude higher TOFs. For practical, biotechnological applications two conclusions can be drawn:
1. It is energetically very favorable, to couple such reactions directly to the light reactions, i.e. to stages A) to C) (with decreasing efficiency), while all processes involving C-fixation and biomass-formation result in a tremendous loss of primary (light) energy.
2. Under optimal light conditions, the primary PS-reactions generate too many reduction equivalents, which overflow the slow dark reactions. As excess electrons generate destructive oxygen species which are dangerous for cell life (this process is generally called "photoinhibition", see also chapter 7), photosynthetic cells degrade their excess energy as fluorescence or heat, corresponding incidentally to more than two thirds of the harvested light energy (Figure 6.3). Introduction of new electron sinks in the cellular metabolism through bioengineering can contribute in regaining the major part of this „lost" energy by bypassing the bottlenecks of PET, thereby considerably increasing the efficiency of photosynthesis-based biotechnological processes. As shown in Figure 6.3, this apparently does not interfere with the energy required for basic cellular processes (Wilhelm & Jacob 2011).

With PS as energetic driving force for various biotechnological processes, cyanobacteria can be harnessed for the production of various products as indicated in Figure 6.1. Energetically, products can be linked directly to photosynthetic electron transport (see, for example, biohydrogen below) with minimized loss of primary light energy, or indirectly via ATP and reduction equivalents (NADPH) as mediators (examples: isoprene, fatty acids etc.). Triggering processes by light

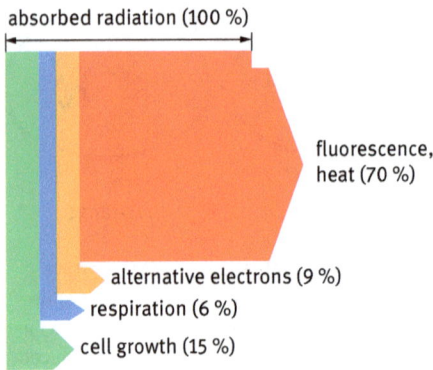

absorbed radiation (100 %)

fluorescence,
heat (70 %)

alternative electrons (9 %)
respiration (6 %)
cell growth (15 %)

Fig. 6.3: (Box 6.1). Fate of the absorbed light energy from a typical microalgal system under saturating (sun) light energy: While the predominant part is lost as fluorescence and heat, only a minor part is used for basic cellular metabolism („housekeeping") (Wilhelm & Jacob 2011). Figure (modified) with kind permission from Springer Science and Business Media.

offers new possibilities for biotechnology (e.g. *via* light dependent promoters, etc.), but their interference with „classical" photosynthetic reactions like state transitions in the short range or circadian processes in the long range have also to be considered.

6.1 Optimization of the "chassis" and principal methods

Engineering of cyanobacteria first requires screening for a suitable host strain and tools for its genetic manipulation. The next step would be a general optimization of the production platform towards the enhanced supply of "energy rich" electrons or reduced carbon for the final product by re-routing of photosynthetic energy pathways (Box 6.1). Usual genetic tools and optimization strategies will be explained by examples with common laboratory cyanobacteria.

6.1.1 Selection of appropriate cyanobacterial strains

The natural diversity and distribution of cyanobacteria is enormous (Hess 2011). They have conquered almost every habitat on earth in which sunlight is available. Extremophilic species have developed thermotolerance to survive in hot springs above 60 °C (e.g. *Thermosynechococcus elongatus*) as well as on the Antarctic shelf ice below −0 °C (e.g. *Nostoc* sp.). Even extreme aridic land areas like the Negev desert in Israel are colonized by specific cyanobacterial species (*Microcoleus* sp.), which form green sand crusts in the upper layer of the desert and play an important role in the local ecosystem. Other species thrive under alkaline conditions (pH ~11) combined with high salinity (1.2 M sodium carbonate) like *Arthrospira maxima*. Most fascinating is the large diversity of marine cyanobacteria with a tremendous

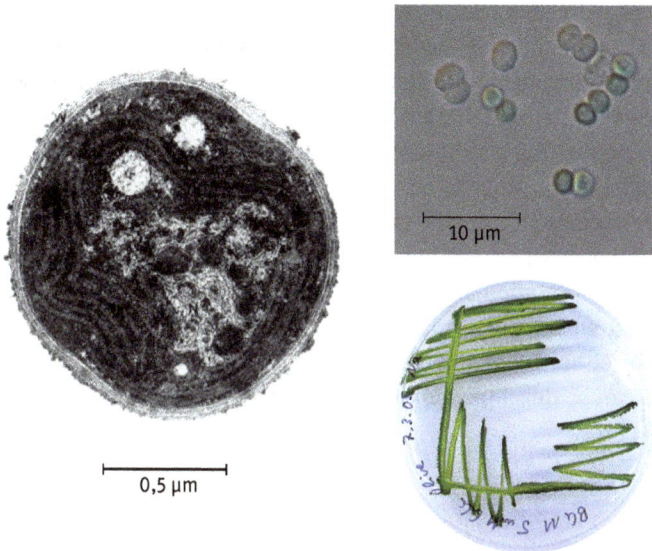

Fig. 6.4: (Box 6.2). Morphology of *Synechocystis* sp. PCC 6803. Cells streaked on agar plate (right, bottom), captured by light microscopy (right, top) and electron microscopy (left).

variety of secondary metabolites. This natural treasure is at the very beginning of being considered as potential biotechnological source, as most of the marine species are largely unknown yet with very few having already been characterized. Precious bioactive compounds include, for example, the antibiotic vancomycin, the immunosuppressive agent cyclosporine, and the anticancer agent bleomycin.

Box 6.2: The model organisms *Synechocystis sp.* PCC 6803 and *Synechococcus sp.* PCC 7002.

Synechocystis sp. PCC 6803 is one of the best-characterized cyanobacterial strains for photosynthesis research and biotechnological applications (Figure 6.4). It was isolated in 1968 from a fresh water lake in California. This mesophilic (growth temp. ~30 °C) organism can be easily grown, is easy to transform, and glucose tolerant strains are available, which can be cultured under photomixotrophic conditions to compensate otherwise lethal mutations of the photosynthetic apparatus. Based on genomic sequencing data, 3,750 potential genes have been annotated on the 3.57 Mbp large circular chromosome and the seven autonomous replicated plasmids. Some cyanobacteria are oligoploid (3 to 10 chromosome copies per cell) and others can be highly polyploid. For *Synechocystis,* up to 218 chromosome copies per cell have been identified experimentally. In contrast to other bacteria, only a few genes are organized in operons.

Synechococcus sp. PCC 7002 was isolated in 1961 from a mud sample of the "fish pens" on Magueyes Island, Puerto Rico. It can grow over a wide range of NaCl concentrations (including sea water) at elevated temperatures (~38 °C), is extremely tolerant to high-light irradiation, and grows much faster than other cyanobacteria. Under optimal conditions, a doubling time of ~2.5 h has been reported. Characteristically, *Synechococcus* 7002 is highly resistant to reactive oxygen and nitrogen species. Like *Synechocystis,* it is naturally transformable, fully sequenced, and capable to grow photomixotrophically. The genome (3,235 genes) is organized in a ~3.0 Mb chromosome and 6 plasmids, which vary in size from 4.8 to 186 kb and in copy number.

The huge natural diversity of cyanobacteria offers several advantages for bioindustrial applications. Usually, they are easy to grow due to their short generation time and minimal requirements (e.g. seawater and sunlight). This offers the possibility to use land areas that are not suitable for agricultural use in order to avoid competition with food production. Additionally, cultures of extremophilic cyanobacteria (e.g. halophilic) are less prone to contaminations by other bacteria. On the other hand, strains can be selected from the large natural diversity that are adapted to a special local environment which facilitates decentral bioindustrial production and could also reduce transportation-associated costs. Such bioactive compounds may be directly isolated from native cyanobacterial strains. Alternatively, the corresponding pathways can either be engineered in other microbial production systems, or cyanobacteria can be developed for a production system through metabolic engineering.

6.1.2 Genetic tools for metabolic engineering of cyanobacteria

Genomic sequencing and the development of genetic tools for the manipulation of cyanobacteria enabled the design of metabolic engineering concepts for the production of specific chemical compounds. Due to the first published cyanobacterial genomic sequence published in 1996, combined with the establishment of numerous tools for genetic manipulation, *Synechocystis* sp. PCC 6803 (Box 6.2) became a well-known model organism and the primary target for metabolic engineering approaches today. Recently, information on the genomic level is complemented by transcriptomic, proteomic, metabolomic, and lipidomic data. *Synechocystis* is easy to handle in the laboratory, and genetic manipulation is facilitated by a natural DNA uptake mechanism.

In contrast to other prokaryotic model systems like *Escherichia coli* (*E. coli*), plasmid based gene expression systems are, in principle, available for cyanobacteria although not yet very common. They are based, so far, on low-copy broad-host-range plasmid vectors like pRSF1010 and introduced into the cyanobacterial cells by conjugation with an *E. coli* helper strain. Although *Synechocystis* bears its own set of plasmids, they have not yet been exploited systematically as genetic tools. Instead, homologous recombination is the main technique for genetic manipulation of cyanobacteria. This method relies on the natural DNA uptake and recombination mechanism, which many cyanobacterial strains utilize for the expansion of their genetic pool. As shown in Figure 6.5, extracellular DNA is linearized, transferred into the cell, and homologous regions are used as starting points for the incorporation into the cyanobacterial genome.

This natural system can easily be used to modify the genomic DNA for a metabolic engineering approach. Existing pathways can be modified by deletion of genes or by changing regulatory elements (e.g. promoters, 5′-UTRs, etc.). Moreover,

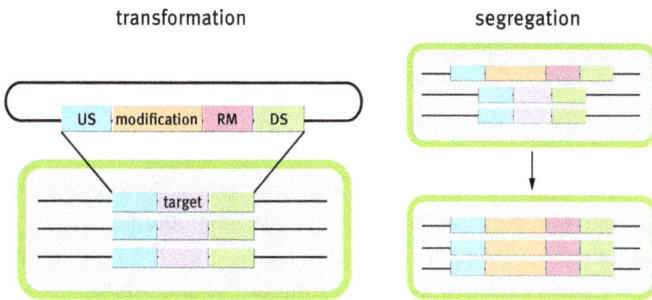

Fig. 6.5: Homologous recombination and segregation of artificial DNA fragments. Transformation: *Synechocystis* cells are incubated with plasmid DNA coding for a resistance marker (RM) and a modified target with upstream (US) and downstream (DS) regions for homologous recombination. Segregation: after the first recombination event, the modified allele has to be transferred in all genome copies by increasing the selective pressure.

novel and/or optimized pathways can be introduced into silent parts of the genome. Major advantage of this system – compared to plasmid based gene expression – is the long-term stability of the genetic information. Once the segregation process (Figure 6.5) is completed successfully, cells keep the introduced information, even if the culture is grown without antibiotic pressure. This offers the possibility to add multiple modifications through recycling of the antibiotic resistance cassette (e.g. by co-transformation of the SacB gene of *Pseudomonas aeruginosa*).

Other genetic tools like antibiotic resistance cassettes (e.g. for chloramphenicol, kanamycin, streptomycin, gentamycin, etc.) and strong or inducible promoters for targeted protein overexpression are also available. It has been shown for *Synechocystis* that the use of phage derived promoters (like the T5-promoter) or the cyanobacterial psbA-promoter yield high level gene expression, although it is not controllable in these systems. In *Synechococcus elongatus* PCC 7942, a *lacI-lacO* repression system similar to *E. coli* was used successfully, while for *Synechocystis* several other inducible systems have been reported: The *petJ* promoter is repressed by increased copper concentration in the media, whereas the *petE* promoter is induced by copper. Another expression system is based on the nitrate-inducible promoter of *nirA*, which can be controlled by the extracellular nitrate concentration. While these systems rely on the exogenous concentration of a specific inducer, cyanobacteria also offer light-controlled elements. Several wavelength dependent photoreceptors that control gene expression in cyanobacteria have been reported. They might also be useful instruments for the development of light quality dependent expression systems for future applications.

These are some emerging examples for the genetic modification of cyanobacteria in dependence of metabolic engineering. However, the decreased costs of large synthetic DNA fragments and the development of novel tools in the upcoming field of Synthetic Biology (synthetic circuits and switches, etc.) will drive the fast development of solar-powered-cell-factories for the production of green chemicals.

6.1.3 Determination of photosynthetic parameters & efficiency

For the utilization of prokaryotic algae as a platform for biotechnological applications, the absorption of light, its conversion into chemical energy, the distribution of the energy in the metabolism, and finally the formation of the target product are key parameters for optimization. Characterization of these parameters require techniques, which target single properties such as net O_2 production, CO_2-utilization and the F_v/F_0 ratio. These allow to draw conclusions on photosynthetic performance and the physiological state. In combination with high-throughput techniques like proteomics, transcriptomics, and metabolomics, they enable a multifactorial in depth analysis of the organisms and provide a holistic view on the biological system. While techniques addressing the physiological state of the algal culture are well suited to quantify the overall productivity and efficiency of a process, high-throughput approaches provide very detailed information on changes in metabolism and regulatory mechanisms of the biological system: Therefore, they can provide clues on metabolic bottlenecks and optimization potential.

The monitoring of chlorophyll fluorescence has a long tradition in photosynthetic research. It is based on the fluorescence emitted by PS2, when the transfer to downstream electron carriers becomes limiting. Chlorophyll fluorescence provides access to parameters, such as PS2 quantum yield, non-photochemical, and photochemical quenching. In addition, light saturation curves show the dependence of these parameters on the light intensity and define the appropriate light regime for the cultivation of different species and mutants of photosynthetic organisms. In cyanobacteria, the contribution of phycobilisomes to the fluorescence emission has to be taken into account, especially if mutations or conditions have an impact on the phycobilisome content.

While the quantum yield of PS2 is often used synonymous with photosynthetic efficiency, it has, however, to be considered that this only takes into account the linear electron transport. For this reason, determination of net O_2 evolution and CO_2 assimilation provide a more solid basis for balancing electron flow with metabolism. Photosynthetic oxygen evolution and respiratory oxygen uptake can be analytically differentiated by membrane-inlet mass spectrometry if photoautotrophic cells are exposed to a gas phase with $^{18}O_2$, i.e. oxygen gas consisting of the stable isotope ^{18}O. This is especially helpful to quantify oxygen uptake induced by the Mehler reaction, i.e. the transfer of electrons to molecular oxygen, which indicates an overload of the metabolism with reducing equivalents.

For an in-depth analysis of cyanobacteria under distinct culture conditions, different high-throughput techniques provide valuable data. The application of modern mass spectrometry for proteome analysis and of next generation sequencing techniques for transcriptomics allow a complete coverage of all relevant proteins and transcripts. They can be used to identify unexpected limitations of the media or the culture environment and also bottlenecks in the metabolism. For the reliable application of these high-throughput techniques, an efficient and reproducible control of the culture conditions is mandatory, especially when targeting factors like light supply, which are strongly dependent on the culture density. Stable and reproducible conditions can be realized very efficiently in a continuous cultivation process (Box 6.3).

6.1.4 Reduction of antenna size and increase of electron transport

In contrast to natural habitats, typically environments with limited light supply, cyanobacteria do not suffer from light limitation in closed photobioreactor (PBR) systems nor in flat open ponds, which usually provide a non-limiting light energy. Rather, they are exposed to too much light, which induces photoinhibition (see Box 6.1), and for this reason, the synthesis of phycobilisomes is usually unwanted

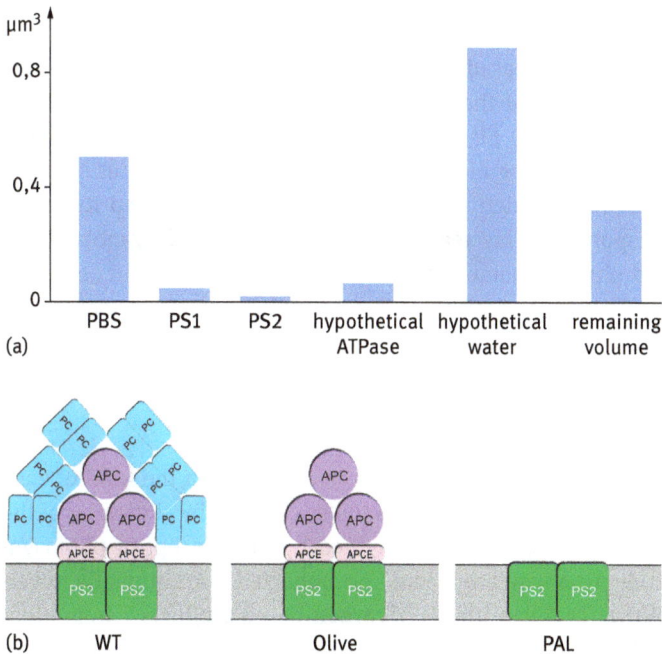

Fig. 6.6: Impact of phycobilisome antenna on *Synechocystis* PCC 6803 WT cell and mutants. Cellular volume required by various components of *Synechocystis* 6803 cells (a). *Synechocystis* 6803 WT phycobilisomes and mutants with reduced (Olive mutant without PC-subunits) or abolished (PAL mutant) phycobilisome content (b). PBS = phycobilisomes, PS1 = photosystem 1, PS2 = photosystem 2, PC = Phycocyanin, APC = Allophycocyanin, APCE = Allophycocyanin linker protein. Fig. 6a (modified) with kind permission of Elsevier B.V. (Moal & Lagoutte 2012).

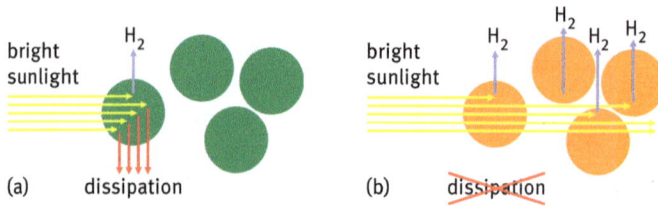

Fig. 6.7: Impact of phycobilisome antenna size on self-shading effect in PBR with high cell densities and consequence on biotechnological product formation – for instance biohydrogen. WT cells with strong shading effect and high dissipation of light energy (a). Cells with truncated PBS antenna resulting in low shading effect and low dissipation of light energy (b).

in biotechnological systems. Besides this destructive potential, especially under high light from an energetic point of view, the synthesis of phycobilisomes is also a waste of energy (Moal & Lagoutte 2012), as phycobilisomes, for instance, constitute up to 63 % of all soluble proteins in *Synechocystis* 6803 WT cells (Figure 6.6a). Reduction of the phycobilisome size (in case of the Olive-mutant by lack of phycocyanin protein subunits, see Figure 6.6b) or their complete deletion (PAL-mutant, see Figure 6.6b) additionally increases photosynthetic linear electron transport considerably (up to 6-fold).

This is mainly due to an increased PS2/PS1-ratio, but also due to a decreased self-shading effect (Figure 6.7a) of the truncated phycobilisome mutants, which results in higher cell densities (see Figure 6.7b). All these combined effects are in favor of more electrons per volume element being available from water splitting for biotechnological reactions, which are dependent from and coupled to photosynthetic electron transport. For this reason, natural or designed cyanobacterial mutants with truncated phycobilisome antenna constitute a very useful „chassis" as starting material for the further design of photosynthesis-powered processes.

6.2 Examples for cyanobacterial biotech products

The photosynthetic capacity of cyanobacteria can be accessed on the level of electrons provided by the photosynthetic electron transfer chain or by skimming of the reduced carbon. The first strategy will be explained by the concept of photosynthetic biohydrogen production with engineered cyanobacteria. Additionally, several examples will be shown for the synthesis of carbon-based products – from proof of principals to industrial scale production.

6.2.1 Non C-based: Photosynthetic biohydrogen production

Biohydrogen is an attractive biofuel product from cyanobacteria (Rögner 2013), as it is released from the cells as gas and, therefore, easy to harvest. Also it is a pro-

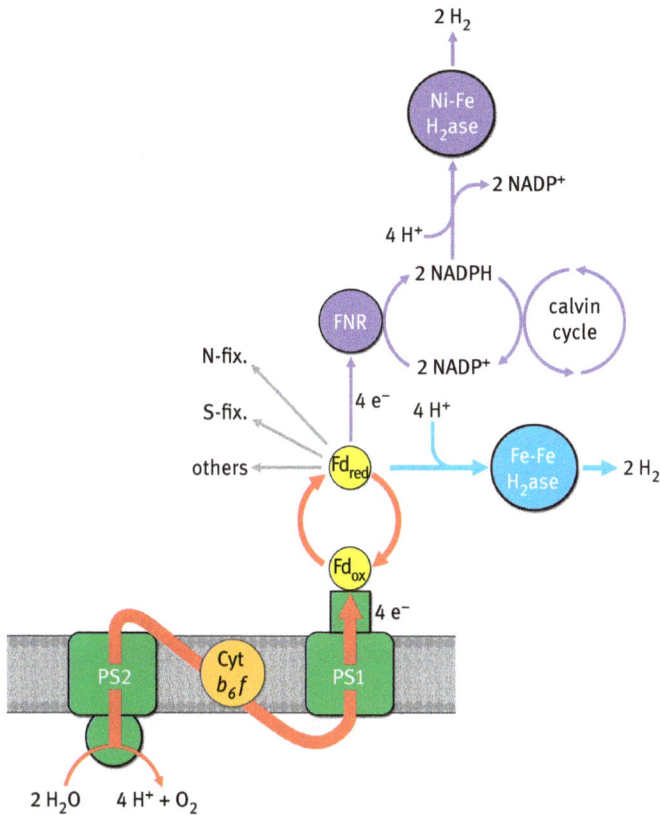

Fig. 6.8: Streamlining cyanobacterial ET for hydrogen production. Electron supply for photosynthesis-based hydrogen production in cyanobacteria (Ni-Fe-H$_2$ase type, purple) and green algae (Fe-Fe-H$_2$ase type, blue) in competition with other metabolic pathways, for instance CO$_2$-, N- and S-fixation. It is apparent that the e-supply via Fd is the most direct path for H$_2$ production, which can be increased by minimizing the interaction of Fd with FNR (e.g. by engineering their binding sites).

duct, which is energy rich and directly formed by using photosynthetic electrons originating from water-splitting, i.e. from a source which is practically unlimitedly available. Although most cyanobacteria contain an internal hydrogenase (H$_2$ase) of Ni-Fe-type, this is rather useless for biotechnological exploitation due to its low activity, its oxygen sensitivity, and its energetic inefficiency with NADPH as substrate sharing various competitive reactions (Figure 6.8). Energetically favorable would be a most direct coupling to photosynthetic electron supply via Ferredoxin (Fd), which functions as electron distributor at the acceptor side of PS1 (see Figure 6.1). This requires the import of a heterologous Fd-dependent H$_2$ase such as the Fe-Fe-H$_2$ase of green algae (see chapter 7) or the engineering of a Ni-Fe-H$_2$ase for interaction with Fd. While the functional expression of a highly active eukaryotic Fe-Fe-H$_2$ase in cyanobacteria could already be shown, the inherent oxygen-sensitivity of this type of hydrogenase has still to be overcome by directed and random mutagenesis approaches; alternatively naturally oxygen-tolerant H_2ases could be used after engineering of a Fd-binding site. Although these approaches are still ongoing, they promise a high potential for the future direct coupling of electron sinks to photosynthetic electron transport before the lower-efficient C-fixation steps are involved (see Box 6.1). The re-design of electron pathways, which are required for this approach and which should re-route electrons for biohydrogen at the expense of biomass-production are illustrated in Figure 6.8.

6.2.2 C-based biofuels and bulk chemicals

Cyanobacteria are ideal solar driven catalysts for the direct conversion of CO_2 into valuable organic compounds, as they are the simplest organisms performing oxygenic photosynthesis (Ducat et al. 2011). The only byproduct formed in this process is oxygen. Metabolites directly downstream of the carbon fixing reactions are used as precursors for the engineered pathways that lead to the desired products (Figure 6.9). This concept of a solar-based microbial cell factory was already realized for several carbon-based chemicals on academic and industrial scale (Rosgaard et al. 2012).

Genetically engineered cyanobacterial strains for the production of carbon-based biofuels (e.g. bioalcohol and biodiesel) may provide viable alternatives to fossil fuels as they offer several advantages in comparison with other biological systems. The direct conversion of CO_2 into the desired product by a prokaryotic system a) maximizes the efficiency of the process compared to microbial fermentation of plant-derived material, b) avoids the land use competition for production of food, and c) relies on the simpler organization and manipulation of the bacterial host strain in comparison with eukaryotic algae. Moreover, cyanobacteria have developed numerous carbon concentration mechanisms (CCMs) that enable a more efficient fixation of carbon by the RUBISCO enzyme compared to other photoautotrophic organisms.

So far, the production of several biofuels with engineered cyanobacteria has already been explored. Ethanol was produced by overexpression of two enzymes – pyruvate decarboxylase (pdc) and alcohol dehydrogenase II (adh) from the bacterium *Zymomonas mobilis* under the control of the strong rbcL promotor. Joule Unlimited Inc. and other companies have developed improved strains with up to 5000-fold increased ethanol production rates (1 mg l^{-1} h^{-1}). They are tested now on pilot plant scale but none of these systems is fully commercialized yet. Isobutyraldehyde production is introduced into cyanobacteria by overexpression of ketoacid decarboxylase (kivD) from *Lactococcus lactis*. The volatile product (6 mg l^{-1} h^{-1}) evaporates from the culture, which allows collection from the headspace by condensation of the efflux gas. Further conversion to isobutanol is achieved by co-expres-

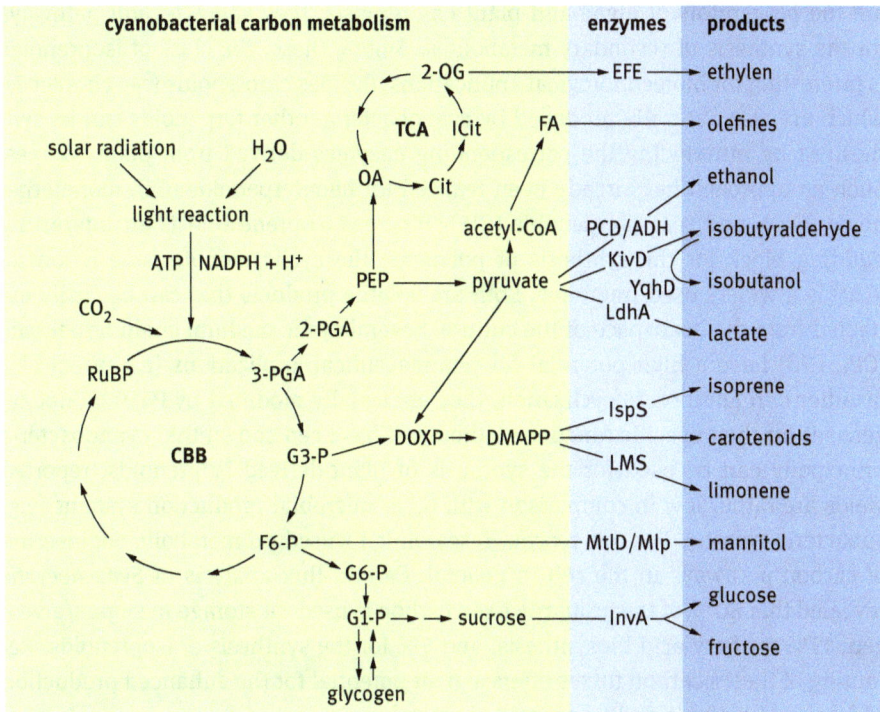

Fig. 6.9: Generation of carbon based products with cyanobacteria by engineered biochemical pathways. Abbreviations: CBB, Calvin-Benson-Bassham cycle; TCA, tricarboxylic acid cycle. Central metabolites: RuBP, ribulose-1,5-bisphosphate; 3-PGA, 3-phosphoglycerate; G3-P, glyceraldehyde-3-phosphate; F6-P, fructose-6-phosphate; G6-P, glucose-6-phosphate; G1-P, glucose-1-phosphate; DOXP, 1-deoxyxylulose-5-phosphate; DMAPP, dimethylallyl-pyrophosphate; 2-PGA, 2-phosphoglycerate; PEP, phosphoenolpyruvate; OA, oxaloacetate; Cit, citrate; ICit, isocitrate; 2-OG, 2-oxoglutarate; FA, fatty acids; Heterologous enzymes: EFE, ethylene-forming enzyme; PDC, pyruvate decarboxylase; ADH, alcohol dehydrogenase; KivD, oxoacid decarboxylase; YqhD, NADPH-dependent alcohol dehydrogenase; LdhA, lactate dehydrogenase; IspS, isoprene synthase; LMS, limonene synthase; MtlD, mannitol-1-phosphate dehydrogenase; Mlp, mannitol-1-phosphatase; InvA, invertase.

sion of alcohol dehydrogenase (YqhD) from *E. coli* but the reported yield was lower (approx. 50 %) for the alcohol. Also, cyanobacterial strains producing and releasing free fatty acids, which can be further converted to biodiesel, were developed by overexpression of an acyl-acyl carrier protein thioesterase (TesA) from *E. coli*. Free fatty acids accumulate in the culture medium with yields up to 4 mg l^{-1} h^{-1}. Interestingly, several cyanobacteria contain enzymes for the direct conversion of intermediates from the fatty acid metabolism into olefins. Overexpression of the corresponding acyl-acyl-carrier protein reductase and aldehyde decarbonylase induces the release of fungible hydrocarbon fuels.

In addition to the production of biofuels, cyanobacterial strains have also been developed to extend their capability to synthesize bulk and fine chemicals. As they are the progenitors of algae and plant chloroplasts, they share several pathways for the synthesis of secondary metabolites. Among these, the class of isoprenoids is interesting for biotechnological applications. Besides carotenoids (see chapter 7), which are also naturally produced by cyanobacteria, other terpenoids can be synthesized by introducing the corresponding enzymes derived from plant sources. Such an approach has already been realized for hemiterpenoids (C5), monoterpenoids (C10), and sesquiterpenoids (C15). Whereas isoprene (C5) is an interesting building block for the synthesis of polymers, the cyclic monoterpene limonene (C10) is a widely used fragrance. Both are volatile products that can be easily extracted from the headspace of the culture. Several cyclic medium chain terpenoids (C15, C20) have a high potential for pharmaceutical applications (e.g artemesin, taxadien). In addition to cyclization, they are usually modified by P450 monooxygenases for their specific function. Although it has been shown that cyanobacteria principally can be used for the synthesis of plant-derived isoprenoids, reported yields are rather low in comparison with other microbial production systems (e.g. eubacteria, yeasts). For this reason, research is focused on metabolic engineering of carbon pathways in the cell in general. Carbon flux analysis of *Synechocystis* revealed that 80 % of the primarily fixed carbon is used for storage in sugars/glycogen, 15 % for fatty acid biosynthesis, and 5 % for the synthesis of isoprenoids. Rerouting of theses carbon fluxes offers a great potential for the enhanced production of fatty acid biofuels and isoprenoid chemicals.

Ethylen is another interesting platform chemical that is used in many industrial relevant processes – for instance as fruit ripening enhancer in the food industry. Therefore, the ethylene-forming enzyme (EFE) from *Pseudomonas syringae* was expressed in *S. elongatus*; however, the transformants were severely affected (decreased growth rate, yellow-green color phenotype) and were genetically unstable with a highest achieved production rate of about 6 % of the total fixed carbon. Sugars are of special interest as they are the main carbon sink in cyanobacteria being produced only by photosynthetic organisms. Although cells usually store glucose as glycogen, freshwater cyanobacteria can accumulate sucrose under salt stress conditions, which can be cleaved into glucose and fructose by a heterologously expressed invertase. These products can be exported into the medium by co-expression of the glucose facilitator (glf) from *Z. mobilis* that functions as a sugar-proton symporter. Usually, such transport systems use the proton motive force at the cytoplasmic membrane of bacteria to facilitate the proton-coupled import of sugar. If expressed in cyanobacteria, these transporters function in the opposite direction due to alkalization (~pH 9) of the medium, with a maximal achieved total concentration of glucose and fructose of 0.045 g/L. A very similar strategy was applied for the production of lactic acid, which serves as building lock for the synthesis of biopolymers. Due to co-expression of lactate dehydrogenase (ldhA) and a lactate transporter (lldP) from *E. coli* in the cyanobacterium *S. elongatus*, lactate was accumulated in the medium up to 0.06 g/L within 4 days.

6.3 Strategies to design mass cultivation in photobioreactors

Photoautotrophic microorganisms require special standards for the construction of bioreactors. Although the vast majority of microalgal biomass is still produced in open ponds, cultivation in closed photobioreactor systems minimizes both the amount of fresh water and of the ground area required for installing the cultivation facility. In addition, closed systems are a prerequisite for the cultivation of genetically modified photoautotrophic microorganisms and for the isolation of volatile fermentation products.

6.3.1 Photobioreactor Design

While the design of bioreactors for heterotrophic micro-organisms is technically mature with stirred-tank reactors being established as common design principle, photosynthetic life style has a strong impact on the development of photobioreactors. Among the various geometries still under discussion, the three most common designs of photobioreactors are column reactors, flat-bed reactors, and tubular reactors (Figure 6.10). All more complex designs can basically be described as a combination or variation of these basic reactor elements.

Central parameters to characterize and compare different reactor designs are the ratio of transparent reactor surface area to reactor volume and the ratio of transparent reactor surface area to the ground area covered by the reactor setup. While the first parameter characterizes the entry of light and its distribution in the culture volume, it is also mandatory to assess the mixing behavior. The light transfer is generally increased with increasing the surface/volume-ratio. This, however,

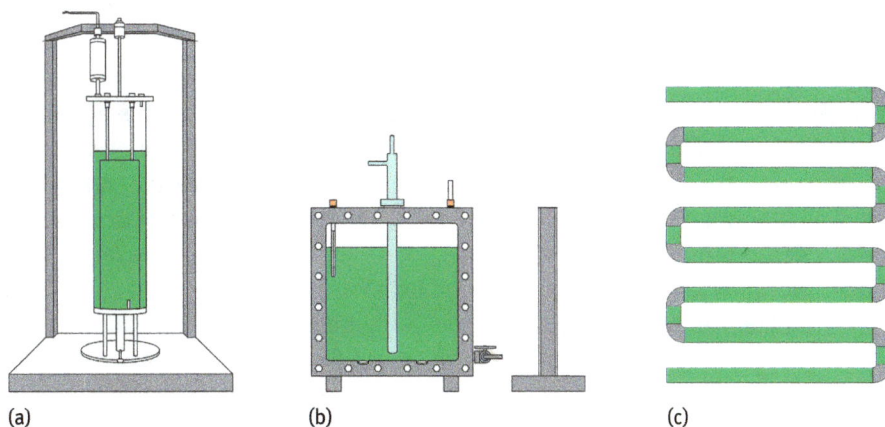

Fig. 6.10: Common designs for closed photobioreactors. (a) Column reactor, (b) flat-bed reactor, (c) tubular reactor.

reduces the mixing efficiency, which is important for mass and heat exchange within the reactor. In contrast to both flat-bed and tubular reactors, which can achieve high surface/volume ratios, column reactors show a good mixing behavior. On the other hand, gas exchange and product inhibition are limiting elements in tubular reactors.

As the principle driving force of photosynthesis, light, and its penetration into the reactor volume, sets strong restraints to the reactor design. Based on light supply and light penetration depth, different volume elements of the bioreactor can be attributed to either productive light zones with sufficient or unproductive dark zones with insufficient light intensities to drive photosynthetic metabolism. For a given reactor geometry, light penetration depends on the density of the light absorbing photoautotrophic microorganism and declines exponentially with increasing culture density due to mutual shading of the cells. For dense cultures, which are required for relevant volumetric productivities, light zones can be limited to a few millimeters depth from the transparent reactor surface. However, as outlined in box 6.1, too much light can have negative impacts such as photoinhibition. Although cyanobacteria have evolved many strategies to protect and recover from light induced damage, photoinhibition in general causes loss of productivity and may also impair product quality. For this reason, high productivity of the entire reactor volume requires appropriate mixing as it transfers cyanobacterial cells from unproductive to productive parts of the reactor.

The handling of surplus light intensities is very important for process engineering: Dependent on the cyanobacterial species, photosynthetic productivity is typically saturated at light intensities above 200 μEinstein m^{-2} s^{-1}. This is very relevant with regard to typical European summer midday light intensities which exceed 1000 μEinstein m^{-2} s^{-1}, i.e. five times the photosynthetic capacity. Such surplus light intensities are converted into fluorescence or heat and are lost energy for the biological system. On the other hand, the excess supply in light can be compensated by a larger transparent surface area of the reactor relative to the ground area. In process engineering, this is achieved by vertical photobioreactor installations with surface areas exceeding the ground area more than 5-fold. This approach, known as „light dilution", spreads light, hitting a given ground area over a large surface area of the reactor, resulting in an irradiance of each reactor surface element with only a fraction of the incident light. The implementation of this concept largely depends on the geographic location of the installation, i.e. the average light intensities, the cyanobacterial species and its photosynthetic capacity.

In addition to irradiation, CO_2 is also a limiting factor for the cultivation of cyanobacteria, as it is the key substrate for carbon fixation and biomass formation. Insufficient supply of carbon dioxide results in a drop of photosynthetic efficiency and in a rise of the unfavorable oxygenase side reaction of the enzyme Ribulose bisphosphate carboxylase (RuBisCo). In column and flat-bed reactors, gas exchange can be realized without significant energy supply; also it is readjustable during cultivation by changing gas flow and gas composition. In contrast, gas exchange in tubular reactors usually occurs in designated devices, i.e. degassers or aeration tanks, where oxygen is removed and saturated with carbon dioxide. Here an appropriate residence time for the illuminated tubular reactor is set by adjusting the flow rate of the culture through the reactor. This should avoid dead zones lacking carbon dioxide, which do not contribute to growth or productivity of the reactor system.

6.3.2 Downstream processing

The term *downstream processing* summarizes all operations applied for separation, isolation, or generation of products from the fermentation broth leaving the bioreactor. Although this processing is strongly dependent on the product and the involved microorganism, some general considerations are important for a successful performance of cyanobacteria-based processes. One of the most significant limitations for photoautotrophic processes is the achievable cell density and the corresponding biomass concentration. The physical limit of cell density is reached for any type of microorganism when the entire reactor volume is filled with densely packed microorganisms, with cell-to-cell contact. For prokaryotic microorganisms with a diameter of about 2 µm, this corresponds to approximately 2.4×10^{11} cells/mL. For aquatic microorganisms with a water content of 90 %, such a cell density is equivalent to a dry mass concentration of 100 g/L. While for industrial fermentation processes with heterotrophic microorganisms typical biomass dry mass concentrations range between 20 g/L and 50 g/L, highest reported dry mass concentrations for photoautotrophic microorganisms are in the range of 20 g/L, and for most applications even significantly lower. For higher densities, the demand for light and CO_2 supply are difficult to satisfy with the existing technologies.

This low biomass content for photoautotrophic microorganisms per reactor volume results in different limitations for products, which are directly proportional to biomass: Growth-linked products on the one side and non-growth-linked products (such as secondary metabolites, which are not directly correlated with biomass) on the other. Due to their direct dependence on biomass itself or combined components (lipids, carbohydrates or carotenoids), the impact on growth-linked products is severe: In this case, the removal of water or the extraction from culture media is energetically and economically only profitable for high-price products like carotenoids and other fine chemicals. In contrast, for non-growth-linked products, their concentration in the media and their separation from the media are the decisive factors. Especially for gaseous and hydrophobic products, which separate spontaneously from the culture and can easily be removed, continuous production systems operating at high cell densities and low growth rates have the potential to target products of medium and low price segments.

6.3.3 Scale up

Factors with impact on the design of photobioreactors and biotechnological processes with photoautotrophic microorganisms involve biological productivity, ecological sustainability, and economic profitability. While the production of a compound in cyanobacteria might be regarded as scientific success, a sustainable process is only achieved when the net energy balance is positive, i.e. when the energy content of the products is larger than the energy required for the production. The decisive factor for the industrial application of a process is the profit and the return of investment, i.e. costs for investment, maintenance, and personnel have to be considered.

One major factor both for ecological demands and economic investments is the size of the reactor system. While larger systems generally reduce the costs per reactor volume unit, in the case of photobioreactors, the light distribution is the most critical parameter for an upscaling as light supply limits their volumetric productivity. Unproductive dark zones within the reactor volume can be minimized by limiting the thickness (diameter) of photobioreactors to <10 cm, which enables a sufficient light penetration depth. Additionally, hydrostatic pressure and mixing parameters limit both height and width of the reactor. For these reasons, flat-panel photobioreactor units are limited to less than 500 L if high technical standards

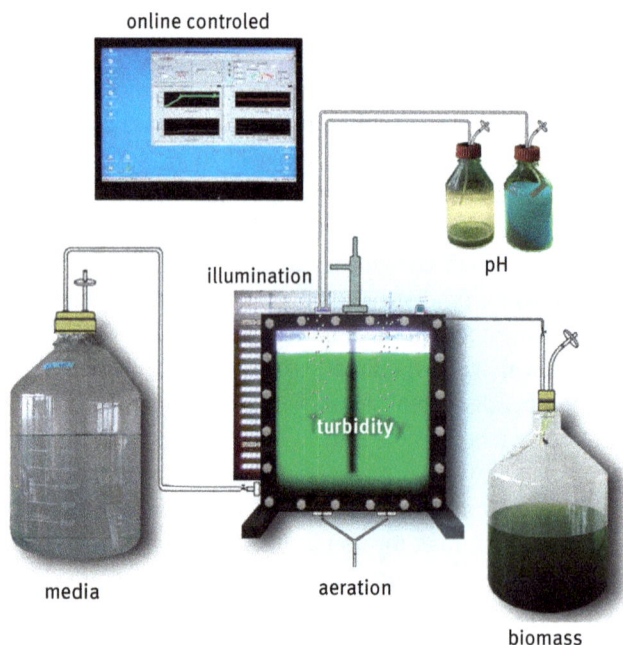

Fig. 6.11: 5 L flat panel photobioreactor run as continuous cultivation system under online control (RUB Plant Biochemistry in cooperation with KSD, Hattingen).

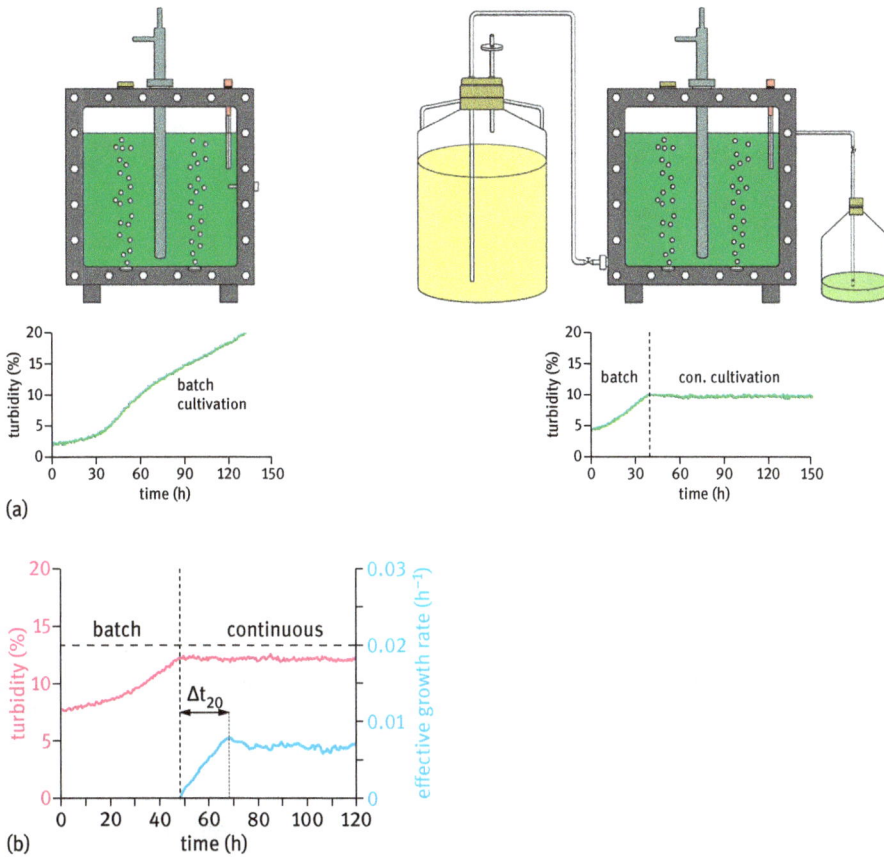

Fig. 6.12: (Box 6.3): Principles of continuous cultivation. (a) Comparison of batch (left) and continuous (right) cultivation. (b) Transition from batch to continuous cultivation. Both constant culture density and a constant growth rate are obtained when a steady-state is reached.

(combined with high costs) have to be avoided. A significant further upscaling can only be achieved if the number of such bioreactor units is increased. On the other hand, investment costs are reduced if facilities for downstream processing are shared for such joined reactor units (see Figure 6.11 for prototype of 5 L flat panel reactor system).

Box 6.3: Characteristics of batch and continuous cultivation systems.

In a batch system (Figure 6.12a), the injection of an inoculum to the medium starts the culturing process. The subsequent microbial growth kinetics can be subdivided into lag-, accelerating growth-, exponential growth-, decelerating growth-, and stationary phase within a steadily exhausting culture medium. In contrast, continuous cultivation is characterized by the continuous addition of fresh sterile culture medium and the continuous removal of biomass, product, and used medium during the cultivation process. A steady state of the culture is realized by balan-

absorbed radiation (100 %)

fluorescence, heat (35 %)

alternative electrons (29 %)

respiration (14 %)

energy for growth (22 %)

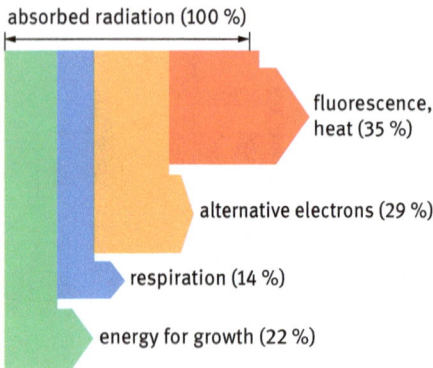

Fig. 6.13: (Box 6.3): Higher efficiency of phototrophic mass culture due to optimization in continuous culture set-up: Minimization of self-shading and optimization of light-intensity results in an about 50 % reduction of energy losses by fluorescence and heat (compare with Figure 6.3 of Box 6.1) (Wilhelm & Jacob 2011). Figure (modified) with kind permission from Springer Science and Business Media.

cing the microbial growth rate and its dilution. In comparison with a batch culture system, the characteristic advantages of a continuous culture (Figure 6.12a) are constant production rates, elimination of down times for cleaning and sterilization, and the realization of defined cultivation conditions with the phenotype fully adjusted to the environmental conditions. Based on the type of process control, chemostat and turbidostat cultures can be distinguished: A chemostat is a microbial cultivation system, to which medium is fed at a constant rate while the total volume of the culture is kept constant by an overflow. This results in a steady state with constant biomass concentration, if the dilution rate is lower than the maximal growth rate, i.e. the specific growth rate corresponds under these conditions to the dilution rate. In a turbidostat, the culture density is monitored and kept constant to a defined setpoint by regulating the dilution rate with fresh culture medium. The specific growth rate of the culture will, therefore, adjust according to the growth-limiting parameter. Figure 6.12b shows the transition from batch to turbidostatic process control. After inocculation, the culture grows exponentially up to the set point of the turbidostatic control (turbidity setpoint: 12 %). When the turbidity of the culture exceeds the setpoint, the culture is diluted by inflow of fresh media and removal of dense culture. After equilibration, both culture density and specific growth rate μ stay constant. Generally, continuous cultivation systems are advantageous to maintain slow growing organisms at high culture densities: These are required for relevant space-time yields, especially when the targets are non-growth linked products such as hydrogen which allow a constant product removal.

In terms of optimal utilization of light energy, continuous cultivation allows the control of all relevant parameters, which is important to minimize losses due to fluorescence and heat (see Box 6.1): The potential of this approach to use alternative electrons for engineered biotech processes is obvious by comparing Figure 6.3 – growth under unoptimized conditions – with growth under conditions of continuous cultivation (Figure 6.13). It should be considered, that in this case, only the „environmental conditions" have been optimized by switching to continuous cultivation conditions. In addition, engineering of the internal cellular metabolism can minimize the energy for „housekeeping", i.e. the energy which is required to keep the organism alive. This will provide another substantial impact on redistribution of harvested light energy for the benefit of biotechnological processes, the limitation of which is still unknown.

(a)   (b)

Fig. 6.14: Large scale photobioreactors. (a) 100 L Photobioreactor illustrating the potential of scale-up of a flat-bed reactor type for future mass fermentation of cyanobacteria (prototype; chair Plant Biochemistry, Ruhr-University Bochum, with KSD Co., Hattingen). (b) Pilot project to use flat bed photobioreactors as cover for buildings.

Most recent examples for upscaling of flat plate photobioreactors are shown in Figure 6.14 – in one case a pilot plant for 100 L, in another the first „algal house" with sun-exposed surfaces being covered by flat plate reactors operated in continuous flow mode. In this case, the generated biomass is harvested in a special set-up in the basement of the building and contributes to the energy balance of the house (IBA Hamburg 2013).

6.4 Outlook – future potential of cyanobacteria

Photoautotrophic lifestyle with its independence from sugars or other reduced carbon compounds as energy source stimulates visions of carbo-neutral production of chemicals and commodities within cyanobacteria. Realization of such processes depends mainly on their economical profit and return of investment. For generation of revenue, cyanobacterial processes have to outcompete both biotechnological processes in heterotrophic microorganisms and classical chemical production. With their virtually unlimited access to reduction equivalents and the ability to fix CO_2, cyanobacterial systems have a considerable advantage on target compounds with high demands in both energy and carbon supply. Currently, as central devices of cultivation and production systems for photoautotrophic organisms are in development, only products and commodities of targeted high-price segments may achieve economic profit. While general cost optimizations for reactors and peripheral devices will arise with the experience gathered from realized production systems, several technical advancements, – such as increasing culture density, sea water compatibility and improvement of the downstream processing for constant product removal – are necessary to target products from mid- and low-price segments.

For this purpose new strains have to be isolated from the giant natural treasure available especially in the oceans. These strains must have the ability to be easily/ naturally transformable, and this competence has to be used for further improvements towards the targeted product. In combination with optimized photobioreactor systems and advanced tools for metabolic engineering provided by synthetic biology this will be an attractive method to harness cyanobacteria for biotechnological purposes using the power of photosynthesis.

Key-terms

batch cultivation, biofuels, biohydrogen, continuous cultivation systems, down stream processing, electron transfer reactions, flat panel photobioreactors, metabolic engineering, photosynthetic efficiency, photosynthesis, streamlining, *Synechocystis* PCC 6803, *Synechococcus* PCC 7002

Questions

- Which primary processes are responsible for photosynthetic energy conversion?
- What is „photosynthetic efficiency"?
- Why is energy lost during photosynthetic energy conversion?
- What is the advantage of cyanobacterial biotechnology in comparison with the biotechnology of eukaryotic systems?
- What is the advantage of cyanobacterial diversity for bioindustrial applications?
- Why are *Synechocystis sp.* PCC 6803 and *Synechococcus sp.* PCC 7002 suitable species for the production of biomolecules?
- What is the genetic basis for metabolic engineering of these organisms?
- Which techniques are suitable to determine photosynthetic parameters and efficiencies?
- What is the rationale for the design of cyanobacterial strains with reduced antenna size?
- What is the potential of photosynthesis-based biotechnology and which products could be profitable?
- Which applications are realistic for light-triggered genes and how can they be realized?

- What are the critical points for the development of a hydrogen producing cyanobacterium?
- Which central carbon pathways/metabolites provide precursor molecules for bioindustrial applications?
- What is the difference between batch cultivation and continuous cultivation?
- What is „light dilution"?
- What is „downstream processing" and why is it important for the production costs?

Further readings

Flores, E. & Herrero, A. 2014. The cell biology of cyanobacteria. *Caister Academic Press.* ISBN: 978-1-908230-38-6
Sharma, N. K., Rai, A. K. & Stal, L. J. 2013. Cyanobacteria: An Economic Perspective. *Wiley.* ISBN: 978-1-119-94127-9
Sarma, T. A. 2012. Handbook of Cyanobacteria. *CRC Press.* ISBN: 978-1-57808-800-3

References

Blankenship, R. E., Tiede, D. M., Barber, J., et al. 2011. Comparing photosynthetic and photovoltaic efficiencies and recognizing the potential for improvement. *Science* 332: 805–809.
Ducat, D. C., Way, J. C., Silver, P. A. 2011. Engineering cyanobacteria to generate high-value products. *Trends Biotechnol* 29: 95–103.
Hess, W. R. 2011. Cyanobacterial genomics for ecology and biotechnology. *Curr Opin Microbiol* 14: 608–614.
Moal, G. & Lagoutte, B. 2012. Photo-induced electron transfer from photosystem I to NADP+: Characterization and tentative simulation of the *in vivo* environment. *Biochim Biophys Acta (BBA) – Bioenergetics* 1817: 1635–1645.
Rögner, M. 2013. Metabolic engineering of cyanobacteria for the production of hydrogen from water. *Biochem Soc Trans* 41: 1254–1259.
Rosgaard, L., de Porcellinis, A. J., Jacobsen, J. H., et al. 2012. Bioengineering of carbon fixation, biofuels, and biochemicals in cyanobacteria and plants. *J Biotech* 162: 134–147.
Wilhelm, C. & Jakob, T. 2011. From photons to biomass and biofuels: evaluation of different strategies for the improvement of algal biotechnology based on comparative energy balances. *Appl Microbiol Biotechnol* 92: 909–919.

Thomas Happe and Anja Hemschemeier

7 Eukaryotic microalgae in biotechnological applications

What is commonly known as "algae" is in fact a very diverse group of organisms that have similar traits, but that are not necessarily closely related in terms of phylogeny. Biologists have still not come to an agreement on the systematic classification of algae. This is partly due to the fact that the evolution of algae did not follow a straight line, so that algae are found in different phylogenetic groups and belong to different taxa (Tirichine & Bowler, 2011). A rather broad definition describes algae as aquatic plant-like organisms that carry out oxygenic photosynthesis. This concept covers unicellular algae that are only a few micrometers in diameter, as well as the giant kelp that can grow to lengths of more than 40 meters. In everyday speech, cyanobacteria also are called microalgae or blue-green algae, but they have to be distinguished from other algae because they are true prokaryotes. Still, it is generally accepted that a cyanobacterium was the progenitor of the plastid of eukaryotic algae and higher plants, and cyanobacteria carry out oxygenic photosynthesis. Therefore, depending on the perspective, cyanobacteria and algae certainly have some traits in common. Nevertheless, in this chapter, an "alga" is defined as a eukaryotic, plastid-containing organism, and the term microalgae describes microscopic, unicellular eukaryotic algae.

Algae have a long tradition of serving human purposes. Macroscopic benthic algae belonging to the taxonomic groups of green, red, and brown algae, and commonly referred to as seaweed, have been used as food for centuries, mostly in Asian countries. Thalloid brown algae are sources of alginic acid, which is employed in food industry, medicine and cosmetics, while agar-agar and agarose, commonly used in life sciences, are derived from red algae. Some cyanobacterial species have also been serving as food for a long time. However, the commercial application of cyanobacteria and microalgae reached a new level in the middle of the 20th century, when the demands of an increasing world population required additional sources of valuable food, high-value chemicals, and energy. Then, large-scale cultivation facilities and the extraction of specific high-value compounds such as carotenoids or polyunsaturated fatty acids (PUFAs) were industrially established (Spolaore et al., 2006). With growing scientific knowledge and genetic technologies, microalgae are furthermore discussed to be used as producers of biofuels or drugs and therapeutic recombinant proteins.

This chapter provides an overview of the biotechnological exploitation of eukaryotic microalgae, which are in many cases regarded as good alternatives to using plants for reasons outlined below. First, the principles of photosynthesis and its biotechnological implementation will be reviewed briefly. Then, products already on the market (mostly algal biomass, carotenoids, and PUFAs), as well as potential future applications based on natural algal traits (photosynthetic production of molecular hydrogen (H_2) or lipids) and the metabolic pathways forming these products will be presented. Also, genetic engineering strategies and techniques aiming to optimize these pathways or to introduce novel metabolisms will be explained.

7.1 Biotechnological exploitation of oxygenic photosynthesis

What makes plants and algae both essential for the biosphere and so interesting for applied technologies is their ability to carry out photosynthesis. Photosynthesis is the most important process on earth, because it uses light energy to convert inorganic molecules such as carbon dioxide, nitrate or sulfate into compounds that build up a living cell or that can be used as energy sources. Organisms that carry out photosynthesis have been used by humans from the beginning on and served as food, shelter, energy, and medicine. The utilization of plants became more organized with the emergence of agriculture, and technical and scientific progress resulted in highly modern intensified cropping systems employing special cultivations or genetically modified plants. Oxygenic photosynthesis is often attributed to be a blueprint for the capture of solar power. Modern science works at understanding the details of photosynthetic light-conversion, one ultimate goal being the artificial reproduction of the principle for biomimetic approaches. However, in its natural form in plants and algae, photosynthesis has been and will continue to be used biotechnologically. All of our fossil fuels were ultimately generated by photosynthetic organisms that convert sun-light into fixed carbon. Today, so-called energy crops are cultivated specifically for yielding biomass to be transformed into energy, either by generating bioethanol from carbohydrates or biodiesel from plant oil, by fermenting plant material to biogas (mostly methane) or by simply burning them to generate heat. Beyond the use of biomass, specific products of photosynthetic organisms are of economic relevance. Here, native compounds such as carotenoids or therapeutic ingredients are harvested, but more and more, plants and microalgae are also used for the production of organic molecules not naturally occurring in these organisms.

7.1.1 Principles

During photosynthesis, the conversion of light- into chemical energy is originally catalyzed by specific molecules, the chlorophylls. These cyclic tetrapyrroles (porphyrins)'have an extensive system of delocalized electrons, so that visible light can be used to raise single electrons of the chlorophyll molecule to a higher energy level. From there, the energy-rich electron is transferred to further electron carriers, and, in the process, releases its energy step by step. These processes are carried out in specialized protein complexes that work in series and form the photosynthetic electron transport chain. During oxygenic photosynthesis, two photosystems are employed. Photosystem II (PSII) is capable of generating the strongest oxidant known in biology. With a redox potential of about 1.3 V, the oxidized chlorophyll reaction center of PSII is capable of oxidizing water, thereby extracting electrons and protons, and releasing molecular oxygen (O_2) (Figure 7.1). From PSII, electrons

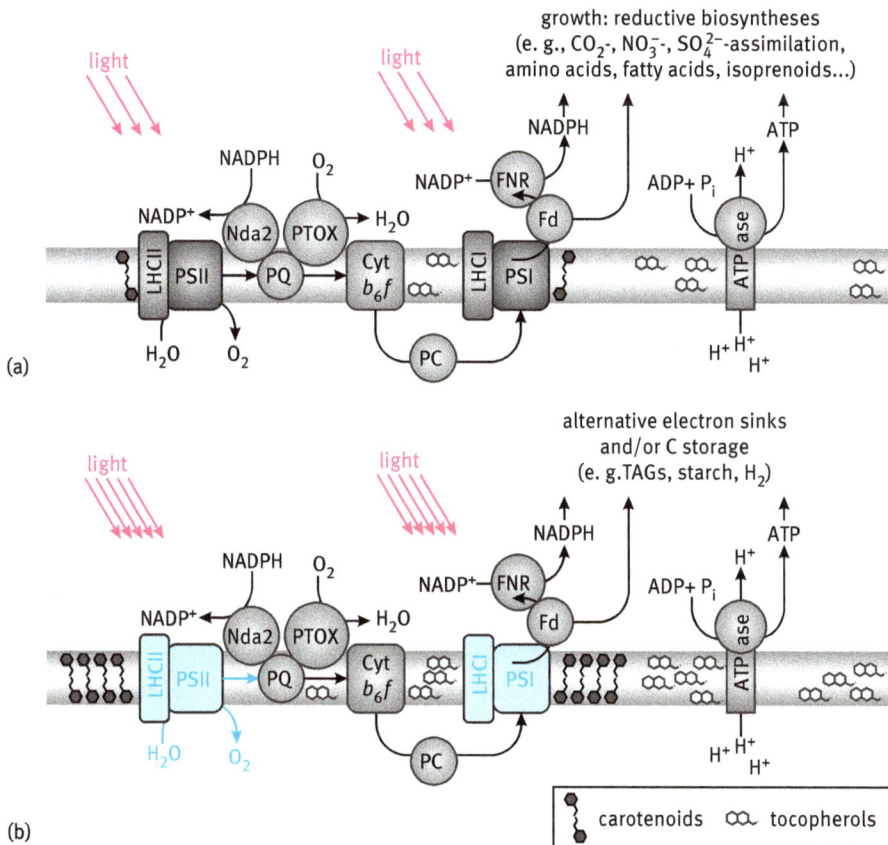

Fig. 7.1: Scheme of the photosynthetic electron transport chain under favorable growth conditions (no light stress) (a) and under light stress (e.g. under nutrient deprivation) (b). (a) Under favorable growth conditions, reducing equivalents, and ATP are used for assimilatory and biosynthetic processes. (b) Under light stress, resulting from nutrient deprivation, alternative electron sinks such as H_2 production or TAG- and starch accumulation are employed. Light stress results in the accumulation of carotenoids and antioxidants such as tocopherols. ATPase: ATP synthase; Cyt b_6f: cytochrome b_6f complex; Fd: ferredoxin; FNR: ferredoxin-NADPH-oxidoreductase; LHCII, I: light harvesting complexes of photosystem II, I; PC: plastocyanin; PQ: plastoquinone; PSII, I: photosystem II, I; PTOX: plastid terminal oxidase; Nda2: monomeric NAD(P)H-plastoquinone oxidoreductase of *C. reinhardtii*; TAG: triacylglycerol.

are transported via the membrane-soluble electron carrier plastoquinone to the cytochrome b_6f complex and from there via the water-soluble plastocyanin protein to photosystem I (PSI) (Figure 7.1). After absorption of light, chlorophylls in the reaction center of PSI create a strong reductant that is able to transfer electrons to the small soluble iron-sulfur protein ferredoxin. Ferredoxin, in turn, is used as an electron donor for various reductive processes. The ubiquitous redox equivalent nicotinamide adenine dinucleotide phosphate (NADPH), generated via ferredoxin-$NADP^+$ reductase (FNR), is ultimately used for assimilatory processes such as carbon dioxide (CO_2) fixation. Reduced ferredoxin is additionally the electron donor needed for assimilatory nitrate and sulfate reduction, as well as various further biosynthetic processes (Figure 7.1). The protein complexes of the photosynthetic electron transport chain are embedded in specialized membranes, termed thylakoid membranes. During electron transport, the released energy is additionally used to build up a proton gradient across the membrane, which is then used by an adenosine triphosphate (ATP) synthase to generate the "energy currency" ATP (Figure 7.1).

Oxygenic photosynthesis is an ingenious process, which not only allows an alga or a plant to grow, but which feeds all heterotrophic organisms. Moreover, it is generally believed that the accumulation of O_2 after the evolutionary invention of oxygenic photosynthesis by early cyanobacteria was the pre-condition for complex organisms to evolve. Not only does aerobic respiration provide the highest energy yield, but network analyses indicate that more than 1000 biochemical reactions in contemporary organisms depend directly or indirectly on the O_2 molecule.

However, for the living cell, oxygenic photosynthesis has its drawbacks and can be a dangerous process. The huge differences of the redox potentials of individual molecules, the generation of radicals and the chemistry of oxygen, altogether the necessary foundation of photosynthesis, are also predestined to cause damage if not carefully balanced. Excited chlorophyll molecules, usually in the electronic singlet state, can go over to the triplet state. Triplet chlorophyll, in turn, can convert the rather harmless and unreactive triplet O_2 molecule to singlet O_2, which is very reactive and can destroy other molecules in the vicinity. Even under the most favorable growth conditions therefore, the PSII reaction center protein D1 (gene designation *psbA*) shows a high turn-over and has a half-life of only about two hours. The D1 protein harbors the electron carriers of PSII and is the direct target of radical site reactions. Another potentially harmful reaction is the reduction of O_2 to the superoxide radical by PSI, the so-called Mehler reaction.

Radical reactions and the generation of reactive oxygen species (ROS) are normal side-effects of photosynthesis, and photosynthetic organisms have evolved several defense strategies to cope with the inherent dangers of photochemical light conversion. Only if energy input, that is absorbed light energy, and energy output, in terms of utilizing redox and energy equivalents, are not balanced does the generation of harmful molecules overcome the stress defense systems of the cell. This is then called light stress, or, because it results in decreased photosynthetic activity, photoinhibition. It should be noted that "light stress" does not only occur at high light intensities. Rather, photoinhibition and stress can occur in low light, given that light absorption exceeds the capacity of the electron transport rate and the intermediate electron carriers, respectively. Expressed quite casually, light stress occurs when electrons "spill over". A common situation in which the light intensity may be low and still result in light stress is nutrient deficiency. When a certain nutrient is not available or strongly limited, essential cell components cannot be built. Sulfate depletion can result in the insufficient biosynthesis of the amino acids cysteine and methionine and additional vital sulfur-containing compounds such as coenzyme A (CoA). This forces the cell to reduce or even stop growth and cell division, and as a consequence, the utilization of NADPH and ATP is decreased, resulting in an over-reduced photosynthetic electron transport chain.

7.1.2 Utilization

Heterotrophic organisms like animals benefit from the inherent stress of oxygenic phototrophic organisms. Their protective measures are traditionally used for human purposes, but are also recognized today to have the potential of true biotechnological applicability. One important defense line is composed of antioxidant molecules that are vitamins for us. Tocopherols and tocotrienols, known collectively as vitamin E, are exclusively synthesized by photosynthetic organisms (Figure 7.1). They are lipid-soluble antioxidants that protect fatty acids from oxidation by ROS. Ascorbic acid, vitamin C, is a water-soluble antioxidant present in quite high concentrations in plants and algae. Carotenoids are also mostly synthesized by photosynthetic organisms, and, besides being additional light-absorbing pigments, play an important role in the quenching of excess light energy within the photosynthetic apparatus (Figure 7.1). Carotenoids are essential for human beings, for example β-carotene, which is the precursor of vitamin A needed for sight as well as for neurological, immunological and many other essential processes in our body. Further carotenoids such as lutein, zeaxanthin or astaxanthin are important antioxidants and additionally protect our retina and our skin from ultraviolet light. While animals including *Homo sapiens* have, by eating plants, benefited from these ingredients from their emergence on, especially carotenoids have gained considerable biotechnological potential in the twentieth century (see 7.2).

7.1.3 Economics

This chapter aims to provide a biological view on microalgae biotechnology, and to give some examples of already economically established processes and possible future applications (Table 7.1), instead of providing a feasibility study in terms of efficiencies and production costs. However, economic aspects should be addressed at least briefly, so that the potential of microalgae can be put into a larger context. For the interested reader, the book "Microalgal Biotechnology: Integration and Economy" (DeGruyter, edited by C. Posten, C. Walter) is recommended.

Although actual numbers and time-frames of future scenarios differ, there can be no doubt that our society has to develop alternative routes for many industrial and agricultural applications. The United Nation estimates that the world population will grow from seven (2011) to nine billion by 2050 and that about 70 % more food will be needed. Also, about 50 % more fuel and 30 % more freshwater will be required. Simultaneously, reserves of fossil-fuels will gradually become depleted, even though here, too, the scenarios differ.

Biotechnology using photosynthetic organisms seems a perfect solution, and for various reasons, microalgae are considered to be more suitable than higher plants for many of the envisioned applications. Several species of microalgae grow faster than higher plants and/or have a higher light-conversion efficiency. They grow in liquid medium and thus can be cultivated in closed photobioreactors (PBRs). These offer the possibility of vertical arrangement, thus saving space and lessening the competition with agricultural land. In fact, cultivating energy crops is not beyond dispute, one of the reasons for that being the competition with available space for plants for food and feed. Closed PBRs furthermore allow to meticulously control growth parameters, which is especially important for the standardized generation of fine-chemicals and drugs. Not least, sealed cultivation vessels represent a strict containment of genetically modified organisms (GMOs). Cultivation of genetically engineered higher plants, in contrast, always bears the risk of unwillingly distributing foreign genetic material, for example through pollen, bacteria or insects.

There has been and still is quite a hype about utilizing microalgae, and many unrealistic numbers on how much biofuel, for example, could be generated by them were published. However, it will take more technical progress before microalgae are economically (and sustainably) usable, also for low-value products such as fuel (Stephens et al., 2013). Ultimately, only if companies can make profits will microalgae reach large-scale marketability. This, in turn, will not least be influenced by the governments, for example by charging for CO_2 emissions. However, also in terms of renewability and sustainability do many aspects have to be considered and true life-cycle assessments provided. In the following, some advantages and challenges of using microalgae will be listed.

The most obvious advantage of microalgae (and other photosynthetic organisms) is their ability to live on and convert the vast excess of energy that we get for free: sunlight. About 1300 ZJ (that is 1300 Sextillion or 1300×10^{21} Joule) of photosynthetically active radiation (PAR) reach the surface of the Earth each year. This amount is huge, especially when comparing it with the momentary global

(a)

(b)

Fig. 7.2: Algal cultivation and harvesting. (a) Test facility of the University of Queensland in Brisbane, Queensland, Australia. From left to right: small-scale raceway-pond, flat-panel PBRs (consisting of specialized bags supported in a metal frame), and a plastic-tubular system. Photographs © T. Happe. (b) Facility of Roquette Klötze GmbH & Co. KG in Klötze, Germany. From left to right: Centrifuge to separate the algae from the culture medium (© Roquette Klötze GmbH & Co. KG), the drying machine, and a view on the factory hall (both © Jörg Ullmann, Roquette Klötze GmbH & Co. KG).

energy demand of 0,5 ZJ. On the other hand, to make use of the light conversion by microalgae, they have to be supplied with ample light, and this is a challenge for the design of cultivation facilities that must provide sufficiently short light paths. Designing PBRs that allow good growth rates of the cells while simultaneously being cost-efficient is an actual topic for research and development (R&D), for example at the University of Queensland (Brisbane, Australia) (Figure 7.2 a). A further advantage that might even be economically translated into profit given that CO_2 emissions become a true cost factor is that plants and microalgae consume CO_2, which is supposed to be one of the major greenhouse gases. Though many processes using the resulting biomass will set this CO_2 free again, this is at least CO_2 neutral. Again, however, as outlined above for light, sufficient amounts of CO_2 have to be introduced in algal cultivation vessels to increase efficiency. This means technical investment in forms of pipes, gas pumps and mixing devices, but also proximity to CO_2 generating industry, if exhaust gas is to be used.

Microalgae can, at least theoretically, be grown in areas that are not used for crops, and vertical PBRs would offer the additional advantage of being space-saving. However, construction costs for such facilities may be large. For example, water has to be transported to ponds that are constructed in dry regions, and water-losses due to evaporation have to be considered. One pronounced challenge of using unicellular microalgae is their high dilution with the aqueous medium. Depending on the product that is to be extracted, harvesting (for example, by centrifugation, Figure 7.2b) and drying are cost- and energy-intensive processes. Nevertheless, technical advances promise to solve many issues of large-scale microalgae cultivation and product extraction. For example, several concepts of microalgae harvesting are more profitable, such as flocculation or sedimentation, though these are not suitable for every application.

The fact that for now, most algal products are not competitive, has not prevented several companies from investing in algae. In many cases, the solution to profitability is regarded to be the combination of low-value products such as fuels with high-value products such as PUFAs. For example, the company Aurora Algae Inc. with headquarters in California, USA, and Western Australia, offers a microalgal product palette that includes fuel, feed for aquaculture and animals, proteins and PUFAs.

Further integrated systems deploy algae for certain tasks and then use the resulting biomass as feed or fertilizers. There are already several wastewater-treatment facilities that introduce microalgae in their systems. In some stages of wastewater processing, which is based on microbial degradation processes, certain bacteria need oxygen. In so-called Advanced Integrated Wastewater Pond Systems (AIWPS), used, for example, by the Californian Company GO$_2$WATER, oxygen is supplied by photosynthetic microalgae, which, in turn, grow on catabolites from the bacteria, including CO_2. The German company Subitec, a spin-off from the Fraunhofer Institute for Interfacial Engineering and Biotechnology (Fraunhofer IGB) in Stuttgart, has developed and patented a vertical so-called Flat Panel-Airlift-Photobioreactor, which has a high productivity at relatively low production costs. 180-liter-pilot reactors have been constructed by order of some German biogas and block-heat and power plants to investigate the potential of microalgae to grow on CO_2 from the exhaust gases.

Many people regard microalgal biotechnology as having a true economic future, especially when remaining technical issues are solved and alternatives (such as fossil fuels) become too expensive. However, most of the more reliable sources predict that algae will not to 100 % save the world, but instead be valuable contributions locally or in niche applications.

7.2 Established applications

Notwithstanding the challenges of large-scale microalgae cultivation, some applications deploying eukaryotic microalgae are already economically established,

Table 7.1: Selected products of photosynthetic microalgae that are already on the market or have future potential. Randomly chosen examples of companies or institutes that have established algal cultivation and sell the respective product or are promising to establish marketability of a certain product.

Species	Product	Application	Companies
Chlorella vulgaris	whole cells (dried)	dietary supplement, feed (aquaculture, farm animals), cosmetics/skin care	Taiwan Chlorella Manufacturing CO., LTD (Taiwan), Roquette Klötze GmbH & Co. KG (Germany)
Dunaliella salina	β-carotene	dietary supplement (vitamin A, antioxidant)	Cognis Nutrition and Health / BASF Personal Care and Nutrition (Australia/Germany)
Haematococcus pluvialis	astaxanthin	aquaculture, dietary supplement (antioxidant)	Cyanotech Corporation (USA)
Phaeodactylum tricornutum	PUFAs	dietary supplements, infant formulas, aquaculture	Fraunhofer Institute for Interfacial Engineering and Biotechnology (Germany)
Chlamydomonas reinhardtii	Mammary Associated Amyloid (MAA)	animal feed (and human food) supplement; supposed to protect intestines	Triton Health & Nutrition (USA)

such as using whole extracts of *Chlorella vulgaris* as dietary supplements, β-carotene production by the halophilic green alga *Dunaliella salina*, the extraction of the carotenoid astaxanthin from *Haematococcus pluvialis* or polyunsaturated fatty acids (PUFAs) from various species (Table 7.1).

7.2.1 Microalgal biomass

As photoautotrophic organisms, microalgae can biosynthesize all compounds needed by a cell, such as all proteinogenic amino acids. In contrast, humans depend on the dietary supply of tryptophan, threonine, phenylalanine, histidine, methionine, valine, leucine, isoleucine, and lysine. Microalgae furthermore contain vitamins and minerals needed by heterotrophic organisms. Some species synthesize certain carbohydrates thought to have beneficial effects on health (such as β-1, 3-glucan), and others synthesize PUFAs not produced by animals or higher plants. According to various studies, the biomass of the unicellular Trebouxiophyceaen green alga *C. vulgaris* (Figure 7.3a) is rich in valuable protein, lipid, carbohydrates, and vitamins, as well as further compounds such as minerals and secondary metabolites (Table 7.2). *C. vulgaris* is cultivated on an industrial scale to use the biomass as dietary supplement for humans, but also to be used as animal feed or for cosmetics

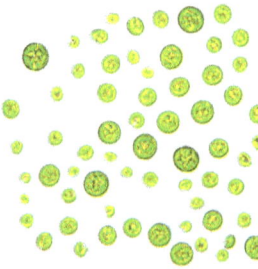

(a)   *Chlorella* sp.

glass tubing system

Chlorella pills

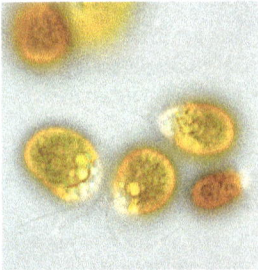

(b)   *D. salina*

Hutt Lagoon colored by *Dunaliella*

β-carotene
containing food

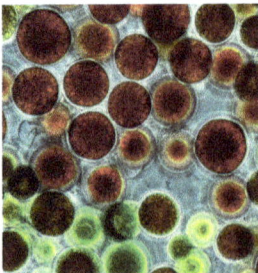

(c)   *H. pluvialis*

H. pluvialis cultivation at Cyanotech

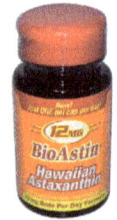

H. pluvialis derived
astaxanthin

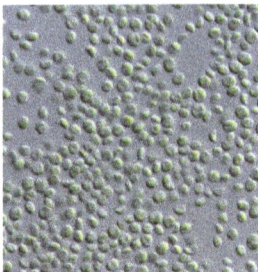

(d) *Nannochloropsis* sp.

Aurora algae cultivation pond in Texas

algal oil

Fig. 7.3: Examples of biotechnologically relevant species of unicellular eukaryotic algae (all to the left), their cultivation by the indicated companies (all in the middle) and the respective product (all to the right). (a) Biomass production from *C. vulgaris* by Roquette Klötze (photographs of *Chlorella* cells and glass tubing system: © Jörg Ullmann, Roquette Klötze GmbH & Co. KG; photograph of capsules: © Roquette Klötze GmbH & Co. KG). (b) β-carotene generation by BASF Personal Care and Nutrition from *D. salina* (photograph of *D. salina* © Dr. Jürgen E. W. Polle; photograph of Hutt Lagoon by Steve Back supplied courtesy of BASF; photograph of foods that can contain β-carotene supplied courtesy of BASF). (c) *H. pluvialis* as an astaxanthin source used by Cyanotech (photograph of *H. pluvialis* © Subitec GmbH, photographs of the *H. pluvialis* cultivation ponds and BioAstin used with permission by Cyanotech Corporation). (c) *Nannochloropsis* is established as a lipid producer by companies such as Aurora Algae inc. Photograph of *Nannochloropsis* cells, of Aurora's Texas cultivation facility and of Aurora's algae derived oil © Aurora Algae Inc. Details to (a)–(d) are explained in the text.

Table 7.2: Nutritional composition of spinach, *Spirulina*, and *C. vulgaris*.

	Unit	Spinach, fresh[1] per 100g	Spirulina, raw[1] per 100g	Spirulina, dried[1] per 100g	Chlorella, dried[2] per 100g
Energy	kcal	23	26	290	360–430
Protein	g	2.86	5.92	57.47	58–68
Lipid	g	0.39	0.39	7.72	6–16
Fiber	g	2.2	0.4	3.6	6–18
Calcium	mg	99	12	120	110–330
Iron	mg	2.71	2.79	28.5	70–200
Magnesium	mg	79	19	195	200–500
Phosphorous	mg	49	11	118	1.64
Potassium	mg	558	127	1363	600–1500
Sodium	mg	79	98	1048	57.5
Zinc	mg	0.53	0.2	2	1.93
Copper	mg	0.13	0.597	6.1	0.53
Manganese	mg	0.897	0.186	1.9	5.05
Selenium	µg	1	0.7	7.2	n. d.
Vitamin C	mg	28.1	0.9	10.1	30–150
Thiamin	mg	0.078	0.222	2.38	1–3
Riboflavin	mg	0.189	0.342	3.67	4–6
Niacin	mg	0.724	1.196	12.82	15–35
Pantothenic acid	mg	0.065	0.325	3.48	4.28
Vitamin B6	mg	0.195	0.034	0.364	1–3
Folate, total	µg	194	9	94	(2.4)[a]

[1] Values are from the United States Department of Agriculture (USDA) Nutrient Database. Spinach raw: Full report 11457, Spirulina raw: Full report 11666, Spirulina, dried: Full report 11667.
[2] Values are from 4Spectrum™ Premium Quality CHLORELLA web page (www.natural-nutrition-ltd.com/chlorella/chlorella.htm) and originally from Analysis Certificate No.103044178-002, dated May 20th, 2003, issued by Japan Food Research Lab.
[a] folic acid.

(Table 7.2: Continued)

Table 7.2: (continued)

	Unit	Spinach, fresh[1] per 100g	Spirulina, raw[1] per 100g	Spirulina, dried[1] per 100g	Chlorella, dried[2] per 100g
Vitamin A (RAE)	µg	469	3	29	19.8
Beta-Carotene	µg	5626	33	342	40–150
Vitamin E	mg	2.03	0.49	5	24.5
Vitamin K	µg	482.9	2.5	25.5	506
Fatty acids, saturated	g	0.063	0.135	2.65	n.a.
Fatty acids, monounsaturated	g	0.01	0.034	0.675	n.a.
Fatty acids, polyunsaturated	g	0.165	0.106	2.08	n.a.
Tryptophan	g	0.039	0.096	0.929	1.18
Threonine	g	0.122	0.306	2.97	2.37
Isoleucine	g	0.147	0.331	3.209	2.23
Leucine	g	0.223	0.509	4.947	5.07
Lysine	g	0.174	0.312	3.025	4.9
Methionine	g	0.053	0.118	1.149	1.3
Phenylalanine	g	0.129	0.286	2.777	2.91
Valine	g	0.161	0.362	3.512	3.23
Histidine	g	0.064	0.112	1.085	1.2

(Table 7.1). One example of *Chlorella* biomass production is its cultivation by the German company Roquette Klötze GmbH & Co. KG in a patented glass tubing system in which temperature, CO_2 content, and pH can be controlled. 500 km of glass tubing holding 700 m^3 of algal suspension are located in a greenhouse (Figure 7.3a). The company mostly produces powder and pills made from dried *C. vulgaris* cells, as well as food that includes algal extract, such as bread and pasta. From the living cells to the edible product, several steps are necessary. The algae are quite dilute and have to be centrifuged to separate them from the liquid, which is re-introduced into the facility. Then, a mild spray drying process is applied until only a residual moisture of about 3% remains (Figure 7.2b). The overall process is quite short in order to preserve the vitamins.

7.2.2 β-carotene and astaxanthin

As already mentioned above, carotenoids such as β-carotene are important for human nutrition, but they are also used in animal feed, in cosmetics or as food colorants. The species mainly used for β-carotene production is the unicellular Chlorophyceaen green alga *D. salina* (Table 7.1), which is halophilic and tolerates very high light intensities. To counteract the osmotic pressure caused by high-salt solutions, the alga synthesizes the osmoprotectant glycerol, and in response to high light intensities or macronutrient deprivation, it accumulates high amounts of β-carotene (up to 10 % of the dry biomass) to protect itself against light-induced damage (Figure 7.3b). *D. salina* is commercially exploited, for example, by Cognis Nutrition and Health / BASF, which uses parts of the natural Hutt Lagoon in Western Australia for large-scale cultivation of *D. salina* (Figure 7.3b). The microalga *H. pluvialis*, also belonging to the class of Chlorophyceae, is utilized by some companies to produce the xanthophyll astaxanthin (Table 7.1, Figure 7.3c). Astaxanthin is often applied in the aquaculture of salmon or salmon trout and is responsible for the pinkish color of their flesh. Additionally, it is supposed to be healthy for humans, mainly due to its antioxidant characteristics. *H. pluvialis*, which appears green under standard growth conditions, accumulates very high amounts of astaxanthin under stress conditions (Figure 7.3c). Companies harvesting astaxanthin from this alga usually work in a two-stage process – the cells are grown under the most favorable conditions to obtain biomass (green stage) and then stressed, mainly by nutrient deprivation, salt stress and/or higher light and temperature to induce astaxanthin production (red stage) (Figure 7.3c).

Carotenoids belong to the chemical group of isoprenoids or terpenoids. Isoprenoids are synthesized by all organisms, but they are especially versatile in plants, where they constitute both primary metabolites (i.e. compounds that are essential for the basic metabolism of each cell) and secondary metabolites (which are mostly used for the interaction of the plant with its environment). The direct precursor of carotenoids is geranylgeranyl pyrophosphate (GGPP), which, in turn, is made from isopentenyl pyrophosphate (IPP) and dimethylallyl pyrophosphate (DPP) (Figure 7.4). IPP is the first product of isoprenoid biosynthesis and can be converted to its isomer DPP by the enzyme IPP isomerase. Today, two pathways of isoprenoid biosynthesis are known, and most organisms use only one of them. The mevalonate pathway, also termed HMG-CoA reductase pathway after its rate limiting step catalyzed by 3-hydroxy-3-methyl-glutaryl- (HMG-) CoA reductase, is active in animals, fungi and archaebacteria. The non-mevalonate pathway (also termed Deoxyxylulose 5-phosphate- (DOXP-) or Methylerythritol 4-phosphate- (MEP-) pathway) is employed by most eubacteria and the malaria parasite (*Plasmodium falciparum*). Plants carry out both pathways. The mevalonate pathway is active in the cytosol and the non-mevalonate pathway operates in the plastid. In *D. salina*, only the plastidic non-mevalonate pathway seems to operate, as inferred from inhibitor studies. This pathway starts with the condensation of pyruvate and glyceraldehyde 3-phosphate to DOXP by DOXP synthase, followed by an NADPH-dependent reduction to MEP by DOXP reductase. MEP cytidyltransferase uses CTP (cytidine triphosphate) to generate 4-diphosphocytidyl-methylerythritol (CDP-ME), releasing pyrophosphate during the reaction. An ATP-dependent phosphorylation by CDP-ME

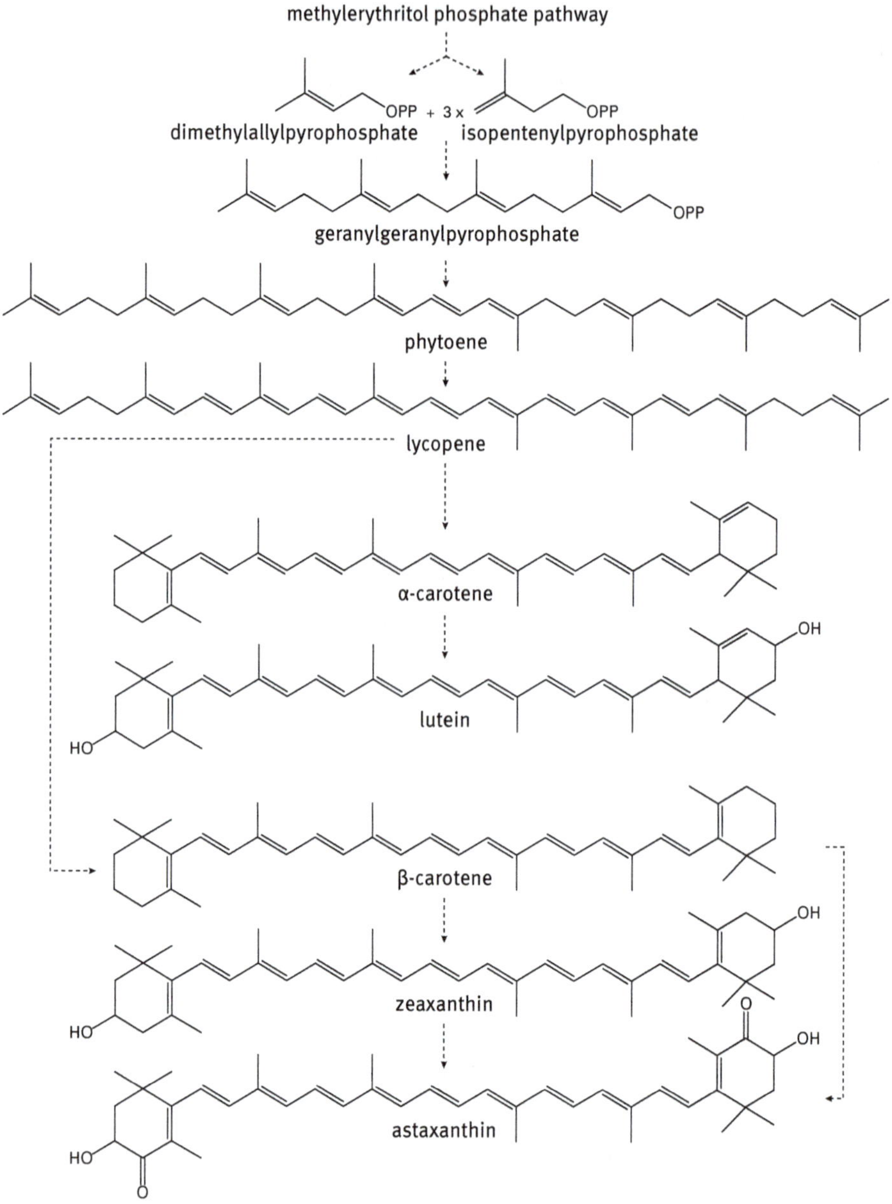

methylerythritol phosphate pathway

dimethylallylpyrophosphate + 3 x isopentenylpyrophosphate

OPP OPP

geranylgeranylpyrophosphate

phytoene

lycopene

α-carotene

lutein

β-carotene

zeaxanthin

astaxanthin

Fig. 7.4: Carotenoid biosynthesis. Only some important intermediates and end products, starting with isopentenyl pyrophosphate (IPP) and dimethylallyl pyrophosphate (DPP), are shown. Details are explained in the text.

kinase results in the production of CDP-ME phosphate, which is cyclized to Methylerythritol 2,4-cyclopyrophosphate (MEcPP) by MEcPP synthase, releasing CMP. MEcPP is reduced and linearized to Hydroxymethyl-but-2-enyl pyrophosphate

(HMB-PP) by HMB-PP synthase, which requires two electrons delivered by two reduced ferredoxins for this reaction. Finally, HMB-PP is converted to IPP or DPP by HMB-PP reductase, using NADPH as an electron donor, and usually generating a $5:1$ mixture of IPP and DPP. One DPP and three IPP are used to generate GGPP (Figure 7.4). First, each one DPP and IPP are condensed to geranyl pyrophosphate (GPP) by dimethylallyl*trans* transferase. (2E,6E)-farnesyl diphosphate (FPP) synthase adds another IPP to GPP, generating FPP, and finally, a third IPP is added by geranylgeranylpyrophosphate synthase, forming the C_{20} molecule GGPP.

The first committed step of carotenoid formation is the condensation of two GGPP to the C_{40} compound phytoene by phytoene synthase (Figure 7.4). Phytoene is colorless – the typical yellow, orange, red or pink colors of carotenoids do not appear before conjugated double-bonds are introduced by sequential desaturation steps. In plants, phytoene (nine double-bonds) is first desaturated to ζ- (zeta-) carotene (eleven double bonds) by phytoene desaturase in two steps via phytofluene. Here, 9,15,9′-tri-*cis*-ζ-carotene is formed, which has to be isomerized (by light or enzymatically by 15-cis-ζ-carotene isomerase) to 9,9′-di-*cis*-ζ-carotene before it can be desaturated by ζ-carotene desaturase to 7,9,7′,9′-tetra *cis*-lycopene or pro-lycopene (13 double bonds), again by two desaturation reactions via neurosporene. Both phytoene and ζ-carotene desaturase are membrane associated and require oxidized plastoquinone as electron acceptor. Pro-lycopene is finally isomerized by carotenoid isomerase to all-*trans* lycopene (Figure 7.4), the red carotenoid that gives tomatoes and rose hips their characteristic color. A branching point in carotenoid biosynthesis is the cyclization of the ends of lycopene. One branch that, via α-carotene, finally results in lutein production includes carotenoids that have one β- and one ε-ring, which differ in the position of the double-bond (Figure 7.4). The other branch leads to the synthesis of β-carotene, which has two β-rings. β-carotene, in turn, is the precursor of the xanthophylls zeaxanthin, violaxanthin, and astaxanthin (Figure 7.4).

For the "β,ε-branch", one end of lycopene is first cyclized by lycopene ε-cyclase to δ-carotene, whose linear end is cyclized by lycopene β-cyclase, forming α-carotene. β- and ε-ring hydroxylases finally convert this carotene into the xanthophyll lutein that has a hydroxyl-group at each ring (Figure 7.4). In the "β,β-branch", the two β-rings of β-carotene are also formed by lycopene β-cyclase, and β-carotene hydroxylase hydroxylates the rings to form the β,β-xanthophyll zeaxanthin. Zeaxanthin is an important carotenoid involved in non-photochemical quenching of excess light energy, because it can absorb light and release its energy as heat. Zeaxanthin formation is, furthermore, involved in spatial re-arrangements of the light harvesting antennae, also resulting in non-photochemical quenching.

Astaxanthin is not commonly produced in higher plants, but by many microalgae and cyanobacteria, as well as other bacteria and fungi. In microalgae, the β,β-ketocarotenoid is generated from β-carotene or zeaxanthin. It is proposed that two routes exist that are both catalyzed, though in a different order of reactions, by β-carotene oxygenase. This enzyme, also termed β-carotene ketolase, introduces the keto-groups to each ring, while β-ring hydroxylase forms the hydroxyl groups as already described above.

7.2.3 Polyunsaturated fatty acids (PUFAs)

Fatty acids (FAs) are carboxylic acids with long aliphatic chains that can be unsaturated or saturated, meaning that they have double-bonds between C-atoms or not (Figure 7.5). In the cell, FAs are major components of the lipids that constitute the plasma membrane. In these lipids, two FAs are usually esterified with a glycerol molecule, whose third hydroxyl group carries a phosphate that, in turn, is esterified to different groups such as choline (forming phosphatidylcholine) or inositol (forming phosphatidylinositol). FAs can also be part of triacylglycerols (TAGs), in which the three alcohol groups of the glycerol backbone are esterified with FAs. TAGs, also termed neutral lipids or oil, constitute a form of storage for FAs, but can also serve re-structuring purposes. FAs are energy rich and can be oxidized by the cell to generate ATP, and humans nowadays use FAs derived from plant oil for the production of biodiesel (see 7.3).

In humans, several FAs, especially unsaturated and polyunsaturated ones, have important neurological and immunological functions. Eicosanoids such as prostaglandin, which derive from oxidation of long-chain (C_{20}) PUFAs (eicosapentaenoic acid (EPA), arachidonic acid (AA) or dihomo-γ-linolenic acid (DGLA)) (Figure 7.5) play roles in inflammatory processes, in the immune system, and also in the central nervous system. The PUFA docosahexaenoic acid (DHA) (a C_{22} fatty acid that is biosynthesized from EPA) (Figure 7.5) is an important constituent of the cell membranes in various human organs, for example in the brain, where it is involved in modulating transport activities. The human body can synthesize these PUFAs only when provided with omega- (ω-) 6 linoleic acid (LA) (a C_{18} fatty acid with two double-bonds), and ω-3 linolenic acid (a C_{18} fatty acid with three double-bonds) (Figure 7.5). In the FA nomenclature, the ω-C-atom is the last atom of the carbon chain, and ω-3 or ω-6 indicate that the first double-bond is introduced after the third or sixth C-atom, respectively, counted from the end. Humans cannot insert double-bonds after the ninth C-atom of a FA (counted from the carboxyl-end) and, therefore, depend on the nutritional uptake of such unsaturated FAs. Additionally, the supplementation of the diet with EPA and DHA seems to have beneficiary effects on health. Often, PUFAs in capsules are derived from oceanic fish, which can synthesize these important fatty acids, but which also accumulate them due to their diet. However, ultimately, the PUFAs in seafood derive from microalgae, and several species of eukaryotic microalgae are already exploited to produce PUFAs for human nutritional supplements (Table 7.1).

In plants and algae, FA biosynthesis takes place in the plastid and is very similar to FA generation in bacteria (Figure 7.5). It starts with the rate-limiting ATP- and

ATP + CO$_2$ ADP+P$_i$

acetylCoA

malonylCoA

ACP

CoA

acetyl-ACP

ACP, CO$_2$

malonyl-ACP

ACP

x n

R

acyl-ACP

S-ACP

acetoacetyl-ACP

buturyl-ACP

S-ACP

NADPH

NADP$^+$

H$_2$O

NADP$^+$

NADPH

O$_2$ + 2e$^-$

2 H$_2$O

OH

β-hydroxybuturyl-ACP

crotonyl-ACP

R

OH

(a)

mono-unsaturated fatty acid

ω6   Δ9

ω1

Δ1   OH

linoleic acid (ω-6 unsaturated fatty acid)

ω3

α-linolenic acid (ω-3 unsaturated fatty acid)

OH

eicosapentaenoic acid (EPA)

OH

arachidonic acid (AA)

OH

dihomo-γ-linolenic acid (DGLA)

OH

(b)

docosahexaenoic acid (DHA)

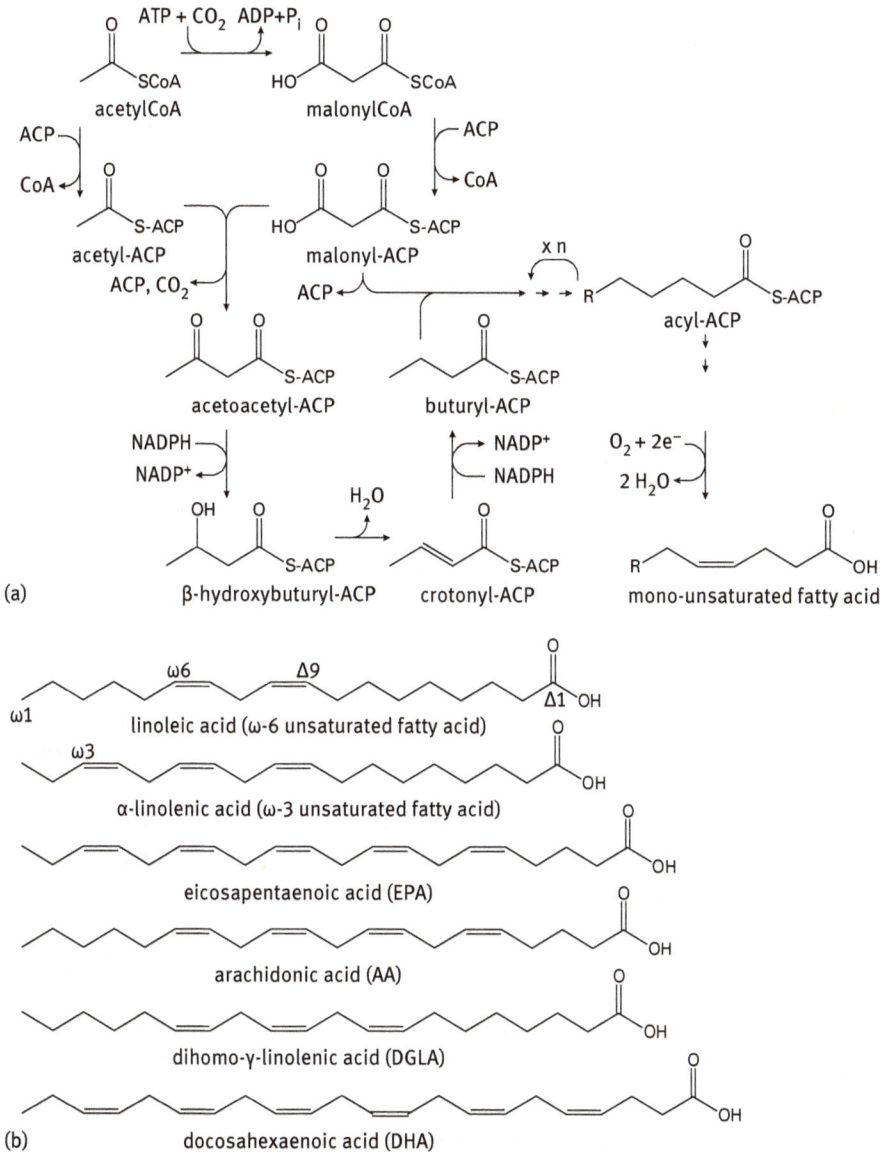

Fig. 7.5: Fatty acid biosynthesis (a) and some important polyunsaturated fatty acids (b). (a) In plants and algae, fatty acids are produced from acetylCoA and CO$_2$ in the plastid. Desaturation occurs both in the plastid and the endoplasmic reticulum by O$_2$-dependent desaturases. (b) The human body can synthesize the PUFAs EPA, AA, DGLA and DHA only when provided with linoleic and linolenic acid, which have to be taken up by the diet.

biotin-dependent carboxylation of acetylCoA to malonylCoA by the enzyme complex acetyl-CoA carboxylase (ACCase). Plastid ACCase is a tetrameric bacterial-type enzyme consisting of biotin carboxylase, biotin carboxyl carrier protein, α-carboxyltransferase, and β-carboxyltransferase. First, bicarbonate (HCO_3^-) is converted to carbonylphosphate, which then carboxylates the biotin co-factor. This "activated" carboxyl-group is transferred to acetylCoA, forming malonylCoA (Figure 7.5). MalonylCoA is passed over to the SH-group of the phosphopantetheine co-factor of acyl carrier protein (ACP) by malonylCoA-ACP transferase. The synthesis of FAs continues with the multimeric enzyme complex fatty acid synthase (FAS). FAS catalyzes the sequential addition of malonyl-units to an acyl-ACP, and because one CO_2 molecule is released during this reaction, two carbons are added in each round. The first step is the condensation of malonyl-ACP and acetyl-/acyl-ACP by β-ketoacyl-ACP synthase accompanied by the release of CO_2 (Figure 7.5). The 3-keto group of the condensation product is reduced by β-ketoacyl-ACP reductase to a 3-hydroxy acyl species (3-hydroxy butyrate in the first reaction with acetyl-ACP). From the 3-hydroxy acyl species, a water molecule is eliminated by β-ketoacyl-ACP dehydratase, and finally, the enoyl-group that resulted during water elimination is reduced by enoyl-ACP reductase. This sequence of reactions results in the addition of two fully saturated (methylene-) carbons per round, and continues until saturated C_{14} to C_{18} acyl-chains are present and released from the FAS complex by acyl-ACP thioesterases (Figure 7.5).

Longer fatty acids are synthesized from a different set of enzyme complexes termed elongases, that utilize acyl-CoA forms of existing fatty acids and add 2-carbon units derived from MalonylCoA. Different elongase systems specific for a certain fatty acid chain length, and number and position of double bonds, exist in most organisms studied to date. The enzymatic steps are similar to those of FAS and involve β-ketoacyl CoA synthase, β-ketoacyl CoA reductase, β-hydroxyacyl CoA dehydratase, and trans-2-enoyl CoA reductase. However, instead of ACP, CoA is usually the FA-carrier. Depending on the FA species that is to be generated, elongation steps often alternate with desaturation steps.

Desaturation of FAs is mostly catalyzed by non-heme iron containing desaturases that can be soluble or membrane-bound and that require molecular oxygen as a co-substrate (aerobic desaturation) and two electrons per introduced double-bond (Figure 7.5). Depending on the desaturase, FAs esterified to ACP, CoA or lipids are used as substrates. In the plastid of plants and algae, ferredoxin is used as the electron donor, so that, in the light, photosynthesis is the source of low-potential electrons. FA desaturation can also occur in the endoplasmic reticulum, where cytochrome *b*, itself reduced by NADH via a flavoprotein, is the electron donor. Some bacteria such as *Escherichia coli* can also anaerobically introduce double bonds. In *E. coli*, this is carried out by a FAS system that has an additional *trans*-2-*cis*-3 isomerase activity so that the double-bond of the enoyl-intermediate is isomerized to a C-C double bond by a dehydrase-isomerase mechanism. Some marine prokaryotes (such as *Shewanella* sp.) and eukaryotes (e.g. *Schizochytrium*) use polyketide synthases (PKS) for the anaerobic biosynthesis of PUFAs. PKS are multimeric enzyme complexes that have similar enzyme activities to FAS (also see Chapter 4 for more details on and products of PKS). However, PKS carry out side-reactions such as the dehydrase-isomerase reaction of *E. coli* FAS described above, or do not complete the full circle that adds two reduced carbons to a precursor as is the case in FAS. Therefore, PKS can synthesize a whole palette of products, PUFAs among them.

To date, mostly PUFAs from heterotrophic dinoflagellates and protists (*Crypthecodinium, Schizochytrium*) are on the market, but several photosynthetic microalgae are being developed for large-scale processes, too. For example, the photosynthetic diatom *Phaeodactylum tricornutum* or the Eustigmatophyceaen species *Nannochloropis* have high contents of EPA and other valuable PUFAs, and the Fraunhofer Institute for Interfacial Engineering and Biotechnology grows *P. tricornutum* in the FPA reactors mentioned in section 7.1.3 (Table 7.1).

7.3 Potential future applications

As mentioned before, microalgae are regarded as promising biofuel producers. Probably the most advanced sector is biofuel generation from algal oil (see 7.3.1), which is being established by several companies, but which is in most cases not profitable yet. Another route with prospects is the photosynthetic generation of molecular hydrogen, a natural trait of some microalgal species.

7.3.1 Triacylglycerols – Biodiesel

Plant or animal lipids are the basis for the so called biodiesel, which is made from triacylglycerols by transesterification with primary alcohols, mostly methanol or ethanol. In Germany, so-termed B7 diesel may contain up to 7 volume-% of biodiesel. Production and usage of biodiesel is controversial. Biodiesel made from plant seed oil is based on renewable resources and utilizes the ubiquitous energy of the sun. It is more bio-degradable than petrodiesel and, upon its combustion, releases less particulate matter and sulfur dioxide. Its production has few waste products, as long as the remaining plant matter is used, for example, as fertilizer or feed, and the glycerol accrued during the transesterification is exploited by the chemical or cosmetics industry. On the other hand, one should not neglect that energy input is needed, too. For example, plants have to be fertilized with nitrate and other nutrients. Industrial ammonia production by the Haber-Bosch process is very energy-intensive and ultimately depends on natural gas. Phosphorous is a limited natural resource. Also, the transesterification needs primary alcohols and energy input in form of heat, and combustion of biodiesel releases more nitrogen oxides. Not least, land that is used for growth of fuel plants competes with land for food and feed plants or has to be reclaimed from nature. Biodiesel made from microalgal

TAGs might alleviate some of the cons of plant-derived diesel, if, for example, the algae are grown on non-arable land or in vertical PBRs and if water and nutrients are recycled.

Triacylglycerols (TAGs) are normal constituents of cells and form energy and carbon reservoirs, but they can also serve as storage for fatty acids used for the biosynthesis of membrane lipids. The *de novo* synthesis of TAGs requires glycerol-3-phosphate and acylCoA and starts with the acylation of glycerol-3-phosphate by acyl transferases to form lysophosphatidic acid. Esterification with a second acyl chain results in phosphatidic acid, which has to be dephosphorylated to diacyl-glycerol (DAG) before the third acyl chain can be attached by an acylCoA-dependent DAG acyltransferase (DAGAT). The third acyl chain of TAGs can also derive from membrane lipids via transesterification, catalyzed by phospholipid diacyl-glycerol acyltransferases (PDATs). TAGs are stored in lipid bodies, which, in plants and microalgae, can be plastidic or extra-plastidic. Lipid bodies are dynamic organelles which not only contain TAGs, but also TAG-, phospholipid- and FA-metabolizing enzymes, as well as structural proteins such as oleosins. It is known from several species of microalgae that they accumulate TAGs in response to stress, often under nutrient- and especially nitrogen starvation. The marine microalga *Nannochloropsis* is discussed to be a promising producer of TAGs, as oil can constitute up to 30 % of its dry weight under standard growth conditions, and up to about 60 % upon nitrogen starvation. The unicellular Chlorophycean green alga *Chlamydomonas reinhardtii*, too, accumulates substantial amounts of TAGs when deprived of nitrogen and sulfur.

7.3.2 Photobiological generation of molecular hydrogen (H_2)

Molecular hydrogen (H_2) is an energy-rich gas with a heating value of about 140 MJ × kg^{-1} compared to about 40 MJ × kg^{-1} for gasoline (though one has to consider that H_2 is a gas with a density of only about 0,09 g × l^{-1}, while that of gasoline is about 750 g × l^{-1}). H_2 is regarded as one potential future energy carrier and can be used either as a fuel, i.e. for combustion, or for operating fuel cells, thereby generating electricity. When just regarding the reaction by which H_2 releases its energy, mostly the reaction with O_2, H_2 is also a very "clean" energy carrier with only water being the "exhaust". There are numerous further applications for H_2, which today are operated by H_2 mostly extracted energy-intensively from fossil fuels. For example, the Haber-Bosch process depends on H_2, but also many chemical syntheses.

Though still in its infancy regarding industrial applicability, a way to renewably produce H_2 might be the utilization of microalgae, because some species possess enzymes that can generate molecular hydrogen. These so-called hydrogenases can be found in all kingdoms of life and are mainly classified as [NiFe]- and [FeFe]-

hydrogenases based on the metal (nickel and/or iron) content of their active sites. In unicellular eukaryotic microalgae, only [FeFe]-hydrogenases have been reported to date. Only few species have been analyzed regarding their capability of producing H_2 light-dependently. However, the H_2 metabolism of *C. reinhardtii* has been intensely studied and is quite well understood. *C. reinhardtii* possesses two [FeFe]-hydrogenases, HYDA1 and HYDA2, with HYDA1 seeming to be the isoform that is responsible for most of the H_2 generation in the light. HYDA1 is located in the chloroplast and accepts electrons from reduced ferredoxin to reduce protons (H^+) to H_2. Ferredoxin, as noted above, is the major electron acceptor of PSI, and under certain conditions, *C. reinhardtii* produces H_2 using electrons provided light-dependently by the photosynthetic electron transport chain (Figure 7.1). In principle therefore, the alga and its chloroplast, respectively, combine solar-driven electrolysis (water-oxidation at PSII) with H_2 production. However, microalgae do not produce H_2 under standard growth conditions, but only in stress situations that make alternative electron sinks necessary. Furthermore, [FeFe]-hydrogenases are very sensitive towards molecular oxygen and cannot operate under conditions in which PSII is highly active and a lot of O_2 is evolved, respectively. To date, the best environmental condition to trigger a relatively high H_2 producing activity in *C. reinhardtii* that lasts for several days is sulfur- (S-) deprivation (Hemschemeier et al., 2009). As noted above, nutrient deficiencies usually result in decreased cell division processes, making alternative electron sinks necessary. However, the absence of S seems to have effects on the cells that are especially favorable for H_2 photoproduction. Probably the main reason for this is that besides the Calvin cycle and further assimilatory processes, PSII is a main target of S limitation. PSII O_2 evolution activity decreases to less than 10 % after two to three days of S-deficiency, while respiratory O_2 uptake rates stay rather constant. In sealed algal cultures, O_2 is thus consumed after a few days, resulting in hypoxic conditions and allowing the activity of the O_2 sensitive HYDA1 enzyme. In the first days of S deprivation, *C. reinhardtii* accumulates starch and TAGs. Besides residual PSII activity, the oxidative degradation of starch seems to be a major electron source for H_2 production in the later stages. Reducing equivalents generated during starch and glucose oxidation are re-oxidized by a thylakoid NAD(P)H-plastoquinone-oxidoreductase that transfers the electrons to the plastoquinone- (PQ-) pool in a process termed non-photochemical PQ reduction (Figure 7.1). From PQ, electrons are transported via the cytochrome $b_6 f$ complex and PSI to ferredoxin. The effect of S deficiency on the activity of this "post-PSII" part of the electron transport chain is weaker and delayed compared to the strong effects on PSII, which might explain its usefulness for H_2 production. In contrast, nitrogen- (N-) deprivation has a major effect on the cytochrome $b_6 f$ complex, which is probably the reason why N-deficient *C. reinhardtii* cells produce much less H_2 than S-deprived cultures.

To date, it is not predictable if and when microalgal photoproduction of H_2 will become a future application. It has been calculated that the process has to become at least 100-fold more efficient to come into the range of cost-effectiveness. Also, the pros and cons of algal biotechnology mentioned above stay true for this process. An additional disadvantage is that H_2 is a very volatile (and potentially explosive) gas, so that PBRs have to be especially tight. On the other hand, the light-conversion efficiency can be better when compared to photosynthetic lipid or starch production, because H_2 is a direct product of the photosynthetic electron transport chain.

7.4 Genetic modification

Nowadays, genetic engineering is a common tool used by scientists to analyze genes and their roles in an organism, respectively, but also to specifically alter or introduce a trait with the aim of creating organisms that are optimized for a certain application. The earliest form of "genetic engineering" was breeding, but with the development of molecular biological techniques, the gene content of many organisms can be purposefully changed. In plant-based biotechnological applications, genetic engineering is commonly employed to optimize plants for cultivation purposes, such as making them resistant to a certain herbicide or to a plant pathogen. A further rapidly developing area of genetic manipulation is so-called pathway engineering, during which metabolic pathways that result in a certain desirable product are altered in a way that the product is synthesized in higher amounts or with altered characteristics. Finally, genes from other species are introduced to let plants or algae produce entirely xenogenic compounds including peptides.

To date, most basic research on microalgal metabolism and genetics is carried out with the unicellular green alga *C. reinhardtii*. After decades of intensive research, this alga has the status of a well understood model organism used in many research fields studying, for example, photosynthesis, nutrient assimilation, circadian and diurnal rhythms, and flagellar function (a comprehensive overview on *C. reinhardtii* metabolism and methods is given in the Chlamydomonas sourcebook, see "further reading"). *C. reinhardtii* has also gained interest to be employed in biofuel production, because it can evolve molecular hydrogen (H_2) (see above) and accumulates significant amounts of starch and TAGs under nutritional stress conditions. In recent years, the microalga has attracted attention regarding the heterologous production of recombinant proteins such as human therapeutic proteins, in the chloroplast. The three genomes (nuclear, chloroplast and mitochondrial) of *C. reinhardtii* have been fully sequenced and are publicly available. On this genomic basis, all kinds of "omics" (Box 7.1) are well established for the organism, and many "omic"-studies of *C. reinhardtii* subjected to various environmental conditions are available. This flood of data has in turn resulted in the discovery of new pathways or candidates that are important for a certain metabolism.

Box 7.1: Omics.

The suffix "-omics" (often used as a proper noun, "Omics") describes technologies of the life sciences which aim to characterize the full set of a certain compound in a cell or an organism. For example, genomics analyze the full genome (instead of examining individual genes), transcriptomics and proteomics aim to capture and analyze all transcripts and proteins, respectively, present in a cell at a given moment or condition. Further common omics are metabolomics which aim to categorize every metabolite, or lipidomics, which examine the whole set of lipids and fatty acids. By providing all-encompassing information, omics are an important tool for systems biology, which aims to understand a cell or an organism as a whole. Systems biology is also important for biotechnological approaches, because in many cases, it is not sufficient to alter a single catalytic function in order to optimize a certain desirable trait. Omics that characterize nucleic acids have made substantial progress due to the development of *second* and *third generation sequencing* technologies, that, often by imaging methods, and by modifications of the original DNA-sequencing technologies, allow to sequence millions of sequences simultaneously and to capture even very low-abundant sequences. Other omics are not that (relatively) easy. For example, because proteins are so versatile and sometimes membrane bound, usually not the whole proteome of a cell can be captured. The same is true for metabolites, which often are unstable or have to be chemically modified before analyses that have to be conducted by various chromatographic techniques. Most omics need bioinformatics that evaluate the huge data-sets.

7.4.1 Pathway engineering

In order to study or alter known and novel pathways of interest, genetic engineering is a valuable technique, and many genetic tools have been developed for *C. reinhardtii*. One drawback is that as in most higher organisms, homologous recombination does not efficiently take place in the nuclear genome of the alga. Therefore, it is nearly impossible to specifically knock out a certain nuclear gene in order to analyze its function or for pathway engineering. Instead, foreign DNA inserts randomly (also termed ectopically) into the ca. 111 mega base pairs large nuclear genome of *C. reinhardtii*, often disrupting structural genes in this process. This random mutagenesis is quite helpful when searching for entirely new gene functions, as long as a certain phenotype can be specifically addressed. For example, genes that are essential for the algae to grow photoautotrophically were often found by transforming *C. reinhardtii* with a selectable marker gene and screening the resulting transformants for photoautotrophic growth. In this regard, *C. reinhardtii* has been an especially valuable organism to study essential photosynthetic genes, because the alga can grow heterotrophically using acetate as a carbon and energy source. The screening systems therefore consisted of replica agar plates of mutant library colonies which were cultivated in parallel in the presence or absence of acetate.

In order to let only algal cells grow that have integrated the piece of DNA into their genome, several selection markers are available for *C. reinhardtii*. In the beginnings, selection was often carried out using mutant strains that are auxotrophic for a certain compound due to a non-functional gene. These mutants were then transformed with the respective intact gene and transformants selected based on their restored prototrophy. The most-used system is the transformation of a strain that is L-arginine-auxotrophic, due to a mutation of the *ARG7* gene encoding argininosuccinate lyase, with a wild type copy of the *ARG7* gene. The cell suspension used for transformation is then plated onto agar-plates containing no arginine, so that only transformants whose arginine-prototrophy was restored by nuclear integration of the wild type *ARG7* gene can grow. Further selection markers commonly used for *C. reinhardtii* nuclear transformation are antibiotic resistance cassettes such as aminoglycoside phosphotransferase genes from *Streptomyces rimosus* (gene *aphVIII*), conferring resistance to paromomycin, neomycin, and kanamycin, or *Streptomyces hygroscopicus* (*aph7″*), allowing *C. reinhardtii* transformants to grow on hygromycin.

To specifically address a certain gene in *C. reinhardtii*, the natural process of ribonucleic acid (RNA) interference (RNAi) has been employed successfully for several years. RNAi in general describes cellular phenomena during which RNA, mostly double-stranded RNA, interferes with and usually inhibits gene expression. Though still under intensive research and exhibiting variations of the individual steps and the participating nucleic acids and proteins, RNAi can be briefly described to occur as follows: Double-stranded RNA is first processed into small fragments of about 20 nucleotides by special ribonucleases (such as "Dicer"). These fragments are then recognized by the RNA-induced silencing complex (RISC) and unwound. Subsequently, one strand binds to a complementary sequence on an mRNA molecule which is then cleaved. Alternatively, translation of the mRNA is blocked. Because this results in a decreased pool of specific mRNAs and/or the encoded protein, the process is also described as post-transcriptional gene silencing. Naturally, RNAi is used by organisms to defend themselves against viruses or transposons, but also for gene regulation. The natural mechanism has been adopted by scientists to interfere with the expression of a certain target gene whose function they aim to analyze. In *C. reinhardtii*, several approaches to generate gene "knock-downs" have been successfully followed. Introduction of engineered gene cassettes from which RNA in *antisense* direction to the target mRNA is transcribed were reported to work, as well as constructs in which parts of the target gene are cloned in an inverted-repeat fashion, resulting in a transcript that forms a hairpin structure. However, *C. reinhardtii* is not a passive recipient of foreign and manipulating DNA and tends to silence inserted constructs, most probably by methylation and making the respective genetic region inaccessible to the transcription machinery. It is still unclear how this "defense mechanism" of *C. reinhardtii* works, but its activity has been recognized by many work groups who reported that, for example, a successful gene knock-down phenotype disappeared after several generations. One method to force the alga to express a certain RNAi-construct and to simultaneously allow picking efficient knockdown phenotypes is the so-called tandem-RNAi, in which the inverted repeats of the target gene are fused to inverted repeats of a selection marker gene. This marker gene encodes tryptophan synthase, an essential enzyme for the biosynthesis of the amino acid tryptophan. Tryptophan synthase, however, also converts 5-fluoroindole to the toxic 5-fluorotryptophan. Thus, transformants in which the tryptophan synthase encoding gene is efficiently suppressed grow on 5-fluoroindole (as long as supplied with tryptophan), while those with normal expression levels die. As the sequence of the gene of interest is included in the same inverted repeat construct, the probability that it is simultaneously down-regulated is high. However, this proceeding includes working with potentially toxic compounds, and the tryptophan auxotrophy of the transformants may have undesired side-effects. A quite stable and efficient RNA interference method was developed later and makes use of a natural microRNA from *C. reinhardtii*. MicroRNAs are encoded in the genome and are first transcribed into non-coding so-called primary microRNAs, then processed into precursor microRNAs that form stable hairpins. This precursor microRNA is cleaved by Dicer and the resulting small double-stranded RNA molecules interfere with their complementary mRNA molecules as described above. To use this natural phenomenon for targeted gene knock-down, a natural microRNA-gene of *C. reinhardtii* was introduced into a *C. reinhardtii* expression vector that allows exchanging the target-specific part of the natural microRNA with a short 21 base pair fragment complementary to the research-specific target gene. Studies using this proceeding reported a good efficiency and stability of the knock-down.

Nevertheless, RNAi seldom results in a complete loss of gene function and may have unwanted side-effects, either by a reaction of the cells to the unnatural microRNA, or by the integration of the miRNA-encoding DNA into another gene. Therefore, clean and stable knock-outs of individual genes are usually favored. Very recently, high-throughput approaches to identify nuclear mutants of specific genes were developed. These techniques are still dependent on random insertional mutagenesis, which might be biased as some genetic regions may be more accessible to the integration of foreign DNA as others. Still, many desired mutants were identified by this proceeding. In its simpler version, about one hundred thousand individual *C. reinhardtii* mutant colonies are created by transformation with an antibiotic resistance cassette. 96 of the colonies are combined and their collective genomic DNA isolated. These DNA-pools are again pooled in groups of ten, creating DNA-"superpools". Using these superpools as templates, polymerase chain reaction (PCR) is carried out with oligonucleotide primers being complementary to the antibiotic resistant cassette sequence and the gene of interest. As mentioned above, the genome of *C. reinhardtii* is available on public internet sites, so that researchers can search for genes that might play an important role in the pathway to be examined or engineered. From these genes, as well as the selection marker, oligonucleotide primers are deduced for both directions. For large genes, several primers are deduced to bind circa every kilo base pair. Finally, individual PCR reactions are performed. Here, the total number of PCR-reactions can be reduced by applying so-called Multiplex-PCR reactions, in which several primers are used simultaneously. Still, in view of the number of annotated gene loci in the *C. reinhardtii* genome (17,741 loci as of August 2015) and the necessity to screen a large excess of mutant colonies to be able to isolate a knock-out of one target gene, this proceeding involves a huge input of labor and time. Here, the technical progress in high-throughput robotic systems, as well as in nucleic acid sequencing allowed the design of new strategies. The generation of mutant libraries remains the same, but both the creation and maintenance of the mutant library with its one hundred thousand individual algae colonies and the isolation of DNA can be performed by a robotic system. Then, as soon as DNA has been isolated, the genomes are sequenced completely and analyzed by bioinformatics approaches to identify and index the genes that are deleted. Thus, indexed libraries of *C. reinhardtii* mutants can be created and made available for scientific or applied purposes.

7.4.2 Expression of transgenes

With new molecularbiological techniques at hand, microalgae are increasingly considered to be valuable platforms also for heterologous products. These may be chemicals to be used in industrial syntheses or as fuels, but also recombinant proteins for chemical or medical purposes. Several organisms have long been used to produce such substances, including bacteria such as *Escherichia coli* or *Bacillus* sp., fungi (*Saccharomyces cerevisiae, Hansenula polymorpha, Pichia pastoris* and others), as well as insect and mammalian cell cultures. Especially for the generation of human proteins, eukaryotes are usually better suited than bacteria, because they produce correctly folded proteins with disulfide bridges and post-translational modifications such as glycosylation. Plants and algae are self-sufficient regarding their energy metabolism, offering an environmentally friendly way of manufacturing specific products and not seldom the direct conversion of photosynthetically fixed carbon into valuable molecules. For economic reasons, this may also be a cost advantage given optimal cultivation facilities, because less energy has to be invested to grow and sustain the organisms. Not least, the chloroplasts of plants and algae allow a high level of gene expression and protein synthesis, respectively, due to the inherent capacity of plastids to express photosynthetic genes at a high level. Tobacco plants, for example, have been used to generate many therapeutic proteins such as Insulin like growth factor-1 or vaccines, which were shown to be functional and exhibit immunogenicity, respectively. However, because of the advantages of microalgae already described above, these are increasingly developed as "factories" for recombinant products, too (Specht et al., 2010). Especially concerning therapeutic products, the feasibility of cultivating algal suspensions in confined and controllable containments is an advantage, because contamination with further organisms or air and water pollutants can be prevented. Using well-examined organisms such as *C. reinhardtii,* that are classified as GRAS (Generally Recognized As Safe by the US American Food and Drug Administration) and that do not produce potentially toxic secondary metabolites, like plants of the genus *Solanaceae* (e.g., tobacco), is of further advantage, because in some cases, the respective pharmaceutical could even be administered orally through whole cell extracts. In this developing field, once again, *C. reinhardtii* is employed, mostly owing to the many tools for nuclear and chloroplast transgene expression already developed and a quite good understanding of its physiology.

For the efficient generation of a heterologous product, the respective transgene has to be expressed properly. As mentioned above, *C. reinhardtii*, like many other eukaryotes, is somewhat recalcitrant to express foreign genes, or even its own genes when provided as copy-DNA, i.e. as intron-less genetic information. Here, basic research has, sometimes by chance, discovered several genetic elements that help to circumvent these issues (Figure 7.6). The most common tools used in *C. reinhardtii* expression cassettes for the nuclear genome are the promoter of the gene *HSP70A* encoding the heat shock protein 70A, the 5'- and 3'-flanking regions of the gene for the small subunit of ribulose-1,5-bisphosphate carboxylase/oxygenase (Rubisco) (*RBCS2* gene) as well as introns of the *RBCS2* gene, and the 5'- and 3'-flanking regions of the *PSAD* gene that codes for the PSI reaction center subunit II (Figure 7.6a). Though various promoters have been used for specific purposes, such as the promoter of the highly and constitutively expressed beta-tubulin 2 gene (*TUB2*), or inducible promoters (for example from the ammonium-repressible and nitrate-inducible *NIT1* gene encoding nitrate reductase), the *HSP70A*-, *RBCS2*- and *PSAD* elements are mostly included in these specific expression cassettes, too, because they confer some global regulatory effects that enhance transgene expression in *C. reinhardtii*. In case of the *HSP70A* promoter, the molecular basis of its beneficiary influence has been elucidated to a large degree and is related to an opening of the chromatin structure by histone modifications and the action of the transcription factor that activates the promoter, heat shock factor 1 (HSF1). In case of *RBCS2*, though a combination of the *RBCS2* promoter with the *HSP70A* promoter is commonly used and seems to be especially effective (Figure 7.6a), it is predominantly the *RBCS2* introns that confer positive effects on transgenes. Most often, only the first intron is used, and it has been shown that it can be cloned in both directions as well as up- or downstream of the promoter or within the cDNA to be expressed. Two copies of this intron, as well as using all the three *RBCS2* introns in their physiological order were reported to have even more enhancing effects. Obviously, there are regulatory and expression stimulating elements within these introns. Likewise, it is not clear how the flanking regions of the *PSAD* gene help a cDNA to be expressed efficiently in *C. reinhardtii*. However, as the *PSAD* gene itself is intron-less, an uncommon feature for an eukaryotic gene, it seems plausible that some regulatory elements are located within the *PSAD* promoter and/or the 5'- or 3'-untranslated regions (UTRs).

expression cassettes for nuclear gene expression

| pHSP70A | pRBCS2 + RBCS2 5' | RBCS2 1st intron | gene of interest | RBCS2 3' |

| pPSAD + PSAD 5' | | gene of interest | PSAD 3' | gDNA 3' of PSAD |

expression cassettes for plastid gene expression

| rbcL 5' | gene of interest | rbcL 3' |

(a) | atpA 5' | gene of interest | rbcL 3' |

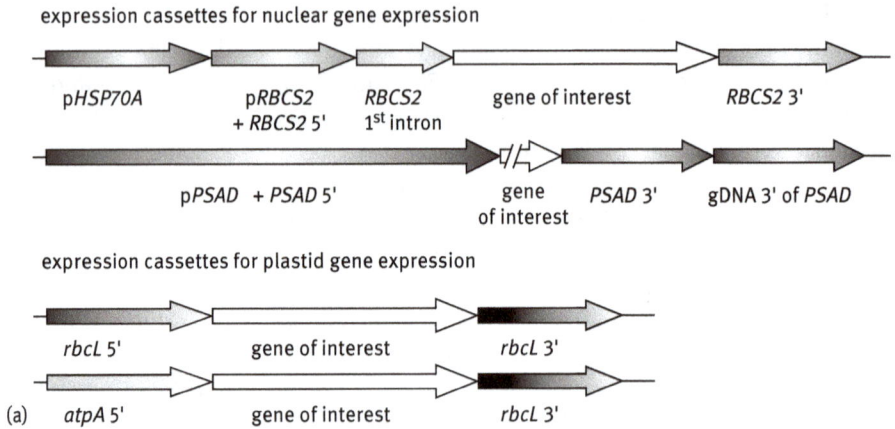

Codon usage and codon optimization

    E. coli (GC content coding regions 52.35%)

CTG	ACC	ATC	CG**T**	GT**A**	TC**T**	GGC	TAC	GCA	GT**A**
L	**T**	**I**	**R**	**V**	**S**	**G**	**Y**	**A**	**V**
CTC	AC**C**	AT**C**	CG**C**	GT**G**	TC**C**	GGG	TAC	GC**T**	GT**G**

C. reinhardtii (GC content coding regions 66.30%)

C. reinhardtii partial CDS versus *E. coli* codon usage

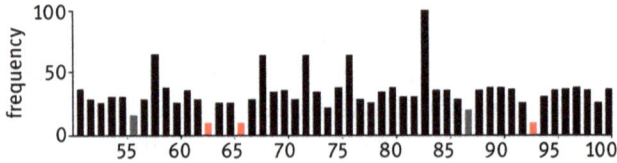

codon-optimized *C. reinhardtii* partial CDS versus *E. coli* codon usage

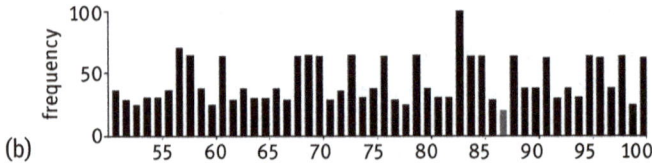

(b)

Fig. 7.6: Examples for expression cassettes used to (heterologously) express cDNAs in *C. reinhardtii* (a) and for codon bias and codon optimization (b). (a) Transgenes can be placed under the control of regulatory elements of other genes to enhance their expression in the *C. reinhardtii* nucleus or plastid. For the nucleus, promoters (p) and 5'- or 3'- untranslated regions (indicated as 5' or 3') of the *HSP70, RBCS2,* and *PSAD* genes, encoding Heat Shock Protein 70A, the small subunit of Rubisco (RbcS) and the PsaD subunit of PSI, have proven useful. Expression from the chloroplast genome works well under the control of 5'- and 3' untranslated regions of the *rbcL-* and the *atpA-*genes, coding for the large Rubisco subunit (RbcL) and a subunit of the ATP-synthase. (b) A protein sequence that is found in both species, *E. coli* and *C. reinhardtii*, is encoded by different codons on the respective mRNA. Upon codon optimization of a *C. reinhardtii* cDNA to be heterologously expressed in *E. coli*, codons that are seldom used in the host (red and gray indicates that the codon frequency in *E. coli* is less than 10 % and 20 %, respectively) have been converted.

Heterologous expression in chloroplasts is somewhat less prone to negative regulatory effects exerted by the host when compared to nuclear expression. The reason for this is the fact that the chloroplast genome, as well as its expression machinery resemble those in prokaryotes. Most plastid encoded genes are intronless, and often, they are transcribed polycistronically, which is helpful when one aims to express multiple genes simultaneously, for example when introducing a completely new metabolic pathway into the host cell. What is more, homologous recombination does take place in the plastid genome, allowing a targeted insertion of an expression cassette. Thereby, undeliberate knock-outs of other genes can be prevented, and if desired, specific genes can be replaced by modified versions.

In the chloroplast, gene expression is mostly regulated at the translational level, and both 5'- and 3'-UTRs of the mRNAs confer regulatory effects. While the 3'-UTR is mostly related to mRNA stability, the 5'-UTR, while also influencing mRNA stability, binds trans-acting factors, whose genes are often nuclear and which are responsible for a concerted expression of nuclear and plastid encoded proteins of the photosynthetic apparatus. In view of the potential of chloroplasts in applied fields, new regulatory elements are searched for or even synthesized *de novo* using information obtained by massive sequencing of chloroplast UTR mutants. To date, only a handful of regulatory regions have been used (Figure 7.6a).

A very important factor for efficient transgene expression both from the nuclear and the chloroplast genome is the codon usage of the introduced gene. There are 64 codons (including three stop codons), but only 20 proteinogenic amino acids, so that most amino acids can be encoded by multiple codons. Codon usage or codon bias refers to the fact that different organisms tend to have certain codon preferences (Figure 7.6b), and the respective tRNA pools match the codon frequency. Therefore, a transgene that has many codons that are not often used by the host may deplete the cell's tRNA pools for these codons, resulting in slow or halting translation of the protein and also in decreased fitness of the host organism. To avoid this, transgenes are often codon-optimized before heterologous expression, i.e. the whole sequence is synthesized *de novo*, using codons preferred by the host (Figure 7.6b).

Using chloroplast expression cassettes, several human antibodies or other therapeutic proteins, as well as viral antigens, have been synthesized by microalgal (mostly *C. reinhardtii*) chloroplasts, and shown to be functional by *in vitro* protein-protein interaction tests (such as Enzyme Linked Immunosorbent Assay, ELISA) and sometimes even in mouse models. The first example was a human IgA anti-herpes large single-chain antibody, whose codon-optimized gene was expressed from the *C. reinhardtii* chloroplast genome using the *rbcL* 3'-UTR and either the *rbcL* or the *atpA* 5'-UTRs as regulatory flanking sequences. The resulting antibody protein is able to bind herpes virus (HSV8) proteins. Besides using purified recombinant proteins, *C. reinhardtii* cells synthesizing a certain heterologous protein might also be applied orally. Some pathogens enter the body through the mucosa of the stomach or the intestines, and oral vaccination is well suited to fight an infection already at the site of pathogen attachment and invasion. A fusion protein of a *Staphylococcus aureus*-specific epitope (fibronectin-binding domain D2) and a cholera toxin B subunit (which helps internalization of the fusion protein in the gut) was heterologously synthesized in *C. reinhardtii* chloroplasts and shown to enhance the immunity of mice when applied orally in form of whole cell extracts.

The company Triton Health & Nutrition in San Diego, CA, USA, has a product line called the PhycoShield™ which includes several proteins that are supposed to help protect against intestinal diseases. The first product, Mammary Associated Amyloid (MAA), is announced to be on sale in 2014, as a result of a strategic partnership between Triton and an algal biotechnology company that develops efficient algal cultivation systems in Gilbert, AZ, USA. MAA is a protein found in the colostrum – the first milk – of mammals and stimulates mucin production in the intestine. The mucin layer, in turn, protects against bacterial infection. Triton aims to orally administer MAA by whole algae extracts that can, for example, be used as feed for livestock, but also as dietary supplementation for humans.

Key-terms

Applied phycology, astaxanthin, biofuels, biomass, carotenoids, *Chlamydomonas reinhardtii*, *Chlorella vulgaris*, codon usage, *Dunaliella salina*, *Haematococcus pluvialis*, high value products, hydrogen, isoprenoids, light stress, lipids, metabolic engineering, microalgae, *Nannochloropsis* sp., nutrient deficiency, *Phaeodactylum tricornutum*, photobioreactors, photosynthesis, polyunsaturated fatty acids (PUFAs), solar irradiance, transgenes, triacylglycerol (TAG), vitamins

Questions

- What is the broad definition of "algae"?
- What are "blue-green algae"?

- Which algae are used to extract agar-agar and which other species are used to extract alginic acid?
- Which molecules carry out the first step of light conversion during photosynthesis?
- Why is the formation of triplet chlorophyll potentially harmful for the cell?
- What is the "Mehler reaction"?
- When do photosynthetic organisms encounter light stress?
- Under which condition can light stress occur at low light intensities?
- What are the functions of carotenoids during photosynthesis?
- Under which conditions does *Dunaliella salina* produce high amounts of β-carotene?
- To which chemical group do carotenoids belong to?
- What is the first committed step of carotenoid biosynthesis?
- In which cellular compartment of plants and algae are fatty acids biosynthesized?
- What is an environmental condition that triggers triacylglycerol accumulation in microalgae?
- How is biodiesel generated?
- What is the direct electron donor of the hydrogen producing HYDA1 enzyme in *C. reinhardtii*?
- What is an environmental condition that triggers a sustainable H_2 evolution in *C. reinhardtii*?
- What does the term 'omics' mean?
- Why is it not possible to specifically knock out a gene in the nuclear genome of *C. reinhardtii*?
- What is a common technique to post-transcriptionally inhibit gene expression?
- What do codon usage and codon optimization mean?

Further reading

Graham, L. E., Graham, J. M. & Wilcox, L. W. 2009. Algae, 2nd ed. Benjamin-Cummings Publishing Company. ISBN-10: 0321559657.

Posten, C. & Walter, C. 2012. Microalgal Biotechnology: Integration and Economy. DeGruyter GmbH, Berlin/Boston. ISBN-10: 3110298279.

Harris, E. H., Stern, D. B. & Witman, G.B. 2009. The Chlamydomonas Sourcebook, 2nd ed., 3-Vol set, Academic Press, San Diego, CA. ISBN-10: 0123708737.

References

Hemschemeier, A., Melis, A. & Happe T. 2009. Analytical approaches to photobiological hydrogen production in unicellular green algae. Photosynth Res 102: 523–540.

Specht, E., Miyake-Stoner, S. & Mayfield, S. 2010. Micro-algae come of age as a platform for recombinant protein production. *Biotechnol Lett* 32: 1373–1383.

Spolaore, P., Joannis-Cassan, C., Duran, E., et al. 2006. Commercial applications of microalgae. *J Biosci Bioeng* 101: 87–96.

Stephens, E., Ross, I. L. & Hankamer, B. 2013. Expanding the microalgal industry – continuing controversy or compelling case? *Curr Opin Chem Biol* 17: 444–452.

Tirichine, L. & Bowler, C. 2011. Decoding algal genomes: tracing back the history of photosynthetic life on Earth. *Plant J* 66: 45–57.

Ansgar Poetsch and Andreas Harst

8 Strain design and -omics technologies

Given the vast diversity of molecules synthesized by (micro-) organisms, nature seemingly possesses an endless metabolic capacity. Harnessing this potential for the biotechnological production of ever-increasing numbers of different molecules with large-scale processes, undergoing continuous improvement in terms of economy and ecology is one of the key aims and challenges in modern industrial microbiology. In particular for the production of fine chemicals (e.g. amino acids, lipids, vitamins; white biotechnology) and pharmaceutical chemicals (e.g. antibiotics, hormones; red biotechnology), biosynthesis has demonstrated to be more favorable than chemical synthesis.

Development of an industrial production process is a multidisciplinary effort involving engineers and biologists, since it requires development and optimization of the production processes and its devices such as fermenters and centrifuges, as well as the used organism. Whereas aspects central to the design and engineering of a production process are exemplified for photosynthetic microorganisms in chapter 6 and 7, this chapter will give an overview of past and present approaches for the screening, optimization, and design of strains for industrial purposes. From classical methods, utilizing random mutagenesis in combination with tailored screening assays, to modern techniques, exploiting -omics data and biological network models, the current arsenal for strain design and optimization will be presented. Favorable traits of such strains are primarily high yield of the desired product concomitant with low amounts of side products – one must keep in mind that *downstream processes*, amount to about 50 % of production costs of a bioprocess. Further, commonly desired strain improvements are: increased stress tolerance (heat, pH, toxic molecules, etc.), extension of substrate range for growth and product formation, and addition of new metabolic activities leading to the production of chemicals new to the organism. Special attention will be paid to the underlying principles and the application of modern -omics techniques, which will be illustrated by strain engineering with *Corynebacterium glutamicum* for the large-scale production of amino acids.

8.1 Classical methods of strain optimization

Prokaryotes play important roles in red, white, and green biotechnology (chapter 7) alike. In medical research and pharmaceutical production, bacteria synthesize proteins used as biopharmaceuticals. An overview of the historic development of strain design can be seen in Table 8.1. The best known example is the production of insulin by *Escherichia coli*, which started in 1978. By now, a large number of other pharmaceuticals, ranging from antibodies to vaccines, are manufactured in

Table 8.1: History of Strain design. The following table lists major milestones and breakthroughs of strain development for biotechnology. In the pioneering years of biotechnology efforts were focused on strain selection and screening. Thereafter identified strains were optimized by random mutation, an approach which in recent years was replaced by more targeted strategies as directed mutagenesis and metabolic engineering.

Technologies	Achievements	Year
strain selection based on environmental screening	Isolation and cultivation of single bacterial colonies on agar plates by Robert Koch	1876
	acetone production with *Clostridium acetobutylicum*	1917
	sorbitol transformation to L-sorbose with *Acetobacter suboxydans* for vitamin C	1933
	microorganisms can be mutated by physical or chemical treatments to achieve improved properties	1930s
	industrial-scale penicillin production	1944
strain optimization using mutation and dedicated screening	new antibiotics and improvement of producer strains	1950s
	amino acid production with auxotrophic *Corynebacterium glutamicum*	1960s
	Gene cloning, recombinant microorganisms	1970s
	metabolite profiling using gas/liquid chromatography separation	1970s
	methods for DNA sequencing	1975
	directed mutagenesis	1978
	concept of synthetic biology based on novel gene arrangements	1978
genome manipulation with molecular biology	recombinant production of human protein (insulin) in bacteria	1978
	recombinant alpha-amylase	1980
	protoplast fusion for strain breeding	1986
	polymerase chain reaction (PCR)	1988
	Invention of MALDI and ESI ionization for mass spectrometry	1990s
	concept of proteome and proteomics	1994
	concept of metabolic engineering	1995
	first complete bacterial genome (*Haemophilus influenzae*)	1995
	DNA microarrays for transcriptomics	1995
systems and synthetic biology approaches	2nd generation DNA sequencing, RNA sequencing	2000s
	genome-based strain reconstruction for *Corynebacterium glutamicum*	2000s
	de-novo synthesis of a bacterial genome	2002
	Artemisinin production with yeast using synthetic biology	2006

microbial processes. Also, involvement of prokaryotes in white biotechnology is of great economic significance as production of chemicals by microbial fermentation is a multi-billion dollar business. The beginning of this success story lies in the isolation of chemical compound producing bacteria from their natural habitats

Historically the first white biotechnology process was the production of acetone from starch by *Clostridium acetobutylicum* in 1917 by Chaim Weizmann. In the following decade, more microbial strains were identified which produced compounds and excreted them. Isolation of these strains enabled manufacturing of large quantities of the desired materials based on sugar as a carbon source.

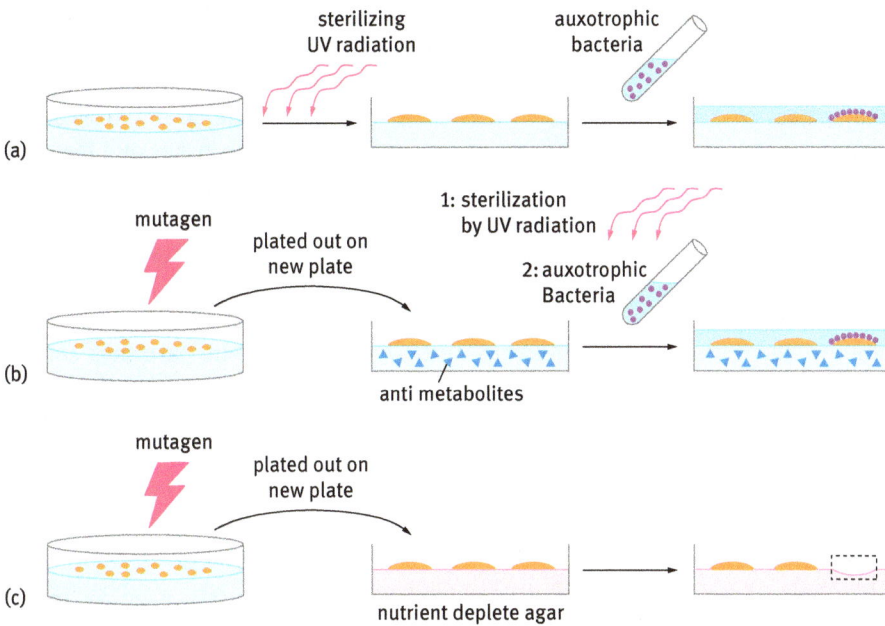

Fig. 8.1: Classical methods of strain screening and optimization. (a) Production screening by growing an environmental sample on medium lacking the targeted compound. The cultivated plate is sterilized by UV radiation and colonies are covered with agar containing auxotrophic bacteria. (b) Screening for strains insensitive to antimetabolites (triangles ▲). Production strains are subjected to mutagenesis by UV radiation and then grown in presence of antimetabolites (triangles ▲). Producing colonies are screened for by auxotrophic bacteria. (c) For generation of auxotrophic production strains, bacteria are mutagenized and transferred to new plates lacking the nutrient. If now colonies cease to grow, they have developed auxotrophies.

The first necessary step of industrial scale production is isolation of microorganisms excreting the desired compound. Therefore, a bioassay had to be used which enables the selection of bacteria from uncharacterized samples based on their ability to synthesize a desired product. For these screening methods, bacterial isolates were grown on plates consisting of defined test media. Chemicals contained in these media are usually also present during production, thereby selecting strains especially resistant to the conditions of production. Inoculated plates were irradiated with UV for sterilization. Subsequently agar containing bacteria auxotrophic for the desired compound were applied to the UV-treated plates. Growth of these bacteria on plate implicated the presence of the desired compounds and thus identified bacteria which are able to secrete them. These bacteria could then be isolated from unsterilized master plates (Figure 8.1a).

Wild type strains only synthesize small amounts of a product and invest more energy into cell division and cell survival. For the design of high yield production strains regulation systems and the competition between pathways had to be diminished. Therefore, one avenue for higher productivity is to delete competing pathways, hence in the beginning of industrial production high yielding strains with amino acid and vitamin auxotrophies were introduced (Figure 8.1c).

Use of these strains is expensive due to the necessary addition of nutrients. A major goal of strain optimization, therefore, was to generate strains resistant to regulation mechanisms as allosteric inhibition of metabolic enzymes and feedback inhibition. To screen for strains which show resistance to this regulation random mutation of cells was performed in presence of compounds inhibiting synthesis. Strains which synthesize the product under these conditions usually harbor enzymes desensitized to feedback and general regulation (Figure 8.1b).

A further approach which served as a substitute for auxotrophic strains was the introduction of so called leaky strains. Leaky biosynthetic pathways contain enzymes characterized by a lower turnover of metabolites. The metabolite's concentration remains below the threshold triggering feedback inhibition, but is high enough to prevent starvation.

Up to now, the generation of mutants in strain design has been performed following two general approaches: random mutation and targeted strain engineering. Random mutation was the method of choice for biotechnological use from the 50s to the 80s of the 20th century. Though this method is well established and generated a number of strains with high production capabilities still used today, there are several drawbacks ranging from high labor intensiveness to the accumulation of undirected mutations. An additional shortcoming of these mutations is their detrimental effect on the process robustness of the strains, as these became more sensitive to changes in pH and temperature and other stresses occurring during fermentation.

In the mid 80s, protoplast fusion has matured for industrial application. Here, the cell wall of microorganisms is treated with specific lytic enzymes, leaving only the plasma membrane as outer barrier intact. Induced by physical or chemical mechanisms, two protoplasts can be fused and genes transferred. This breeding approach allows further improvements by combining strains, commonly obtained by random mutagenesis, with differing positive traits.

Targeted engineering of genes is based on technologies as PCR and heterologous gene expression, which only became available in the early 90s of the 20th century. This approach avoids the pitfalls characteristic for random mutagenesis. By introducing and changing genes in a defined manner, no undesired mutations occur in most cases, and existing production strains can be improved in a targeted way. Before applying genetic engineering, a target gene has to be identified, therefore, the genome of the targeted organism should be sequenced and functionally annotated.

8.2 Genome engineering

In the previous paragraphs, strain optimization was limited to genetic engineering of single genes and random mutagenesis. But the ability to completely sequence genomes also provides new avenues for improving biotechnological processes. The sequences can be used to identify and eliminate prophage-like sections from the genome as well as to engineer the genomic structure of an organism.

The enhanced knowledge of bacterial genomic structure provided by next generation sequencing (introduced in section 8.5) can be used to increase the efficiency and production volumes of biotechnological processes. These would be much more economic if bacteria would act as biocatalysts only containing an essential set of genes, required for the production of the desired compound. Deletion and remodeling of large parts of an organism's genome occurs often during evolution, thus theoretically, it is possible to edit the genome of an organism without harming its ability to survive. By now, many studies have shown that it is possible to delete regions which strongly contain transposons and bacterial phages, and engineering of chassis organisms harboring a reduced genome is ongoing.

A new strategy to genetically modify biotechnological organisms is the use of transposons. Transposons are mobile genetic elements, which can be excised from the genome by transposases at one position, and then be reintegrated at another position. Insertion sequences are genetic elements, which also undergo a transposition, but do not contain genes. Use of these sequences provides a method to generate random mutations in an organism offering a higher degree of control than in previously used methods. Insertion sequences displaying unspecific target sequences, high insertion frequency, and a high stability of the mutants are most suitable for this method. Besides generation of mutants, transposons are able to transfer large fragments of genetic material, an application especially interesting for biotechnology, as it allows the integration of new proteins and pathways into an organism. On the other hand, identification of a transposition's location is difficult and, thus, complicates the application of transposon mutagenesis.

Recombination techniques have shown great promise for genome modification. They are especially versatile in deletion of large genetic segments and also can replace genetic elements. Recombination technology is based on the application of the FLP/FRT and the CRE/Lox recombination systems. FLP and CRE are enzymes with recombinase activity binding to double strand recognition sites called FRT or Lox. When attached to these sites the enzymes catalyze a recombination reaction. If two recognition sites show the same orientation on the DNA strand, the segment flanked by the sites will be excised during the recombination reaction (Figure 8.2). A disadvantage of this reaction is that a recognition site will stay behind in the genome of the organism. Another system utilizes DNA double strand breaks (DSB) for recombination, exemplified in Box 8.1. These approaches can be used for the knockout of genes as well as the random deletion of large genome segments to generate minimal genome strains.

As we have seen in the previous sections, many methods have been developed for the integration and deletion of large genetic segments. One application was the introduction of the gene cluster (14.3 kb) for nisin biosynthesis into *Bacillus subtil*

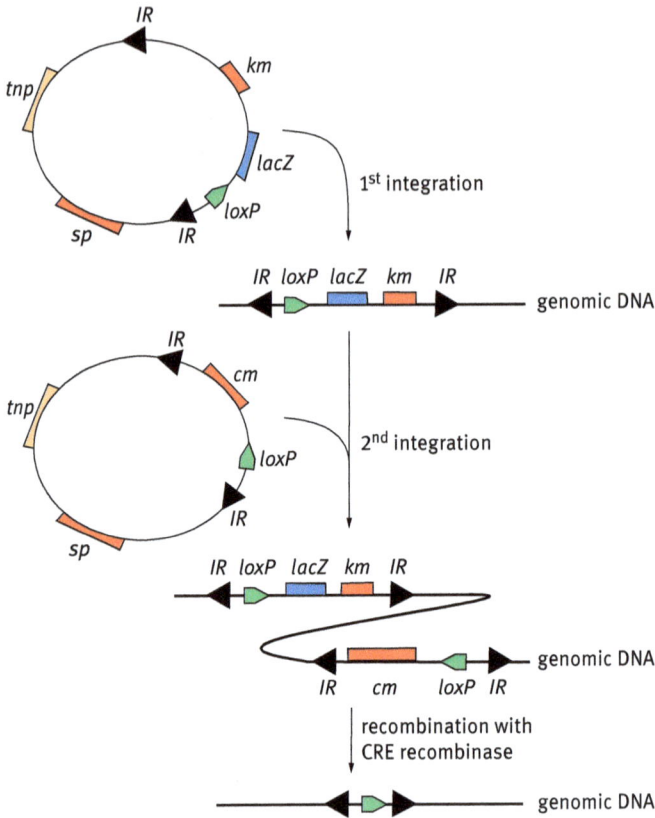

Fig. 8.2: Use of the Cre Lox system for deletion of genetic segments. From the first mini transposon vector, the *loxP* gene and a gene coding for kanamycin resistance are integrated into the genomic DNA. After integration of a second *loxP* element with the same orientation as the first one and expression of CRE recombinase, the nucleotide sequence between both *loxP* elements is removed by recombination.

Box 8.1: DNA Recombination by double strand breaks.

DNA is cut by a mega nuclease at a desired position and intramolecular recombination takes place leading to the deletion of the desired gene. This method is quiet advantageous considering that no alien genetic material is integrated into the genome. Via homologous recombination, genetic segments can also be replaced in the genome. A genetic element containing a homologous sequence, a marker gene, and double strand break sites are necessary for this approach. If a DSB occurs in the DNA, the recombination machinery of the cell integrates the homologous segment accompanied by the marker gene into the genome while the target sequence is excised from the genome.

is, using high fidelity polymerases. Nisin is an antimicrobial peptide inhibiting the growth of gram positive bacteria naturally occurring in *Lactococcus lactis.* Experiments like this show that remodeling of a prokaryotes genome is feasible. But still, recombination technology has to be improved and optimized.

8.3 Engineering metabolism: from enzymes to metabolic networks

With interactions between metabolites and enzymes being at the heart of metabolic networks, works towards their experimental determination and mathematical representation have been initiated more than hundred years ago. The simplest mathematical abstraction of enzymatic catalysis is realized in the Michaelis-Menten equation: $\frac{v}{vmax} = \frac{[S]}{Km + [S]}$, with v = reaction rate, $vmax$ = maximal reaction rate, $[S]$ = substrate concentration, Km = Michaelis constant.

However, kinetics for the majority of enzymes are much more complex, and involve, for instance, multiple reaction steps, substrates, products, and inhibitors. Hence, sophisticated experiments and instruments were developed to investigate reaction mechanisms, mostly for purified enzymes *in vitro*. A thorough understanding of enzyme mechanism inferred by *in vitro* studies, can indeed be important for optimizing metabolic networks, yet the main benefit still lies in the identification and quantitative description of substrates and inhibitors for a given enzyme. Hence, *in vitro* analysis of enzymes delivers direct experimental evidence about metabolite and enzyme interactions, which are much harder to infer from techniques operating at the metabolic network level. Furthermore, experimentally determined kinetic parameters (*Km*, *vmax*, etc.), which can be nowadays retrieved from specialized electronic resources (e.g. BRENDA database) are invaluable for the development of kinetic models used in metabolic engineering with the caveat that the conditions (concentration, ionic strength, etc.) used for *in vitro* analysis do not correspond in all aspects to the *in vivo* situation.

Organisms must be viewed at the network level as highly complex systems of interacting DNA, mRNA, proteins and metabolites. Despite tremendous progress in experimental and computational approaches, even for well-characterized bacteria like *E. coli* these networks are far from being comprehensively understood. In practice, simplification and reduction of network representation, in an appropriate manner for the question at hand, are inevitable for model-based predictions to pose a feasible tool in strain design. Of all elements within a molecular network of the organism, models for metabolism reached the highest level of sophistication, and are usually the most important for strain improvement. For the representation of large metabolic, even genome-scale networks, stoichiometric models are currently preferred. Various modeling approaches have been developed and applied to metabolic engineering, whereby flux balance analysis (FBA), metabolic control analysis (MCA), and kinetic modelling are frequently applied. Metabolic network

Metabolic network analysis.

In the following, common approaches for metabolic network analysis will be discussed: For the representation of large metabolic networks, stoichiometric models are currently preferred. The model requires information about all participating enzymes and metabolites, all occurring chemical reactions, and all stoichiometric coefficients, i.e. the proportion of substrate and product molecules involved in each reaction. Reconstruction of network topology can be used to predict the theoretical metabolic capacity of an organism, such as operation of glycolysis. Genome-scale models have been realized for several industrially relevant microorganisms as *B. subtilis*, *C. glutamicum*, with the most recent model for *E. coli* containing 2,251 reactions, 1,136 metabolites, and accounting for 1,366 genes. Flux balance analysis (FBA) builds upon stoichiometric models and introduces further constraints on the metabolic model. The cardinal constraint is assumption of a steady state, meaning that all fluxes in and out of the network must be balanced, thus concentrations of all metabolites remain constant. Other common constraints are definition of irreversible reactions (based on thermodynamics) and boundaries for the magnitude of individual metabolic fluxes. FBA is used to predict the flux distribution in the network under a given objective, which may be biomass formation, maximum yield of a metabolite, utilization of a carbon source, aerobic or anaerobic growth. Although constraints minimize the range of flux distributions fulfilling the objective, it is not uncommon that more than one possible flux distribution is found, and further experimental work or model refinement are required to narrow down the number of flux solutions. The fact that FBA does not rely on (experimentally determined) kinetic parameters is an advantage, because these are often not available, but also a disadvantage, because it cannot predict metabolite concentrations. FBA is a powerful tool to predict on the network level effects of introducing or deleting genes, such as re-routing of pathways or occurrence of auxotrophies.

How does change in flux at one point of the pathway affect flux distribution in the network under a given objective? Which reaction is most sensitive to changes in parameters (enzyme kinetics, substrate/product concentration)? These important questions in metabolic engineering can be addressed by metabolic control analysis (MCA). Essentially, MCA provides a framework that quantitatively connects properties of the network with properties of each reaction. Mathematically, this is achieved by calculating changes of steady-state variables in response to small changes of one or more parameters. In MCA, the sum of all these coefficients amount to 1 and denote the different local and global levels of control. An example for a local coefficient is elasticity, which links the properties of an enzyme to its potential for flux control through variation in metabolite concentration. For instance, assuming Michaelis-Menten kinetics, an enzymatic substrate conversion with reaction rate v in the range of Km has a relatively high, yet with v in the range of vmax has a relatively low elasticity, since for the latter small changes in the substrate concentration only affect little the reaction rate v. The global flux-control coefficient denotes the effect of changes in steady-state flux at one step in a pathway on relative changes in pathway flux. Furthermore, MCA provides a theoretical framework for the observation that the first reactions of a metabolic pathway and branching points often underlie the highest level of regulation.

Kinetic metabolic models incorporate biophysical properties (enzyme/transport kinetics, substrate concentrations etc.) allowing to interrogate temporal phenomena (e.g. oscillations) and non-equilibrium states of a metabolic network. Obviously, such models hold great predictive power, but their realization on the genome scale is only emerging and significantly hampered by the scarcity of required input parameters that must be experimentally obtained. Because of this, kinetic models are so far used less frequently for strain improvement.

models can support and guide strain design in a rational manner. For the development and refinement of models, -omics techniques, in particular genomics and fluxomics (section 8.5) are becoming increasingly important. Among the many purposes of modeling, introduction of new metabolic reactions, prediction of metabolic fluxes for different physiological conditions, and identification of most promising intervention points for increasing fluxes towards a desired metabolite, should be mentioned.

As soon as promising interventions for a given metabolic networks have been identified by experimental and/or modeling techniques, they must be experimentally verified in the organism by genetic manipulation. Common strategies thereof for a simple theoretical network are summarized in the following list and in Figure 8.3:

1. Overexpression of rate limiting enzymes for a metabolic pathway eliminates metabolic bottlenecks.
2. Overexpression and genetic engineering of branch point enzymes can redirect metabolite flows to enhance a compound's production.
3. Insertion of (heterologuous) enzymes insensitive to regulation.
4. Modification of an active center resulting in more effective catalysis of a reaction.

The convergence of genetic engineering, metabolic network analysis, and -omics technologies has profoundly changed strain design. Whereas classical methods of strain optimization were untargeted and screened for random increases in production, modern methods enable researchers to directly engineer towards a rise of production. Due to these methods, strain design has by now moved closer to engineering and away from random screening.

A recent application of metabolic engineering is the production of non-native chemicals in organisms. For synthesis of a compound, which is not produced by an organism in its natural environment, enzymes producing it have to be cloned into a host organism from a heterologuous source (Figure 8.3). The production of substances which are not naturally produced by biological pathways demands altering of an enzyme's substrate specifity (chapter 2). The modulation of substrate specificities and targeted manipulation of an active site reaction by enzyme evolution can be used to introduce synthetic metabolic pathways, with non-natural substrates as well as products. Integration of these pathways might lead the way to design microbial cell factories. These metabolic engineering approaches can only be efficient when they are assisted by models of enzyme kinetics and metabolic networks. An example for application of this approach is the synthesis of polyamides in *C. glutamicum*.

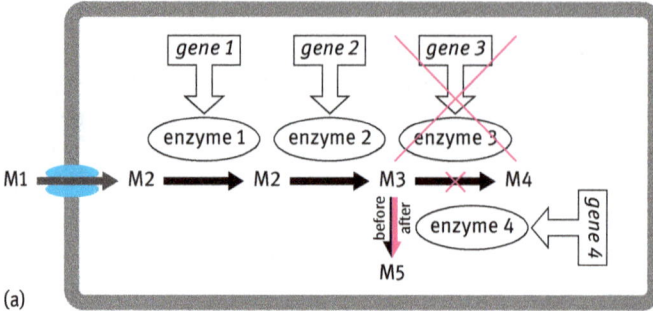

(a)

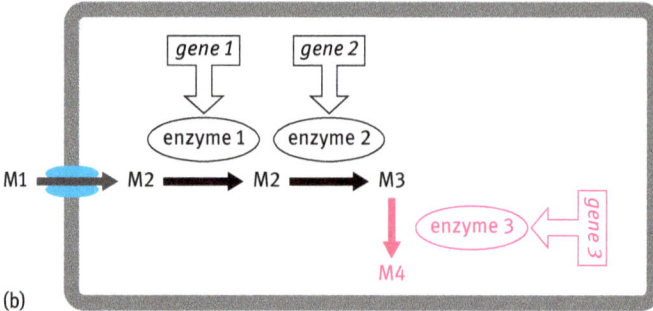

(b)

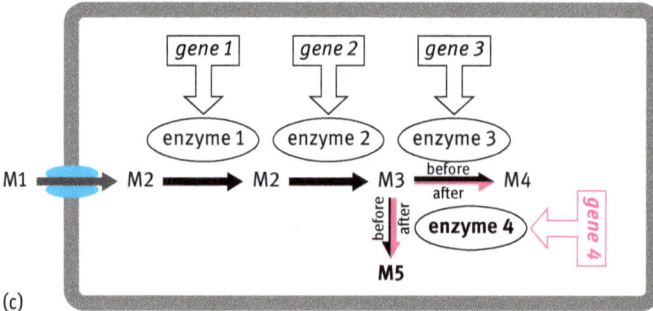

(c)

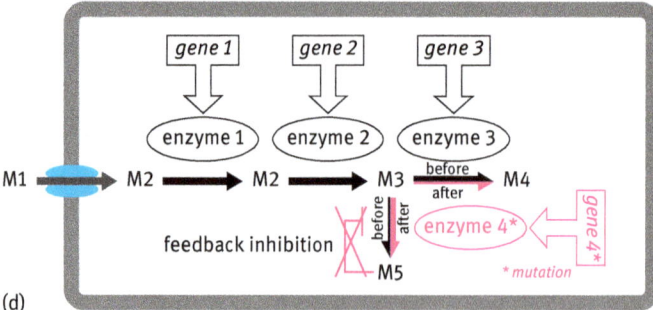

(d)

Fig. 8.3: Illustration of the main strategies of metabolic engineering. (a) The flux to metabolite M5 after mutation (red arrow) is increased in comparison to the flux to M5 before the mutation (black arrow). This is due to the knock out of the competing reaction catalyzed by enzyme 3. (b) An organism is enabled to synthesize a new metabolite M4 from M3 by insertion of the heterologuous gene 3. (c) After increasing the copy number of gene 4, the flux from metabolite M3 to M5 (red arrow) is increased, while the flux from M3 to M4 decreases (red arrow). (d) The Wildtype form of enzyme 4 is inhibited by metabolite M5, decreasing fluxes from M3 to M5 (black arrow). Targeted mutagenesis of enzyme 4 leads to desensitized enzyme 4*, resulting in an increase of metabolic flows from M3 to M5 (red arrow) and a decrease of flux from M3 to M4 (red arrow).

8.4 -Omics technologies for strain characterization and optimization

Strain characterization and optimization was, for a long time, based on costly methods such as random mutation and manual clone screening. With the onset of the -omics age, the biotechnological attributes of a strain can be investigated on the gene, transcript, protein, and metabolite level. Thus, evaluation of a strain's fitness for biotechnological production has become a mainstay of strain development. Additionally, the identification of gene and protein targets for genetic engineering is an important application.

The defining feature of an organism is its genome, which determines its characteristics and properties. To understand the genome, the genetic code it contains has to be sequenced. Classical sequencing (Sanger sequencing) is based on the termination of DNA chain polymerization with dideoxy nucleotides. Though this method is very robust and accurate, increases in throughput are only possible by use of additional sequencers, limiting scale and speed of the method. To overcome this bottleneck in throughput, the polymerization reaction was miniaturized so that a large number of reactions could take place in parallel. The technologies introducing these improvements were summarized under the name Next generation sequencing (NGS). Two sequencing platforms established early in the development of NGS were SOLiD sequencing by ABI and 454 pyrosequencing by 454 Life Sciences. In both platforms, DNA amplification was miniaturized by emulsion PCR; here, chain elongation reactions take place on beads contained in small aqueous droplets inside an oil water emulsion. Though gene amplification is similar, there are strong differences in sequencing. Pyrosequencing establishes the nucleotide sequence by detecting the release of diphosphates after incorporation of a nucleotide during the DNA elongation reaction. For detection, the luciferin oxyluciferin reporter system is used, and sequencing takes place in parallel to synthesis. The SOLiD platform sequencing approach is based on ligation of four fluorescently labelled dibase probes, which interrogate every 1st and 2nd base per ligation step and takes place after gene amplification. These sequencing technologies are only able to sequence shorter segments up to 600 bp, therefore a considerable bioinformatics effort has to be invested to assemble genome sequences from this data.

The main approach for genome sequencing of bacteria relevant in biotechnology is sequencing based on ordered libraries of genome fragments. To generate a genetic library for an organism, the complete genome has to be fragmented, and the resulting chunks are then cloned into plasmids, which are introduced into bacteria. Using hybridization, the cloned sequences are screened for overlapping regions to indicate neighboring sequences. By performing this procedure for a large number of clones, these can be assembled into only a few contiguous genetic segments representing a large part of the genome. The remaining gaps can also be mapped using other library approaches as bacterial artificial chromosomes (BAC). With next generation sequencing devices, whole genomes were assembled using smaller fragments, and gene libraries were used to provide a scaffold to order the acquired sequence information. This strategy is known as shotgun sequencing of a whole genome. Assembly of whole genomes from smaller sequences requires a considerable bioinformatics effort (chapter 14).

In the previous paragraph, we have seen the role which genomics can play in the prediction and manipulation of an organism's physiological traits and phenotype. Besides that, a central process of living organisms is the translation of genes into proteins. The first step of this process is the transcription of DNA sequences into mRNA. Therefore, quantification of gene transcription rates can give insight into biological processes. Several ways for transcription quantification are available. In microarray experiments, a specific sequence of each gene is converted to an oligonucleotide, which is immobilized in a single dot on the micro array chip. Quantification is performed by labeling the extracted RNA with fluorescent dyes. To compare conditions or strains, they are labeled with different dyes. This RNA is hybridized with the micro array probes, gene expression is quantified by measuring the response signal following laser excitation. This signal is used to calculate the ratio of a gene's expression between two conditions. A disadvantage of this method is that it only provides relative quantification for a small segment of the mRNA and is not able to report on real time dynamics of gene expression.

The technologies introduced for DNA next generation sequencing approaches are now also used for transcriptomes. Here they provide high throughput and high dynamic ranges. For sequencing of proteinogenic RNA sequences in eukaryotes, non coding introns have to be excised, therefore spliced mRNAs harbouring poly-A tails are converted to cDNA, which is then sequenced in the same manner as genomic DNA. RNA quantification is performed by comparing the read number per gene to the read number of known quantities of sample RNAs. If further information as, for example, RNA content of the cells is available, absolute quantification of transcripts can be performed, i.e. the number of transcripts per cell can be calculated.

The number of an enzyme's transcripts can have a strong impact on biotechnological production. Especially rate limiting steps of metabolic pathways show a strong dependence on transcription. When comparing two production strains with similar genomic background, different transcription of genes can lead to increased production. The knowledge about these differences, and how to manipulate transcription can be put to good use in biotechnological production. Using the next generation sequencing and RNA microarray approaches, a number of transcriptional phenomena leading to increased production have been elucidated.

Transcriptional and genetic regulation are important players in the control of metabolic processes. But many processes of cells are regulated on the protein level. Information as presence of proteins in a cell, posttranslational modifications, and synthesis rates are very important. Proteomics is attempting to detect, quantify, and investigate the proteome of a biological condition, and provides many methods to unravel these questions. In contrast to DNA or RNA, proteins cannot be sequenced in a replication based approach, but have to be digested into smaller peptides to generate sequence information. Sequencing is based on fragmentation of peptides in electro spray ionization mass spectrometry (ESI-MS), or matrix-assisted laser desorption ionization (MALDI) mass spectrometry. Mass spectrometry is a physical method to detect ionized atoms or molecules. For shotgun (bottom up)proteomics, proteins extracted from a cell are first digested by enzymes cleaving after specific amino acids. The peptides originating from these digests are then ionized in gas phase and fragmented by collision with e.g. helium atoms. For this fragmentation, a collision energy is selected which only breaks the amino bonds of the peptide. Fragments are detected by first accelerating them in an electrostatic field and by impact on electron multipliers. By measuring the movement of these ions in the electrostatic field, the mass to charge ratios, i.e. fragment masses, can be determined. Since fragmentation preferably occurs at the peptide bonds, these masses are then used to deduce the sequence of amino acids. MS-proteomics can give a more detailed insight into the physiological state of a cell by directly investigating the state of its tools, proteins, including their quantity and modifications.

Metabolomics is based on the comprehensive measurement of metabolites and their concentrations. A bacterial cell contains around 700 core metabolites of which the most common is 10,000 times more highly abundant than the metabolites with the lowest concentration. Extraction has to be performed with precautions to stop metabolic reactions and to guarantee abundant extraction of compounds. To stop, i.e quench metabolic reactions, cells are usually shock frozen by low temperature extraction media. Composition of these media has to be such selected that they abundantly dissolve ionic and neutral metabolites. Identification and analysis of the analytes can be performed using conventional liquid chromatography mass spectrometry (LC-MS) as well as gas chromatography mass spectrometry (GC-MS) and nuclear magnetic resonance (NMR). Whereas GC-MS is used for identifying uncharged metabolites, modern high resolution LC-mass spectrometers can identify polar compounds in positive and negative mode with high sensitivity as well as high resolution. Quantification of metabolites is based on isotope labeled internal standards which are used as calibrants for quantification. NMR gives direct access to the 3D structure of a metabolite and its concentration (Reaves et al. 2010).

molecules	DNA	RNA	proteins	metabolites	fluxes
technology	high throughput whole genome sequencing	high throughput RNA sequencing	1D/2D gel based MS/MS proteomics	identification of small molecules by GCMS /HPLC MS	flux balance analysis or metabolic control analysis
target	sequencing of complete genomes	sequencing and quantification of all transcribed RNAs in a sample	identification and analysis of all proteins expressed in a sample	identification of all metabolites in an organism	calculation of all fluxes inside a metabolic network

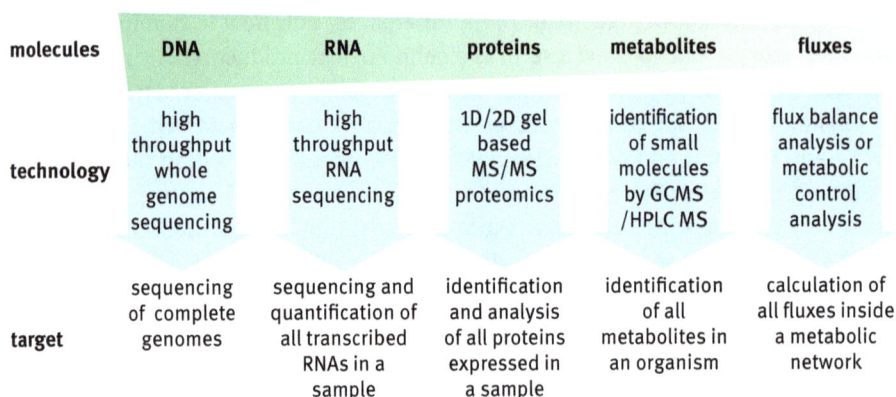

Fig. 8.4: Schematic representation of several -omics technologies, their target molecules, and their objectives.

The quantitative metabolic state of a cell cannot be interrogated by the previously introduced technologies alone. These can only give a limited insight into the concentration and regulation of enzymes and metabolites. To establish the concentration changes of metabolite pools, called fluxes, mathematic models are combined with metabolomic measurements. For these approaches, first the in vivo kinetic parameters of the involved enzymes and the concentration of the metabolites have to be established. Mathematic models are then used to derive all fluxes in a metabolic network. Isotope fluxes are measured by feeding cells isotopically labeled substrates and subsequent determination of isotope enrichment during the following steps. This enrichment can be measured by mass spectrometers and nuclear magnetic resonance.

Systems biology is the field of biology striving to understand biological systems by analyzing the complex networks they are made of. Important tools of systems biology are quantitative models incorporating the information from different biological levels. The qualitative and quantitative elucidation of an organisms metabolism, succinctly called metabolomics, combined with genomic, transcriptomic, and proteomic information will pave the way to an integrated systems biology of biotechnological processes (Figure 8.4).

8.5 Synthetic biology's impact on strain design

Synthetic biology is the manipulation and engineering of organisms using modern molecular biology techniques. Synthetic biology strives to plan and construct artificial networks, cells or even whole organisms from scratch using synthesized DNA sequences. Another major goal is the construction of biological modules which resemble control circuits and function as switches. The design of these tools is informed by systems biology and mathematical models. Introduction of heterologuous pathways and pathways leading to non-natural products into microorganisms can also be considered as synthetic biology. Despite this, there is a controversial discussion going on if these experiments should be defined as part of metabolic engineering because they aim more at establishing new synthesis routes for chemicals than to achieve the goals put forward by synthetic biology.

Construction of biological modules by DNA synthesis led to the collection and dissemination of genetic elements, which can be combined to form synthetic pathways or genetic circuits. Public collections of these genetic elements were founded to publicize and advance the use of synthetic biology. Though a large range of these modules are available, they alone cannot lead to emulation and reconstruction of biological systems as many factors leading to the functions of biological processes are still not understood. Only with input from other branches of biology will synthetic biology be able to attain its promise: the design and creation of biological systems de silico. Still, the experience gathered by these tools will add valuable information about the way biological processes work.

While assembly of small genetic units might lead to replication of single biological processes or cellular compartments, the design of artificial cells is based on synthesizing the genome according to digital data. Whole prokaryotic chromosomes can by now be synthesized and inserted into cells as the synthesis of a *Mycoplasma Mycoides* genome and subsequent integration into a *Mycoplasma capricolum* cell free of DNA has shown. Further steps might enable the creation of living artificial cells, this would be a very significant event entailing a range of ethical as well as practical questions. Issues as the impact of artificial cells on biotechnology are still subject of a controversial debate. Many researchers argue that artificially created cells will not automatically lead to increased yields in biotechnology, because natural evolution is a more efficient process than human engineering.

A further field of synthetic biology has been the design of switches and control circuits based on genetic elements. Genetic elements and sensor proteins can be coupled in such a way that expression of genes is dynamically regulated by the intracellular conditions. Gene regulation networks can also be used to design control circuits. Such circuits can be designed using transcription regulators, as promoters and repressors. These can be arranged in such a way that they perform a logical operation. For example, they can respond to the presence of two compounds with a binary output as occurrence of fluorescence or its absence. Such simple networks can be used to implement complex functions. Therefore, the design and assembly of such functional networks is aimed at defining the response of a cell to a certain stimuli.

Advantages of marrying synthetic- and systems biology are obvious. Omics technologies can be used to detect targets of synthetic biology. And the global changes in artificially designed cells or biological processes can be investigated by systems biology approaches. This information can be used to guide synthetic biology. For example, adaptations to stress brought forward by evolutionary engineering can be determined using systems biology. These adaptations can then be introduced to other organisms or strains increasing their resistance to adverse environments.

Synthetic biology is a field of research bearing considerable promise as well as significant risks. Creation of artificial life and application of artificial biological control circuits might transform modern biology and biotechnology. For example, genetic control circuits and development of biological sensors in organisms might innovate control and online monitoring of fermentative cultures in biotechnology. The generation of organisms based on synthesized DNA is a first step to artificial cells, but as long it is not completely understood how genes, proteins, and metabolism work together to form living cells, this goal seems to be unattainable Also, the consequences and implications of these endeavors, if they were to be successful, cannot all be predicted.

8.6 The amino acid producer *C. glutamicum* as an example for strain engineering

Of all the bacteria used in the biotechnological industry, *Corynebacterium glutamicum* is the one with the highest economic significance. It produced amino acids at an estimated value of 7 billion dollars in 2007. Amino acids derived from fermentative processes are used in a number of fields ranging from human nutrition and pharmaceutical products to additives in animal nutrition. The best known example is glutamate, an amino acid that is used for seasoning of meals. Historically, glutamate was produced by decomposition processes of natural raw materials or chemical syntheses. Chemical synthesis leads to occurrence of racemates awhile humans and animals are only able to utilize the L-enantiomer. Decomposition processes, on the other hand, are energy intensive and make use of harsh means of extraction as hydrochloric acid or organic solvents. After the Second World War, Japan was plagued by food shortages, and it was of utmost importance to find a cheap alternative for glutamate production. In a screening experiment, *C. glutamicum* was isolated in a soil sample from the Tokyo zoo. It showed the highest glutamate yield and was selected for amino acid production.

Soon after its potential for glutamate production was discovered, experiments started to find strains able to produce other amino acids. Random mutagenesis by UV radiation led to the emergence of homoserine auxotrophic strains producing L-lysine. To improve lysine yield, further random mutations were performed, leading to strains which showed more auxotrophies, increasing the biochemical flow into lysine production. To mitigate the effects of these auxotrophies, the before mentioned leaky strain strategy was used. In a study by Hilliger and Hertel, a homoserine auxotrophic lysine production strain of *C. glutamicum* was reverted into a prototrophic form and grown in presence of a threonine analogue which inhibits the homoserine dehydrogenase. The homoserine prototrophic mutant, harbouring a deregulated homoserine dehydrogenase, showed a high lysine production capability.

These random mutation approaches were soon to be replaced with genetic engineering. In 1990, the genes coding for aspartate kinase, the main regulator of lysine biosynthesis, could be identified. The structure of the enzyme is very complex and the comparison of sequences from several strains was necessary to find the regulatory subunit encoded by *lyscβ*. When a variant of this subunit insensitive to feedback regulation was expressed in wildtype (WT) *C. glutamicum,* the resulting mutant would start to produce lysine. Decreased susceptibility of aspartate kinase to lysine feedback regulation was also achieved by nucleotide substitution of codon 279, where nine mutations showed decreased sensitivity. Especially substitution of the threonine residue with proline had an impact on regulatory activity and showed that targeted genetic engineering can lead to similar results as the random mutation approach.

By the 90s, genetic modification had reached the enzyme level. Taking the step to modification on the genome level afforded the comprehensive elucidation of the *C. glutamicum* genome. In 2003, two laboratories, the Tauch group in Germany and the Ikeda group in Japan, published the first complete sequences of the *C. glutamicum* genome. Both studies used different approaches for sequencing. The German group sequenced an ordered cosmid library (Kalinowski et al. 2003), while the Japanese group used shotgun sequencing of unordered plasmid and BAC libraries (Ikeda et al. 2003). Though the former method is coupled to high labour intensity, it leads to a lower number of redundant sequence reads than the second method. Information from gene sequencing was used to identify mutations which introduce the lysine production ability. These mutations were then integrated into a WT strain and resulted in a lysine production rate of 3 g per hour. As main benefit of this approach, termed "genome-based strain reconstruction", the introduction of detrimental mutations unavoidable during classical random mutagenesis, was obviated. Also, knowledge of the complete genetic sequence enables identification of transposons and prophages. As could be seen here, the elucidation of an organism's genome can strongly help in generating more productive strains for biotechnology.

Being the major producer of amino acids, many metabolic flux studies have been performed on *C. glutamicum*. These studies showed that metabolite fluxes through anaplerotic pathways are important for providing precursors to lysine production, while tricarboxylic acid cycle (TCA) activity competes with lysine biosynthesis. Genetic engineering informed by these results, aimed at mutating the glucose-6-phosphate dehydrogenase and the phosphogluconate dehydrogenase, both enzymes of the pentose phosphate pathway (PPP). The introduced mutations aimed at reducing feedback inhibition and increased metabolic flows through the PPP, and hence increased lysine production. A similar effect was observed when fructose-1,6-bisphosphatase was overexpressed. These examples show that metabolite flux analysis is a valuable tool, if one wants to increase product yields in microorganisms (Pfefferle et al. 2003).

Transcriptome analysis identifies regulators of expression and the impact of mutations on the expression profile of a strain. When deriving an arginine producing strain from the WT using genetic engineering, it was observed that arginine biosynthesis proteins were less induced in this strain than in strains generated by random mutation. Induction of these proteins could be enhanced when a mutated *leuC* gene was introduced to this strain, increasing product accumulation. Lysine production could also be increased by deleting the *ilvB* gene in a production strain. The deletion affected amino acid biosynthesis in an indirect manner by changing the transcription of 49 genes. The advantage of strain design based on transcriptomic analysis is the ability to manipulate cells on the more subtle level of expression regulation.

Recent metabolic engineering approaches were aimed at inserting heterologous pathways into *C. glutamicum*. Progress in genetic engineering has expanded the range of compounds which can be produced by *C. glutamicum*, as well as the list of compounds the bacterium can use as a carbon source. Insertion of *E. coli* enzymes made it possible for *C. glutamicum* to metabolize glycerol, a common byproduct of oil seed processing. Also, the rate of glucosamine utilization was increased by mutating the promoter of the *nagAB* gene, resulting in higher expression rates of the gene. The production of compounds as 1,2-propanediol and gamma amino butyric acid was enabled in *C. glutamicum* by inserting enzymes as methyl glyoxal synthase and glutamate decarboxylase from *E. coli*. Polyamide, better known as Nylon, is a polymer of high industrial significance. A precursor for this polymer, diaminopentane, can be synthesized from lysine by decarboxylation. To synthesize diaminopentane in *C. glutamicum*, a lysine decarboxylase from *E. coli* was inserted into a *C. glutamicum* lysine production strain and flows towards diamino pentane were increased by deleting a lysine exporter (Figure 8.5) (Kind et al. 2014).These first steps to establish new products in this organism with high biotechnological significance lead the way to the use of *C. glutamicum* as microbial cell factory, producing a vast range of unnatural or nonnative products for the biotechnological industry.

As we have seen, *C. glutamicum* went through an extensive development. In the beginning, there were the first glutamate producing strains, followed by strains producing other amino acids and optimized for biotechnological growth. Also, the methods for strain development changed with time; random mutagenesis was the first method of choice and was later replaced by genetic engineering. Today, genome breeding and synthetic biology are the hallmarks of a new era in biotechnology, leading the way to more efficient and versatile microbial cell factories.

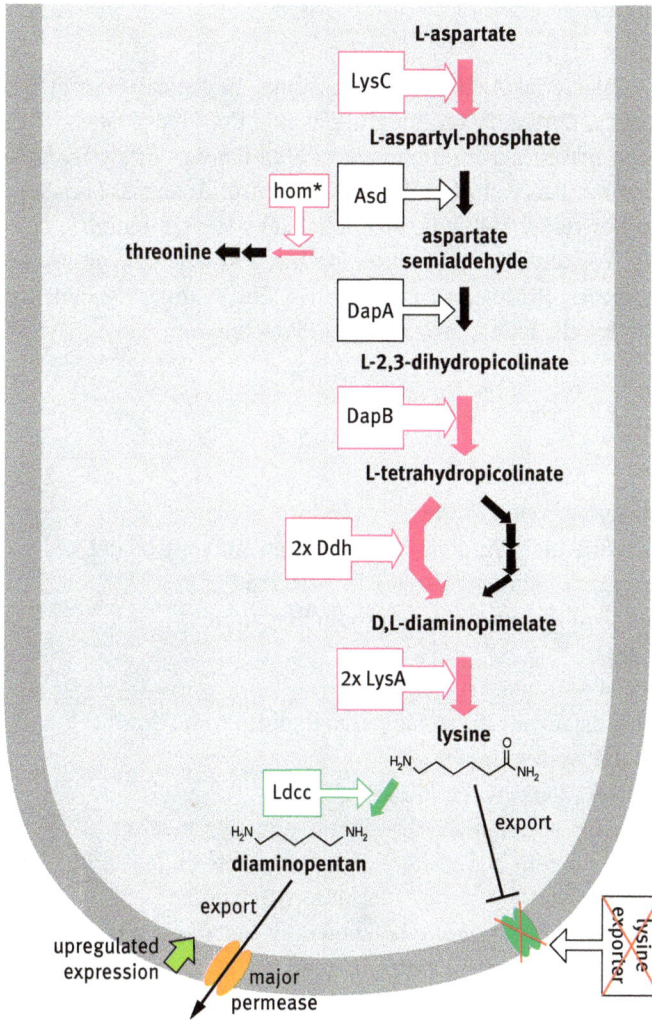

Fig. 8.5: Production of Diaminopentane by engineered *C. glutamicum*. Here, the lysine biosynthesis pathway was engineered to produce a polyamide. Diaminopentane is synthesized by decarboxylation of lysine catalyzed by decarboxylase Ldcc from *E. coli*, to increase flow through the decarboxylation pathway, the *C. glutamicum* lysine exporter *lysE* was deleted, while the permease P_{sod} *cg2893* was overexpressed. Also, the genes coding for the enzymes LysC, Dapb, DdH, and LysA were overexpressed, while the *hom* gene was attenuated. Abbr. of enzymes involved in lysine biosynthesis: Aspartate kinase (LysC); Aspartate semialdehyde dehydrogenase, dihydropicolinate synthase (DapA); dihydropicolinate reductase (DapB); meso-diaminopimelate dehydrogenase (DdH); diaminopimelate decarboxylase (LysA); homoserine dehydrogenase (hom) (Kind et al. 2014).

Key-terms

Amino Acid production, Bacterial Artificial Chromosome, Biological Control Circuit, Bottom up Proteomics, Corynebacterium glutamicum, Cre/LoxP system, Double strand break, Enzyme engineering, Fermentation, Flux Balance Analysis, Fluxomics, Genome engineering, leaky strains, Metabolic Control Analysis, Metabolic engineering, Metabolic Networks, Metabolomics, Michaelis-Menten-kinetics, Next Generation Sequencing, Polyamides, production strains, Propanediol, protoplast fusion, random mutagenesis, RNA Micro Arrays, RNA Seq, Sanger sequencing, strain screening, Synthetic cell, Transposons, White Biotechnology

Questions

–   What are the characteristics of a production strain?
–   How can the terms white biotechnology and red biotechnology be defined?
–   How can a glutamate producing prokaryote be selected?
–   What is the advantage of auxotrophies in production strains?
–   What is a leaky strain?
–   Explain the concept of protoplast fusion?
–   Why are prophages a disadvantage to biotechnological production?
–   How does the Cre/LoxP system function?
–   Explain the advantages of minimal strains?
–   What do the components of the Michaelis-Menten-equation mean?
–   Are in vitro and in vivo kinetic parameters of enzymes interchangeable?
–   How can metabolic networks be changed for flux optimization?
–   What are the parameters which need to be known for metabolic flux analysis?
–   Describe the main strategies of metabolic engineering!
–   What are the differences between Sanger DNA sequencing and next generation sequencing?
–   What additional information has proteomics to offer in comparison to transcriptomics?
–   How are the different -Omics approaches combined in systems biology?
–   In which ways do Fluxomics and metabolomics complement each other?
–   What are the goals of synthetic biology?
–   How can transcription regulators and genetic elements form a control circuit?

Further readings

Yukawa, H. & Inui, M. 2013. Corynebacterium glutamicum. *Biology and Biotechnology*. Springer Verlag, Heidelberg, New York, London.

Baltz, R. H., Davies, J. E. & Demain, A. L. 2010. Manual of industrial microbiology and biotechnolo-
gy. american society of microbiology press, Washington.
Ali, N. (ed.). 2012. Methods in molecular Biology. *Microbial Systems Biology*. Humana press, New
York.
Barh, D., Zambare, V. & Azevedo, V. (ed.). 2013. OMICS: Applications in Biomedical, Agricultural
and Environmental Sciences. CRC press. Boca Raton.

References

Hilliger, M. & Hertel, W. 1997. Regulation of L-lysine biosynthesis in prototrophic revertants of
Corynebacterium glutamicum. *J Basic Microbiol* 37(1): 29–40.
Ikeda, M. & Nakagawa, S. 2003. The *Corynebacterium glutamicum* genome: features and impacts
on biotechnological processes. *Appl Microbiol Biotechnol* 62: 99–109.
Kalinowski, J., Barthe, B. 2003. Bartels D., et al. The complete *Corynebacterium glutamicum* ATCC
13032 genome sequence and its impact on the production of L-aspartate-derived amino acids
and vitamins. *J Biotech* 104: 5–25.
Kind, S., Neubauer & S., Becker, J., et al. 2014. From zero to hero – Production of bio-based nylon
from renewable resources using engineered *Corynebacterium glutamicum*. *Met Eng* 25:
113–123.
Lee, Y. W, Kim, Y. T., Jang, Y. S., et al. 2011. Systems metabolic engineering for chemicals and
materials. *Trends in Biotech* 29(8): 370–378.
Pfefferle, W., Möckel, B., Bathe, B., et al. 2003. Biotechnological manufacture of Lysine. In:
Scheper T, ed. Advances in Biochemical Engineering/Biotechnology, Heidelberg, Germany,
Springer Verlag 79: 59–112.
Reaves, M. L. & Rabinowitz, J. D. 2010. Metabolomics in Systems Biology. *Curr Opin Biotech* 22(1):
17–25.
Stephanopolous, G. 2012. Synthetic Biology and Metabolic Engineering. *ACS Synth Biol* 1:
514–525.
Suzuki, N. & Inui, M. 2013. Genome Engineering of *Corynebacterium glutamicum*. In: Yukawa, H.,
Inui. M. (ed.) Microbiology Monographs *Corynebacterium glutamicum. Biology and
Biotechnology* 23: 90–104, Heidelberg, Germany, Springer Verlag.

Markus Piotrowski

9 Biotechnology of higher plants

Life on our planet as we know it is totally dependent on plants. Not only do they produce the oxygen that we are breathing and form a part of our daily diet, also the animals that we are eating were raised, giving them plants as fodder. Plants consume CO_2, and thus are an important sink for this greenhouse gas, keeping our planet cool. They produce fibres for our clothing (e.g., cotton, jute, linen, hemp), sweeteners and spices for our meals (e.g., sugar, pepper, cinnamon, mustard, saffron, ginger), luxury foods (e.g. coffee, cacao, tea, tobacco), and medicine (e.g., digoxin, quinine, codeine). Their wood is used in paper manufacture and as construction material. Plants provide fuels, either as coal or, nowadays, as bio-fuels. Finally, plants, especially their flowers, are used for decoration.

This list can easily be extended beyond the pages of this chapter. The reason why life without plants is not imaginable is that plants are the main primary producers on our planet. As photoautotrophic organisms, they do not rely on products of other organisms but can satisfy all their needs from an inorganic environment and with the use of light energy. They fetch minerals and water from the soil and, by the process of photosynthesis, they can fix gaseous CO_2 into carbohydrates. With these building blocks and chemical energy, which is also produced during photosynthesis, plants can produce all molecules (e.g., amino acids, nucleotides, sugars, lipids, vitamins) that they need for their life. Finally, land plants produce 98 % of the global biomass of approximately 560 billion tonnes carbon.

This chapter starts with a description of several breeding techniques: conventional breeding techniques making use of biotechnological methods or not, and transgenic breeding. After this follows a description of the main transgenic traits in todays' commercially available genetically modified crops and other traits were are worth mentioning for several different reasons. This part is closed by an overview on the current status of commercialized transgenic plants. While most of these topics deal with plants that are used as food or fodder, the final part of this chapter shows an additional example of plant biotechnology: the use of plants as production systems for pharmaceuticals.

9.1 Conventional breeding techniques

Domestication of plants by humans started after the last ice age, approximately 10,000 years ago. Since then, people have practiced breeding to grow plants with desired traits. The basis of breeding is diversity, either naturally occurring or artificially produced, and the selection by humans. Conventional breeding techniques are those techniques that do not use genetic engineering. These include "traditional" techniques like selective breeding and interbreeding, but also modern, biotechnological methods like micropropagation or mutational breeding.

9.1.1 Traditional breeding

Selective breeding

The oldest form of breeding is selective breeding, where only seeds of plants with positive traits (e.g., many or big seeds, resistance against pathogens, easier processing, better tasting) were used for further propagation. This is positive selective breeding, opposed to negative selective breeding, were most plants were propagated, while only those with clear negative traits were excluded. An example for the successful application of selective breeding is corn (maize, *Zea mays*). The wild progenitor of corn is teosinte, a highly branched grass with small cobs and only few but very hard-shelled kernels (Figure 9.1). Actually, maize looks so different to teosinte that scientists for a long time did not consider teosinte as a possible ancestor of maize. Genetic studies have shown that only five principal mutations were necessary to convert teosinte into a plant that looks similar to today's corn.

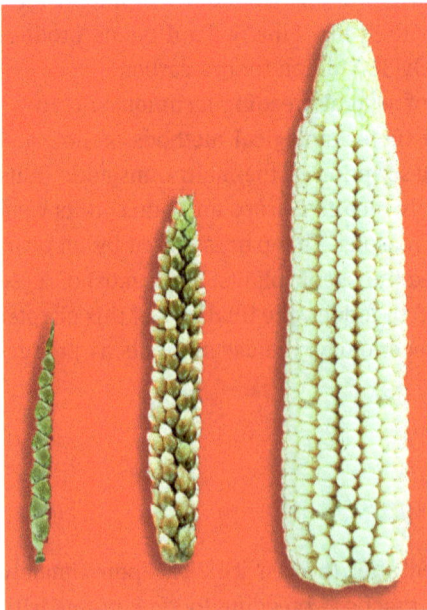

Fig. 9.1: Ears of teosinte (left) and modern maize (right). Teosinte and maize can still be crossed; in the centre is an ear of a F_1 hybrid from both plants. Image courtesy of John Doebley, University of Wisconsin.

Introgression

Crossing/interbreeding is often used for introgression of a certain trait into a cultivated species. This is done by crossing the cultivated variety (elite line) with, e.g., a wild variety (donor line) that possesses a certain trait, e.g., a resistance gene against a certain pathogen. While the offspring of this crossing will contain 50% of the genes of the wild variety, and may show many unwanted characteristics, repeated backcrossing with the elite line results in out-crossing of these donor genes. Crossing-over events allow that the gene of interest that confers the wanted

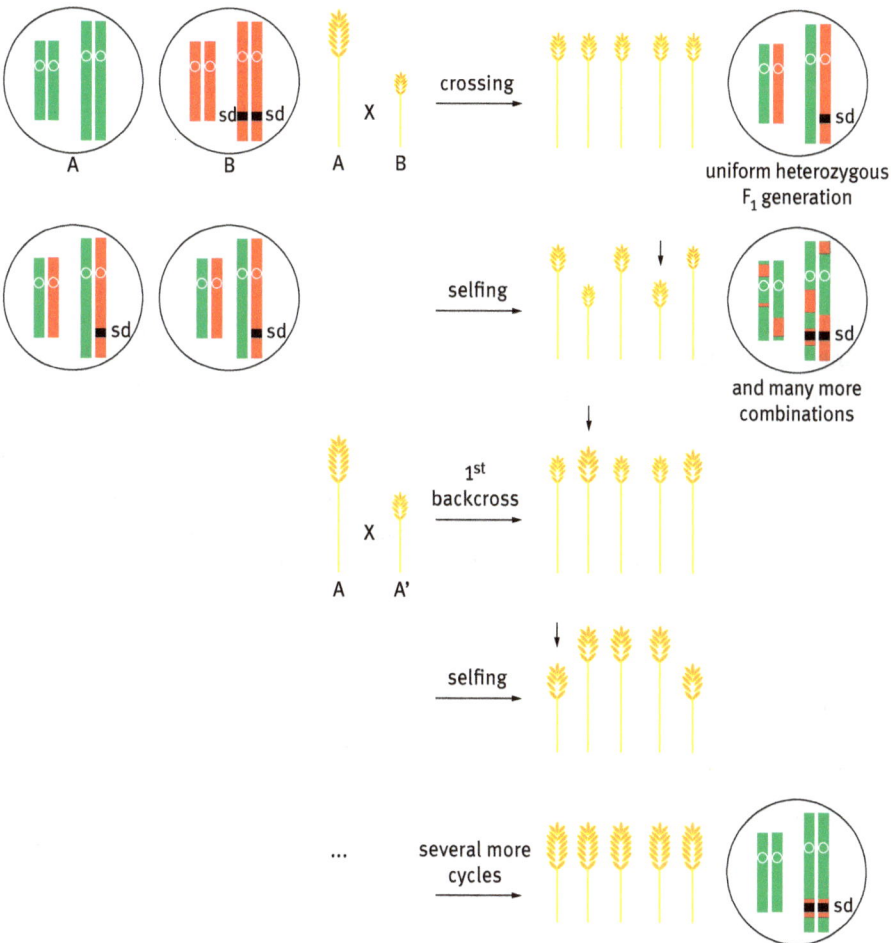

Fig. 9.2: Principle of introgression. For simplicities sake, a genome with only 2 chromosomes is shown. An already existing elite line (A, green genome) is crossed with a donor line (B, red genome) which contains the desired trait, in this case, semi-dwarfism (sd), which is recessive. By repeated cycles of selfing (to get the trait homozygous) and backcrossing to the elite line, more and more of the donor genes are removed. Finally, only a small part of the donor genome containing the trait is left in the elite line.

trait (in this case, resistance against a pathogen) is exchanged into the respective chromosome of the donor line (Figure 9.2). Introgression was used to introduce "shortness" into plants, and thus formed the basis for the so-called green revolution between 1960 and 1980, in which the yields of wheat and rice could be more than doubled. At the beginning of this period, high yielding varieties in combination with the use of synthetic fertilizers lead to long growing plants with heavy seed heads that tend to fall over (plant lodging) which resulted in large losses of the harvest. Short-growing varieties, so called semi-dwarfs, were used to introduce this trait – semi-dwarfness – into the existing high-yield varieties. Not only were the produced varieties more resistant to lodging, but they also had a better harvest index (the harvest index is the ratio of harvested grain to total shoot biomass).

It is important to keep in mind that the basis of introgression is the exchange of chromosome segments by crossing-over. Thus, what is exchanged actually is not "a gene", but a chromosome segment that may contain several genes.

Hybrid breeding

Hybrid breeding makes use of the so-called hybrid vigour or heterosis, the "greater constitutional vigour of crossed plants", as Darwin has put it. Today, heterosis is defined as the superiority of hybrid offspring to each of their pure-line (= homozygous) parents. The increase in yield ranges from 10 % to more than 100 % but often is around 25–50 %. Besides their higher yield, hybrids are also more uniform. Heterosis is only evident for the hybrids of the first generation (the F_1 generation) and gets lost in subsequent generations. Thus, to make use of heterosis, seeds must always be produced from the pure-line parents. This is the reason why farmers cannot use seeds of hybrids for the next generation, and instead have to buy them each year. For a successful production of a hybrid, self-fertilization must be avoided. This can be done by manually removing (emasculation) or covering the stamens (the organs that produce the pollen) in the line that serves as the female parent. Another way is to use plants with cytoplasmic male sterility (see below). Hybrid breeding was first developed for maize, and today, most of the corn varieties are hybrids. However, hybrid breeding was successfully transferred to many other crops, cereals and non-cereals, and often hybrid varieties outnumber the normal varieties.

9.1.2 Breeding techniques using biotechnology

Mutation breeding

In mutational breeding, plants were exposed to chemical, physical or biological mutagens, e.g., ethyl methanesulfonate (EMS), nuclear radiation, UV light, x-rays, and transposons to increase the natural mutation rate. Although widely unknown,

Fig. 9.3: Gamma-radiation field of the Institute of Radiation Breeding in Hitachiomiya, Japan. The field has a radius of 100 m and is shielded by walls of 8 m height. The centre is equipped with an 88.8 TBq Cobalt-60 source. Image © National Institute of Agrobiological Sciences (NIAS), Japan.

Table 9.1: Examples of plants derived from mutation breeding. In total, more than 3200 mutagenized varieties are officially released.

Variety	Trait	Mutagenizing Agent	Comment
Grapefruit Rio Red	Pulp with red color	Thermal neutrons	Mutagenesis on existing natural red mutant
Rice Calrose 76	Semi-dwarfism	γ Radiation	First semidwarf table rice released in the USA
Barley Golden Promise	Shorter straw, good yield, good malting properties	γ Radiation	Used in beer and whisky brewing
Pear Gold Nijisseiki	Resistance to black spot disease	γ Radiation	
Chrysanthemum Bronze Charmette	Bronze flower color	x-Rays	
Canola Clearfield	Herbicide resistance	EMS	EMS mutagenesis of pollens grains

EMS: ethyl methanesulfonate

more than 3,200 crop varieties produced by mutation breeding have been officially registered. The red-colored Rio Star grapefruit, for example, originates from a spontaneous mutant ("Ruby Red"), which was further mutagenized by irradiation to obtain an even more reddish fruit. Approximately one half of the registered mutant varieties are cereal crops, one quarter non-cereal crops, and the final quarter are ornamental plants. Figure 9.3 shows the y field of the Institute of Radiation Breeding in Japan. Table 9.1 gives some examples of cultivars that were produced by mutation breeding.

Crossing over species borders

Crossing is usually only applicable for organisms of the same species. However, in some cases, species borders may be overcome and fertile offspring of parents from different, although related species can be produced. One example is Triticale (Figure 9.4), which is a cross between wheat (*Triticum aestivum*) and rye (*Secale cereale*). Crossing rye (as male parent) with wheat (as female parent), however, only

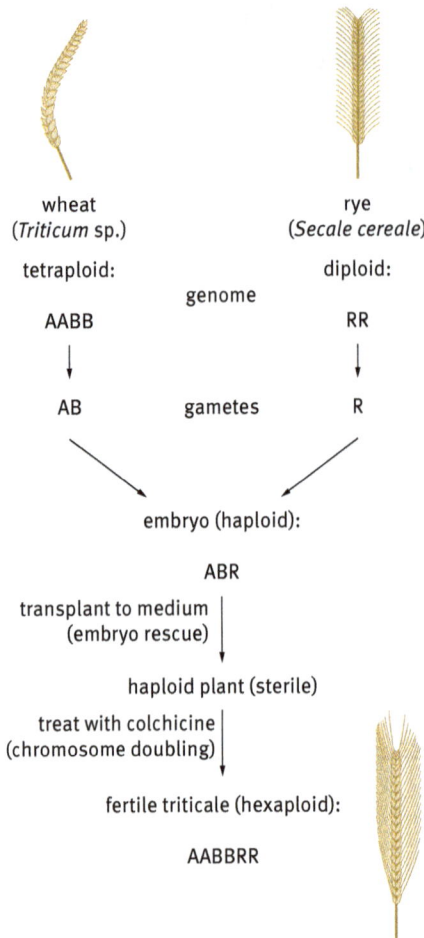

Fig. 9.4: Making of Triticale, a cross between wheat and rye. For details, see text.

seldomly results in the formation of fertile Triticale. Only a very small number of seeds are produced, many of them may have a degenerated endosperm (and thus will not develop) and the plants that survive are sterile, since they contain a single copy of each of the parents' genome. Only occasionally spontaneous genome duplication occurs, leading to an amphidiploid (two diploid genomes) plant which is fertile. Triticale breeding was brought forward by two techniques: 1. embryo rescue allows raising plants from defective seeds by transplanting the plant embryo from the seed on culture medium, and 2. treatment with colchicine, the poison of the autumn crocus, results in chromosome doubling. Colchicine prevents the formation of microtubules, thus during mitosis, chromosomes are doubled, but the mitotic spindle cannot be formed and the cells do not divide. By applying these techniques, the production of primary Triticale was much improved. Secondary Triticale is a cross of two primary Triticale plants. Although Triticale is a very young plant, it is grown in approximately the same amounts as rye (14 million tonnes per year). Yearly global production of wheat, however, is 670 million tonnes.

Somatic cell hybridization/protoplast fusion

Plants can also be "crossed" asexually, by fusion of somatic cells (Figure 9.5). Single plant cells are released from plant tissues by the use of pectinases (see Chapter 4.3.2), enzymes from fungi that cleave pectin, the main component of the plant cell's middle lamella which connects the cell walls of neighbouring cells. Another fungal enzyme, cellulase, is then used to degrade the cell wall. The remaining plant cell without cell wall is called a protoplast. It has a spherical form and must be kept in iso-osmotic medium to prevent bursting or collapse. Protoplast from different plants can be fused in the presence of certain chemicals, e.g., polyethylene glycol or by applying an electric field, thus forming a somatic hybrid cell. Since plant cells are generally totipotent, it is possible to regenerate whole plants from such a hybrid cell (see Chapter 9.2). Although the process of protoplast fusion circumvents the problems of 'natural' crossing and allows the fusion of cells from only distantly related plant species, it is often not possible to regenerate whole plants from such hybrid cells or they develop abnormal and usually, they are sterile. However, fertile plants have often been regenerated from somatic cell hybrids of closely related plants from the same genus or the same tribe. A possible outcome of somatic cell hybridization is cytoplasmic male sterility (CMS). CMS is the result of an incompatibility of the nuclear genome of one partner with the mitochondrial genome of the other partner. Although at first glance counter-intuitive, CMS is an interesting trait, since plants that cannot self-pollinate are highly demanded in hybrid breeding (see above).

Plants produced from protoplast fusions are an in-between regarding gene technology. According to European and German laws these plants are not regarded as genetically modified if the donor plants can be crossed by conventional breeding

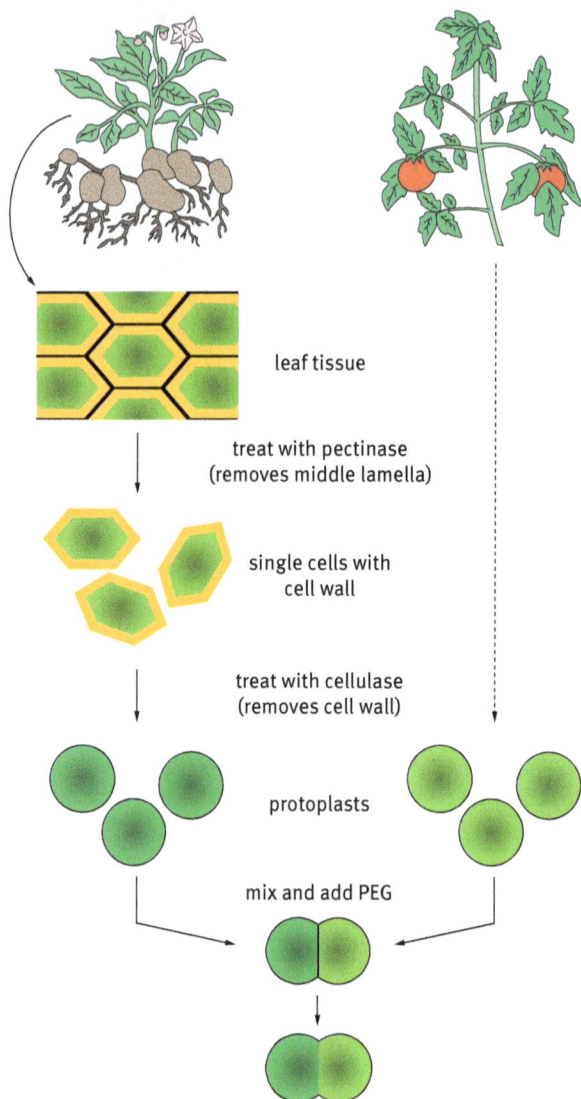

Fig. 9.5: Generation of a somatic hybrid cell.

techniques. If, however, the fused cell contains a new combination of genetic material that does not occur under natural circumstances, it is classified as transgenic (i.e., genetically modified).

Vegetative propagation and micropropagation
There are well-known examples of plants that can be propagated vegetative (asexual); e.g., potato plants are usually grown from potato tubers, but not from potato

seeds. Strawberries can reproduce themselves through runners (stolons). In gardening, propagation of plants by cuttings is common practice. Vegetative propagation is often used when sexual propagation is difficult: Cultivated bananas, for example, do not form seeds or trees usually have a long juvenile phase when growing from seeds. But vegetative propagation is also a method to clonally amplify an individual plant which has superior characteristics. The increased vigour of hybrids can be maintained in the same manner. In micropropagation, a small part of plant tissues (sometimes a single cell) is excised and transplanted to sterile medium. These explants are usually taken from parts of the plant that contain undifferentiated and actively dividing (= meristematic) cells. They are grown as *in vitro* culture and develop into cell cultures, calli (which is an aggregation of undifferentiated plant tissues) or plantlets (a small, vegetatively developed plant). These will be divided (e.g., calli can be mechanically divided in smaller parts) and further propagated over several rounds. By this procedure, the original explant can be amplified several thousand to several hundred thousand fold. Finally, whole plants are regenerated by the external application of plant hormones and planted to soil (see Chapter 9.2 and Figure 9.8). More than 150 plants species are micropropagated, among them banana, apple, pears, strawberry, ornamental plants like orchids and roses, but also crops like sugarcane.

Anther and pollen culture
Cells derived from anthers or single pollen can be used to regenerate haploid plants (since pollen does only contain the male chromosome set). Haploid plants are useful in mutation breeding, because otherwise recessive mutations will show their full effects. Haploid plants are also used for somatic cell hybridization. As described above for Triticale, colchicine can be used to generate a diploid plant (which is homozygous for all loci).

SMART Breeding
SMART stands for "selection with markers and advanced reproductive technologies". Another term describing the same technique is "marker assisted selection" (MAS) or precision breeding. In MAS, the gene of interest or at least a locus for a certain trait is known, and molecular techniques to detect this gene or locus are available, thus it can serve as marker. With such markers, the success of a certain crossing can be analysed long before the trait becomes visible in the phenotype. The markers can already be used to select the parents for a crossing. There exists several types of molecular markers, e.g., cleaved amplified polymorphic sequences markers (CAPS) are based on the presence or absence of a recognition site for a restriction endonuclease in a PCR-amplified genomic region and amplified fragment length polymorphisms (AFLP) use PCR to detect loci of different length (due

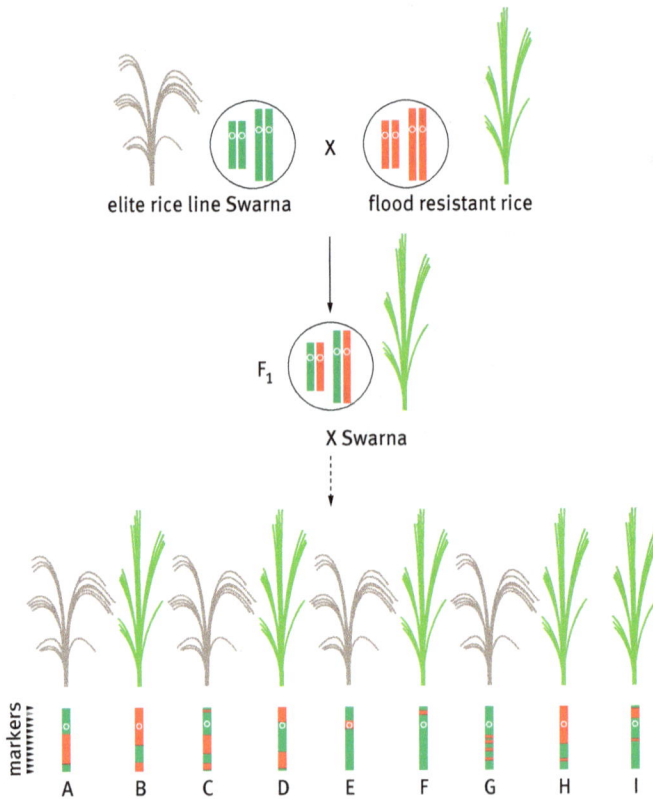

Fig. 9.6: Marker-assisted selection in the generation of flood-tolerant rice plants. A flood-tolerant cultivar is crossed to an existing elite line. The F_1 generation is backcrossed several times with the elite line and the progeny is analysed. Comparing the phenotype (flood tolerance) and the genotype (using molecular markers) show that all resistant plants (B, D, F, H, I) carry a chromosome fragment from the donor plant that co-segregates with a certain marker (shown in grey). The markers can also be used to select that line for further backcrossing which contains the largest portion of the elite genome (marker-assisted backcrossing); in this case the plant F.

to repetitions, deletions or insertions). The marker must not reside within the gene causing the trait but should be in close proximity so that separation of the marker and the gene by crossing-over becomes unlikely. On the other hand, the gene(s) responsible for the trait must not necessarily be known, if marker exist that are tightly linked to the trait (Figure 9.6). For breeding flood-tolerant rice in India, markers were first used to identify a genomic region responsible for submerging tolerance in a certain rice cultivar. This cultivar was then crossed with the popular rice variety Swarna. During this process, different markers were used to identify a) the progeny plants containing the "*Sub1*" locus, where b) the *Sub1* locus was introduced into a Swarna chromosome (by crossing-over) and c) which contained only few genes from the donor cultivar. By this strategy, the number of backcrosses could be reduced and the final Swarna-*Sub1* variety only contains the *Sub1* locus in an otherwise Swarna genetic background.

Tilling

Targeting induced local lesions in genomes (TILLING) is a method to identify point mutations in a gene of interest (Figure 9.7). Plants that were chemically mutated in mutational breeding (see above) can rapidly be analysed for a mutation in a certain gene by this method. The gene of interest is amplified by PCR from a DNA mixture from several plants (5–8). The PCR products are melted into single strands by heating and re-annealed by slow cooling. If the gene of one plant contains a mutation, heteroduplexes of non-mutated and mutated-strands will form. Such heteroduplexes can be identified by several methods. The actual standard procedure is to amplify the gene with differently labelled forward and revers primers and treat the re-annealed mixture with the mismatch-specific endonuclease CEL I. If the mixture contains only wild-type genes, a single PCR product with both labels will be present. If, however, one plant contained a mutated gene, the resulting

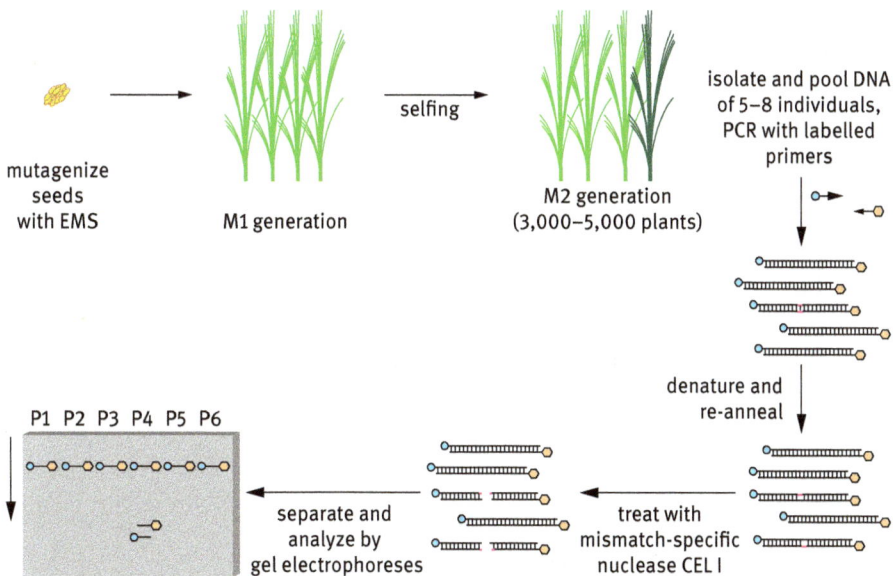

Fig. 9.7: Principle of TILLING. Seeds are mutagenized and plants are propagated to the M_2 (mutant) generation. DNA of these plants is isolated and pooled, and the gene of interest is amplified with differently labelled primers. A cycle of denaturing and re-annealing results in heteroduplex formation if a mutagenized copy of the gene is present. These heteroduplexes are specifically cleaved by the nuclease CEL I and can be identified by, e.g., gel electrophoresis. In this example, one pool of plants (P4) contains a mutagenized gene copy, which results in three bands: The wild-type homoduplex carrying both labels and the cleaved heteroduplex fragments, each carrying one label.

heteroduplex will be cut by CEL I and in addition to the wild-type PCR product, two fragments, each carrying only one label will appear. The size of the fragments can be used to determine the region, where the mutation has occurred. Finally, the mutation must be determined by sequencing.

Once a useful mutant has been identified (e.g., a mutation in a gene that causes dwarfism), it must be backcrossed several times with the wild-type line to get rid of the numerous other mutations that are present in the mutant.

9.2 Regeneration of whole plants from plant tissue or single cells

As already mentioned in the last chapter, plant cells usually stay totipotent. Thus, even fully differentiated cells can be de-differentiated, and so it is possible to grow a complete plant from a single cell. This characteristic of plants is not only an

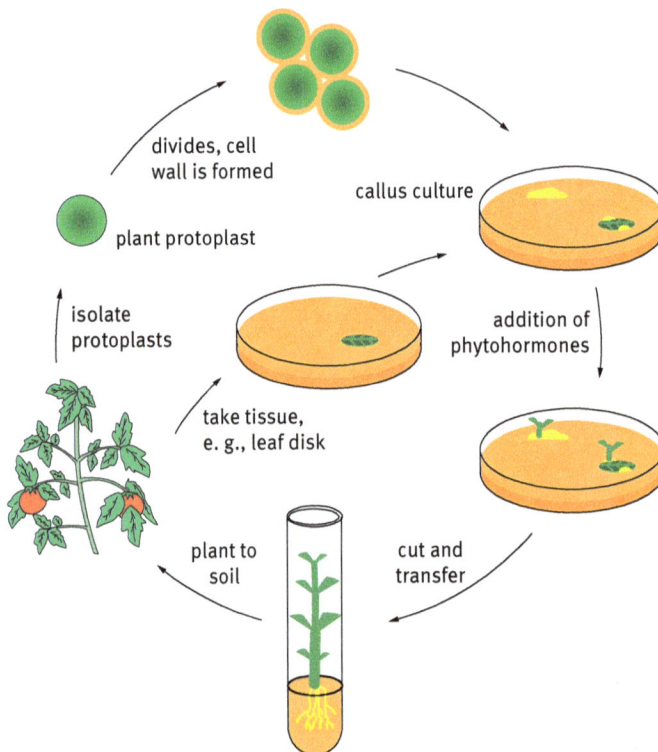

Fig. 9.8: Regeneration of a whole plant from a single cell. An isolated protoplast (see Figure 9.5) is grown in media where the cell divides and forms a callus. By addition of the phytohormones auxin and cytokinin, shoots and/or roots emerge from the callus and develop into a whole plant which can be planted to soil. Regeneration can also be done from tissues, e.g., leaf disks.

important prerequisite for many of the biotechnological breeding techniques mentioned above (like somatic cell hybridization, micropropagation, and cell culture) but also for the generation of transgenic plants, because transformation of plants is mostly done on plant tissues or single cells.

Regeneration may start with a single protoplast (see above, protoplast fusion and Figure 9.8) or a tissue explant, e.g., a leaf that was cut into small pieces and placed on sterile medium. At the wound edges, calli (see above, micropropagation) will develop. Addition of an auxin, a plant hormone, causes rooting, while addition of a cytokinin, another plant hormone, will lead to the formation of shoots. The application of both hormones in the correct concentrations and ratio will result in the development of a complete plant.

The described procedure must be optimized for each plant (e.g., which tissue, cultivation medium, growth conditions, concentrations, and ratio of phytohormones) which may be quite laborious, and although success has often been achieved, there are other examples where regeneration of a complete plant was not successful yet.

9.3 Transgenic breeding

A genetically modified (transgenic) plant is a plant in which the genetic material has been altered in a way that does not occur naturally by mating or natural recombination. Usually, one or few genes (often two: the gene of interest and a marker gene that is used for selecting the transformed plant) under the control of foreign promoters were introduced into the nuclear genome. These genes are typically obtained from organism that cannot be crossed with the transformed plant, either other plants or completely different organisms, for example bacteria. These are the two main advantages of "transgenic breeding" in comparison to conventional breeding: 1. The possibility to introduce a single selected gene into an otherwise unchanged genome (introgression in conventional breeding allows only for exchanging chromosomal fragments which may contain many genes) and 2. the unlimited access, at least theoretically, to the complete gene pool of live.

9.3.1 Direct gene transfer methods

In direct gene transfer, 'naked' DNA is directly inserted into plant cells. This is done mainly by three different methods, 1. protoplast transformation, 2. electroporation, and 3. biolistic transformation. The DNA construct that is transformed can be a usual cloning plasmid which carries the genes of interest (Figure 9.11a). So cloning the genes and their promoters is done in *Escherichia coli* and once the cloning is finished, the plasmid DNA is isolated from the bacterium and used for transformation. However, to avoid transformation of unwanted plasmid components the part of the plasmid that carries the genes to be transformed can be isolated from the plasmid, e.g. by using restriction enzymes. Alternatively, PCR may be used to amplify this section. Direct gene transfer methods are usually carried out on single cells or tissues, thus regeneration of whole plants (see Chapter 9.2) must be carried out.

Protoplast transformation and electroporation
In protoplast transformation, protoplast (see above: protoplast fusion) are mixed with DNA in the presence of polyethylene glycol (PEG) and calcium ions which destabilize the plasma membrane and makes it permeable for the DNA. Nowadays, protoplast transformation is not widely used for the generation of transgenic plants; however, it is an often used method for transient transformation (Figure 9.9). Transient transformation is a useful method, e.g., to study the localization of proteins within a cell. For this purpose, genes are constructed that encode the protein of interest coupled to a fluorescent protein (like the green fluorescent protein GFP), and this constructs are transiently transformed in protoplasts. There, the genes are expressed but not necessarily incorporated into the genome.

In electroporation, cells are made permeable for the uptake of DNA by a short electric pulse of high voltage (usually between 1 and 7 kV/cm). The advantage in comparison to protoplast transformation is that it is not necessary to prepare protoplast; intact cells and even tissues like callus cultures or inflorescence tissue can be used for transformation.

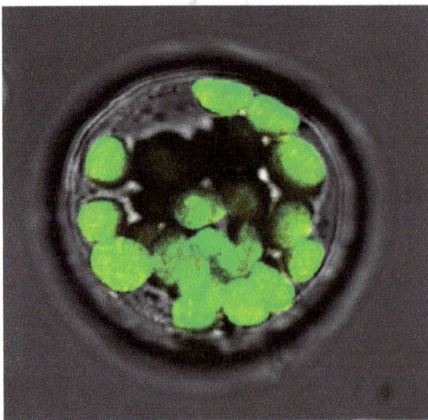

Fig. 9.9: Leaf cell protoplast from *Arabidopsis thaliana* transformed with a *GFP* gene construct that is targeted to the chloroplasts, therefore, they appear green under UV light. Photograph courtesy of Beatrix Dünschede, Ruhr-Universität Bochum.

Biolistics (particle bombardment)

Biolistics is a coinage derived from "biological ballistic". By this method, DNA is shot into the cells. The bullets are small particles (~1 μm diameter) made from tungsten or gold, which are covered with the DNA to be transformed. One of the prototype devices for such particle bombardment, a "gene gun", indeed used gun powder, but nowadays, such apparatuses are operated with compressed gases (e.g. helium) to build up the pressure for shooting (Figure 9.10). Particle bombardment or biolistics is the most widely used direct transformation method. Especially cereal crops are often transformed by biolistics, since the other widely used transformation method, *Agrobacterium*-mediated transformation (see below), is less efficient in these species. Biolistics can also be used for transient transformation and it is the only method which allows the transformation of plastids (e.g., chloroplast) if a DNA-loaded particle hits the organelle. One drawback of the method is that the cells which will be transformed are inevitable injured, so many of them are severely damaged or die. Also, multiple insertions in the genome of sometimes rearranged transgenes are a known issue.

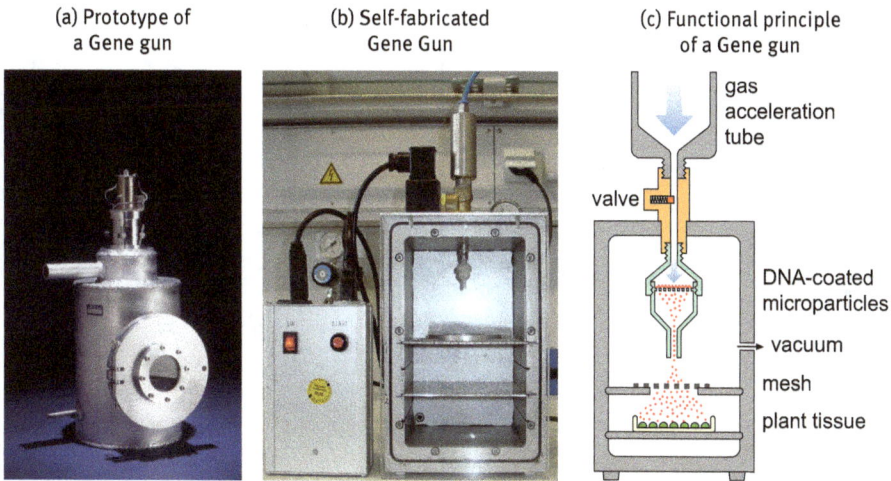

Fig. 9.10: Gene Gun. (a) Prototype III of a Gene Gun constructed by John Sanford, Ed Wolf and Nelson Allen at Cornell University, New York. (b) A self-fabricated Gene Gun. The function principle is shown in (c). Gas from a gas cylinder is used to build up a high pressure. An electric valve is used to release the pressure suddenly, and the DNA-coated particles are accelerated and hit the cells. The chamber containing the specimen is put under vacuum to reduce drag on the particles from air resistance. Photograph in (a) courtesy of the Smithsonian National Museum of American History.

9.3.2 *Agrobacterium*-mediated transformation

Agrobacterium tumefaciens (meaning "tumor-producing soil bacterium") (Figure 9.12a) is a plant pathogen that is able to transfer part of its genetic material into the genome of a plant cell. This transfer DNA (T-DNA) is located on a large plasmid (200–250 kBp), called the tumor-inducing plasmid (Ti plasmid, see Figure 9.11b) and is defined by two border sequences (left border and right border), which are nearly perfect repeats of 25 base pairs. Outside the T-DNA, the Ti plasmid harbors the *vir* region, a cluster of approximately 30 virulence genes which are

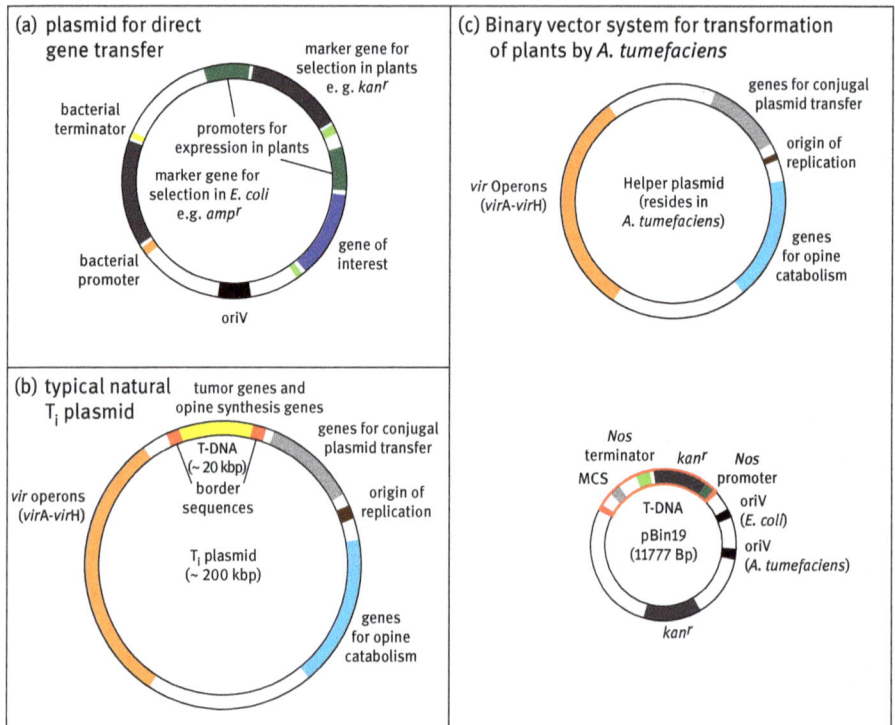

Fig. 9.11: Plasmids for genetic transformations of plants. Note that the maps are not to scale. (a) A typical plasmid construct for direct gene transfer in plants. Besides the usual plasmid elements like selection marker and origin of replication (oriV) for amplification in *E. coli*, it contains the gene of interest and a marker gene for selection in plants, both under the control of promoters (shown in dark green) and terminators (light green) that are active in plants. (b) A typical T₁ plasmid from *Agrobacterium tumefaciens*. Several different T₁ plasmids occur in nature. Details see text. (c) Binary vector system. The T-DNA is removed from the original T₁ plasmid which is now called helper plasmid. pBin19 is a first generation binary plasmid, carrying the T-DNA borders, two kanamycin resistance genes (for selection in bacteria and in plants) and two origins of replication (oriV), thus cloning can be done in *E. coli* before the construct is put into *A. tumefaciens*. For easier cloning of the transgene construct, a multiple cloning site (MCS) is integrated in the T-DNA.

(a)

(b)

(c)

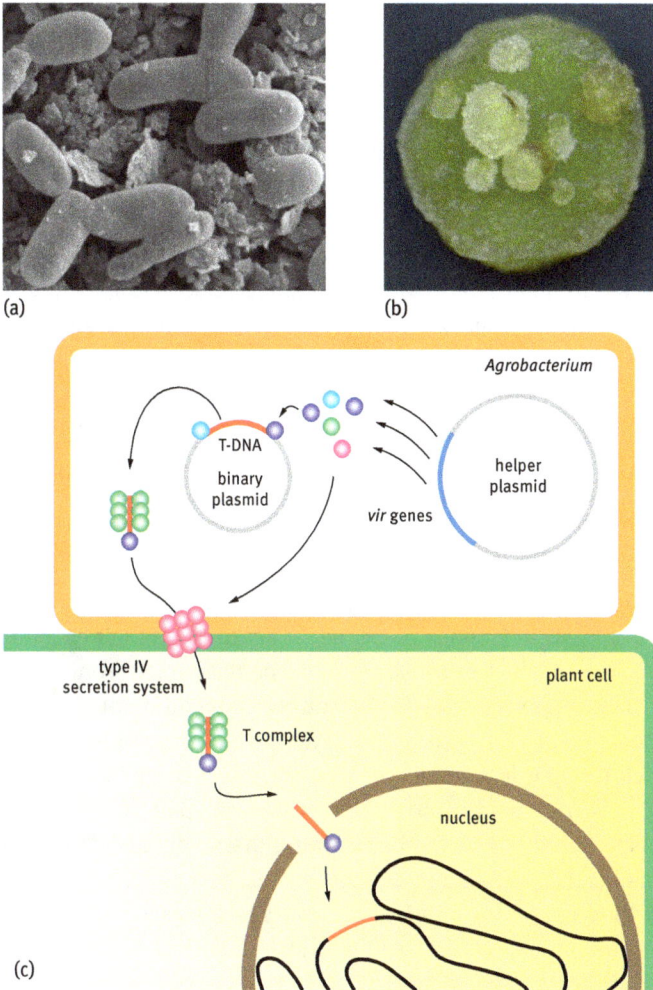

Fig. 9.12: *Agrobacterium tumefaciens* and *Agrobacterium*-mediated transformation. (a) Electron micrograph of *Agrobacterium tumefaciens* cells. (b) Tumors on a potato disc after infection with *A. tumefaciens*. (c) Transformation of plants by *A. tumefaciens*. Vir proteins cut the T-DNA as single-stranded DNA form the plasmid and bind to it, forming the T-complex. Others form a secretion system through which the T-complex enters the plant cell. Finally, the T-DNA is integrated in a random position into the plant genome, most probably by non-homologous end joining in double-stranded breaks (illegitimate recombination). Photographs in (a) and (b) courtesy of Philip Möller, Ruhr-Universität Bochum.

responsible for the transfer of the T-DNA into the plant cell. Two of them (*vir*A and *vir*G) encode a bacterial two-component receptor responsible for detecting a wounded plant cell and activating the expression of the other *vir* genes. Other vir proteins cut the T-DNA as linear DNA out of the plasmid (virD1 and virD2), cover the DNA to protect it (virE2), and build a type IV secretion system (several virBs) which connects the bacterial and plant cells. After transfer to the plant cell, the T-DNA is transferred into the nucleus where it is inserted at a random position in the plant genome by processes which are usually involved in repair of double strand breaks. This type of integration, which is also observed in direct gene transfer methods, is called illegitimate recombination and does not involve sequence homology between the transferred DNA and the integration site.

The transferred T-DNA contains genes for the biosynthesis of the plant hormones auxin (inducing cell elongation) and cytokinin (inducing cell division), which is the reason for the formation of tumors at the infected sites (Figure 9.12b). Other genes of the T-DNA are responsible for the formation of opines; these are either conjugates of certain amino acids with certain keto acids or with the sugar mannose. The plant cell is unable to make use of these conjugates but the Agrobacteria possess genes for opine catabolism, also localized on the Ti plasmid (but outside the T-DNA), and thus, can utilize these compounds a C-, N- and energy source.

Two important findings allowed for the successful application of *Agrobacterium* as 'natural genetic engineer' in the production of transgenic plants: 1. The T-DNA is only defined by the border sequences; this means that the tumor- and opine-producing genes can be removed (resulting in a 'disarmed' Ti-plasmid) and replaced with any DNA of interest. 2. Since the *vir* genes act through their encoded proteins on the T-DNA, i.e., they act *in trans*, they can be decoupled from the T-DNA. This means that the *vir* region and the T-DNA can reside on different plasmids (in the same *Agrobacterium* cell). This led to the development of so-called binary vector systems, in which the *Agrobacterium* contains the helper plasmid, which is a Ti plasmid carrying the *vir* genes but no T-DNA. The other part is the binary plasmid, a much smaller plasmid (5–15 kbp) containing the modified T-DNA and additional genetic elements that makes handling of this plasmid much easier (Figure 9.11c). The T-DNA may contain a multiple cloning site for easier cloning of transgenes, but more importantly, the binary plasmid is a shuttle vector which allows for replication in *Agrobacterium* and *E. coli*. Thus, construction of the T-DNA can be done by standard molecular biology techniques with *E. coli* and once the cloning is finished, the resulting binary plasmid is transformed in *Agrobacterium* that already contains the helper plasmid. The process of transforming a plant cell by *A. tumefaciens* carrying a binary vector is shown in Figure 9.12c.

Generation of transgenic plants using *Agrobacterium* usually involves incubation of plant tissue, e.g., sliced pieces of leaves, with the modified *Agrobacterium* strain, which requires regeneration of whole plants from callus culture (Figure 9.8). However, in some cases, a germ-line transformation is successful, in which flowering plants are dipped into an *Agrobacterium* solution. If cells of the germ line are transformed, genetically modified seeds will form and transgenic plants can be grown directly from such seeds, making the time consuming and tedious regeneration procedure dispensable.

9.3.3 Transformation of plastids

As mentioned above (Chapter 9.3.1), biolistics allows the generation of transgenic chloroplast. "Transplastomic" plants (plants with a genetically modified plastid genome = plastom) are interesting for several reasons. Since the chloroplast is of bacterial origin, modification of the plastom offers some advantages over modification of the nuclear genome. First of all, gene integration into the plastom runs through homologous recombination, thus genes can be inserted precisely at a certain position. Several genes can be organized in an operon, needing only one promoter (in the nuclear genome, each gene needs its own promoter). Transgene silencing, an epigenetic effect sometimes observed with transgenes in the nuclear genome, does not occur in the chloroplast. Besides these differences caused by the different genetic properties, chloroplast allow for a very high expression and accumulation of proteins. A protein that is heterologously expressed in chloroplasts may account for up to 45 % of the total protein of a cell, making transplastomic plants interesting for the production of protein-based pharmaceuticals like antigens or antibodies (see also Chapter 9.6.2). Finally, transplastomic plants may be an instrument for transgene containment; plastids are usually only maternally inherited, thus the risk of transgene escape via pollen spreading is negligible. On the downside, it sometimes proves difficult to obtain homoplastomic plants; these are plants where all copies of the chloroplast genome in all chloroplasts of the plant carry the transgene. This is because a single plant cell contains several (up to 100) chloroplasts (or other plastids) and each plastid contains approx. 100 copies of its genome. Therefore, homoplasy can only be reached by a strong selection of the transformed cells by the use of an appropriate marker gene (see Chapter 9.3.5).

9.3.4 Architecture of the transgene construct

For a gene to be expressed efficiently in a plant cell, certain requirements must be fulfilled. Most importantly, besides the coding sequence, is the promoter which regulates the expression of the transgene(s). However, also the coding sequence may require optimization, especially if the transgene is not derived from plant.

Promoters
In transgenic plants, very often constitutively active promoters are used which drive the expression of the transgene at every time in every tissue. The most commonly used promoter is the 35S promoter of the cauliflower mosaic virus (CaMV 35S promoter). This promoter drives the transcription of the whole viral genome (approx. 8 kBp) as a single large RNA, which has a Svedberg coefficient of sedimen-

tation (a measure of size) of 35S, thus the name for the promoter. In plants, this promoter is highly active at all times in virtually all tissues. Genetic studies allowed to separate the promoter in different functional units, like a minimal promoter (necessary but not sufficient for expression) and enhancer elements. In plant biotechnology, often enhanced versions of the CaMV 35S promoters are used which carry multiple copies (usually 2–4) of the enhancer elements.

In monocot plants, to which cereal crops belong, this promoter is less active compared to dicot plants. Here, promoters of housekeeping genes are often in use, e.g., the *actin-1* promoter of rice and the *ubiquitin* promoter from maize. Both genes contain an intron in their 5′ untranslated region, which are also included into the promoter constructs since it has been shown that such 5′-UTR introns increase the expression levels.

Promoters of the *Agrobacterium* opine synthase genes, e.g., from nopaline synthase (nos), octopine synthase (ocs), or mannopine synthase (mas) have also been frequently used in transgenic crops, although it is nowadays known that their activity is organ specific and developmentally regulated.

As knowledge about promoters and their activity in plants is increasing, tissue-specific, temporally-active, and regulated promoters are gaining more interest. Why express a bacterial resistance gene all the time, even when the plant is not challenged by bacterial pathogens? Why express a gene to keep tomatoes longer

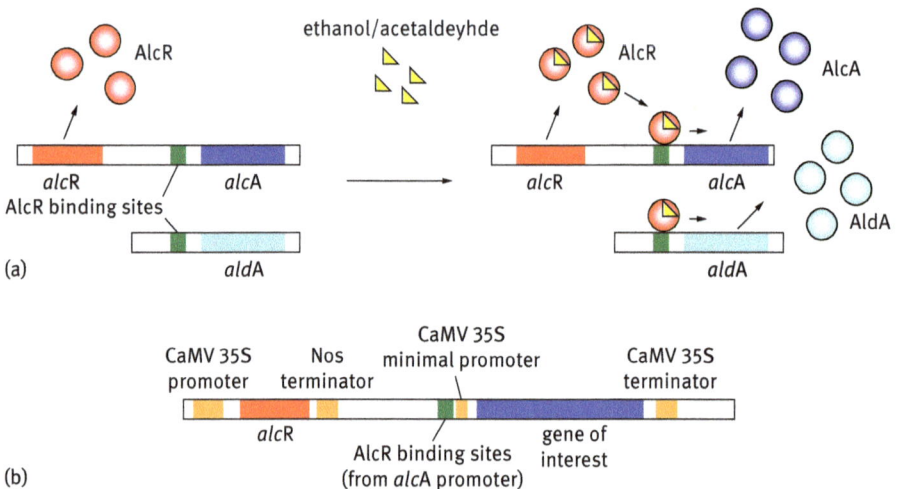

Fig. 9.13: The ethanol switch. (a) Regulation of genes involved in ethanol utilization in *Aspergillus nidulans*. Upon binding of acetaldehyde, the transcriptional activator AlcR binds to specific binding sites in the promoters of the *alcA* and *aldA* genes (coding for alcohol dehydrogenase and aldehyde dehydrogenase, respectively) and induces their transcription. (b) Use of the ethanol switch in transgenic plants. The gene of interest is placed under the control of an artificial promoter consisting of a CaMV 35S minimal promoter and the AlcR binding sites. The *alcR* gene is constitutively expressed from the (full length) CaMV 35S promoter. Nos: Nopaline synthase.

Table 9.2: Chemical-inducible promoter systems.

System	Inducer	Transcription factor	Promoter	Comment
Ethanol induction	Ethanol or Acetaldeyhde	AlcR from *Aspergillus nidulans*	Minimal promoter plus AlcR binding sites	Figure 9.13
Tetracycline de-repression	Tetracycline	TetR from transposon Tn*10* of *E. coli*	Full promoter plus *tet* operator sites	TetR is a transcriptional repressor. Binding of tetracycline releases it from the promoter
Tetracycline inactivation	Tetracycline	Fusion protein of TetR and the activation domain of the VP16 protein from *Herpes simplex*	Minimal promoter plus *tet* operator sites	
Steroid induction	Dexamethasone	Fusion protein of the GAL4 DNA- binding domain (*Saccharomyces cerevisiae*), the VP16 activation domain (*Herpes simplex virus*) and the steroid binding domain of an animal steroid receptor	Minimal promoter plus GAL4-binding sites	The steroid-binding domain is bound by HSP90 in the absence of the inducer. Binding of the steroid releases the transcription factor and allows binding to the promoter

fresh in the whole plant and not only in the fruit, where it exerts its function? Tissue-specific promoters have been used in the production of Golden Rice (see Chapter 9.4.4) to establish carotenoid biosynthesis in the rice kernel. Inducible promoters are especially interesting in basic research, if the gene to be expressed has fatal consequences for the plant. One such system, which has been ported from a fungus to plants, should shortly be introduced here: the alcohol-regulated promoter from *Aspergillus nidulans* (Figure 9.13a). In *A. nidulans*, the *alcR* gene codes for a transcriptional activator that is, however, inactive in the absence of its inducer acetaldehyde. In the presence of acetaldehyde, which is produced by alcohol dehydrogenase from ethanol, AlcR binds to AlcR binding sites in promoters of genes that are subsequently activated. In transgenic plants, the *alcR* gene has been introduced under the control of the CaMV 35S promoter (Figure 9.13b). In addition, the gene of interest, which expression is to be regulated, is placed under the control of an artificial promoter, consisting of the CaMV 35S minimal promoter (necessary but not sufficient for expression, see above) combined with alcR binding sites. Expression of the gene can then be induced by application of ethanol. Other examples of chemical-inducible promoters are given in Table 9.2. Chemical-inducible promoters have not been used so far in the generation of cultivated transgenic crops.

Terminators

Terminators are important for a regulated termination of gene transcription by RNA polymerase. Failure to stop transcription accurately may lead to the formation of aberrant mRNAs which are prone to induce transgene silencing by posttranscriptional gene silencing mechanisms. Very often, the transcriptional terminator of the nopaline synthase gene (*nos* terminator) from *Agrobacterium* is used.

Adaptation of the coding sequence

The differences in transcription, transcript processing and translation between prokaryotes and eukaryotes can make the expression of a genuine bacterial coding sequence in plants ineffective. Bacterial genes do not contain introns, however, they may contain sequences which are recognized in a plant cell as splice sites (so called cryptic introns). The same applies to polyadenylation signals. In addition, mRNA stability in eukaryotes is regulated by ATTTA sequences. Short-living mRNAs often contain several copies of this sequence in their 3′ untranslated region, and removing them increases the life time of the mRNA. Finally, different organisms may have a different codon usage; e.g., of the six possible codons for arginine, the *E. coli* strain K12 mostly uses only two, CGT and CGC (in 78 % of the cases), while AGG is nearly never used (2 %). Rare codons are often accompanied by a low amount of the corresponding tRNAs; thus if a bacterial gene contains many codons that are rare in plants, transcription of this gene in plants may be slowed down or inhibited. Since the degenerated genetic code allows changing the base composition of a coding sequence without changing the encoded protein, bacterial genes can be codon-optimized for the expression in plants. In this process, not only will the codon usage be adapted, but also cryptic splice sites and polyadenylation signals and ATTTA sequences can be removed. Such a plant-optimized gene may be expressed 100-fold higher compared to the genuine bacterial gene.

Finally, if the cellular location where the transgene should exert its function is an organelle, the coding sequence must be equipped with the appropriate targeting sequence. A gene from *A. tumefaciens*, for example, that confers resistance to the herbicide glyphosate (see Chapter 9.4.2), must be modified, so that the encoded protein contains an N-terminal chloroplast targeting sequence. This is done by adding the coding sequence for a known plant chloroplast transit peptide at the 5′ end of the coding sequence of the transgene.

9.3.5 Marker genes

The successful transformation of a plant cell, i.e., the stable integration of the transformed DNA into the nuclear (or plastid) genome, is a rare event; therefore, it is necessary to effectively select these cells and distinguish them from the majority of untransformed cells. For this reason, a marker gene is tightly coupled to the gene of interest, which allows for such a selection. Often used marker genes are either resistance genes against antibiotics, nearly always against kanamycin or hygromycin B, or resistance genes against herbicides, often against the herbicides glyphosate or phosphinothricin. Herbicide resistance is also sometimes the new trait which is introduced; in this case, the transgene can also be used as marker gene. The use of herbicide resistance genes bears the risk of outcrossing the resistance to wild relatives, rendering the herbicide less effective. Thus, other marker systems have been developed which would not provide an advantage to the plant under farming conditions. One such marker is the phosphomannose isomerase (PMI) from *E. coli* (Figure 9.14a). Plants cannot utilize mannose as carbon source; indeed, mannose is converted within the plant cell to mannose 6-phosphate, which, upon accumulation, is an inhibitor of glycolysis. Phosphomannose isomerase converts mannose 6-phosphat to fructose 6-phosphate, a natural intermediate in glycolysis. Plants containing the *pmi* gene can thus use mannose as sole carbon source. There are several reports that using such markers like PMI which do not need the application of toxic substances (like antibiotics and herbicides) also results in a higher rate of identified transformation events.

Especially the use of antibiotic resistance genes is criticized by opponents of genetically modified plants. Notwithstanding the question if such critics are justified, the marker gene is only necessary in the process of transformation, once the transgenic plant is established it is dispensable. Therefore, methods have been developed to remove the marker gene before the transgenic plant is released on the market. This would have the additional benefit that there is one gene less to be evaluated during the registration process for releasing the genetically modified plant. One method which allows the removal of the marker gene is co-transformation (Figure 9.14b); this is transforming the gene of interest and the marker genes independently during the transformation process. This results in independent integration of both genes in the genome, so that they can later be separated by segregation. In biolistics, two separate plasmids (or linear DNA fragments) can be shot at the same time, one carrying the marker gene, the other the transgene to be transferred. For *Agrobacterium*-mediated transformation, there exist three alternatives to deliver transgene and marker gene independently: simultaneous transformation with two *Agrobacteria* strains, transformation with one *Agrobacterium* strain that carries two separate Ti plasmids, and transformation with one *Agrobacterium* strain that carries one Ti plasmid that contains two separate T-DNA regions.

Drawbacks of this co-transformation strategy are that transformation frequencies are usually lower (but still may reach 85 % compared to the standard single transformation), and that integrations into the genome often occur at one locus, although the transgenes were delivered separately. Finally, segregating the marker gene requires crossing, which is not applicable for vegetatively propagated plants.

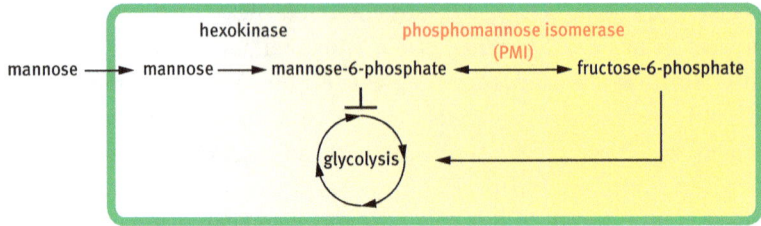

(a)

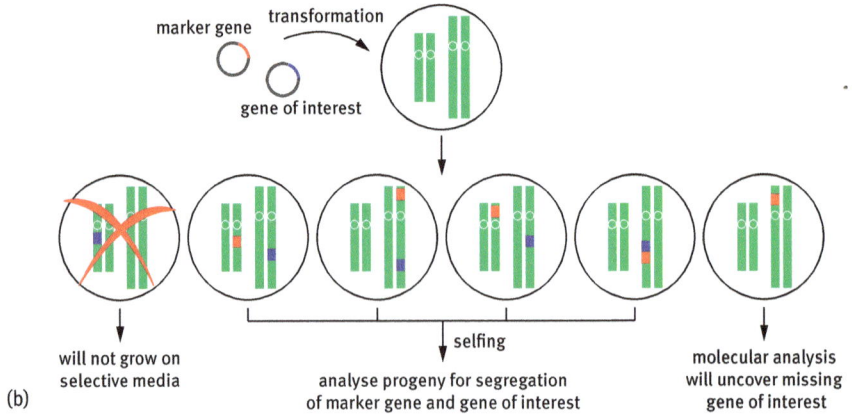

(b)

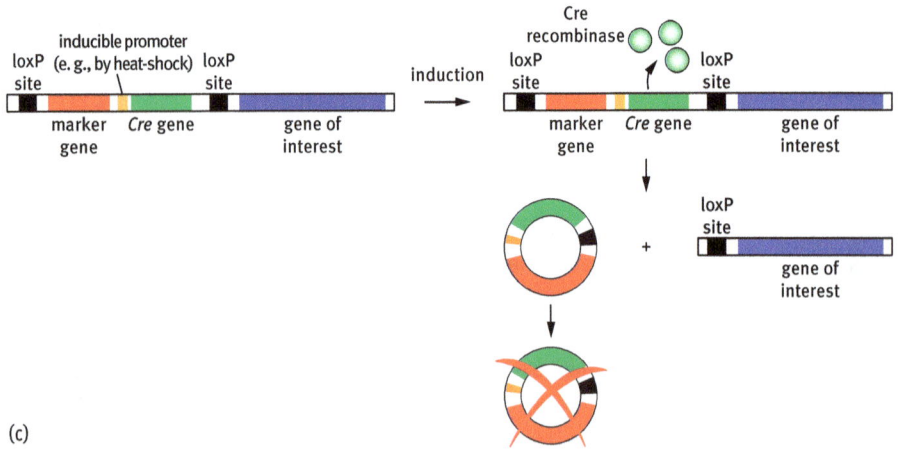

(c)

◄ Fig. 9.14: (a) Phosphomannose isomerase as alternative marker gene. (b) Co-transformation of marker gene and gene of interest on separate constructs (for details, see text). If both genes are delivered separately, they may integrate at different positions in the genome and the marker gene can afterwards be removed by segregation. Before, the plants have to be analysed for the presence of both genes. (c) Removal of the marker gene by a site-specific recombinase. In this example, the marker gene and the gene for the recombinase Cre under the control of an inducible promoter are placed between the two recombination sites (loxP). When production of Cre is induced, it will remove the marker gene and its own gene from the genome. (For the sake of simplicity, other genetic elements like promoters and terminators are not shown)

Another method to remove the marker gene is the use of site-specific recombinases (Figure 9.14c). The Cre recombinase of the *E. coli* bacteriophage P1, for example, recognizes specific sequences called *loxP* sites, and catalyses the recombination of two such sites. By this process, a DNA segment flanked by two *loxP* sites is cut out of the genome, and the flanking *loxP* sites were re-ligated. To produce marker-free transgenic plants, the marker gene is first inserted between two *loxP* sites. The Cre recombinase, which is needed to remove the marker gene, may be delivered to the transgenic plant by several ways. The transgenic plant can be crossed with a Cre recombinase-expressing transgenic plant and the *Cre* gene can later be removed by segregation. The Cre protein can also be expressed only transiently, e.g., by using viral vector systems that spread systemically through the whole plant. Most elegant would probably be to put also the *Cre* gene between the *loxP* sites, under the control of an inducible promoter. Once the *Cre* gene is activated, Cre recombinase is produced and removes the marker gene and its own gene from the genome.

Box 9.1: Antibiotic resistance genes for selection of transgenic plants.

Antibiotics are usually active against microorganisms, but may also be effective against plant cells if the compound targets the plants' plastids, which have their origin in a bacterial ancestor.

Kanamycin, Hygromycin B
Kanamycin and hygromycin B are both aminoglycoside antibiotics derived from different *Streptomyces* species. They bind to the 30S subunit of the 70S ribosome, thus inhibiting protein synthesis. Resistance is accomplished by expressing the *E. coli* genes *neomycin phosphotransferase* II (*nptII*) against kanamycin and *hygromycin-B phosphotransferase* (*hph* or *hpt*) against hygromycin B. The encoded enzymes phosphorylate the respective antibiotics, rendering them inactive.

Chloramphenicol
Chloramphenicol is also an inhibitor of protein synthesis but binds to the 50S ribosomal subunit. Resistance against this compound from another *Streptomyces* strain is conferred by chloramphenicol acetyltransferase (cat) from *E. coli*.

9.3.6 Silencing genes in transgenic plants

The typical method for turning off a certain gene is to replace the gene or parts of it by another sequence via homologous recombination (gene knockout, KO). By placing homologous regions of the gene to be knocked out on either sites of a marker gene and transforming this construct, the targeted gene may be replaced by the marker gene. In the yeast *Saccharomyces cerevisiae* homologous recombination is so efficient that flanking regions of 20 base pairs are sufficient for a successful gene replacement while in other organisms, longer flanking regions are required (e.g., 5–10 kilobase pairs for a gene knockout in mice) (see also Chapters 4.5.1, 8.2 and 10.6.5 for additional examples). Homologous recombination also takes place in plants, of course; however, illegitimate recombination (see Chapter 9.3.2) is the major process by which foreign DNA is inserted into plants genomes, even if extended homologous regions are present in the transgene construct. This means that a transformed DNA is inserted randomly into the genome, and thus homologous recombination cannot effectively be used to target a specific gene for a knockout. Insertional mutagenesis using transposons or *Agrobacterium* (to insert a T-DNA into a gene) can be used to randomly knock out genes, but this requires the generation of several thousand of individual plants (each plant will only have one or few genes knocked out) and a laborious screening process to identify those plants in the collection, in which the gene of interest was effected.

Instead of a classical gene knockout, plant scientists use methods which silence genes post-transcriptionally or transcriptionally. In this case, the term "knockdown" is often used, since the gene is still functional and expression is not always completely inhibited, albeit strongly reduced. These methods are cosuppression, antisense expression, RNA interference (RNAi) and expression of artificial micro RNAs (amiRNA) (Figure 9.15). Cosuppression describes the effect that sometimes, when a plant gene is transformed into a plant and overexpressed, both the transgene and the endogenous gene are silenced. This effect was first discovered in petunia plants, in which genes for the production of the flower color pigments were overexpressed. Unexpectedly, many of the transgenic plants showed less or no pigments in the flowers, indicating that not only the transgene but also the plant's own biosynthesis gene was silenced. In antisense expression, the transgene consist of the sequence of the gene to be silenced (or parts of the gene) in reverse complementary orientation (i.e., in *antisense*). RNA interference is a more recent method, based on the observation that the presence of a double-stranded RNA (dsRNA) can inhibit the expression of a homologous endogenous mRNA. Thus, to silence a plant gene, a transgene is constructed which consists of a part of the target gene in sense and antisense orientation, separated by an intron. During transcription and mRNA processing, a dsRNA is formed, which triggers degradation of the target gene's mRNA. Finally, micro RNAs are small, non-coding endogenous RNAs, which also can form dsRNAs and serve gene regulation by triggering the

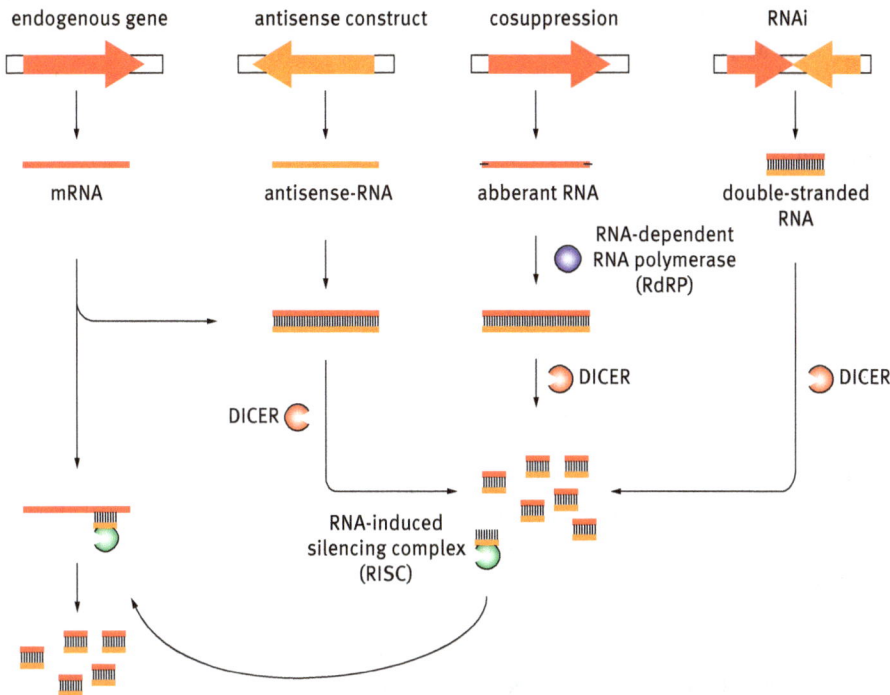

Fig. 9.15: Post-transcriptional gene silencing by antisense expression, cosuppression, and RNAi. Central to all three methods is the formation of a double-stranded RNA which is digested into small fragments of 20–25 bp by the enzyme DICER. These fragments are bound as single-stranded RNA by the RISC, which subsequently binds to the target mRNA and degrades it.

degradation of several plant mRNAs. In artificial miRNAs, the region of a miRNA that is responsible for the identification of the target gene(s) is modified to match a gene of interest.

All these processes share common mechanisms which are usually regarded as post-transcriptional gene silencing (PTGS). Although the details vary in the different processes, here a general description is presented (Figure 9.15): Prerequisite for all four methods is the formation of a dsRNA. In amiRNAs and RNAi-constructs, sense and antisense sequences are encoded in the same gene, thus, upon transcription, a dsRNA is readily formed. In antisense expression, the dsRNA is formed between the endogenous target RNA (sense) and the antisense RNA. For cosuppression, it has been found that strong overexpression of a gene can result in the formation of so-called "aberrant" mRNA, lacking either the 5′ cap or the 3′ poly-A tail or both. Such aberrant RNA is recognized by a RNA-specific RNA polymerase (RdRP) which transcribes them into dsRNA. In the next step, the dsRNAs are cleaved into small pieces of 20—25 base pairs by a dsRNA-specific RNAse called DICER (or DICER-like in plants). In the last step, these small dsRNAs are bound by the RNA-incuded silencing complex (RISC). RISC unwinds the dsRNA and uses the short single-stranded RNA as a guidance to the target mRNA, which is then cleaved by the RNAse activity of RISC.

Expression of a dsRNA homologous to a promoter region of a gene can also lead to transcriptional silencing (TGS) due to hypermethylation of the promoter. However, this method is usually less successful than the methods targeting the mRNA.

9.4 Main traits in commercialized transgenic plants

9.4.1 Input traits and output traits

In plant breeding, new traits can be classified depending on whether they serve to decrease the input (costs) needed in production, these are called input traits, or if they alter the harvested product, so-called output traits. Nearly all genetically modified plants that are actually cultivated carry input traits, either herbicide resistance, which allow effective weed control, or insect resistance, which decreases the use of pesticides. In general, input traits are for stabilizing or increasing the yield. Output traits increase the quality of the product, e.g., fruits with longer shelf life or increased vitamin content, flowers with different colors but also products that are better suited for downstream processing, e.g., potatoes without amylose for amylopectin production. However, although such output traits may attract more attention, in comparison to input traits, they play virtually no role regarding their acreage.

9.4.2 Herbicide resistance

One factor that largely influences agricultural yield is the presence of weeds. Weeds are in concurrence with the crops for space, water, nutrients, and sunlight, and global crop losses due to weed competition have been estimated to be approx. 10 %, although for some crops they may be higher than 50 %. Weeds do not only decrease crop yield, they may also influence harvesting and downstream processing, e.g., oilseed-rape harvested as weed in sugar beet fields may clog the machines for beet processing. Weeds are often wild growing plants but can also be, as in this last example, a crop plant from the former year's cultivation (so-called volunteer plants). Weed control is one of the aims of ploughing, in which weeds are uprooted or cut. The main procedure to control weeds is, however, the application of chemical weedkillers = herbicides. Herbicides may be classified as selective, when they target only specific plants, or as non-selective when they kill all plants. Non-selective herbicides are usually used to keep larger areas, e.g., railway tracks, completely free of plants, but they are difficult to use in agriculture, since they should not come in contact with the crop. On the other hand, selectivity of selective herbicides is dependent on several factors, like amount of herbicide applied or developmental stage of the crop plant. Thus, a wrong application of a selective herbicide may also kill the crop plant.

Resistance against a certain herbicide may be a natural characteristic of the plant. Dicotyledoneous plants (many weeds are dicots), for example, are more sensitive to mimics of the phytohormone auxin than monocotyledoneous plants (cereals are monocots). But herbicide resistance can also be generated by breeding: resistance may be transferred or introgressed from plants in which herbicide resistance has occurred spontaneously. Herbicide resistance can also be induced by mutation breeding (see Chapter 9.1.2). In transgenic breeding, known resistance genes against a certain herbicide can directly be transformed into the crop plant.

Glyphosate resistance

Glyphosate (sold as Roundup) is the most sold herbicide, and most commercialized transgenic plants carry one or two genes conferring resistance against this herbicide. Glyphosate is a very effective non-selective herbicide with very low toxicity against animals and humans. The compound inhibits the plant's shikimate pathway, which is responsible for producing the three aromatic amino acids tryptophan, phenylalanine, and tyrosine, by binding (and thus blocking) to an enzyme of this pathway, EPSP synthase (Figure 9.16a). The shikimate pathway is absent in animals, which have to take up aromatic amino acids with their diet, explaining the low toxicity of glyphosate for these organisms. From some microorganisms, e.g., the *Agrobacterium tumefaciens* strain CP4, EPSP synthases have been identified that are not inhibited by glyphosate, and an EPSP synthase from maize has been made insensitive by site-directed mutagenesis. Expressing these enzymes in transgenic plants makes them resistant against glyphosate. Often, a second transgene is expressed, a bacterial glyphosate oxidase that converts the glyphosate into glyoxylate and aminomethylphosphonic acid.

Phosphinothricin resistance

Resistance against another herbicide, phosphinothricin (sold as Basta or Liberty), is only rarely used in transgenic crops but regularly used in basic research. Phosphinothricin is the toxic degradation product of bialaphos (L-alanyl-L-alanyl-L-phosphinothricin), a natural herbicide from some *Streptomyces* species. Phosphinothricin inhibits glutamine synthetase, which results in the accumulation of toxic ammonia (Figure 9.16b). Resistance against phosphinothricin can be accomplished by expressing the *bar* gene (<u>b</u>ialaphos <u>r</u>esistant) or *pat* gene (<u>p</u>hosphinothricin <u>a</u>cetyl<u>t</u>ransferase), also from some *Streptomyces* species, which both code for an acetyltransferase that renders phosphinothricin inactive.

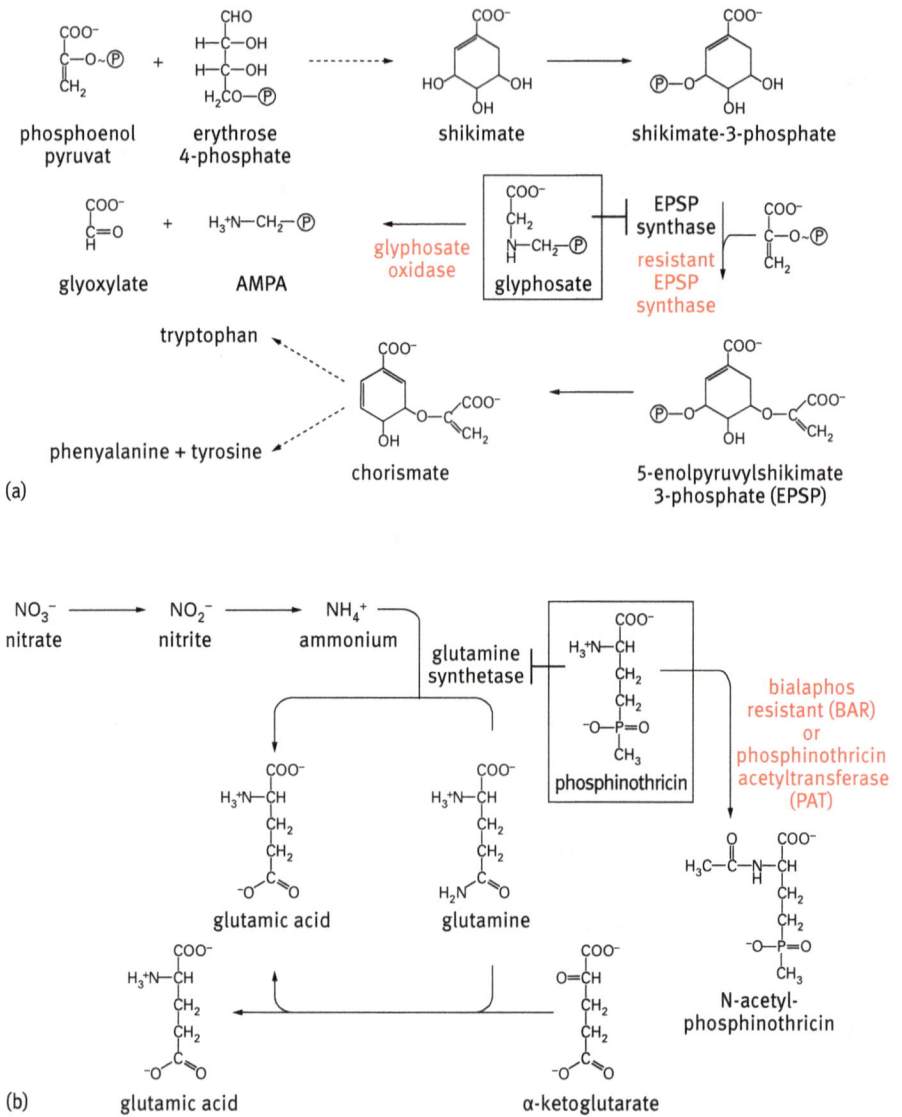

(a)

phosphoenol
pyruvat

erythrose
4-phosphate

shikimate

shikimate-3-phosphate

glyoxylate

AMPA

glyphosate
oxidase

glyphosate

EPSP
synthase

resistant
EPSP
synthase

tryptophan

phenylalanine + tyrosine

chorismate

5-enolpyruvylshikimate
3-phosphate (EPSP)

(b)

NO$_3^-$
nitrate

NO$_2^-$
nitrite

NH$_4^+$
ammonium

glutamine
synthetase

phosphinothricin

bialaphos
resistant (BAR)
or
phosphinothricin
acetyltransferase
(PAT)

glutamic acid

glutamine

N-acetyl-
phosphinothricin

glutamic acid

α-ketoglutarate

Fig. 9.16: Herbicides and herbicide resistance in transgenic plants. (a) Glyphosate inhibits the EPSP synthase in the shikimate pathway, an essential pathway for the biosynthesis of aromatic amino acids. Resistance can be achieved by expressing a glyphosate-resistant EPSP synthase and/or by expressing a glyphosate-catabolizing enzyme, here glyphosate oxidase. (b) Phosphino-thricin inhibits the glutamine synthetase responsible for the incorporation of nitrogen (as ammo-nium) into amino acids. The inhibition leads to a toxic increase in ammonium. Resistance is accomplished by expressing BAR or PAT, which convert phoshpinothricin into the inactive com-pound N-acetyl-phosphinothricin.

9.4.3 Herbivore resistance using Bt toxin

The second most transgenic trait in commercialized transgenic plants is insect resistance. Estimated yield losses by herbivory in major crop plants lies between 8 % and 16 %. Herbivores cause direct losses by eating the product, or indirect losses by diminishing the plant's performance. In addition, herbivores may be vectors for pathogens. Insect resistance is accomplished by expression of toxic proteins from *Bacillus thuringiensis*. This is a spore forming bacterium that produces crystals consisting of so-called crystal proteins (cry proteins), also known as δ-endotoxins or "Bt toxin" (Figure 9.17a). When eaten by an insect, the crystals are dissolved, the cry proteins become activated by proteolytic cleavage, and they form pores in the plasma membrane of the gut's epithelium, finally leading to the death of the insect. Bt toxins are mostly active against larvae of the three insect orders Lepidoptera (butterflies and moths), Coleoptera (beatles), and Diptera (flies and mosquitoes). More than 500 different cry proteins are known which show some specificity against the different insect orders; e.g., Bt toxins belonging to the Cry1, Cry3, and Cry4 classes are specific against Lepidoptera, Coleoptera, and Diptera, respectively. Solutions or powders containing spores and crystals from different *B. subtilis* strains are used as natural insecticides since 1938, and Bt formulations are approved for the use in organic farming.

(a)

(b)

Fig. 9.17: (a) Electron micrograph of protein crystals (Bt toxin) from *Bacillus thuringiensis*. (b) Leafs of transgenic (left) and wild type (right) peanut plants challenged with larvae of the Lesser Cornstalk Borer. The transgenic plants contains Bt toxin due to the expression of a *cry* gene from *B. thuringiensis*. The image in (a) is from P. R. Johnston and Jim Buckman. The photographs in (b) are from Herb Pilcher, and were released by the Agricultural Research Service of the United States Department of Agriculture (USDA ARS).

Since the toxic principle is a protein, it was easy to transfer Bt toxins into transgenic plants (first accomplished in 1987 with tobacco and tomato) (Figure 9.17b). Today, transgenic plants carrying Bt toxins are mostly cotton (Bt toxins against the cotton bollworm) and maize (Bt toxins against the European corn borer and Western corn rootworm). While the transgenic plants of the first generation carry only one *cry* gene, it is now usual to transfer genes for two or more Bt toxins into one plant ("stacked traits"). This broadens the protection of the plant but also makes the emergence of Bt toxin resistance in insects less likely. "SmartStax" maize, released 2010 in the US and Canada, and approved in the EU in 2013, contains 6 genes for different Bt toxins (in addition to two genes for resistance against glyphosate and phosphinothricin).

9.4.4 Other traits

The two transgenic traits described above, herbicide resistance (nearly exclusively against glyphosate) and insect resistance through expression of Bt toxins, account for 99 % of the worldwide acreage of transgenic plants. The following traits, although insignificant regarding their acreage on a global scale, are listed due to their historical or future importance or as outstanding examples of plant biotechnology.

Resistance against viruses: the papaya ringspot virus
In 1998, papaya cultivation in the Puna region of the Hawaiian main island suffered severely from the spread of the papaya ringspot virus, decreasing the yield to 65 % compared to 1992, the first year when the virus was detected in Puna. Approximately 50 years before, the same virus ruined papaya cultivation on the Hawaiian island Oahu, which lead to the transfer of papaya production to the main island Hawaii. Scientists were aware that this would only be a temporary solution; thus, in the late 1980s, they developed two transgenic papaya varieties called *Sunup* and *Rainbow*. The primary transgenic plant was produced by biolistics and introduced the gene for the viral coat protein under the control of the strong CaMV-35S promoter (see chapters 9.3.1 and 9.3.4). It has been shown before that overexpression of viral coat proteins in plants may confer resistance against the respective viruses, although at that time, the reasons for the resistance were unknown. Today, it is known that cosuppression due to posttranslational gene silencing (see Chapter 9.3.6) leads to silencing of the original viral gene. In the same year when the papaya ringspot virus was detected in Puna, a two year field trial was started on the heavily infected island Oahu and the transgenic varieties remained free from disease, while the yield of the non-transgenic papaya dropped to 13 % due to infestation in the same time. In 1997, the transgenic varieties were approved for use in

food and feed production, and in 1998, seeds were given to the Hawaiian papaya growers. Within 4 years, papaya production in Puna recovered and reached levels comparable to that prior to the virus outbreak.

The first transgenic vegetable: the Flavr Savr tomato
The Flavr Savr tomato was the first transgenic plant that was sold as whole food to the public marketplace. It was designed to reduce and delay softening, thus allowing the fruits to grow longer on the vine and resulting in a prolonged shelve life. This was achieved by antisense expression (see Chapter 9.3.6) of a fragment of the polygalacturonase gene. Polygalacturonase is responsible for the breakdown of pectin, which forms the middle lamella of the plant's cell wall. Since the middle lamella is responsible for connecting the cell walls of neighbouring cells, its breakdown results in softening of the tissue (see also Chapter 9.1.2, protoplast fusion). The polygalacturonase levels in the Flavr Savr tomato were reduced to below 1% compared to normal tomatoes. The tomato had been developed by Calgene in California since the mid 1980s and was ready for marketing in the early 1990s. In 1992, the United States Food and Drug Administration (FDA), for the first time, faced to deal with a genetically modified food, decided that genetically engineered foods would not be regulated differently than conventional foods ("[The FDA is] not aware of any information showing that foods derived by these new methods differ from other foods in any meaningful or material way, or that, as a class, foods developed by the new techniques present any different or greater safety concern than foods developed by traditional plant breeding"). In 1994, the FDA approved the use of neomycin phosphotransferase (the product of the kanamycin resistance gene used as selection marker, see Chapter 9.3.5) as a food additive. In the same year, the Flavr Savr tomato was brought to the market.

At the same time, a "twin" of the Flavr Savr tomato was developed for use in production of tomato puree by the British company Zeneca Seeds. Because this tomato, like its Flavr Savr counterpart, contained higher amounts of pectin, which is also a gelling and thickening agent, production of double concentrated tomato puree is easier, since the paste has a higher viscosity. In addition, heat treatment to inactive polygalacturonase activity and to concentrate the paste could be reduced.

In the beginning, both the Flavr Savr tomato as well as the tomato puree "made with genetically modified tomatoes" (marketed from 1996) were quite successful. The Flavr Savr tomato could be sold as premium product at 2½ –3½ times the price of gas-green tomatoes. The tomato puree, due to the lower production costs, could be sold less expensively than "conventional" tomato puree and reached an estimated share of 60% of the canned tomato market in the United Kingdom. Calgene fell into several severe problems and was finally taken over by Monsanto in 1997. Following production problems, this was the end of the Flavr Savr tomato. The transgenic tomato puree was withdrawn in 1999 due to public concern about genetically modified foods.

Box 9.2: Induced ripening of harvested tomato plants.

Ripe tomatoes are too soft for transportation. Thus, it is usual practice to harvest and transport tomatoes when they are still green and hard. Ripening is induced by spraying with ethylene, which is the natural plant ripening hormone, when the tomatoes approach their final destination. However, this induced ripening of isolated tomatoes differs in some aspects from the natural ripening on the vine; therefore, "gas-green" tomatoes taste different (less flavourful) compared to fresh tomatoes.

The starch potato Amflora: A long story with a rapid end

Starch, the typical storage compound for the photosynthetically produced glucose, is composed of two polysaccharides, the unbranched amylose, consisting of $\alpha(1\rightarrow4)$ coupled D-glucose molecules, and the highly branched amylopectin, in which additional $\alpha(1\rightarrow6)$ bonds are formed. Plant's starch usually contains 20–30% amylose and 70–80% amylopectin. Besides in food industry, starch is used as thickener, e.g., in cooking, but also as glue or to stiffen garments before ironing. As a technical product, it is used in paper, adhesives, building, and textiles for thickening, binding and sizing. Thickening is mainly due to the amylopectin constituent while amylose mediates the (often) undesirable property of gelling (Figure 9.18). Since separation of amylopectin and amylose is expensive, and thus economically not viable, starch is often chemically modified to reduce the gelling property of the amylose constituent. At the end of the 1980s, BASF Plant Sciences developed a potato in which the gene for the enzyme responsible for amylose biosynthesis,

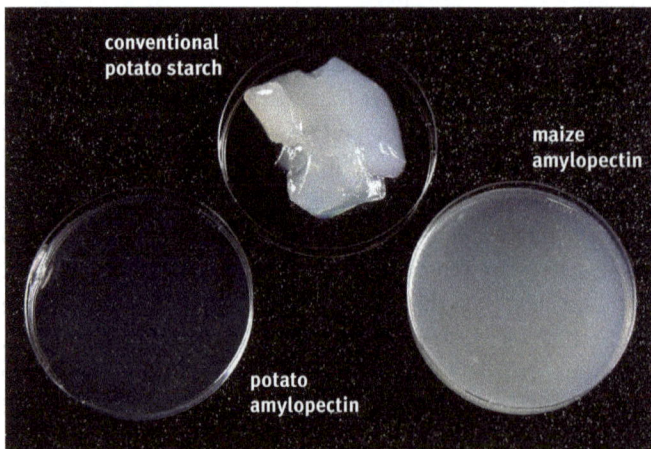

Fig. 9.18: Gelling properties of different starches. Starch isolated from conventional potatoes (top) show strong gelling due to the presence of amylose. Pure potato amylopectin (as present in Amflora potatoes) shows no gelling (left). The opaque appearance of maize amylopectin (right) is due to contaminations with proteins (which are absent in potato amylopectin). Photograph courtesy of Lyckeby Starch AB, Sweden.

the granule-bound starch synthase, was silenced by an antisense approach (see Chapter 9.3.6). The transgenic plants (named *Amflora*) were produced by *Agrobacterium* mediated transformation of leaf discs, using a kanamycin resistance gene (*nptII*) as marker. Tubers of these plants contained less than 2 % of amylose in their starch. An application for approval of Amflora potatoes in the EU was filed in August 1996 and finally gained authorization in March 2010. In the same year, Amflora was grown in the Czech Republic (150 ha), Sweden (80 ha), and Germany (15 ha). In Germany, approx. 1 ha was destroyed by opponents of genetically modified plants. In Sweden, a contamination of an Amflora field with approx. 50 non-approved transgenic Amadea potatoes (another amylopectin potato) occurred due to an accidental commingling. In 2012, BASF announced that they would stop the development of transgenic plants solely targeted at cultivation in the European market, due to the lack of acceptance of transgenic plants in many parts of Europe. The headquarters of BASF Plant Sciences was moved from Germany to the US. In 2013, the General Court of the European Union annulled the authorization for the Amflora potato because of procedural errors occurring during the authorization process.

Golden Rice, the "rice that could save a millions kids a year"

According to estimates of the World Health Organization (WHO), approx. 250,000 to 500,000 children become blind every year due to vitamin A deficiency, and half of them die within 12 months after losing their sight. Vitamin A is the name for a group of compounds including retinol, retinal, and retinoic acid. Retinal is the prosthetic group of rhodopsin, the light receptor in photoreceptor cells responsible for vision. However, vitamin A is not only important for vision, but vitamin A deficiency can also cause anaemia and weakens the immune system. The main reason for vitamin A deficiency is prolonged dietary deprivation with little animal sources of preformed vitamin A. Vitamin A can also be produced from β-carotene derived from plants (thus called provitamin A). Rice, however, which is the staple food in southern and eastern Asia, does not contain β-carotene in the grain. This is not because rice cannot produce β-carotene at all, indeed carotenes are photosynthetic pigments present in photosynthetic tissues of all plants, but because two of the genes encoding enzymes responsible for the biosynthesis of carotenes, phytoene synthase, and phytoene desaturase are turned off in the endosperm tissue of the grain (Figure 9.19a).

Golden Rice is a transgenic plant, in which two transgenes, a plant derived phytoene synthase gene and a bacterial carotene desaturase gene, both under the control of an endosperm specific promoter, have been introduced. Grains of the first transgenic plants contained 1.7 µg of carotenoids per gram. In further studies, the used phytoene synthase which originated from daffodil was identified as bottleneck. An improved version, Golden Rice 2, contains the phytoene synthase gene

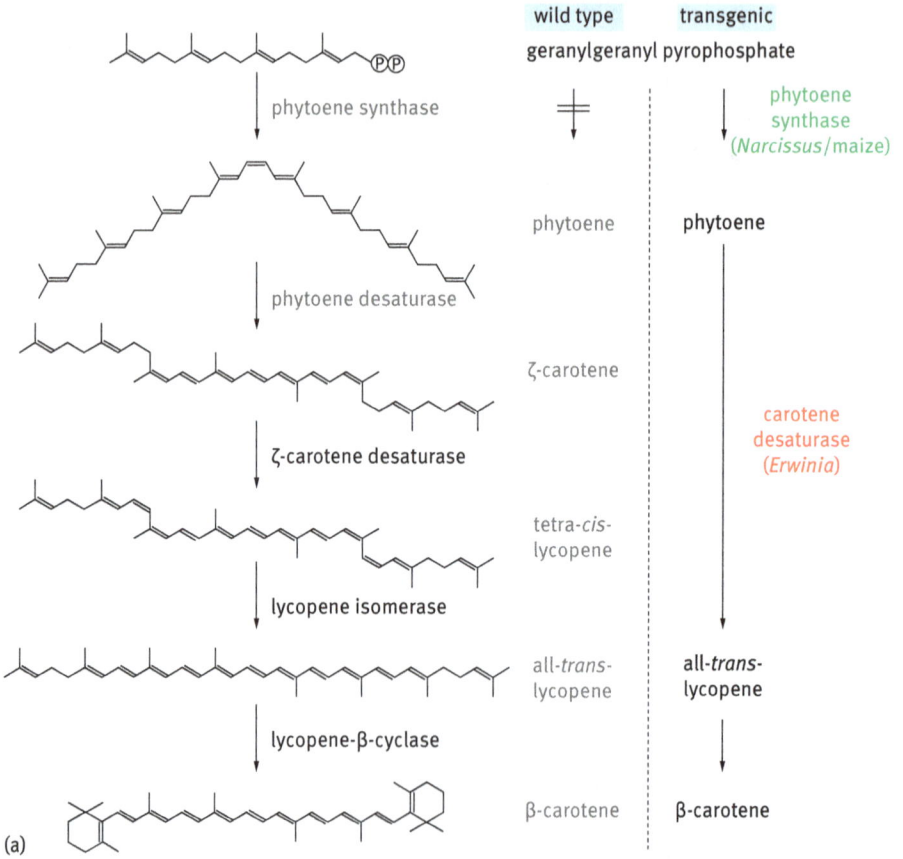

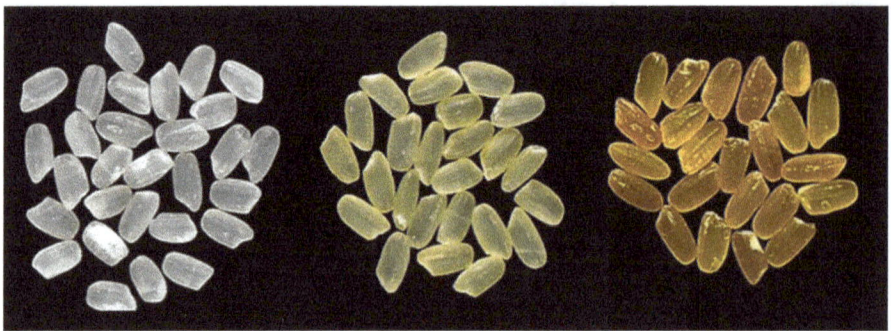

wild type   transgenic

geranylgeranyl pyrophosphate

phytoene synthase

phytoene synthase (*Narcissus*/maize)

phytoene   phytoene

phytoene desaturase

ζ-carotene

carotene desaturase (*Erwinia*)

ζ-carotene desaturase

tetra-*cis*-lycopene

lycopene isomerase

all-*trans*-lycopene   all-*trans*-lycopene

lycopene-β-cyclase

β-carotene   β-carotene

(a)

(b)   wild type   golden rice 1   golden rice 2

◀ Fig. 9.19: β-Carotene biosynthesis in rice. (a) Biosynthesis in wild type rice kernels and transgenic kernels. In wild type kernels, the genes encoding for phytoene synthase and phytoene desaturase are not expressed. Thus, biosynthesis stops at the stage of geranylgeranyl pyrophosphate.
(b) Polished rice kernels from wild type plants, Golden Rice 1 (expressing phytoene synthase from
daffodil) and Golden Rice 2 (expressing phytoene synthase from maize). Photograph courtesy of
the Golden Rice Humanitarian Board (www.goldenrice.org).

from maize and produces up to 37 µg carotenoids per gram (Figure 9.19b). Given a
daily intake of 200 g of Golden Rice 2, this would fully provide the daily vitamin A
needs of children from 1–3 years of age.

Drought resistance, an upcoming trait

One of the most important factors affecting crop productivity is the availability of
water. The impact of drought as environmental stress for agriculture will increase
due to the change in the global climate, as well as the shortage of freshwater due
to the increasing world population. Drought resistance is a complex trait and involves many processes. Plants may cope with drought stress by reducing water loss
by transpiration or by improving water uptake due to an increased root system.
Tolerance may be developed by accumulation of osmoprotectants. Some crops with
increased drought resistance have been developed by conventional breeding techniques. A transgenic approach proved to be successful by expressing cold shock
proteins (Csp) from E. coli and Bacillus subtilis in several plants. These cold shock
proteins are RNA chaperones that can resolve misfolded RNA and are thought to
positively influence transcription and/or translation. Although the involvement of
cold shock proteins in tolerance against several stress conditions is proven, the
molecular mechanism(s) for this effect is still unknown. However, transgenic maize
plants expressing CspB from B. subtilis under the control of the rice *actin1* promoter
(see 9.4.4) performed 10–15 % better under temporal water-limiting conditions than
its non-transgenic predecessor. This maize is available (and grown) in the US since
2013.

Transgenic ornamental plants

Ornamental plants and nursery plants have a global production value of approx.
20 billion Euros per year, with the EU having a share of some 40 %. Each German
spends approx. 100 Euros per year on ornamental plants. There are actually only
two traits that are addressed by transgenic approaches: vase life and flower color.
Withering of cut flowers, like fruit ripening, is controlled by the plant hormone
ethylene. Withering can be (and usually is) delayed by adding certain chemicals
like silver thiosulfate, which inhibits the perception of ethylene or amino-oxyacetic
acid, which inhibits 1-amino-cyclopropane-1-carboxylic acid synthase (ACC syn-

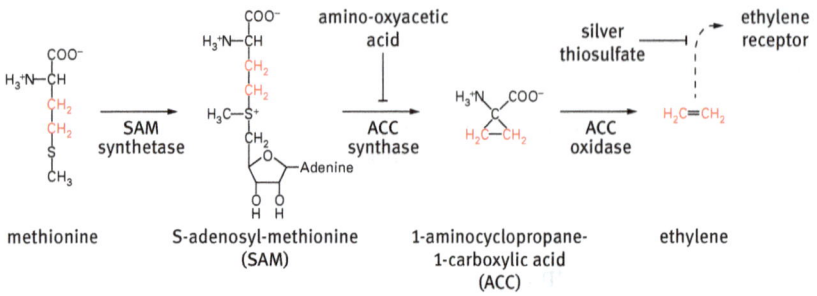

methionine   S-adenosyl-methionine   1-aminocyclopropane-   ethylene
(SAM)   1-carboxylic acid
(ACC)

(a)

control   transgenic plant
non-treated   treated   non-treated
(silver thiosulfate)

(b)

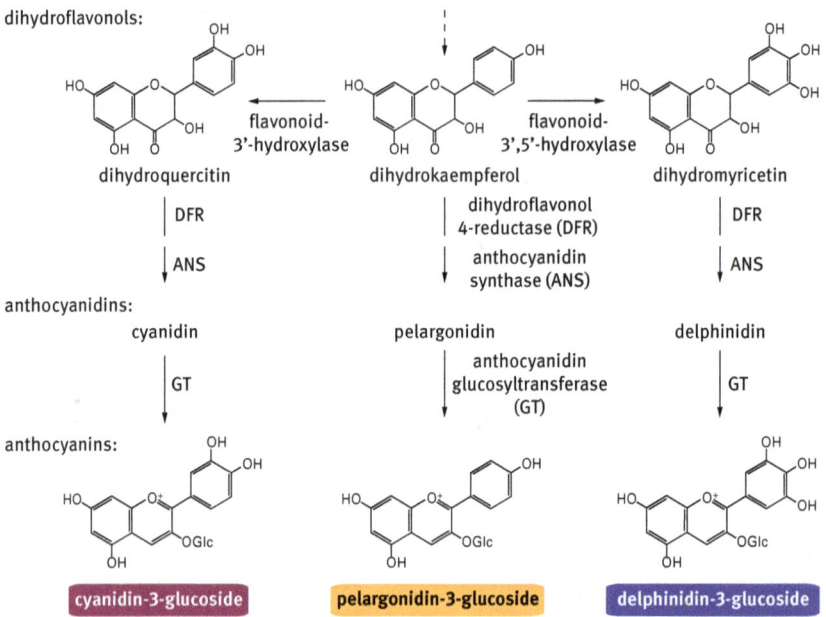

(c)

(d)

(e)

◄ **Fig. 9.20:** Important pathways for transgenic ornamental plants. (a) Biosynthesis of the plant hormone ethylene from methionine. Ethylene causes wilting of flowers, thus inhibition of ethylene biosynthesis results in a longer vase life. (b) Control plants (treated and non-treated) and non-treated transgenic plants 10 days after cutting. The non-treated control plants have already wilted. (c) Last steps in the biosynthesis of anthocyanins, typical pigments of many flowers and fruits. (d) Wild-type white progenitor and transgenic blue carnation. (e) Transgenic "blue" rose (left) and its mauve-colored progenitor (right). Photographs courtesy of Suntory Limited.

thase), the penultimate enzyme in ethylene biosynthesis (Figure 9.20a). Withering has been successfully delayed in transgenic carnation (*Dianthus caryophyllus*) by overexpressing a truncated version of the ACC synthase gene by cosuppression (see 9.3.6), increasing the vase life from 10 days (with chemical treatment) to 22 days (without chemical treatment).

Carnation is also the first commercialized transgenic plant with a changed flower color. Flower pigments often belong to the group of anthocyanidins which range in color from orange to blue. One factor controlling the color of anthocyanidins is the degree of hydroxylation in one of its rings of the core structure; increasing the number of hydroxyl groups changes the color from orange (pelargonidin) to red (cyanidin) to blue (delphinidin) (Figure 9.20c). Carnations are usually red or pink due to the presence of the anthocyanidins cyanidin and/or pelargonidin. Mutations in genes of anthocyanidin biosynthesis may result in white flowers; one such white carnation cultivar, White Unesco, lacks activity of dihydroflavonol reductase (DFR), which catalyses the first step in the conversion of the direct anthocyanidin precursors. Thus, this cultivar still contains the precursors necessary for anthocyanidin production. Blue/mauve colored carnations (Figure 9.20d) were generated by introducing two genes, one from pansy (*Viola sp.*) or petunia (*Petunia hybrida*), encoding a flavonoid 3',5'-hydroxylase (F3'5'H) and the second from petunia, encoding the missing DFR. Both genes are involved in the biosynthesis of the blue colored delphinidin.

Only recently (2009), have transgenic blue roses been released to the market. The generation of these plants was more difficult than the production of blue carnations, since no white *DFR* mutant was available. The commercialized cultivar also contains two additional genes, the same *F3'5'H* gene from petunia as mentioned above, and an *anthocyanin 5-acyltransferase* gene from torenia (*Torenia hybrida*). This plant has a pale purple color (Figure 9.20e), and although possessing significant amounts of the blue colored delphinidin (up to 95%), it still contains pelargonidin and cyanidin. In the laboratory, it was possible to turn off the rose's own anthocyanidin biosynthesis by knocking-down the rose's *DFR* gene by RNAi (see 9.4.6). Introducing the *F3'5'H* and *DFR* genes from pansy and iris (*Iris hollandica*), respectively, resulted in roses with 98% delphinidin. Still, these roses are pale purple ('blueish', at best). The reason for this is that the actual color of anthocyanidins is controlled by many factors besides their chemical structure, like the pH value in the vacuole (where the anthocyanidins are stored), complexation with metal ions, and the presence of co-pigments (like flavones and flavonols).

Although the "blue rose" does not have the same dark blue as the blue carnation, this plant represents a milestone in rose breeding. Roses are the most important plants on the flower market and breeders have tried for centuries to breed a blue rose, becoming a synonym for the impossible.

9.5 Current status of commercialized transgenic plants

The first commercialized transgenic crops (virus resistant tobacco) were grown in 1992 in China. In 1994, the Flavr Savr tomato was the first transgenic food which was approved (see Chapter 9.4.4) for commercial sale. The breakthrough of transgenic crops occurred in the years 1995 and 1996, in which more than 40 transgenic crops were approved for sale. Since then, the global acreage of transgenic crops has always increased; today (2014) transgenic crops are grown on approx. 180 million hectares, representing more than 10 % of the global acreage which is estimated to be 1.5 billion hectares. Most of them (40 %) are grown in the USA; however, although the USA is still by far the country with the highest acreage of transgenic plants, 8 of the 10 countries with the highest acreage of transgenic crops are developing countries (Table 9.3). Since 2012, developing countries grow more genetically modified crops than industrial countries. Of the roughly 30 countries that are grow-

Table 9.3: Top ten countries and European countries growing the largest amounts of genetically modified crops in 2013. Data from James (2013).

Rank	Country	Acreage (Million hectares)	Main transgenic crops
1	United States of America	70.1	Maize, soybean, cotton, canola, sugar beet, alfalfa, papaya, squash
2	Brazil	40.3	Soybean, maize, cotton
3	Argentina	24.4	Soybean, maize, cotton
4	India	11.0	Cotton
5	Canada	10.8	Canola, maize, soybean, sugar beet
6	China	4.2	Cotton, papaya, poplar, tomato, sweet pepper
7	Paraguay	3.6	Soybean, maize, cotton
8	South Africa	2.9	Maize, soybean, cotton
9	Pakistan	2.8	Cotton
10	Uruguay	1.5	Soybean, maize
16	Spain	0.1	Maize
22	Portugal	< 0.05	Maize
24	Czech Republic	< 0.05	Maize
26	Romania	< 0.05	Maize
27	Slovakia	< 0.05	Maize

Table 9.4: Main transgenic crop species in 2013. Data from James (2013).

Crop	Acreage (Million hectares)	%	Adoption rate (%)
Soybean	84.5	48	79
Maize	57.4	33	32
Cotton	23.9	14	70
Canola	8.2	5	24
Alfalfa	0.8	< 1	–
Sugar beet	0.5	< 1	–
Papaya	< 0.1	< 1	–
Others	< 0.1	< 1	–
Total	175.3	100 %	–

ing biotech crops, approx. 20 are developing countries. In addition to the countries that are actually growing transgenic crops, there are approx. 35 countries (including the EU counting as one) in which transgenic crops are allowed for use as food or feed or for growing. The EU plays only a minute role; due to the high opposition of politicians and most of the EU countries' citizens against genetically modified plants, the portion of transgenic crops on the global acreage is below 0.1 % (between 100,000 and 150,000 hectares).

Nearly all transgenic crops (99 %) are represented by only 4 species: soybean (approx. 50 %), maize (approx. 30 %), cotton (approx. 15 %), and canola (oilseed rape, approx. 5 %) (Table 9.4). The adoption rates (the percentage of transgenic varieties) are 80 %, and 70 % for soybean and cotton, respectively, and 30 % and 25 % for maize and canola, respectively.

As mentioned above (Chapter 9.4.4), these plants carry only two transgenic traits: herbicide resistance, in most cases against glyphosate (Roundup), and insect resistance due to Bt toxins. Stacked traits, i.e. the combination of several herbicide resistance genes or several different Bt toxins, or both, are becoming more important; approx. 30 % of the transgenic crops (measured in acreage) contain more than one transgenic trait.

9.6 Plants as producers of pharmaceuticals

9.6.1 Pharmaceuticals isolated from plants

Plants produce a wide range of so-called secondary metabolites; these are low-molecular weight compounds that are not important for the normal development of the plant but which are indispensable for the survival of the plant in its environment. Often they are toxins targeted against pathogens and herbivores. However, following Paracelsus' principle "*dosis sola facit venenum*" (the dose makes the poi-

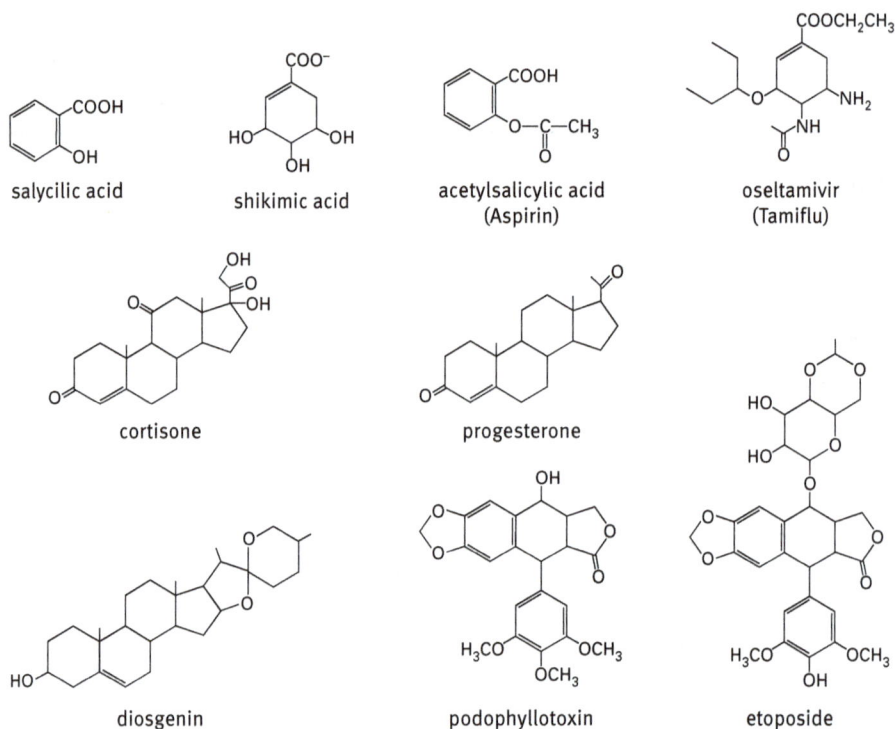

Fig. 9.21: Plant-derived precursors (salicylic acid, shikimic acid, diosgenin and podophyllotoxin) of selected pharmaceuticals (Aspirin, Tamiflu, cortisone, progesterone, etoposide). For details, see text.

son), many of these are used as pharmaceuticals and have been used for centuries as such. Examples of the use of plants as medicine in ancient times is the white willow (*Salix alba*), the bark of which contains the pain- and fever-relieving salicylic acid, and the quina tree (Cinchona tree, *Cinchona officinalis*), the bark of which contains quinine, which was (and still is) used against malaria. The earliest records of usage of plants as medicine date back to 1500–2000 BC. Approximately one quarter of todays' pharmaceuticals contain at least one compound which is (or was) directly or indirectly derived from plants (Figure 9.21). Plant-derived pharmaceuticals may be classified in the following categories:

– The compound was originally identified from plants but today it is synthesized chemically or is produced in microbial systems. E.g., most of the pharmaceutically used salicylic acid, the precursor of acetylsalicylic acid (Aspirin), is now chemically synthesized and no longer extracted from the bark of willow trees. Chemical synthesis, however, is becoming difficult, and thus often no longer feasible if the molecule of interest becomes more complex. This is especially true for molecules with many chiral centres, since chemical reactions are usually not stereospecific.

- Another example is shikimic acid, a precursor for the anti-influenza drug Tami-flu (oseltamivir phosphate). Shikimic acid is still mainly obtained from seeds of the Chinese star anise (*Illium verum*). However, since shikimic acid is an intermediate in the biosynthesis of aromatic amino acids (see Chapter 9.4.2), and this pathway is also present in most microorganisms, it is also increasingly obtained from genetically modified *E. coli* cells.
- The compound is made from a plant-derived skeleton. Diosgenin, isolated from the roots of certain *Dioscorea* species (yam), has a steroid skeleton which is used for the production of different steroid drugs, e.g., corticosteroids and con-traceptives. The large amounts of diosgenin available from yam, and thus, the cost-effective production of steroid hormones is one of the foundations for the success of the oral contraceptive pill.
- The compound is a semisynthetic derivative of a natural plant product. In con-trast to the above-mentioned category, these precursors do already possess the wanted pharmacological activity, but other factors prevent them from being used directly, e.g., they are too toxic, have low solubility, a short half-life, or they show a low resorption into the human body. Podophyllotoxin, derived from *Podophyllum* species (mayapple), although exhibiting strong antineoplas-tic activity is too toxic for use as chemotherapeutic agent (it is used, though, locally as antiviral agent against warts). Etoposide and teniposide are glyco-side derivatives of podophyllotoxin used in chemotherapy.
- The compound is isolated in its active form from the plant. Examples are the potent anti-cancer drugs vinblastine and vincristine, which are indole alka-loids isolated from *Catharanthus roseus* (Madagaskar periwinkle).

A major problem is often the limited availability, since both compounds are only present in low amounts in the plant and/or the plants themselves are limited. E.g., Taxol (generic name: paclitaxel), a very effective drug against certain types of can-cer, is present only in tiny amounts in the bark of the yew tree (*Taxus brevifolia*), and calculations were that approx. six 100-year-old trees must be killed to treat one patient, which would bring the Pacific yew in danger of extinction and made the drug very expensive. Fortunately, a suitable precursor was found in much high-er amounts in the needles of the European (or English) yew (*Taxus baccata*), which could be chemically converted to either Taxol or docetaxel, which is similarly effec-tive as Taxol. Today, most of the world's production of Taxol and docetaxel (trade name Taxotere) is produced in plant cell cultures.

Several strategies can be applied to increase the availability of the compound of interest (Figure 9.22):
- The use of cell cultures instead of intact plants, since the latter may take sever-al years before they can be harvested.
- Addition of elicitors, these are compounds that activate the plant defense reac-tions and may result in a higher production of the compound of interest.

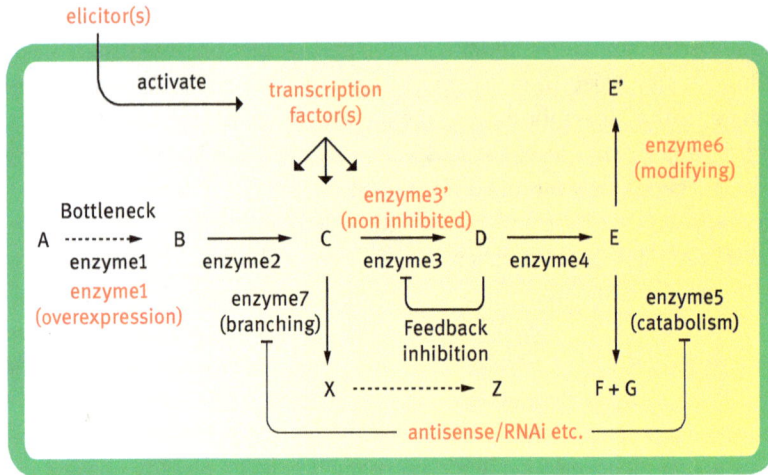

Fig. 9.22: Strategies to increase the production of a compound of interest. A whole pathway may be activated by overexpression of transcription factors, which may also be activated by exogenously applied elicitors. Limiting enzymes may be overexpressed, thus overcoming bottlenecks. Feedback inhibition can be removed by expressing non-inhibited versions of the enzyme. The expression of catabolizing and branching enzymes can be inhibited by antisense expression, cosuppression or RNAi. Finally, the product may be already modified in the cell by expressing the respective transgenes.

–   The genes for the biosynthesis of a certain secondary metabolite are often regulated in a coordinated manner by one or few transcription factors, which may be overexpressed by a transgenic approach.
–   Expression of rate-limiting enzymes ("bottlenecks") may be enhanced by a transgenic approach. Similarly, feedback-inhibition of key enzymes may be overcome by expressing non-inhibited enzymes.
–   Inhibition of catabolic pathways or efflux of precursors and intermediates into competitive pathways by transgenic approaches (e.g., silencing catabolic and branching enzymes).
–   Finally, the additional expression of modifying enzymes by a transgenic approach may convert the compound into a pharmacologically more valuable compound.

9.6.2 Plants as producers of proteinaceous pharmaceuticals

Many pharmaceuticals like vaccines, antibodies, hormones, inhibitors, and enzymes are proteins which, in principle, can be produced in many different organisms. Roughly 100 proteins of human therapeutic value are available to date. Bacteria like *E. coli*, represent a well characterized, convenient, and inexpensive system for protein expression, and human insulin used against diabetes is derived from them since 1982. However, proteins that require posttranslational modifications (very often: glycosylation) for their function can often not be expressed in active form in bacteria and instead must be expressed in eukaryotic systems capable of performing these reactions. Yeasts like *Saccharomyces cerevisiae* and *Pichia pastoris* may be used (see Chapter 4.5.2); however, their glycosylation patterns are very different from the pattern in humans. The current alternatives are mammalian or human cell cultures but these are very expensive and bear the risk of contamination with human pathogens. Thus, since approx. two decades, efforts are being made to produce proteinaceous pharmaceuticals in genetically modified animals (see Chapter 10) or plants, a process termed as "pharming" (a portmanteau of farming and pharmaceutical). Plants offer several advantages over the other production systems: they are even more inexpensive than microbial fermentation systems, they possess a posttranslational machinery which is very similar to the human ones, it is easy to modify them genetically (in contrast to animals), and they are intrinsically free of human pathogens. The feasibility of plant-produced proteinaceous pharmaceuticals was demonstrated in 1995, when it was shown that mice that were fed with transgenic potatoes, expressing the heat-labile enterotoxin from *E. coli*, developed an oral immune response. This gave rise to the idea of "edible vaccines", where the plant is not only the production system for the vaccine but also its container. Already in 1989, the expression in tobacco plants of a functional antibody from mice was demonstrated. Antibodies expressed in plants are sometimes referred to as "plantibodies".

A problem with using plants as protein expression system is that the levels of the recombinant proteins are often low compared to bacteria or animal cells. This problem may be overcome in certain cases by expressing the protein in the plastids of the plant (see Chapter 9.3.3). In recent years, a process called "Agroinfection" was developed in which *Agrobacteria* that contain DNA from a plant virus in their T-DNA are infiltrated into the plant, e.g., by pressing an *Agrobacterium* solution into the apoplast by use of a syringe (not equipped with a needle) or by immersion of the plant in a *Agrobacterium* solution and applying a vacuum. The viral DNA in the T-DNA is modified to contain the gene of interest. When the viral DNA is transcribed, a viral RNA is formed which replicates and produces viral proteins including the recombinant protein of interest. Depending on the actual system, the viral RNA may spread to the whole plant or stays in the infiltrated leaves. In these transient systems, the proteins are produced within a few days and may accumulate up to 80 % of the total soluble protein in the plant.

Today, a number of plant-made pharmaceuticals are under investigation in clinical trials. In 2006, the first plant-made pharmaceutical was approved for its use as a veterinary drug in poultry: a vaccine against Newcastle disease virus consisting of the haemagglutinin neuraminidase of the virus that is produced in tobacco cell cultures. Finally, in 2012, the first plant-made pharmaceutical for use in humans was approved (site note: it is also the first prescription medication certified

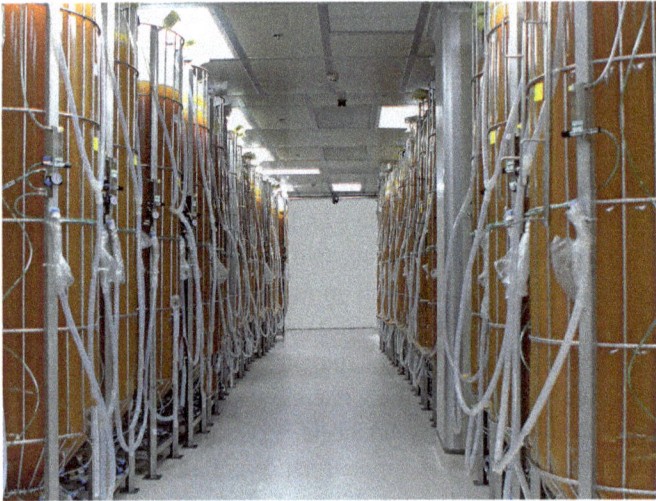

Fig. 9.23: Bioreactors containing cell cultures of genetically modified carrot cells that produce an enzyme to treat Gauchers disease, an inherited disorder. Photograph courtesy of Protalix Biotherapeutics.

as kosher); Elelyso is produced in cell cultures of carrots and consists of the enzyme taliglucerase alfa, a recombinant form of glucocerebrosidase, to treat Gaucher disease. Patients with Gaucher disease possess a defect of this lysosomal enzyme, which may be more or less pronounced.

It is significant that the only two approved plant-made pharmaceuticals are produced in contained systems, cell cultures, which grow in fermenters similar to microbial production systems (Figure 9.23), thus losing one of the main advantages of plants as protein production platform: its low costs. The initial euphoria about edible vaccines (e.g., a vaccine banana) has given way to disillusionment, mainly due to the great reservations against genetically modified plants for human use. Today, the vision of an open field full of maize plants which contain a vaccine in their grains seems unimaginable.

At the time of this writing (summer 2014), a massive break out of Ebola virus disease occurred in West Africa. Due to the critical circumstance, the World Health Organization endorsed the use of experimental drugs. One of these was ZMapp, a mixture of three different antibodies produced by Agroinfection in *Nicotiana benthamiana*.

Key-terms

Agrobacterium tumefaciens, Amflora, *Bacillus thuringiensis*, biolistic transformation, breeding, Bt toxin, Flavr Savr, flower color, Golden Rice, herbicide resistance, heterosis, hybrid, input traits, insect resistance, marker genes, output traits, pharmaceuticals, plants, protoplast, regeneration, totipotency, transformation, Triticale, vitamin A

Questions

– Name ten products of plants which are vital or important to humans.
– Name several breeding methods and categorize them.
– Why can't a farmer keep seedlings of hybrids for growing them the next season?
– Is mutation breeding regarded as transgenic breeding? Why (not)?
– What are genetic markers?
– What is the totipotency of plant cells?
– Name three methods to produce a genetically modified plant.
– Which genetic elements are on a typical plasmid used in biolistic transformation?
– What is a binary vector?
– Which genetic element, besides the coding region, is of vital importance for the expression of a transgene in a plant cell?
– How can the production of a protein that is encoded by a bacterial gene be increased in a transgenic plant?
– What is the function of a marker gene? Name three and explain their function.
– How can gene silencing be accomplished in a transgenic plant? What is the underlying mechanism?
– What are input traits and output traits? Give examples.
– Name the most common transgenic traits in genetically modified crops.
– What is "pharming"?
– What are advantages when using plants as producers of pharmaceuticals?

Further readings

Kempken, F. & Kempken, R. 2012. Gentechnik bei Pflanzen: Chancen und Risiken, 4th ed., *Springer Spektrum*, Berlin, Heidelberg, New York.
Slater, A., Scott, N. W. & Fowler, M. R. 2008. Plant Biotechnology: The genetic manipulation of plants, 2nd ed., *Oxford University Press Inc.*, New York.

References

James, C. 2013. Global status of commercialized biotech/GM crops 2013. ISAAA brief no. 46, ISAAA, Ithaka.

Stefan Wiese, Dennis Stern and Alice Klausmeyer

10 Transgenic animals

This chapter will focus on the history and achievements of transgenic animals. There will be insights in the methodologies and different strategies for foreign gene expression in living animals. Apart from this, the production of genetic clones from living animals is explained.

Stories about transgenic animals have lately come into society's focus when first reports were made on cloning pets for home use. Cloning a beloved pet might be a by-product of a technology that originally had the intention to clarify the function of RNA and proteins in living animals. Nowadays, due to problems like bovine spongiform encephalitis (BSE) in cattle, scientists even try to generate knock out cows that miss a functional Prion protein to exclude the transmission of BSE in cattle. Apart from such attempts, the development of transgenic and knock out technologies for living organisms has brought deep insights into the function of many different RNAs and proteins. Especially the medical sciences use animal models to mimic human diseases with the hope of generating a cure in the long term. This is, of course, the silver lining of this technology. Criticisms and ethical concerns for the use of animals have generated control as well as restrictions for the use of animals which are absolutely necessary. Such restrictions hopefully help to keep Pandorás box closed – the cloning of humans.

10.1 History of transgenic animals

Manipulating the genome of any species had to fulfil a lot of prerequisites. The first, and of course, most important was the identification of DNA as a "code for life". Together, with the elucidated molecular structure of the DNA, published by James D. Watson and Francis Crick in 1953, in the following years (1961–1971), the identification and purification of specific restriction enzymes first identified by Daniel Nathans, Werner Arber, and Hamilton Othanel Smith led to the establishment of a new field in biology and medicine, the molecular Biology. This period was followed by the first report on integration of foreign DNA into a genome by Rudolf Jaenisch and Beatrice Mintz in 1974. They injected the simian virus 40 (SV40) DNA into the blastocoel of a mouse embryo and subsequently established the first transgenic mouse line, published in 1976, by Rudolf Jaenisch. For this mouse line, the technique for the generation changed to the pronucleus injection of DNA into the male pronucleus, in the period where the pronuclei of egg and sperm coexist in the fertilized egg. It took 5 years until the term "transgenic animal" was established within the scientific community by the laboratories of Frank Ruddle, Frank Constantini, and Elizabeth Lacy. In general, animals which carry a transgene, independent from whether it is a random integration or a specific integration, are called "transgenic animals". Knock out animals generated by homologous recombination are also included in this terminus. To distinguish between random-integrated transgenic animals and knock out animals many people make a subdivision as such that "transgenic animal" means only the animals produced by pronucleus injection and knock out animals which are generated via homologous recombination in embryonic stem cells. The laboratory of Richard Palmiter first detected a functional expression of a transgene, the human growth hormone (HGH), stably expressed in transgenic mice, published in 1982. Within a shorter period of only three years, the generation of other transgenic species followed (e.g., sheep, pig, and rabbit). In 1989, for the first time, homologous recombination as a method for targeted inactivation and mutation of specific genes was published by Simon Tompson et al. Mice expressing a mutant hypoxanthine-guanine phosphoribosyltransferase (HPRT) enzyme were generated by partial gene deletion and subsequent insertion of a shorter gene version into the open reading frame. This new achievement also enclosed the use of target-derived mutation in embryonic stem (ES-) cells. For a longer time period, mice were the only species where the stem cell culture from the inner cell mass of a blastocyst was possible including the successful reintegration into a foster blastocyst (Figure 10.4). Finally, in the year 2000, the generation of cloned sheep (e.g., Dolly) opened the possibility to manipulate the genome of somatic cells and subsequently generate transgenic animals by nuclear transfer into a fertilized enucleated egg (Figure 10.7). This procedure only has the disadvantage that the generated clone from a somatic cell lacks the normal refilling of the telomeric ends (see Box 10.1), and therefore, has the genetic age of the nucleus used for this nuclear transfer.

Nowadays, due to subsequent integration of new tools to the genetic manipulation and stem cell technology, the targeted cell type-specific inactivation using the Cre recombinase system and/or the Flp recombinase system makes it possible to use a modular construction system by which the mice carrying the e.g. Cre recombinase under a cell type-specific promoter are bred to Cre recognition side carrying genes in mice. This procedure can generate various cell type-specific inactivation

Box 10.1: Telomers and the telomerase.

Telomers represent the ends of linear chromosomes. They are highly repetitive and prevent the linear chromosome from fusion and degradation by nucleases. The telomeric ends need a special enzyme for their replication, the telomerase. This enzyme guarantees the maximum quantity of the telomeric ends. The telomerase is expressed in highest levels in oocytes and spermatogonia and assures the refilling of these ends during cell division of these germ cells. During development of an embryo, the activity of the telomerase strongly decreases and is finally only detectable in the germ line cells. The somatic cells constantly lose parts of their telomeric ends during cell cycling as the DNA polymerase can only partly duplicate the repetitive telomeric ends of the chromatids. This missing activity of the telomerase is one major reason for aging leading finally to cell death, and for the organism to perish. The reexpression of the telomerase in any tissue is one of the most prominent reasons for the development of cancer.

of one gene just by using different Cre mouse lines. Another new invention are the induced pluripotent stem cells (IPSCs). With this method, somatic cells are reprogrammed by transient expression of up to four transcription factors (KIF4, c-Myc, Sox2, Oct4). This generates a cell line that has nearly identical properties when compared with ES cells. The method solves the problem of the nuclear transfer method described in brief above because the telomeric ends are refilled during this reprogramming process.

10.2 Isolation of fertilized eggs, morulae and blastocysts

Ovulation of a female is the prerequisite for the isolation of eggs. The number of maturing follicles is species dependent but can be increased by hormone supplementation. While the general mitotic divisions for the later oocytes in the female mice already occur between day 12 and 13 of gestation, these cells stop at the prophase I of meiosis. Mice have cycle duration of 4–5 days. The follicle stimulating hormone (FSH) initiates the follicle growth. This can be additionally stimulated by treatment with FSH. The reason for more mature follicles generated by this treatment is similar to that for humans. The natural FSH level leads to a follicle maturation of more than the finally mature ones. Increasing amounts of FSH subsequently lead to maturation of more if not all initialized follicles. The reason for this bereavement of follicles during maturation is a kind of quality control. Only the "best" of the best should reach maturity. The pituitary gland subsequently releases the luteinizing hormone (LH), which leads to the maturation of the follicles, while the Theka interna cells are primed to produce estrogens. Addition of β-human corticogonadotropin (βHCG) 12 hours before ovulation can synchronize the mice in this stage so that the later vaginal plug time is nearly identical. Additional LH release from the hypophysis leads to the follicle rupture and ovulation. The time line for isolation of fertilized eggs starts with mating. For mice, this is usually annotated to midnight due to the fact that mice display a night active mating behavior. The vaginal plug is visible up to 16 hours after mating and during this time, the chance for the isolation of fertilized eggs in the one-cell-stage is highest (Figure 10.2). The earlier the next morning the isolation starts, the higher the rate of fertilized eggs that show up the male and the female pronucleus. This is the ideal time point for pronucleus injection of DNA into the male pronucleus (Figure 10.4). The two-cell stage is present the same day in the later evening but the usual laboratory procedure, in this case, might be to isolate the fertilized eggs and, subsequently, cultivate them until they reach the two-cell stage. This stage can be used for fusion to generate tetraploid embryos (Figure 10.3). At day 2,5 after fertilization, the isolation of 8-cell stage embryos from the oviduct guarantees development of blastocysts in short term cultures. These morulae subsequently can either be used for morula aggregation or can be cultured to be used for blastocyst injection. Blastocysts can

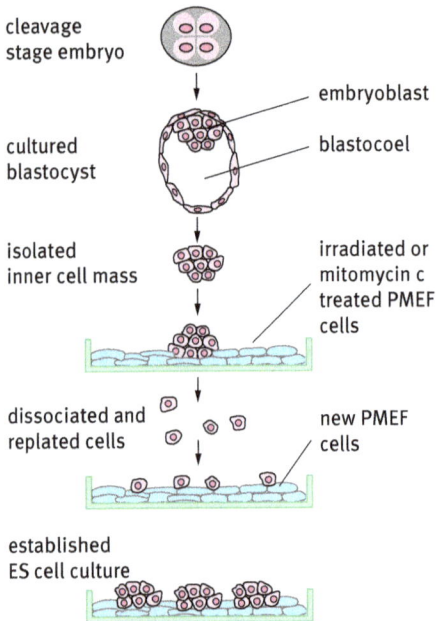

Fig. 10.1: Establishing a new embryonic stem cell line. Embryos can be isolated as cleavage stage embryos or as a blastocyst. The cleavage stage embryos can be cultured until they reach the blastocyst stage. Prolonged culturing leads to complete filling of the blastocoel with cells (not shown here). The isolated inner cell mass can be cultured on irradiated or mitomycin-treated primary mouse embryonic fibroblasts (PMEF), which build up a monocell layer, provide factors but cannot divide. The first cell colonies are dissociated from the plate and replated on new PMEF cells. ES cells grow as 3-dimensional colonies on the monolayer of PMEF cells. A new ES cell line is established by this procedure.

also be isolated from the uterus at day 3,5 after fertilization. The blastocysts can be used for blastocyst injection of embryonic stem cells (Figure 10.1) or isolation of the inner cell mass for establishing a stem cell line. While the morulae are an undefined cell mass, the blastocysts can anatomically be subdivided into three different parts. The trophoectoblast, which later generates the embryonic part of the placenta, the inner cell mass, which gives rise to the embryo and represents the pluripotent stem cells, and the blastocoel, which is a liquid-filled area within the trophoectoblast (Figure 10.4). The metamorphosis of a morula to the blastocyst in mice is accompanied with the onset of transcription and translation of the embryonic cells. At the end of the morula stage, the maternal transcripts are degraded and the transcription machinery of the embryo is initiated. So even if there were severe chromosomal disarrangements, the fertilized egg could possibly develop to morula stage but not further (Green 2007).

Why are mice so special? Why are they ideal for the isolation of stem cells? Mice have a special behavior in nature. When life changes to worse and the animals do not have enough to eat, the females have the possibility to retain implanta-

light/dark day time
0:00
5:00
0.5 –12 h – β-HCG treatment
19:00
0 h
5:00
0.5 –12 h – vaginal plug, single cells
19:00
0 h – 2-cells stage
5:00
1.5 –12 h
19:00
0 h – 4-cells stage
5:00
2.5 –12 h – 8-cells stage
19:00
0 h – morula
5:00
3.5 –12 h – blastocyst
19:00
0 h
5:00
4.5 –12 h – implantation to the uterus
19:00

ovary
infundibulum
swollen Ampulla
sperm
oviduct
utero-tuberal boundary
blasto-cysts
uterus

Time

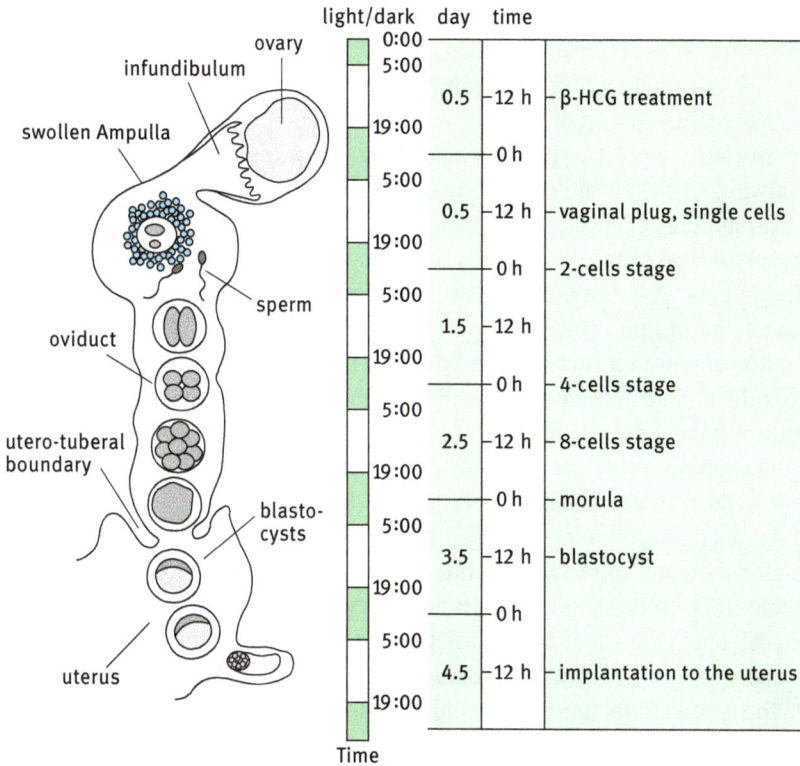

Fig. 10.2: Scheme for the isolation of fertilized eggs, morulae, or blastocysts from the uterus of pregnant mice. Mice are stimulated with b-HCG (beta human corticogonadotropin) 12 hours before fertilization occurs. This leads to an increase in fertilized eggs and synchronizes the female animals. Next morning, the vaginal plugs prove that breeding has taken place. Fertilized eggs subsequently develop in the oviduct until they reach the morula stage on day 2,5. The morulae transfer through the utero-tuberal boundary and transform to blastocysts in the distal uterus until day 4. Implantation occurs on day 4,5.

tion, which usually occurs between day 4 and 4,5 after mating, for up to 14 days. In this time period, the blastocysts stay within the uterus. The cells of the inner cell mass further divide and finally fill the whole blastocoel. In contrast to this, many other species immediately interrupt a preimplantation pregnancy when life circumstances change for the worse. This natural retaining effect in mice can even be mimicked when the ovaries are removed after the vaginal plug occurred. This might be an indication that the embryos in a preimplantation stage in mice are somewhat more robust. The second point is that the cultured cells of the inner cell mass only slightly change their size.

10.3 Tetraploid embryos

The use of mouse stem cells for generation of chimeric mice, that later can give rise to a new genetically modified mouse line, has been in use since it was first published in 1982 (see 10.1). The prerequisite for a successful germ line transmission was always that the stem cell line was male by genotype. The injected blastocysts for the chimera generation were either male or female. So the resulting chimera could also be a mixture of male cells from the stem cells and female cells from the blastocyst inner cell mass. The resulting animals might have a male phenotype as long as the germ line cells are in part male, but the mating behavior of these chimeric mice quite often turns out to be a real bottle neck as they have a lower testosterone level and accompanied with this, they might display a reduction in mating and/or have reduced amounts of viable sperm cells. This problem led to an invention that circumvents the problem of chimeric mice with a mixture of male and female cells. The use of tetraploid embryos leads to the generation of 100 % chimeric animals, and therefore, the resulting offspring from the blastocyst injections are already equal to the F1 generation offspring of conventionally produced chimeric mice (Figure 10.3). For generation of tetraploid eggs, the isolated embryos in the two cell stage are used. An electric pulse leads to the fusion of the two cells and the subsequent reunion of the separated nuclei. Due to the internal stores for RNA from the mother side, these tetraploid cells develop as an embryo and successfully generate a blastocyst in culture. The tetraploid cells which form the inner cell

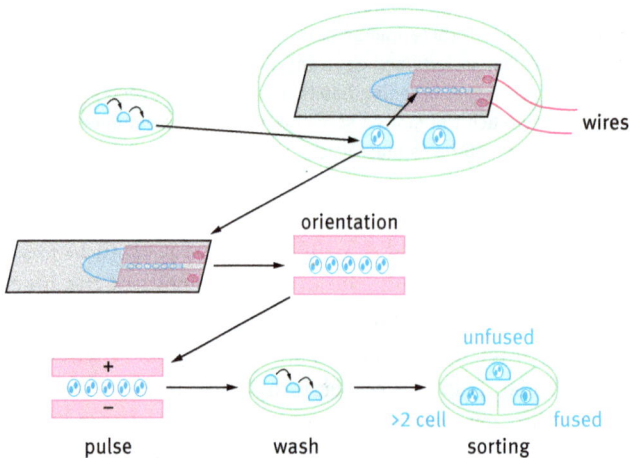

Fig. 10.3: Generation of tetraploid embryos. 2-cell stage embryos of mice are transferred to a mannitol-buffer mixture by subsequent washes. The embryos have to settle down in the solution. Equilibrated embryos are transferred to the interspace of two electrodes within the buffer. A pre-pulse leads to correct orientation of the embryos. The subsequently following pulse then leads to fusion. Buffer washes eliminate the mannitol. The embryos are sorted. Fused embryos have only one cell. Multicell embryos and 2-cell stage embryos are sorted and put to waste.

mass cannot develop any further. Now, the genetically modified embryonic stem cells are injected into the blastocoel. They integrate into the inner cell mass, and they now exclusively generate the embryo. While the tetraploid cells are unable to generate the later embryo, they can generate the placental cells of the resulting embryo after implantation. This procedure ensures 100 % chimeric offspring generated by the embryonic stem cells. A disadvantage of this method might be that this procedure needs increased numbers of fertilized eggs as there might be a greater number of embryos that either do not match the criterion for a two cell stage embryo, and not all two cell stage embryos fuse after the pulse has been given (Figure 10.3). One should further notice that this blastocyst injection procedure with tetraploid embryos has been used successfully to generate mice from induced pluripotent stem cells (IPSCs) (see chapter 10) (Yamanaka 2012). The resulting offspring were 100 % clones of the donor mouse for the IPSCs. This method circumvents the age problem of nuclear transfer-derived clones as with e.g. the cloned sheep Dolly.

It remains to be mentioned, that the combination of the IPSC technology for humans together with the generation of tetraploid embryos might even be a possible tool for generating human clones. Therefore, we need worldwide restrictive rules that prohibit such use in humans.

10.4 Pronucleus injection of fertilized eggs

Preimplantation embryos, in general, appear susceptible to foreign DNA, when they get into direct contact. The first transgenic animal report was on SV40 DNA injected to a blastocyst. This injection resulted in an animal that showed integration of this foreign DNA into its genome. The rate of integration for this kind of injection was low at that time. The change to pronucleus injection of DNA into the male pronucleus led to offspring that showed integration in almost all cells of the body, accompanied with germ line transmission to the next generation. The pronucleus stage exists for up to 12 hours after fertilization. While the female pronucleus is still busy finishing the last steps of the meiosis II division, and the freshly fertilized egg giving then rise to the last pole body, the male pronucleus has to be decondensed as the DNA is extremely packed by protamine proteins instead of histone proteins. This difference in condensation is visible through light microscopy. While the male pronucleus has an oval shape, the female pronucleus, even after finishing the meiosis II division, is bigger and more round-shaped. Due to massive rearrangements in the male pronucleus (the DNA has to be decondensed before the two chromosome sets fuse to one nucleus), this difference made it possible to inject the linearized DNA directly into the male pronucleus, which has a much higher capacity to integrate foreign DNA than the DNA of the female pronucleus.

The DNA integration for this method is random and the way of integration is not really controlled. This means that the foreign DNA could be integrated in

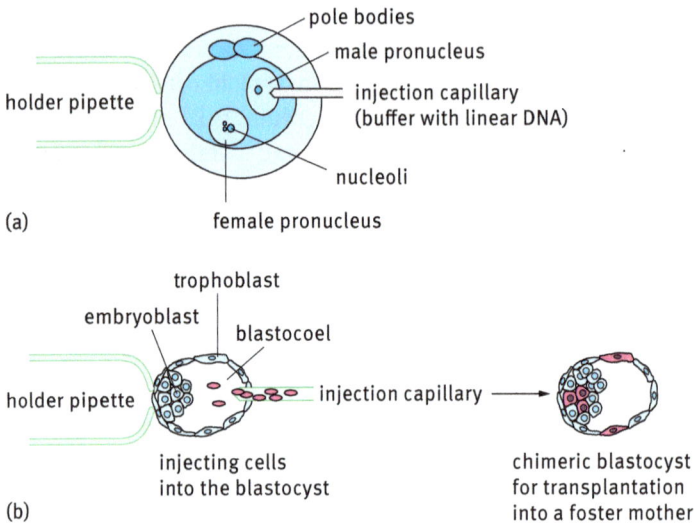

(a)

(b)

Fig. 10.4: Principles for pronucleus injection and blastocyst injection. (a) Pronucleus injection is used for isolated linearized DNA. DNA in saline is injected into the male pronucleus of an isolated fertilized egg of a mouse before fusion of the female and the male pronuclei has taken place. The injected egg is subsequently lastly retransferred in the morula stage. (b) Blastocyst injection is performed using transgenic ES cells (red) which are injected into the blastocoel of the blastocyst. The injected cells integrate either into the trophoblast, and therefore participate in the placenta development, or integrate into the embryoblast and participate in the further development of the embryo, possibly also including the germ cell lineage.

euchromatic as well as in heterochromatic parts of any chromosome. Depending on the place of integration, this might result in a low expression, or even complete shutdown of the transgene. Integration into the X-chromosome also shows up variable effects depending on the sex of the offspring. Another problem is the way of integration. While multiple integration sides could be reduced to single by mating, the transgenes usually integrate as tandem repeats, which could be in fewer (1–3 copies) or higher numbers. This might result in a massive expression of the transgene because each copy has its own promoter and expresses the mRNA.

These uncertainties make it necessary to determine the transgene by southern blot for at least 3 generations, and to produce at least 3 independent and expressing lines. The southern blot should not only determine the stability of the transgene transmission to the next generation but also determine the amount of integration. This can be managed by serial dilution of the transgene vector with wild type DNA. The densitometric determination gives a linear correlation for the number of vector copies and the amount of DNA used. The importance of this determination might be estimated with the following example.

Some effects shown up in the next generations might be due to the expression level of the transgene. In some cases, low and high expressing lines might be inves-

Box 10.2: Spinal muscular atrophy.

Spinal muscular atrophy (SMA) is a human disorder with recessive trait. The disease has different severity grades, type I, II, III, and IV with type I with very early onset, and type IV with a very mild outcome. Nearly all humans carry a duplication at the telomeric end of chromosome 5. This duplication includes the hSMA1 gene. The hSMA2 gene is expressed but carries an 840C to T mutation, affecting the splicing of exon 7. This exon is skipped in 90 % of all transcripts from the SMA2 gene. The approximately 10 % complete mRNA only produce approximately 10 % functional protein. The severity of the disease depends on the copy numbers of SMA2. The region around SMA2 appears unstable so that SMA2 can appear in more than 1 copy. There are even cases with a deletion of the telomeric hSMA1 copy which carry 10 or more copies of SMA2. These individuals appear healthy.

tigated even on purpose when e.g. earlier experiments have shown that the protein level is important for the disease level. This is e.g. the case for the mouse model mimicking the spinal muscular atrophy (see Box 10.2) in humans (hSMA). The level of functional SMA protein determines the onset and even the severity of the disease. The mouse model is a mixture of a conventional mouse (m) SMA knock out (homozygous knock out animals do only exist until the morula stage) which is bred to a transgenic animal carrying the human SMA2 transgene. When the hSMA2 gene, a splice variant of the fully functional SMA1 gene, is present in those mice which lack the murine version of SMA1, the hSMA2 transgene rescues the phenotype. But if this hSMA2 transgene is only present in 1–3 copies in the genome, it can only rescue the phenotype until day 3–7 postnatal. In contrast, hSMA2 high copy transgene carriers can rescue the SMA knock out phenotype completely.

Transgenic approaches can be used to overexpress a gene of interest or to modulate its function, by either interferance with the RNA synthesis (antisense RNA) or interferance with the protein function by introducing a transgene with e.g. a dominant-negative mutation (see Figure 10.6). Dominant-negative mutations can either interfere because the protein is produced predominantly by overexpression of the transgene, or they can interfere in protein assembly when the protein itself can only function as dimers or tetramers. The mutated region, in this case, usually involves the active center of the protein.

10.5 Embryonic stem cells and induced pluripotent stem cells

ES cells can be newly generated from blastocysts (Figure 10.1). Digestion of the zona pellucida following digestion of the cell aggregate and culture of single cells on irradiated primary mouse embryonic fibroblasts (PMEFs) will lead to growth of single colonies. The cultured cells are sensitive to differentiation unless they are treated with differentiation inhibiting factors like leukemia inhibitory factor (Keller 2005). This cytokine is, in part, provided by the irradiated fibroblast but should be supplemented additionally. The irradiated or chemically division-inactivated fibroblasts cannot divide in culture. Apart from this, they display a contact inhibition feature which makes them grow in a monolayer. The freshly isolated and cultured ES cells lack this contact inhibition. They grow as three-dimensional colonies, and the single cells are much smaller than the fibroblast. This growth behavior not only makes it easy to distinguish between PMEF cells and ES cells but also makes it easy to pick single colonies as new clones. Therefore, single colonies can easily be picked up by sucking it up with a pipette tip. Transfer to e.g. a 96-well plate, and digestion with trypsin to generate a single cell suspension is, therefore, used as a tool to analyze single colonies after electroporation of a target vector. The target vector can either generate a knock out or conditional knock out with Cre-LoxP flanking sites to one or more exons. This usually disrupts the open reading frame of the gene of interest. In the first years of use, one problem for the Cre-LoxP system was the neomycin resistance cassette, which was present in the constructs. Even if placed in an intron with its own promoter, this presence could lead to disturbances in gene expression. The solution was to flank the neomycin resistance cassette with a third LoxP site or to flank it with Flp sites. The secondary excision of the neomycin resistance cassette leaves back a single LoxP site which is alternatively used to cut out the respective exon or in case of Flp flanking sites the Flp site is latest cut out when the cell expresses the Cre recombinase and the respective region is deleted.

Mouse embryonic stem cells have been generated since the late 70s in the 20th century. One of the first lines was the E14Tg2a cells which were derived from an F1 intercross of Sv129 mice. This mouse line has been used for a longer time period to generate new ES cells, as it appeared that they are easier to generate and easy to handle. The problem with the genetic background of the stem cells became more apparent when people tried to elucidate the role of specific genes in neurological and neuroimmunological diseases like e.g. multiple sclerosis (Aguzzi 1994). One of the experimental models for multiple sclerosis is the experimental autoimmune encephalitis (EAE). The SV129 mouse line turned out not to be susceptible to this model. In contrast, the inbred mouse line C57Bl/6 is highly susceptible to this multiple sclerosis model in mice. In case a conventional knock out mouse has been made with SV129 derived ES cells, despite several back-crosses to C57Bl/6, the knock out allele itself will always be SV129 derived as there are no cross over events possible during meiosis. A difference in the genetic background in a certain gene might also influence the outcome of planned experiments as point mutations might give rise to protein variants that are less susceptible for the disease. Therefore, a check for the genetics of the ES cells, as well as a check for the genetic backcrosses of the investigated transgenic mouse line, is highly recommended.

IPSCs can be generated from any cell type but the efficiency is quite diverging so that some cell types appear much more susceptible to this procedure. Earliest procedures involved the use of viruses which express the four transcription factors KIF4, c-Myc, Sox2, and Oct4. Meanwhile, especially because of problems for the

(a)

(b)

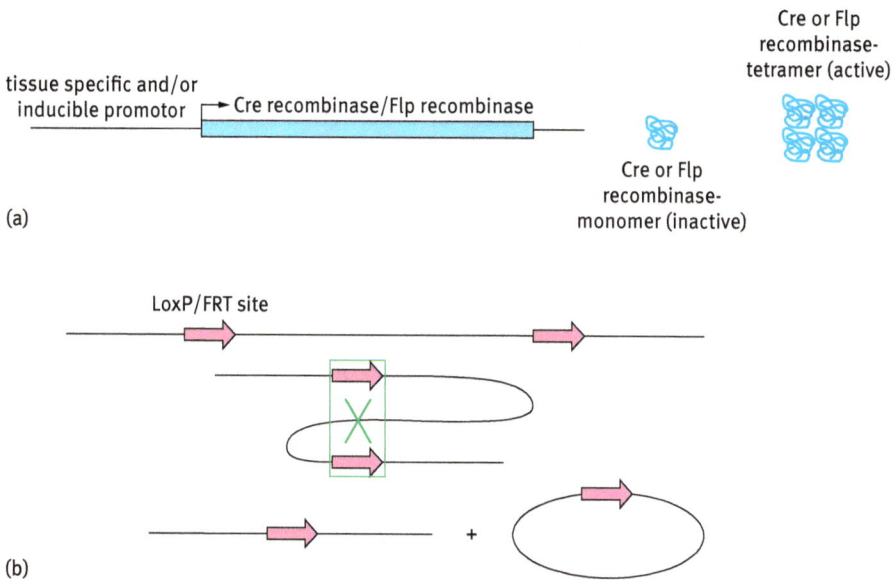

Fig. 10.5: Cre and Flp Recombinases. (a) Cre and Flp Recombinases are expressed under a tissue-specific and/or an inducible promoter. The enzyme is generated, depending on the tissue and time point. The monomeric proteins are inactive, and they have to form a tetramer to be active. Therefore, the starting point of expression might not be the starting point of activity, and the excision within a tissue is very often less than 100 %. (b) The active Cre or Flp enzymes recognize their specific binding sequences and form a loop that leads to excision of the DNA in between. The recognition sequences have to be in parallel and not in antiparallel formation as this would cause an inversion of the sequence in between instead of excision.

Box 10.3: Cre- and Flp Recombinases.

The causes recombination (Cre) enzyme was first described by Nat Sternberg and David Hamilton in 1981. This enzyme is produced by the bacteriophage P1 and is natively used for integration of the phage DNA into the genome of the host bacterium. The enzyme recognizes a specific 34 bp sequence with two 13 bp palindromic sequences and an 8 bp intervening sequence. Two of these sites are necessary for recombination events, and the sequences in between are therefore called floxed. The LoxP sites have to be in parallel for correct excision events (for more details see Figure 10.5).

The Flippase (Flp-) recombinase, also called Flp-FRT (Flippase recombination target) is a system derived from a *Saccharomyces cerevisiae* plasmid. The recognition sequence for the Flp is also 34 bp. A 13 bp inverted identical sequence is spaced again by an 8 bp intervening sequence. Two of these sequences are necessary for excision of the DNA lying in between. The Flp enzyme exists in several variants which recognize different sequences. Therefore, the flanking Flp sequences have to be identical (Figure 10.5).

use of human IPSCs in possible personalized medicine, the transcription factors are transiently expressed. This omits the problem of viral DNA in the newly generated IPSCs. These cells can alternatively be used for generating transgenic mice when the IPSCs are handled like the ES cells described before. The combinatory use of tetraploid embryos and IPSCs in mice already generated fully viable cloned mice so that the question of stemness of these cells has been proven (see also Chapter 11). Nonetheless, the ES cells will serve as a standard to prove the quality of the IPSCs.

10.6 Strategies for gene transfer

From the very beginning of genetic manipulation of fertilized eggs and blastocysts, the type of manipulation was determined by the type of scientific question – but in the end the outcome was sometimes vice versa as the phenotype of the genetically modified animal quite often raised concerns towards the original thought on the role of the gene, RNA or protein looked at. In general, one can subdivide the different types of manipulation either by the method, or more precisely, by the "target". These different "targets" are a) transgene expression, b) insertion mutagenesis, c) antisense RNA, d) dominant-negative mutation, e) homologous recombination, and as a subtype here, f) genetic ablation of specific cell types in the adult or within development (Figure 10.6). Cell type-specific approaches, in general, can be performed by use of specific promoters and development-specific expression by use of specific promoters or exogenously activated promoters like e.g. the estrogen-promoters.

10.6.1 Transgene expression

For transgene expression, a conventional pronucleus-injected mouse is generated. The construct includes the open reading frame of a specific gene of interest and a separate promoter. The transgene might be expressed celltype-specific depending on the specificity of the promoter. The readout for this approach is the analysis of the function for the overexpressed protein.

10.6.2 Insertion mutagenesis

A DNA fragment introduced by pronucleus injection can also disrupt an endogenous gene. This approach is random as the injected DNA can introduce in any part of the genome. The use of transposons and respective activators can help to minimize the amount of transgenic animals, which have to be made as one transgenic line can be used multiple times. The readout of this approach is the functional analysis unknown genes.

10.6.3 Antisense-RNA

In this approach, the antisense direction of a messenger RNA is cloned under control of a separate promoter. This construct is then introduced into a mouse by pronucleus injection. The antisense RNA is independently expressed in the cells where the promoter is active. It can form a hybrid with the internal sense mRNA, which should normally be exported from the nucleus to the cytoplasm after correct editing and be transcribed. The nucleus of the cell recognized the double-stranded RNA as incorrect and marks it for degradation. Therefore, the expression level of the endogenous sense mRNA is down-regulated. The expression of an antisense RNA usually does not erase the complete sense-mRNA, so that there might be some remaining level of expression for the protein. This method's aim is the suppression of the function of a protein without complete deletion. A dose-dependent phenotype might be the consequence (Figure 10.6).

10.6.4 Dominant-negative mutation

While the former method is aimed to reduce the level of RNA here, a transgene inserted to the genome by pronucleus injection has the expressed protein as a target. A mutant inactive version of the protein is made because the DNA contains a site-directed mutagenesis version of the wild type DNA, leading to a non-functional protein. If the protein acts as a monomer, the endogenous wild type version and the dominant-negative version of the protein will compete for their targets within the cell. If the dominant-negative version of the protein acts as a dimer or tetramer, it can inhibit the function of the endogenous wild type protein by formation of heteromeric complexes. The aim is to study the function of the endogenous protein or to inhibit the endogenous protein without complete deletion.

10.6.5 Homologous recombination

Gene targeting by homologous recombination can delete a specific gene, and with this, its expression, or it can modify it by defined insertion of mutations into the endogenous gene. This approach requires the use of embryonic stem (ES) cells – or – induced pluripotent stem (IPS) cells. The targeting vector includes the mutation surrounded by two homologous sequences. These sequences should be at best 10 kb or more apart from each other within the genome because this increases the frequency for double recombination events. The classical targeting vectors contain a marker and selection cassette, which usually expresses a neomycin resistance under control of an independent promoter. This ensures that the target vector – once it has been integrated to the genome – expresses the selection marker under all circumstances. Next to one of the homologous flanking regions, the targeting construct includes a HSV-TK. The herpes simplex virus thymidine kinase gene produces a toxic product in the cell in case it is integrated into the genome of the stem cell. This negative selection marker should preclude random integration stem cells so that only the ES cells with a homologous recombination can survive. Therefore, after electroporation of the construct first, the positive selection starts by applying neomycin. In a second step, the remaining colonies are treated with gancyclovir. The remaining stem cell colonies are subsequently picked as single clones and further analyzed. The correct stem cells are then injected to a blastocyst and the resulting chimeras hopefully mate to transmit the mutated allele. Transmission of the mutated allele by the founder, even in the case of 100 % chimeric animals, is only 50 % because the stem cells are heterozygous for the mutation, and they can also, therefore, transmit the wild type allele to the next generation.

10.6.6 Genetic ablation of specific cell types

The genetic ablation of specific cell types is performed as a transgene approach using pronucleus injection. The construct includes a toxic gene product under control of a cell type-specific promoter which is additionally controlled by a regulatory element, that either can be induced or repressed exogenously by applying the activator or repressor (in case of constant feeding of the animals during pregnancy and weaning). The ablation of specific cell types aims to elucidate the role of specific cells throughout development or in the adult.

Fig. 10.6: Strategies for gene transfer. (a) Integration of an *in vitro* designed construct (regulatory ▶ unit in blue and gene of interest in red) into the genome. The open arrow indicates the step of pronucleus injection. (b) Integration of a foreign DNA can disrupt an endogenous gene. The loss of gene function of unknown genes can subsequently be analyzed in the transgenic animals (open arrow). (c) The transgene codes for an antisense RNA (vertical stripes) which is complementary to the mRNA (sense RNA, horizontal stripes). Intracellular hybridization of both RNAs leads to the disruption of the mRNA translation process by degradation. (d) The transgene codes for a mutant version of the endogenous protein. This mutant protein can interfere with the endogenous protein and neutralize its function on protein level. (e) The transgene (in grey) carrying construct injected into the pronucleus, is a toxin which is under control of a specific promoter (in blue). The toxin expression should be regulated by this promoter, and thereby, only be expressed in the targeted cells that upregulate this promoter. (f) The genomic sequence of a gene (red) is interrupted and/or partially deleted *in vitro*. The neomycin resistance (Neo) serves as a positive selection marker, and the thymidin kinase (TK) gene as a negative selection marker that should be excluded by the double homologous recombination event. The construct is electroporated to ES cells where homologous recombination takes place (crosses indicate the recombination event). The resulting embryonic stem cells are heterozygous for the targeted gene modification and are subsequently injected to blastocysts (see also Fig 10.4).

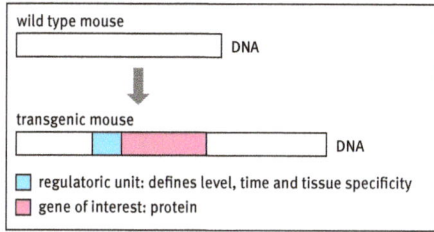

(a)

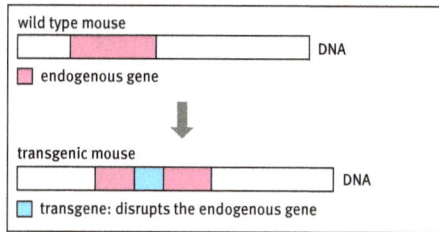

(b)

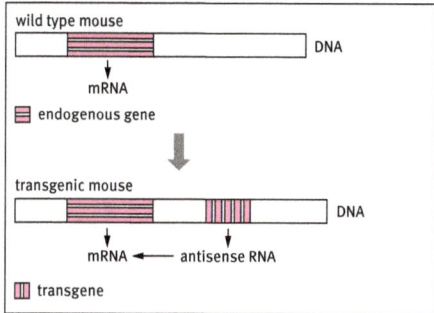

(c)

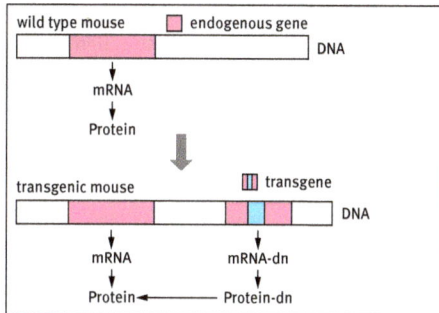

(d)

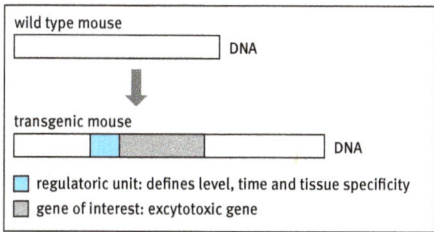

(e)

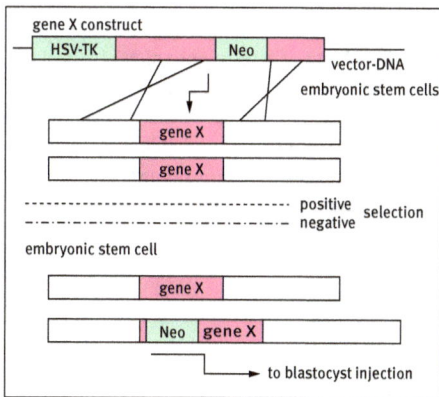

(f)

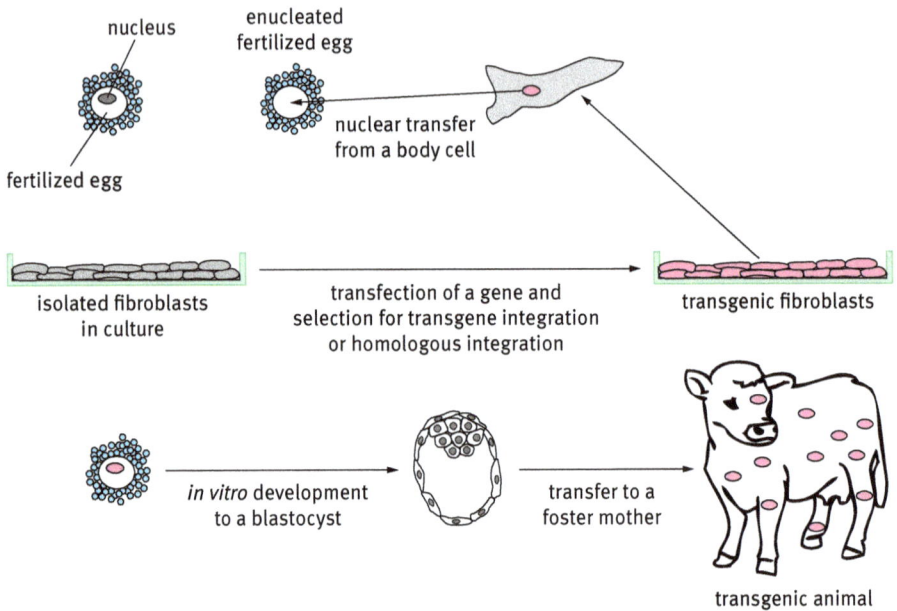

Fig. 10.7: Generation of transgenic animals using enucleated fertilized eggs.
A fertilized egg of a recipient animal is used for enucleation. The nucleus of another cell is subsequently injected into the recipient egg. This donor nucleus can be taken from a freshly isolated
cell of a body. In the case of the first cloned sheep, it was e.g. an udder cell which was isolated
and cultured. These cultured cells served as a donor cell line. The cells can be genetically modified prior nuclear transfer to the recipient enucleated egg. This technique can be used to produce
transgenic animals of many mammalian species.

Key-terms

antisense, blastocyst, dominant, embryo, cells, Flip recombinase, homologous recombination, induced pluripotency, knock out, LoxP-Cre, morula, mouse, protein,
tetraploidy, transgenic

Questions

- Who generated the first "transgenic animal" and which kind of DNA was used
  for this approach?
- How do you generate a transgenic mouse and with which method do you gen-
  erate a knock out mouse?
- Explain the different preimplantation stages of a mouse with time frame.
- How can these different stages of a preimplantation embryo be used?
- Explain how a tetraploid embryo is generated.

- What is the difference between a chimeric mouse generated by blastocyst injection of a diploid embryo and by blastocyst injection of a tetraploid embryo?
- How can you achieve the genetic ablation of a specific cell type using transgenic mice?
- Which are the different types of genetic approaches for the analysis of gene, RNA, and protein function?

Further readings

Fallini, C., Bassell, G. J. & Rossoll, W. 2012. Spinal muscular atrophy: the role of SMN in axonal mRNA regulation. *Brain Res* 1462: 81–92.

Mason, P. J., Perdigones N. 2013. Telomere biology and translational research. *Transl Res* 162(6): 333–42.

Bouabe, H. (1) & Okkenhaug, K. 2013. Gene targeting in mice: a review. *Methods Mol Biol* 1064: 315–36.

References

Green, E. L. 2007. Biology of the Laboratory Mouse. 2nd ed., adapted for the Web by Mouse Genome Informatics (MGI), http://www.informatics.jax.org/greenbook/index.shtml.

Yamanaka, S. 2012. Induced pluripotent stem cells: past, present, and future. *Cell Stem Cell* 10: 678–684.

Keller, G. 2005. Embryonic stem cell differentiation: emergence of a new era in biology and medicine. *Genes Dev* 19: 1129–1155.

Aguzzi, A., Brandner, S., Sure, U., et al. 1994. Transgenic and knock-out mice: models of neurological disease. *Brain Pathol* 4(1): 3–20.

Andreas Faissner, Jacqueline Reinhard and Ursula Theocharidis

11 Stem cell biology and applications in neuroscience

In the last decades, many new achievements and methods have widely expanded the knowledge and use of stem cells in research and clinical trials. Beginning with the fertilized egg, the present chapter defines different types of stem cells in the developing and mature organism, and discusses ethical questions and potential therapeutic application strategies.

11.1 Embryonic stem cells

The fusion of a female and male germ cell enables the development of a whole organism. Based on this great potential, it has come into the focus of research to identify the main cell types and molecular players, which orchestrate and enable this developmental process.

11.1.1 Fertilized egg cell and totipotency

The fertilized mammalian egg, the zygote, is the source of all cell types that have been described in mammals and is, therefore, considered totipotent. Totipotency can be proven in the mouse by inserting cells endowed with a traceable marker into early embryonic stages, that is the blastocyst. The resulting organisms are named chimeras, because the emerging embryos contain genetic information from two distinct genomes that distribute into discernable cell lineages (see Chapter 10). In case progeny of the transplanted cells constitutes the germline, the next generation will be genetically homogeneous, and each cell will be a descendent of the transplanted ancestor. The marker associated with the transplanted cell can serve a proof of totipotency. This strategy has successfully been used in mice, but is obviously not applicable to human beings. Based on transplantation studies, it has been concluded that beyond the fertilized egg cell, also the early stages of the embryo up to the eight cell stage may be totipotent, as well as embryonic stem (ES) cells derived from the inner cell mass (epiblast) (Figure 11.1). The inner cell mass represents an early stage of mammalian embryogenesis prior to the process of gastrulation and has been used as a source of cells that could be established as permanent cell lines *in vitro*. In the mouse system, ESCs served as a source of cells for controlled genetic recombination and constitute the basis of the generation of genetically modified mouse lines (see Chapter 10).

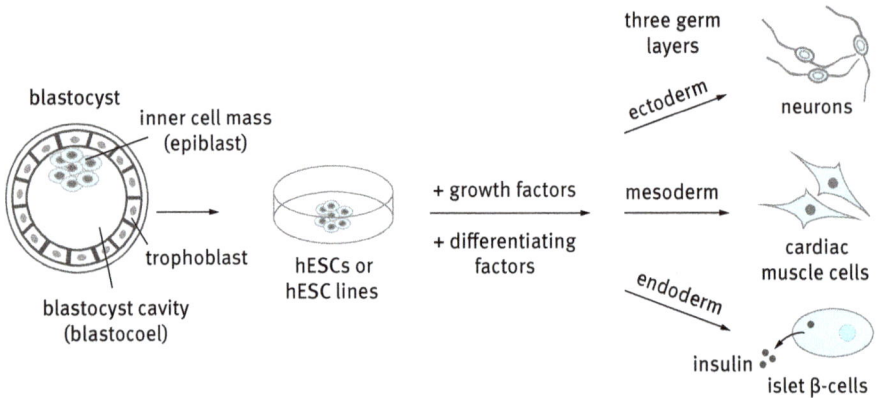

Fig. 11.1: Embryonic stem cells.
Human embryonic stem cells (hESCs) and stable hESC lines can be derived from the inner cell mass (epiblast) of the blastocyst. As these lines have been derived from early embryonic stages, they can be considered at least pluripotent, meaning that these cells can form derivatives from all three major embryonic layers, namely the ectoderm, mesoderm, and endoderm under defined conditions. These three germ layers give rise to neurons, cardiac muscle cells or insulin-producing islet β-cells as examples.

11.1.2 *In vitro* fertilization and ES-lines

Beginning in the 1970s, the development of techniques of *in vitro* fertilization in humans by Robert G. Edwards (Nobel Prize 2010) permitted the cultivation of human embryos in the culture dish. These embryos were destined for implantation, but in the context of the procedure more embryos were cultivated than needed for implantation, which led to so-called supernumerary embryos. In 1998, James Thomson described a protocol that allowed the cultivation of human embryonic stem cells (hESCs) obtained from the inner cell mass, the epiblast of human embryos grown in the process of *in vitro* fertilization. Stable lines could be derived from these cultures, the so-called hES-lines (Figure 11.1). As these lines have been derived from early embryonic stages, they can be considered at least pluripotent, meaning that these cells can form derivatives from all three major embryonic layers, that is ectoderm, mesoderm, and endoderm. The three germ layers are formed during gastrulation of the embryos of amphibians, birds, and mammals. Many hES-lines have been generated that are of clonal origin, and thereby, homogeneous. Analysis of epigenetic modifications, however, indicates that these lines may not be completely identical, which may have an impact on their differentiation capacity. It is possible that some of the hES-lines are totipotent, but this cannot formally be proven, as it would imply an experimental strategy that touches human cloning. This objective is forbidden by strong legal regulations in most countries in Europe and in the USA, where stem cell research is being pursued. However, the teratoma formation assay can be viewed as a substitute for the forbidden procedures and serves as an appropriate tool for monitoring pluripotency in stem cell research (Box 11.2). When transplanted into recipient immunosuppressed mice, hESCs, in many cases, form teratomas at a quite high frequency, emphasizing their pluripotent character. On the other hand, this assay also illustrates the dangers intrinsically associated with any treatment based on the use of hESCs.

Box 11.1: Ethical issues raised by stem cell research.

The generation, maintenance, and application of hESCs have been deeply controversial from the very beginning. The major reason for a conflict-ridden debate is the fact that the removal of individual ESCs from early human blastocysts derived by *in vitro* fertilization in general goes along with the destruction of the embryo. For this reason, strong pressure groups oppose the generation of ES line and favor the protection of the embryo instead. This raises serious issues concerning the handling of supernumerary embryos that accumulate in fertility cliniques. There are estimates that several hundred thousand embryos are stored in nitrogen-cooled deposits in the United States alone. A general debate in many societies has been engaged concerning the aims, limitations, and prospects of stem cell research, and numerous legal rules have been developed that differ in dependence of cultural and legal traditions of the country concerned. In Germany, the derivation of human ES lines is prohibited. Investigators may use ES lines that have been produced before a defined date (01. 05. 2007) in other countries, provided they have submitted an application that has been approved by the institution concerned, the Bundesgesundheitsamt (BGA, Berlin). This regulation, which seems reasonable at first sight, has some problems attached. For example, many projects are carried out in international collaborations. German investigators would have to withdraw if colleagues in countries with other legal regulations would use hES lines that have been generated after the exclusion date voted in the German Parliament. The development of induced pluripotent stem cells (iPSCs) (see 11.2.4) may contribute in bypassing some of these restrictions.

Box 11.2: Teratoma formation assay.

Studying pluripotent stem cells includes the analysis of the differentiation potential. Several *in vitro* methods can be used to detect cells with different characteristics in stem cell derived cultures, which include immunostainings for tissue specific antigens, formation of embryoid bodies or the analysis of the gene expression profile. A widely used *in vivo* method for the detection of derivatives from all three germ layers is the teratoma formation assay, which is used for quality control purposes in stem cell research. Teratomas are long known in human pathology and represent benign tumors of very unusual cellular composition. In many cases, tissues originating from the major germ layers can be observed, for example, connectives including bones, teeth, epithelia, and neural components. In this sense, teratomas can be viewed as revealing a pluripotent potential of the cells of origin. The transplantation of pluripotent stem cells into immunodeficient host animals leads to the rapid formation of teratomas. The presence of tissue components from ectodermal, mesodermal, and endodermal origin proves the differentiation capacity of the transplanted stem cells.

The use of pluripotent stem cells for therapeutic applications bears the potential risk of tumor formation by transplanted cells. The unlimited self-renewing capacity of the stem cells involves the possible tumorigenicity of their therapeutic progeny. The guaranty for safety in stem cell therapy sets the limits in future stem cell applications.

11.1.3 Tissue-specific and organ-specific stem cells

A wealth of studies in the fields of embryology and developmental biology has provided evidence for tissue-based stem cells. These cells are multipotent, because they can form all cell types of a given organ. Multipotent stem cells have been described in the hematopoietic system, where they form all the cells of the blood, including the cells of the general and the adaptive immune system. It has been proven in the mouse that all the cells of the hematopoietic system can be regenerated by the transplantation of a single hematopoietic stem cell, which forms the basis of bone marrow transplantation therapy in the hematology departments. Further examples for multipotent stem cells are the stem cells of the nervous system (NSCs), of the mesenchyme (MSCs), and of the skin. Multipotent stem cells reside in each organ, where they provide a constant supply of progenitors to substitute dying cells, or for repair processes after small lesions. It is believed that these cells form the basis for physiological organ regeneration, which secures safe physiological function over the life span of an organism. Because these cells are not derived from embryos, it has been suggested that they should be favored over hESCs for the study of potential uses in regenerative medicine. However, this strategy is hampered by specific obstacles, as explained below (see 11.1.5).

11.1.4 General features of stem cells

The most salient and constitutive property of stem cells is the ability to self-renew with each division cycle. This signifies that with each division step, at least one of the daughters must be a stem cell (self-renewal), while the sister cell may be another stem cell (mode of symmetric division), or a progenitor cell that is destined to generate one or several types of precursor cells (mode of asymmetric cell division) (Figure 11.2). Because the progenitor cell is not capable to resume the stem cell

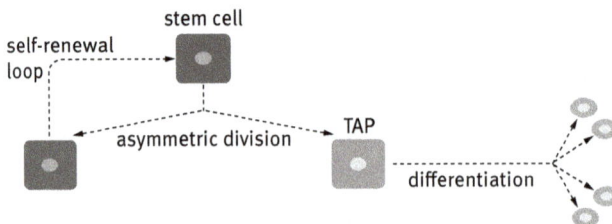

Fig. 11.2: Stem cell concepts.
A stem cell has two cardinal features, which are self-renewal and multipotency. Division of a stem cell can lead to self-renewal when one of the daughter cells has the same characteristics as its mother. The second daughter cell can be a more committed precursor cell giving rise to differentiated cells. These precursors often show a high proliferative activity and are termed transit amplifying precursors (TAP).

properties, at least under normal conditions, it is considered fate-restricted. As the progenitor cell differentiates further to produce precursors with distinct developmental destinies, the fate restriction progresses. Finally, the offspring of the precursor cell lineages will differentiate to specialized cell types of given organ. During early embryogenesis, when a large number of cells needs to be generated in a short time period, the pattern of symmetric division prevails, while at later stages the asymmetric division mode is dominating (see stem cells of the nervous system as an example, 11.5). In view of its biological importance, the switch from the symmetric to the asymmetric division pattern is being intensely investigated. Thus, evidence has been provided that the asymmetric distribution of intrinsic cellular determinants to the progeny within a division cycle distinguishes the self-renewing stem cell from the differentiating sister cell. In this context, the orientation of the mitotic spindle and, consequently, of the cell division plane are important regulatory elements. Beyond this cell-autonomous mechanism, another mode has also been discovered, namely the impact of the cellular microenvironment on the differentiation pathway of the progeny. Micro-heterogeneity of the environment drives self-renewal or differentiation in this perspective, termed environmental asymmetry. This second principle highlights the importance of the environment for stem cell biology.

11.1.5 Adult stem cells

In view of the ethical controversy that was spurred by the discovery that hESCs can be cultivated and propagated as permanent cell lines, the use of adult stem cells for research and application has been considered. With aging, the number of multipotent stem cells in a given organ decreases. This may be a reason for the diminishing ability of aging organs to regenerate, and explain reduced recovery potential in aged populations. Another direct consequence of diminishing numbers of stem cells in the adult organism is that these cells are difficult to detect and isolate from adult organs. Beyond the fact that true stem cells are rare objects, it is very difficult to identify them because stem cell specific markers are missing in many cases. Therefore, specific antibodies cannot be used to enrich for stem cells since the specific surface-cues that could serve for these protocols are not known. Finally, the adult stem cells show a much slower division rate than the embryonic counterparts. For this reason, the stem cell compartment cannot rapidly expand in response to lesions or other stimuli. To circumvent this problem, transient amplifying precursor (TAP) cells have evolved that are generated by asymmetric division (Figure 11.2). Thus, a slowly dividing stem cell can self-renew at a low rate and at the same time, give rise to a rapidly dividing precursor cell. The latter will form a large number of committed precursors in a reasonable amount of time that will serve to compensate cell numbers in the organ concerned. The type-C cell in the adult subventricular zone of the CNS represents a well-studied example of a TAP (see 11.5.2).

11.1.6 The stem cell niche

Adult stem cells are rare and confined to discrete areas in the organism. In some cases, the regions where adult stem cells reside have been localized. For example, two neurogenic zones have been identified in the adult CNS, the subventricular zone of the lateral ventricle and the subgranular zone of the hippocampus (see 11.5.2). The strong topological confinement has led to the concept, that stem cells require a privileged environment for their maintenance and expansion. This environment has been designated with the term "stem cell niche". A stem cell niche contains several cellular elements, the stem cell, -derived progenitors and, at least in several cases, blood vessels and their constituents, the endothelia, and the pericytes that interact by paracrine factors. The blood vessels provide glucose, oxygen, and many other crucial factors that are important for the metabolome of the stem cell compartment. Furthermore, the extracellular matrix (ECM) environment of the stem cells is important, that consists of basal lamina structures and macromolecular assemblies of glycoproteins and proteoglycans. ECM glycoproteins contain signaling peptide motifs that are recognized by specific receptors, e.g., the integrins that interact with the RGD-motif. The proteoglycans expose N- and O-linked carbohydrates, in particular the glycosaminoglycan chains. These are composed of long sequences of repeating carbohydrate dimers that are additionally modified by sulfate groups. Sulfation of the glycosaminoglycans creates specific docking and recognition sites that serve as binding sites for various factors. For example, it is known that heparan sulfate structures are important for the exposure of FGF-2, a cytokine that is involved in the proliferation of stem cells, to the FGF-receptor. While it is undisputed that the ECM is a supportive constituent of the stem cell microenvironment, it has been questioned whether it may have an instructive function. Recent experiments strongly suggest that this may indeed be the case. Thus, when de-cellularized trachea scaffolds that consist basically of the three-dimensional organization of the ECM of that organ were seeded with an enriched population of blood and/or bone-marrow-derived cells that contain hematopoietic stem cells, the trachea scaffold was colonized by the cell preparation and the differentiation of tracheal epithelia was observed. The reconstructed organ could successfully by transplanted into recipients with severely destructed airpipes. These achievements of experimental surgery highlighted the roles of the ECM and have strongly stimulated research concerning the construction of artificial scaffolds.

11.2 Reprogramming stem cells

The use of human embryonic stem cells was highly debated over several years and researchers were drawn between promising capacities of human embryonic stem cells and the ethical problems inherited in their generation. Research was focused on reliable alternatives, which are discussed in the following chapter.

11.2.1 Generation of distinct cell populations from human embryonic stem cells

Lines of hESCs generated from early stage embryos have been studied with regard to their differentiation potential. In this field of research, the wealth of data that had been accumulated in basic research in developmental biology could be successfully applied. Indeed, a variety of morphogens, cytokines, and their receptors had been discovered, that play crucial roles in the specification of distinct cell lineages in developing organs. These discoveries and results have successfully been applied to the derivation of specified cell populations from hESCs in culture. Thus, protocols have been developed that permit for the generation of human motoneurons, a cell population that is degenerating in the fatal disease of amyotrophic lateral sclerosis. Another example concerns the generation of cardiomyocytes that are crucial for the biological function of heart tissues. As a result of these efforts, a large spectrum of specialized cells can, nowadays, be generated from hESCs for the purpose of application.

11.2.2 Human embryonic stem cells as a tool for repair

The primary aim of generating specialized cell types from hESCs has been to obtain material for cell replacement and organ repair. Although most organs supposedly contain adult stem cells, their pace of proliferation and expansion may not suffice to replace large cell numbers that vanish in the context of disease processes of lesion. The hope of the new discipline of regenerative medicine is to provide tools to cross this gap. However, the application of hESC-derived populations of cells for organ repair faces the well-known problem of immunological compatibility. Any hESC-derived line carries the HLA-profile encoded by its genome, which raises the general problem of the immune barrier elevated by recipients to transplanted cells or organs. The hESC lines are derived from a limited number of embryos. In many countries, law prohibits the production of hESCs. This principally restricts the number of HLA-genotypes available in hESC line collections and has the potential to exclude inappropriate recipients from potentially beneficial treatments. This situation has led to the question whether it may be possible to reprogram the hESC in a way that they may be compatible for potential recipients. Several key-experiments have suggested potential strategies to reach that aim.

Box 11.3: Key experiments of reprogramming the Oocyte.

Sir John Gurdon carried out the first experiments aiming at clarifying the question how strong the genetic determination of specialized cells of an adult organ is (Nobel Prize 2012). In his thesis from the 60s, Gurdon addressed the question whether the commitment of the nucleus of an intestinal epithelial cell was reversible or not. This issue touches the broader problem of developmental biology, whether the differentiation process that can be observed during embryonic development is reversible, or not. In his experiment, Gurdon enucleated an egg cell of an amphibian, which was subsequently implanted with the nucleus of a differentiated amphibian intestinal epithelial cell. Gurdon could show that the implanted nucleus was sufficient to allow for the development of an amphibian embryo, showing that all the necessary information could be obtained from the nucleus of a specialized cell. Gurdon concluded from this experiment that the cytoplasm of the recipient oocyte contains factors that reprogram the genetic information of the nucleus of a specialized cell to totipotency. The technique of enucleation and subsequent nuclear transplantation was subsequently used in veterinary medicine and culminated in the generation of cloned mammals, e.g., the sheep Dolly. In this experiment, the oocyte of a given sheep was enucleated and the nucleus replaced, resulting in an offspring that was genetically identical to the donor. The question was raised whether this strategy could not serve to bypass the immunological barrier.

11.2.3 Concepts for the reprogramming of hESCs

Based on the cloning experiment with the sheep Dolly, the concept was addressed whether it may be possible to reprogram human oocytes. One strategy that has been considered has been termed "therapeutic cloning". Conceptually, this proposal follows the paradigm of the sheep Dolly. A human oocyte would be collected from a donor and enucleated. Subsequently, the nucleus-free oocyte would be refurbished with the nucleus of a somatic cell of a donor, for example a skin cell. The resulting oocyte would be stimulated to divide, and the resulting early embryo could serve as a source of hESCs with the genotype of the donor cell. Using established protocols, various precursor cell lines could be generated from the hESC line for the purpose of tissue repair or other objectives. In Germany, the strategy of "therapeutic cloning" is legally prohibited (Embryonenschutzgesetz). Because the need for human oocytes represents a potential restriction, alternative oocyte donors have been explored. In this context, the application of bovine oocytes as an alternative has been tested. These could be used to obtain ESC lines which express the genotypes of the donor's nuclei gained from somatic cells, for example fibroblasts. Personalized ESC lines could thus be obtained for the development of transplantation strategies in the context of regenerative medicine. The generation of interspecies chimeras is, however, not permitted in Germany while it is allowed in Great Britain. This provides another example for the considerable legal divergencies in the stem cell field. Motivated by these problems, investigators have sought for alternative routes to obtain pluripotent cells.

11.2.4 The discovery of induced pluripotency

The studies in basic research had shown that it is possible to reprogram the nuclei of somatic cells by inserting them into enucleated oocytes, suggesting that cytoplasmic factors in the oocytes could be the responsible reprogramming factors. As the genome encompasses about 25,000 genes in humans, it appeared a daunting task to identify the proteins necessary and sufficient for reprogramming. It is true that research in developmental biology had already pointed to several candidate genes implicated in the self-renewal circuit of stem cells. Yet, nevertheless, it appeared a tantalizing task to decipher the responsible candidates by an experimental candidate gene approach. This "tour de force" has successfully been achieved by the team of Prof. Shinya Yamanaka in Japan. The team has systematically infected genetically tagged primary mouse fibroblasts with sets of factors encoded in retroviruses, which insert into the genome of recipient cells and tested the resulting lines for self-renewal and stem cell properties. The combination of the genes Sox2, Oct4, Klf4, and c-Myc ("Yamanaka cocktail") was successful in generating pluripotent lines. Pluripotency was assessed by inserting the transformed fibroblasts into early blastocyst embryos and breeding the resulting mice. Based on the genetic marker, the genome of the transformed fibroblasts had been successfully transferred to the germ-line. In conclusion, the transformed mouse fibroblasts behaved like embryonic stem cells of mice (Takahashi and Yamanaka 2006). The phenomenon was termed induced pluripotency and the discovery honored with the attribution of the Nobel Prize in 2012 (Gurdon and Yamanaka).

11.2.5 Induced pluripotency as a novel, broadly applicable experimental strategy

The discovery by Yamanaka and his team published in 2006 stimulated an intense experimental activity in many laboratories worldwide. Most importantly, it could soon be shown that induced pluripotency also works with primary human fibroblasts and the resulting lines of cells were termed iPSCs (induced pluripotent stem cells) (Takahashi et al. 2007). Human iPSCs (hiPSCs) in many important features resemble hESCs, that is self-renewal, the potential to form defined subpopulations of tissue specific cell types and the capacity to generate teratomas upon transplantation into recipient mice in the teratoma assay (see 11.1.2). It has now become clear that hiPSCs can be obtained from primary human fibroblasts collected by biopsy, but also from other primary human cell types, for example human umbilical cord blood cells (a potential source of early postnatal human cells), bone marrow preparations, periodontal fibroblasts (that are accessible during dental treatment by the dentist), or cellular constituents that can be obtained from human urine. Depending on the cell population of origin, slight variations in the composition of the factors used for induction may be possible. Biotechnology companies have developed kits for the generation of induced pluripotent cells that are based on the canonical factors identified by S. Yamanaka, and the procedure can now be performed in any life science laboratory that is equipped for cell culture and has access to primary human cells. Another branch of activities concerns the mode of application of the canonical factors. In the original protocol, the factors were delivered via retroviral constructs, but it soon became clear that the permanent expression of the factors is not required for efficient induction. As retroviruses in the genome of hiPSCs may themselves represent a health hazard, alternative routes of application have been explored. These include the use of non-integrating viruses that persist in an episomal state (e.g., the Sendai virus), the delivery of the factors as cell membrane permeant proteins, or – as the most recent achievement – the use of small chemicals instead of proteins to achieve induced pluripotency. The latter observation opens the perspective on a new type of pharmacology, aimed at the modulation of the pluripotency circuit of genes. In consequence of these efforts, a large number of hiPSCs is continuously being generated. Several incentives have been started to collect these lines in centralized deposits and to render them accessible for scientific analysis.

11.2.6 Applications of induced pluripotent stem cells

The development of hiPSCs holds the promise to circumvent one of the most important obstacles for the treatment of diseases with engineered cells, the immunological incompatibility. Indeed, ESC-derived progenitors or tissues possess the HLA-genotype of the original donor, leading to the major problem of transplantation medicine, the incompatibility of the major histocompatibility complex. As therapeutic cloning (see 11.2.3) is forbidden in many countries, the application of ESC-derived therapeutics would require immunosuppression of the host and a plethora of bioethical concerns and problems. In this regard, the invention of hiPSCs appears as a long-sought solution. Indeed, the induced cells exhibit the genotype of the donor. These cells and all derivatives are hence immunologically compatible. Furthermore, as the cells carry the genotype of the donor, they represent ideal models to study physiopathological processes on the cellular level, and to test the response of patient cells to drug treatments. This opens the perspective on a scenario of personalized medicine. In a first step, cells would be obtained from a patient and induced to pluripotency (Figure 11.3). These cells could subsequently be amplified and aliquots could be stored for future uses. The given hiPS cell could be taken as starting point for the differentiation of progenitors and for specified tissues using established protocols, and the latter could be studied *in vitro* for cell behavior and potential treatments. In one line of investigations, the hiPSCs could be investigated with respect to pathology, as an *in vitro* disease model. This has been achieved already for several types of neurodegenerative diseases. Analogous

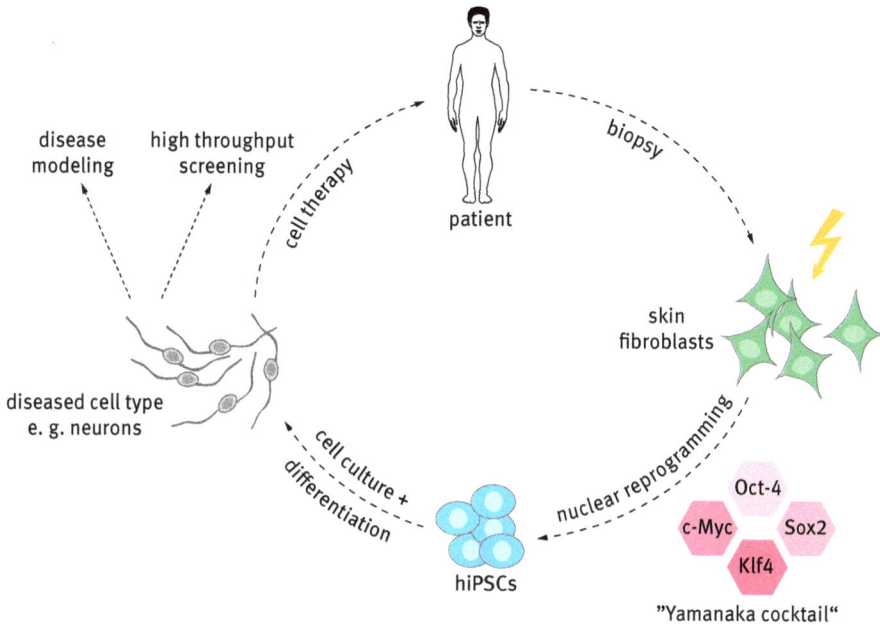

Fig. 11.3: The concept of personalized medicine.
The combination of the genes Sox2, Oct4, Klf4, and c-Myc ("Yamanaka cocktail") is successful to reprogram human skin fibroblasts into hiPSCs. This hiPSCs can be cultivated and differentiated *in vitro* to various cell types e.g. neurons. These cells exhibit the same genetic background as the patient what makes specific disease modeling and drug development possible. Immunological rejections should be minimized upon transplantation of manipulated cell cultures or tissue components when therapies are applied to the same patient.

strategies are pursued for any cell type where a gene mutation is known and suspected as a disease cause, e.g., in diseases based on mutations in ion-channels, the so-called channelopathies. Distinct cells or tissues with specified genetic deficits could be used as a tool for high-throughput screening of drugs designed to modify or even halt the disease process. Alternatively, the genetic deficit could be engineered using molecular biology approaches, and cell lines with corrected genetic deficits could be expanded in order to replace deficient cell populations in recipient organs. On a conceptual level and in order to expand our knowledge on disease processes, genetic mutations can be introduced into iPSCs for any given gene of interest using the TALEN-technology, or analogous strategies to generate directed mutations in stem cell lines to create *in vitro* models of disease. As explained above, these could be used for high-throughput screening of drugs, or other purposes. Finally, the generation of different specialized cell types from iPSCs could serve as starting point for toxicity testing of pharmacological agents. Indeed, many drugs that have been developed with a considerable capital investment have to be withdrawn from the market because of unforeseen toxic side effects. These may be avoided thanks to toxicity testing on crucial tissues, for example, the heart muscle.

11.2.7 Themes of regenerative medicine

A wealth of activities has unfolded that aim at developing new therapeutic approaches to medical problems concerning large populations in advanced industrialized countries. In several areas, medical progress has provided tools to treat previously fatal diseases, mainly using transplantation. Thus, transplantation of hearts, of blood vessels, liver, lung, bone marrow, and skin are, nowadays, clinical routine in large centers. However, the practice of transplantation is often hampered by the lack of suitable organs. The propensity to donate organs stalls or is even regressive in many societies. Furthermore, the availability of HLA-matched donors is often limited, and hence, immuno-compatible organs are difficult to collect. For these reasons, the development of isogenic autologous hiPSCs is viewed as a promising alternative by many clinical investigators. A variety of research lines has been inaugurated that aim at the generation of tissue replacements from hiPSCs. Examples include the derivation of pancreatic islet organs for the treatment of type-1-diabetes, of liver tissue for the replacement of liver cirrhosis, or of dopaminergic neurons for the treatment of Parkinson Disease. From the point of view of incidence, the diseases of the cardiovascular system represent another potential area of application of iPS cells.

11.2.8 The cardiovascular system

Cardiovascular diseases represent an important issue, and heart infarct and its treatment are focus areas of stem cell research. Heart infarct is a consequence of insufficient blood supply to the heart via the coronary arteries, often resulting from reduced vessel diameter caused by the deposition of lipid plaques (atherosclerosis). The immediate consequence is the necrosis of specialized heart muscle and its replacement by connective tissue. The scars that form after elimination of heart muscle may cause mechanical malfunction and reduced efficiency of heart activity and result in serious functional deficits including disability to exert professional activities. A replacement therapy for lost heart cells is, therefore, highly desirable and intense experimental activity has been devoted to this objective. Based on decades of research on heart progenitors and their roles in the regulation of heart development, protocols have been developed that allow for the generation of cardiomyocytes (Moretti et al. 2013). These cardiomyocytes form monolayers of contractile cells, whose contractions can be monitored on multi-electrode arrays (MEAs). The heart progenitors can be driven to the fate of cardiomyocytes by modified RNA that codes for the vascular endothelial growth factor A (VEGF-A), a treatment that also promotes vascular regeneration after infarction *in vivo*. The long-term aim of this research is to provide heart muscle cells for replacement that might be delivered to the injured tissue by catheter-based interventions. Another application of induced pluripotency to cardiovascular diseases is the analysis of arrhythmias, e.g. the long QT-syndrome. Several cell lines have been produced from patients suffering from this disease to test pharmacological treatments or gene therapy approaches *in vitro*.

11.3 Direct conversion as an alternative to iPS cells

Based on decades of successful research in cell and developmental biology, protocols for the derivation of various restricted progenitor cell populations from ES or iPS cells could be derived in a relatively short time period. Yet, even if a large number of technical problems can be solved, a major obstacle to convenient and rapid application of induced pluripotency is intrinsic to the method and the resulting cell populations, namely the time required for efficient differentiation. Indeed, the formation of distinct cell populations from ES or iPS cells, for example the derivation of neuroectodermal cells and neurons, takes months of culture time.

11.3.1 Induced cells

Inspired by the work of S. Yamanaka, investigators probed combinations of transcription factors to test whether selected cell types can be obtained directly from fibroblasts, without transiting via stem cells. Combining different sets of proneural transcription factors, the team led by Marius Wernig was able to show that three factors, Ascl1, Brn2 (also known as Pou3f2), and Myt1 are sufficient to directly convert primary fibroblasts into neurons. The induced neuronal (iN) cells express a variety of neuronal proteins and proved able to form synapses and to generate action potentials (Vierbuchen et al. 2010). The very same three factors proved sufficient to generate neurons from human pluripotent stem cells as early as 6 days after activation of the transgenes, which represents a dramatic acceleration in comparison to the standard differentiation protocol. When these factors were combined with the helix-loop-helix transcription factor NeuroD1, the forced expression of this group of four factors was sufficient to induce rapid conversion of primary embryonic or postnatal human fibroblasts into primary human neurons. These human iN expressed a variety of neuronal markers, were electrically active, and received functional synapses when co-cultivated with primary mouse cortical neurons. These experiments clearly demonstrated the concept that direct lineage conversion is possible, based on the forced expression of transcriptional regulators. These data raise the hope that large numbers of specified cells can be obtained in a relatively short time-window, possibly without bearing the risk of teratoma formation that is associated with the induced pluripotency. Meanwhile, a broad spectrum of directly induced cell types has been reported, namely induced neurons, induced motoneurons, induced dopaminergic neurons, induced cardiomyocytes, induced hepatocytes, and induced oligodendrocytes (Sancho-Martinez et al. 2012).

11.3.2 Epigenetic landscapes of induced cells

The number of ESCs, iPS cells and forced conversion-derived cells is rapidly increasing. This raises several questions. First of all, it is important to assess whether the lines obtained by differentiation of ES or iPS cells are comparable, that is have comparable differentiation potential and otherwise comparable properties. Indeed, the cells are obtained from different individual genetic backgrounds. Furthermore, the starting material often differs with regard to age, and also with respect to tissue of origin. Hence, it is conceivable that the lines, although endowed with similar biological properties at first sight, are different in details of their transcriptomes and the corresponding regulatory states of their respective genomes. In a classical view proposed by the embryologist Waddington, the differentiation pathways starting with the totipotent zygote towards pluripotency, multipotency, and finally, restricted precursor cells of specialized organs reflect equilibrium states of gene expression patterns that are regulated by epigenetic factors. For a long time, it was believed that the corresponding developmental pathways, which find their expression in cell lineage itineraries, are unidirectional. Contrasting this view, the discoveries of S. Yamanaka (see 11.2.4) clearly showed that it is possible to revert cellular specialization towards a pluripotent state, indicating that the epigenetic states are reversible. Furthermore, the discoveries related to forced conversion showed that it is even possible to switch directly from one state of differentiation to another, that is to obtain trans-differentiation without moving back the differentiation pathways to a more embryonic state and to re-differentiate from that situation. Therefore, the view of a hierarchy of differentiation states as proposed by Waddington is, nowadays, questioned (Ladewig et al. 2013). The fact that pluripotent stem cells can be obtained from ESCs, as well as from primary fibroblasts and other specialized cell types, raises the question how comparable the different lines are. On the level of gene expression, current technology allows for the comparison of transcriptomes, either by gene array analysis or by extensive RNA-sequencing. On the level of gene regulation, the epigenetic mechanisms are addressed by genome-wide annotations of patterns of histone methylation. These efforts, in particular high-throughput sequencing, resulted in DNA-methylation maps at the single-nucleotide resolution and genome-wide chromatin maps. Current data sets support the idea that hESCs and hiPS cells resemble each other, but more extensive analysis is definitely required. Another issue concerns the genetic make-up of specialized cells that are obtained from different cellular origins. For example, motoneurons can be generated from ESCs, from iPSCs, and also by direct conversion from fibroblasts. It remains to be established, however, whether these motoneuron populations are truly equivalent. This is relevant with regard to the application of stem cell technology, be it with regard to disease models, to the screening of pharmacological agents, or in the context of disease-oriented transplantation strategies.

11.4 Stem cell technology and standardization

The horizon of numerous studies relating to induced pluripotency clearly is the application in translational research. This certainly requires a robust level of standardization that permits comparison of procedures between centers and countries.

11.4.1 Generation of hiPSCs

Originally, the introduction of reprogramming factors was based on the random insertion of retroviruses. Subsequently, other viruses have been tested, including very efficient Lentiviruses and non-integrating Adenoviruses or Sendai mRNA viruses. In order to avoid viral genes in the procedure, also the transfection with plasmids has been attempted, as well as the direct application of proteins, mRNA or miRNA. These approaches differ with respect to efficiency and research in this field is ongoing. The recent discovery that pluripotency can be induced by a combination of seven small chemical compounds, that is without adding regulatory proteins or genes, has opened the perspective on a pharmacology of pluripotency induction. In parallel to these promising developments, biotechnology companies have begun to market kits of pluripotency factors for the derivation of lines in laboratories equipped with the standard infrastructure of cell biology. Thereby, the Yamanaka protocol has become widely accessible and increasing numbers of laboratories in life sciences have begun to study induced pluripotency using primary cells from patients or experimental animals. After their production, the lines will have to undergo a stringent quality control. This will have to include the routines to exclude contaminations with human viruses, e.g. HIV-1 or hepatitis and with bacteria. On the level of cell and developmental biology, the differentiation potential of the cells will have to be assessed. Indeed, not all induced lines embody the complete pluripotency repertoire and may need additional handling for the derivation of desired terminally differentiated cell types.

11.4.2 Technical challenges of stem cell applications

Beyond the requirements of standardized procedures for the generation of hESC- or hiPSC lines and quality control the translation of stem cell research faces a considerable challenge, the number of cells needed when it comes to application. This can be illustrated by the example of the pancreas, the organ that harbors the β-cells that release insulin. In patients suffering from diabetes, the insulin has to be substituted by regular administration using local injections. Therefore, a replacement strategy using terminally differentiated β-cells would be highly appealing. The human pancreas contains about 800,000 islets, which means that about 10^9

cells would be required for the therapy of one patient. Assuming that one large tissue culture flask can accommodate up to 10^7 cells, this would mean the culture of 100 bottles to produce the required cell population. This clearly transgresses the limits of traditional cell culture laboratories. A solution to this problem could be the use of three-dimensional cultures in suspension. Using spinner flasks, it may be possible to grow about 0.5×10^9 cells in about three weeks. Obviously, spinner flasks or other devices of large-scale cultures require a set of technological supports and controls. Thus, the concentrations of the critical nutrients, of oxygen and CO_2, as well as the pH-values of local micro-environments have to be maintained in controlled and acceptable ranges. To this end, a sophisticated sensor technology has to be developed, as well as computerized feed-forward and feed-back mechanisms for ongoing control and correction of growth conditions. The culture media and conditions themselves have to be adjusted to the requirements of certified medical treatments. This generally excludes the use of animal-derived components, beginning with serum, a standard additive to the media at earlier stages of research in the stem cell field. The general trend in this regard goes to the production of completely xeno-free formulations, where animal-derived compounds are excluded altogether. In this context, only genetically engineered and heterologously expressed and subsequently purified growth factors and/or cytokines are permitted. Along the same line, the use of animal feeder cells in co-culture with ESCs, a necessary standard in the first culture protocols, is strictly prohibited. Currently, several biotechnology companies offer xeno-free, completely defined media for the cultivation of hiPSCs in the perspective of therapeutic applications. Finally, efficient and stringent quality control procedures have to be implemented to charac-

Box 11.4: The automated cell factory.

One aim of stem cell technology is the generation of precursor or differentiated cell populations for repair of organ deficits in patients suffering from disease. Examples are the treatment of Parkinson's disease or of type-1-diabetes, where distinct and clearly defined cell populations are affected, e.g. dopaminergic neurons and β-cells of the islet organ of the pancreas, respectively. In order to be acceptable for therapy in humans, however, this strategy requires highly reproducible techniques for production, cultivation, maintenance, and differentiation of stem cells, be these derived from early embryos or from induced pluripotent stem cells. This supposes strict quality control, as well as highly standardized cell handling and storage protocols. In order to render cell culture procedures independent from variations inherent to human workforce, high-technology machinery is currently being developed. The vision of these endeavors is the construction of fully automated production lines that should guarantee for products meeting the highest quality standards. The compliance with Good Manufacturing Practice (GMP) standards indeed is a prerequisite for acceptability by the Federal Drug Administration (FDA in the USA), the Bundesgesundheitsamt (BGA in Germany) or comparable institutions in other countries that decide about the implementation of new therapeutic tools. The computer aided design figure shows an integral production line for iPS cell lines (Figure Box 11.4). We thank the HiTec Zang GmbH for providing the image of the stem cell factory.

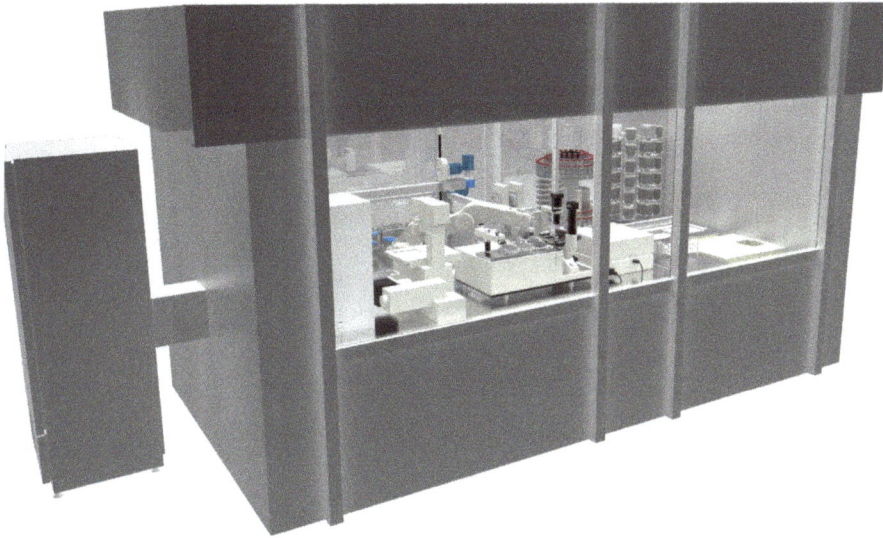

terize the cells thus obtained. Quality control will have to consider the growth properties of the cells, as well as the stage of terminal differentiation reached and the absence of potentially tumorigenic residual ESC populations. Taken together, an extensive high end production line and highly qualified cell and developmental biologists will be required to operate the stage for stem cell-based, personalized regenerative medicine.

11.5 Stem cells of the nervous system

The central nervous system (CNS), which is the most complex organ system in the body, arises from the ectodermal part of the developing embryo. Neuroepithelial cells form the neural tube and elevate their number by symmetric divisions. These early neural stem cells (NSCs) have the potential to give rise to neurons that are integrated in the growing tissue.

11.5.1 The radial glia cell as stem cell of the nervous system

The neuroepithelial cells change their characteristics to a new cell type that has a long process spanning the complete thickness of the growing nervous system. Because of their morphology and the appearance of glial characteristics, this cell type is called radial glial cell (Figure 11.4a). They share characteristics with glial cells,

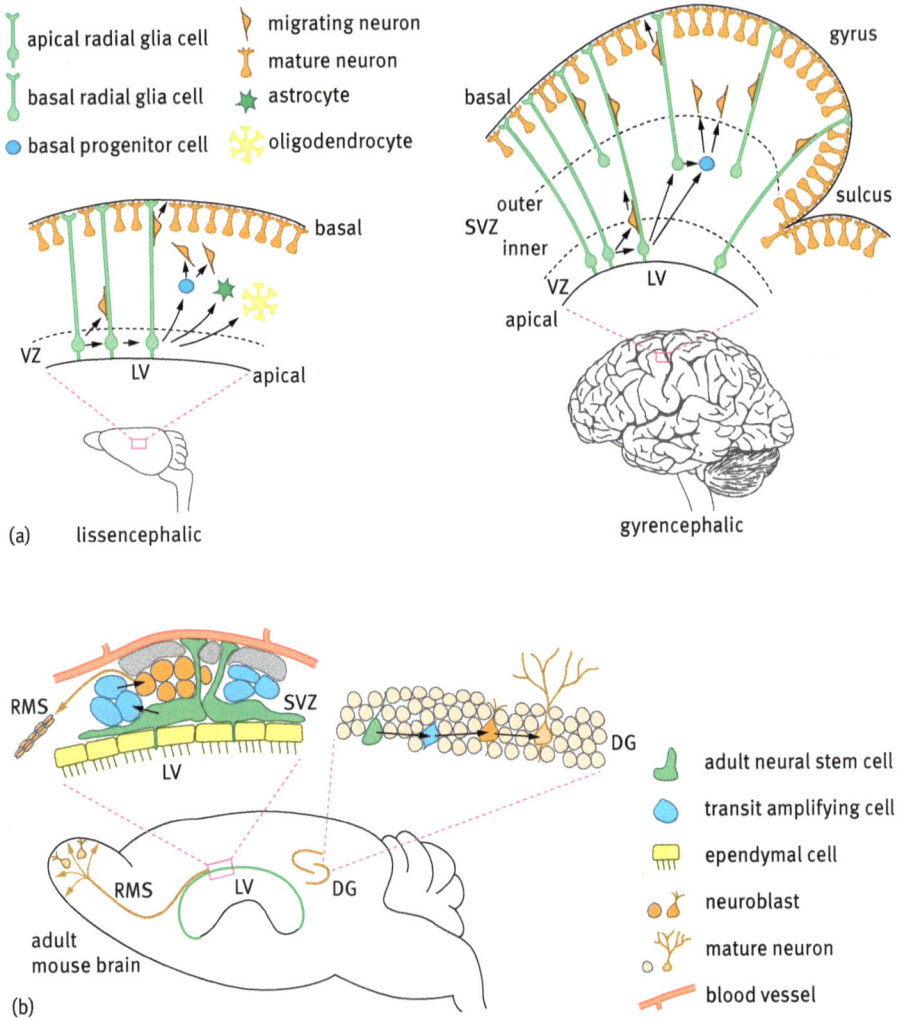

Fig. 11.4: Neural stem cells. (a) Cortical development. Radial glia cells (green) reside in the ventricular zone (VZ) along the lateral ventricles (LV) in the developing cerebral cortex. During development of lissencephalic species (left), which exhibit a smooth brain surface, radial glia cells give birth to neurons (red) and transit amplifying precursors (TAPs, basal progenitor cells, blue) that migrate outwards to more basal positions in the developing tissue. Neurons migrate along processes of radial glia cells before they leave these tracks in their destined layer. The development of the enlarged area of folded (gyrencephalic) cortices as seen in the human brain (right) includes the generation of cells by basal radial glia cells in the outer subventricular zone (OSVZ). (b) Adult neural stem cell niches. In the adult rodent brain, neurogenic neural stem cells (green) can be found in the SVZ lining the lateral ventricles of the forebrain and the dentate gyrus (DG) of the hippocampus. The stem cells (green) divide slowly in these regions, giving rise to faster dividing TAPs (blue), which in turn generate neuroblasts (red). These newborn cells migrate from the SVZ towards the olfactory bulb (OB) throughout the rostral migratory stream (RMS) before maturing to olfactory neurons and integrating in the system. Newborn neurons in the hippocampus mature and integrate into the hippocampal network.

like the expression of glutamine synthetase, S-100β, the astrocytic glutamate transporter GLAST, and the extracellular matrix glycoprotein tenascin-C. Initially, radial
glia cells were described by Pasco Rakic in the 1970s as guiding new-born neurons
along their long processes radially through the developing cerebral cortex. The
expression of glial markers led to the assumption that they transform into astrocytes after the period of neuronal migration. Therefore, they were considered as
glial precursor cells until their capacity to generate neurons as well was detected
(Malatesta et al. 2000). Radial glia cells divide symmetrically or asymmetrically in
the ventricular zone (VZ) lining the lateral ventricles (LV) of the developing forebrain, giving rise to daughter cells with the same characteristics or more differentiated cell types. Progeny of these cells can be transit amplifying precursors (TAPs,
see 11.1.5), giving rise to a higher number of offspring cells. Radial glia cells are
able to generate neurons and the glial cells of the central nervous system, which
are astrocytes and oligodendrocytes. Therefore, they can be identified as stem cells
fulfilling the cardinal stem cell characteristics, namely self-renewal and multipotency. The mechanisms how neural stem cells divide and differentiate have been
cleared in a lot of aspects but many interactions and players in this highly regulated system are still unknown.

The tissue grows outwards and builds layers of cells, which contact other cells
via processes that can reach over long distances to connect different parts of the
nervous system. Neuronal cells are the first ones to differentiate from stem cells
and migrate through the tissue. In later developmental stages, astrocytes and oligodendrocytes, which arise from radial glia cells, find their positions and interaction
partners to take over their specific functions in the system.

Neural stem cells in lissencephalic species, like mice or rats, give rise to columns of progeny cells, which assemble vertically in the tissue layers. This is not
sufficient to generate the larger surfaces of folded brain tissue as found in gyrencephalic species like ferret, sheep or human. Here, an additional cell type was
identified that is only very rare in non-gyrified brains, the basal radial glia cell
(bRGC) (Figure 11.4). These cells localize in the outer part of the relatively thick
subventricular zone (SVZ), the outer SVZ (OSVZ). These cells show characteristics
and functions resembling their apical equivalents but missing the apical contact
to the ventricle. By the addition of another progenitor cell type, the number of
progeny is largely enhanced. Neurons emerging from and migrating along processes of this basal type of radial glia cells build the horizontally enlarged areas that
are folded into gyri and sulci.

11.5.2 Adult neural stem cells

As development proceeds, neural stem cells become restricted to specific areas in
the adult brain (Figure 11.4b). The SVZ lining the lateral ventricles in the rodent

forebrain inherits glial cells with the potential to generate new neurons as well as astrocytes and oligodendrocytes throughout adulthood. These cells share some characteristics with astrocytes and are called type-B cells (Doetsch et al. 1997; Doetsch 2003). They divide slowly and give rise to TAPs with higher proliferation rate, the type-C cells. These type-C cells in turn generate neuroblasts that migrate

Box 11.5: *In vitro* culture of neural stem cells.

Neural stem cells can be grown in cell culture as so-called neurospheres (Figure Box 11.5). There-fore, neural tissue from the region of interest is dissected and dissociated to single cells, which are cultivated in defined serum-free medium under the addition of growth factors. The epidermal growth factor (EGF) and the fibroblast growth factor (FGF) have been shown to promote the proliferation of stem/progenitor cells in the culture. They generate neurospheres, which contain each up to several hundred progenies from single neural stem/progenitor cells. Detailed analy-ses have shown that neurospheres are a heterogeneous mixture of stem and progenitor cells as well as more differentiated cells. The spheres can be dissociated and grown under identical conditions giving rise to secondary neurospheres developing from the stem cells among neuro-sphere cells. Repeated capacity to generate neurospheres can reciprocally hint to the presence of "real" stem cells in the tissue. The fraction of stem cells can be enriched in passaged neuro-sphere cultures and variations of the culture protocol can enforce the homogeneity of pure neural stem cell cultures.

Withdrawal of growth factors, and especially the addition of serum to the neural stem cell culture, lead to the differentiation of neural stem and precursor cells to neurons, astrocytes, and oligodendrocytes, which are the main cell types of the nervous system. The influence of intrinsic and extrinsic regulators of stem cell characteristics can be analyzed using this *in vitro* model.

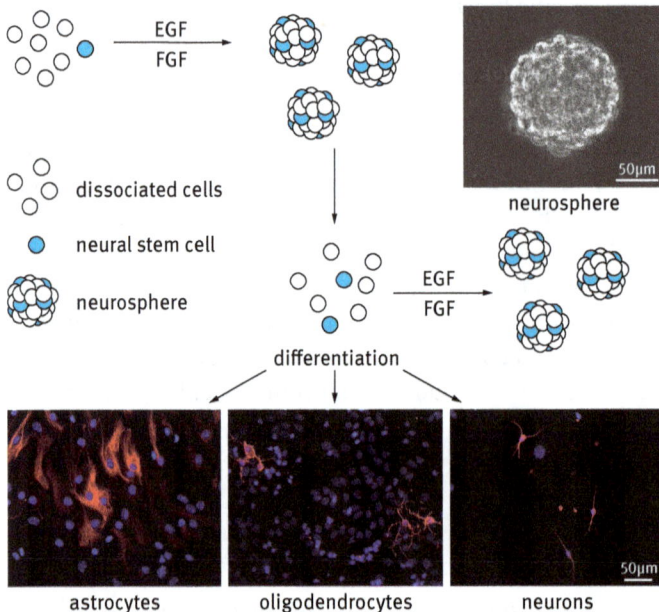

Fig. Box 11.5

through the rostral migratory stream (RMS) into the olfactory bulb to integrate into a complex network as new olfactory neurons. In mouse and rat, this neural stem cell niche is well studied, whereas the population of neural stem cells in the adult human SVZ has not yet been clearly defined and characterized.

The second area where neural stem cells can be found in adult rodents, and also in the human brain, is the subgranular layer in the dentate gyrus of the hippocampus. This brain region is mainly responsible for memory and learning. Stem cells in the hippocampus have glial characteristics and display a radial process. Hippocampal stem cells divide and generate TAPs which transform into neuroblasts. New neurons integrate in the functional networks of the hippocampus. Neurogenesis in the hippocampus is strengthened by physical exercise or learning influenced by an enriched environment but impaired upon high stress levels or the application of antidepressant drugs or alcohol.

11.5.3 Applications of stem cells to neural repair

Brain functions can be massively impaired when the tissue is damaged upon lesion or traumatic situations, or when the oxygen supply via the blood system is impaired by a stroke. Such pathological situations lead to an increased proliferation rate in the stem cell niches and migration of cells towards the lesion site. Nevertheless, the replacement of damaged or injured tissue or the functional integration of stem cell derived progeny cannot be observed yet. An intensive search for factors leading newly generated cells to sites of injury and making a functional replacement possible is ongoing. Maintaining the innate adult stem cell function in the hippocampus by the application of nutrients or growth factors is an attractive idea for the revitalization of impaired hippocampal functions. For the transplantation of human fetal NSCs, first success in clinical trials was reported, e.g., the improved myelination, which is impaired in children suffering from Pelizaeus-Merzbacher disease. Since the development of iPSC protocols for human cells, many efforts were made to apply these for the modeling of nervous system disorders, like ALS (amyotrophic lateral sclerosis), Alzheimer's, Huntington's and Parkinson's disease, schizophrenia, familial dysautonomia or Rett syndrome. In some disorders, specific cell types are affected, like dopaminergic neurons in the substantia nigra of Parkinson's disease patients or motoneurons in the motoric system of ALS patients. The direct conversion of cells to the respective cell type and their functional integration is the aim of numerous preclinical studies. Inappropriate differentiation and short migration potential of transplanted cells are the main problems of stem cell based therapies. Intensive research is going on using animal models with different neurological disorders for the development of cell replacement strategies. Finding the right way of manipulating hiPSCs for the treatment of such neural disorders would offer an ethically unproblematic, patient specific therapy and is, therefore, one of the major challenges of stem cell technology today.

11.6 Development and stem cells of the eye

In early vertebrate embryonic development, the retina originates from the developing brain. Therefore, the retina is considered as part of the CNS. The retinal neuroepithelium serves as an excellent model system for the characterization of neural stem/progenitor cells because of its easy accessibility and the limited heterogeneity of retinal cells. Moreover, retinal stem/progenitor cells are well characterized in regard to the intrinsic and extrinsic factors that regulate their proliferation and differentiation capacity.

11.6.1 Development of the neural retina – Retinogenesis

The development of the neural retina (retinogenesis) starts at embryonic stages with the emergence of the optic vesicles, which form bilateral evaginations of the diencephalon. Retinal cell types arise from proliferating stem/progenitor cells located at the most apical part of the outer neuroblastic layer (ONBL). During a process called interkinetic nuclear migration, dividing mitotic cells (M-phase cells) migrate from the apical part of the ONBL to the basal part of the ONBL and enter the G2-phase at the most basal part of the ONBL. Later on, G2-phase cells migrate back to the apical area, re-enter the M-phase and undergo a new division step or exit the cell cycle to become postmitotic. Based on experimental birth dating and lineage tracing studies, it is well known that retinal stem/progenitor cells undergo different competence stages during retinogenesis and give rise to all retinal cell types in a highly conserved and chronological manner (Figure 11.5a) (Cook 2003; Cepko et al. 1996). Following further differentiation and migration processes, which are accompanied by a synaptic "fine-tuning" period, the retina exhibits a

Fig. 11.5: Retinogenesis and stem/progenitor cells of the eye for treating chronic eye diseases. ▶
(a) During retinogenesis, all seven main retinal cell types are generated from multipotent retinal stem/progenitor cells during embryonic and postnatal developmental states (adapted from Cook 2003). Based on the "competence" model, retinal stem/progenitors pass through a series of competence states, while they are capable to generate a limited number of early- and late-born cell types in a highly conserved and chronological manner over the course of retinal development (Cepko et al. 1996). (b) A schematic drawing of the mature mouse retina illustrates the localization of the main cell types (see also legend (e)) in distinct layers. (c) The adult eye and neural retina contains different types of potential stem cells, such as Müller glia, cells of the ciliary marginal zone, the ciliary body, the iris, the RPE, and the corneal limbus (adapted from Locker et al. 2009). (d) Several chronic eye diseases affect different retinal cell types. Age-related macular degeneration causes the degeneration of RPE cells and rod photoreceptors. *Retinitis pigmentosa* causes also rod photoreceptor degeneration, whereas RGCs are mainly destroyed in glaucoma. RPE = retinal pigment epithelium; OS = outer segments of photoreceptor cells; ONL = outer nuclear layer; OPL = outer plexiform layer; INL = inner nuclear layer; IPL = inner plexiform layer; GCL = ganglion cell layer.

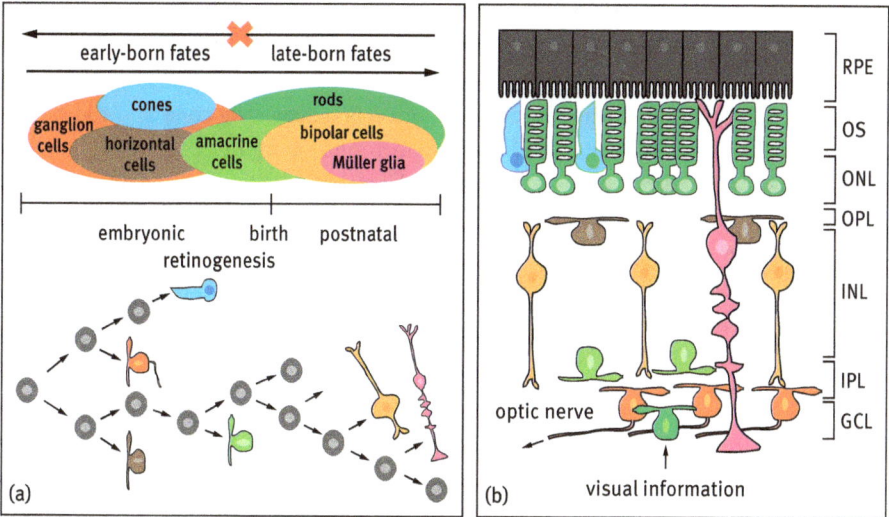

(a)

(b)

RPE

OS

ONL

OPL

INL

IPL

GCL

early-born fates late-born fates

cones

rods

ganglion
cells

horizontal
cells

amacrine
cells

bipolar cells

Müller glia

embryonic birth postnatal

retinogenesis

optic nerve

visual information

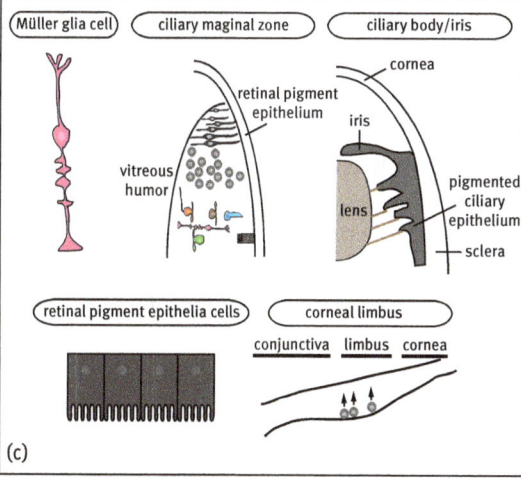

(c)

Müller glia cell

ciliary maginal zone

ciliary body/iris

cornea

retinal pigment
epithelium

iris

vitreous
humor

pigmented
ciliary
epithelium

lens

sclera

retinal pigment epithelia cells

corneal limbus

conjunctiva limbus cornea

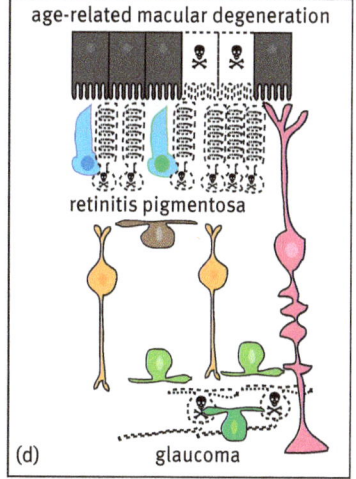

(d)

age-related macular degeneration

retinitis pigmentosa

glaucoma

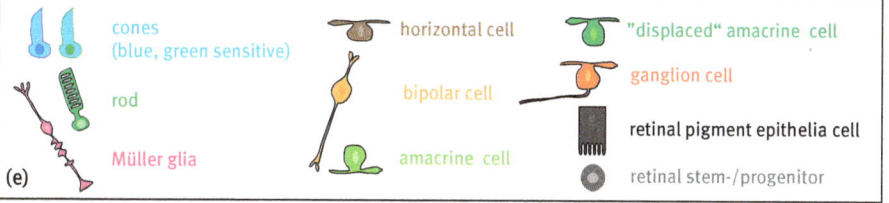

(e)

cones
(blue, green sensitive)

rod

Müller glia

horizontal cell

bipolar cell

amacrine cell

"displaced" amacrine cell

ganglion cell

retinal pigment epithelia cell

retinal stem-/progenitor

mature morphology during postnatal development with its seven main retinal cell types: cone and rod photoreceptors, horizontal-, bipolar-, amacrine-, ganglion-, and Müller glia cells (Figure 11.5a, b, and e).

During visual processing, the visual information is received by cone and rod photoreceptor cells and transmitted to bipolar and retinal ganglion cells (RGCs). There are also laterally connecting horizontal and amacrine cells as well as displaced amacrine cells and Müller glia cells. Rod photoreceptor cells process mainly lower-resolution dim light vision (monochromatic, scotopic vision) and provide black-white vision, whereas different types of cones (in mice blue and green light spectral sensitive) support day light vision and high-resolution color perception (photopic vision). RGCs transmit the visual information along the optic nerve to higher sub-cortical and cortical visual processing centers of the brain.

11.6.2 Stem/progenitor cell populations of the eye and neural retina

The eye serves as a well-defined niche for different types of stem/progenitor cells (Figure 11.5c), although their proliferation and differentiation capacities vary during ongoing development and across different species. At embryonic stages, multipotent retinal stem/progenitor cells are located at the most apical ventricular surface of the retina and give rise to all retinal cell types. During adulthood, retinal neurogenesis is severely limited and retinal regeneration capacity disappears progressively from lower to higher vertebrates. In higher vertebrates, Müller glia represent the inner retinal cell type which is described as resting but limited in regard to stem/progenitor cell characteristics. *In vivo*, upon lesion or injury in several species, or *in vitro*, these resting Müller glia can become re-activated, re-enter the cell cycle and de-differentiate into neuronal cell types (see Figure 11.5c and Box 11.6a, b). In lower vertebrates, such as fish and amphibians, adult neurogenesis arises from persistent cells of the ciliary marginal zone. Quiescent, potential stem/progenitor cells were found beneath the aqueous humor-producing ciliary body ("*pars plana*") at the most peripheral part of the retina. Cells of the iris, retinal pigment epithelia cells, and cells of the corneal limbus also exhibit stem/progenitor characteristics (Locker et al. 2009).

11.6.3 Müller glia as potential stem/progenitor cells

Müller glia cells, named after their discoverer, the anatomist Heinrich Müller in 1851, are the main intrinsic glia of the retina. They fulfill several homeostatic, metabolic, supportive and protective functions. In addition, recent findings indicate that Müller glia cells act as living optical fibers, which guide light throughout the retinal tissue in order to enhance signal-to-noise ratio of visual processing (Franze et al. 2007). Several findings indicate that, under pathological conditions, such as after lesion or injury, a subset of the Müller glia population re-enters the cell cycle and undergoes de-differentiation (Figure Box 11.6a).

Box 11.6: Müller glia reactivity and *in vitro* cultivation of Müller glia.

Upon severe lesion or injury, resting Müller glia may become re-activated, proliferate, and de-differentiate into neuronal cells such as photoreceptors or retinal ganglion cells (Figure Box 11.6a, adapted from Belecky-Adams et al. 2013). Müller glia can be cultivated *in vitro* (Figure Box 6b). For the *in vitro* cultivation of Müller glia, whole retinal tissue is dissected and enzymatically dissociated into a mixed single cell suspension. The dissociated cells are then transferred into cell culture plates with defined serum-enriched culture medium. The cells will grow at 37 °C for a few days. After the mixed culture has grown confluent, the cells are subject to mechanical shaking to remove loosely attached microglia. The cells, firmly attached to the bottom, are consisting of > 90 % Müller glia. Over the next days, Müller glia divide rapidly as adherent cells in culture plates with serum-enriched medium. A pure (> 98 %), neuron-free, proliferative Müller glia culture is observed following several enzymatic dissociation and re-plating steps ("passaging"). The purity can be confirmed by immunocytochemical detection of Müller glia-specific proteins e.g. the intermediate filament vimentin (see insert in Figure Box 11.6b; blue represents nuclear, green represents vimentin staining). Using defined differentiation and transfection protocols Müller glia can be de-differentiated into retinal cells for later transplantation. Thomas Reh and colleagues demonstrated that the viral-mediated expression of the basic helix-loop-helix transcription factor Ascl (achaete-scute complex homolog) in mouse Müller glia reprogrammed them into progenitor cells (Pollak 2013).

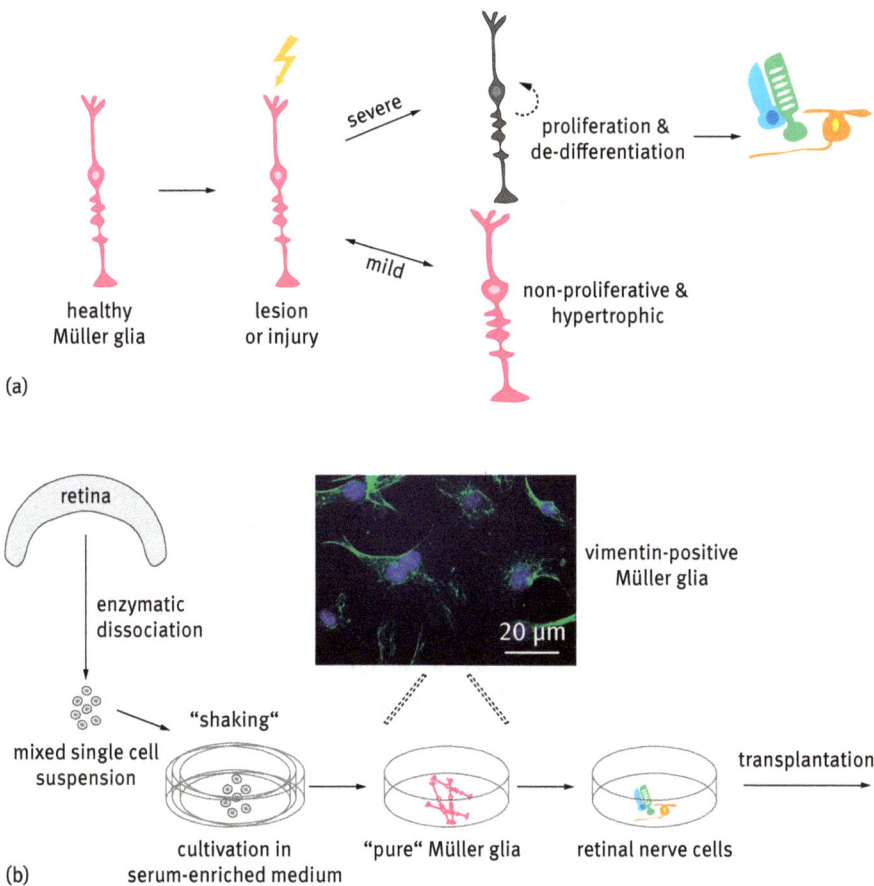

(a)

(b)

Fig. Box 11.6

The stem cell potential of Müller glia has been recognized more than 20 years ago. During regenerative processes, in the fish or chick retina, following injury or after neurotoxin injection, Müller glia become re-activated, re-enter the cell cycle, and migrate to the site of injury. Here, they lose their glial marker expression, begin to re-express retinal precursor markers, and differentiate into new neurons and glia or retain the undifferentiated state (reviewed by Fischer and Reh, 2003 (Fischer and Reh 2003)). The enforced knowledge of Müller glia function and behavior in the healthy and diseased retina will have a great impact on the development of future cell replacement strategies in clinical applications to overcome retinal diseases. Several *in vitro* protocols describe the cultivation of Müller glia cells (see Box 11.6b).

11.6.4 Stem cell based therapies of chronic eye diseases

Several recent and ongoing studies focus on the development of new drugs to treat retinal diseases in humans. Gene therapy holds a high potential to cure retinal diseases. In gene therapy a non-infectious virus e.g., an adeno-associated virus is used to shuttle a specific gene into a part of the retina. For instance, in a human clinical trial, gene therapy was described to be effective and safe in patients suffering from Leber congenital amaurosis (see also ClinicalTrials.gov. numbers NCT00481546).

Cell replacement strategies by transplantation are relevant to several chronic diseases of the eye that lead to irreversible blindness. There is an urgent need to find new therapeutic strategies for retinal cell survival, repair, and replacement. Over a decade, various stem cell-based approaches were investigated for the treatment of diseases such as age-related macular degeneration (AMD), glaucoma or *Retinitis pigmentosa*. The aforementioned diseases affect different cell types of the retina or adjacent tissues (Figure 11.5d).

AMD is the most common cause of visual impairment and blindness worldwide. The disease originates from defects of the retinal pigment epithelia (RPE) cells and overlying tissues, namely Bruch's membrane and the choroid. Many patients still suffer from the secondary progressive photoreceptor loss in the *Macula* ("yellow spot") due to the lack of nutrient support of RPE cells and blood vessel damage. Also genetic diseases, such as *Retinitis pigmentosa*, lead to a severe damage and loss of photoreceptor cells. Glaucoma, also called the "sneak thief of sight", represents the second major cause of blindness worldwide and is characterized by progressive retinal ganglion cell (RGC) degeneration and loss.

Several studies have focused on retinal transplantation using various types of stem cells. While retinal-derived stem/progenitor cells differentiate into retinal cells after transplantation, these cells often fail to migrate and integrate into the host retina. In contrast, several studies indicate that transplanted neural stem cells integrate into host retinal tissue, but these cells fail to differentiate into retinal phenotypes. Recent studies support the theory that committed photoreceptor cells can successfully integrate and migrate after retinal transplantation. Therefore, the retina serves as an attractive target for stem cell therapy.

A directed differentiation of retinal stem cells *in vitro* is extremely tricky and several laboratories demonstrated varying efficiencies and a variety of generated cell types. A promising strategy would be to differentiate all stem/progenitor cells into a single cell type of choice. In addition, generation efficiency needs to be high enough for a successful transplantation and integration of retinal cells. Some mouse and human *in vitro* culture protocols include for example retinoic acid and taurine in order to produce photoreceptor cells. Several *in vitro* protocols for the cultivation of three-dimensional eye-like structures from hES and hiPS cells were also reported. These structures displayed retinal layer formation as well as interkinetic nuclear migration and represent appropriate model systems to study eye formation and therapeutic applications. Due to the risk of the tumorigenicity of remaining stem cells, pure cultures with high reproducibility are necessary.

Several retinal degenerative diseases such as glaucoma or *Retinitis pigmentosa* affect millions of people worldwide. Until today, there are no treatments available to reverse vision loss in these patients. Several clinical trials exist to overcome retinal degenerative diseases using human stem cells. Due to the easy access and the potential to monitor transplanted cells by non-invasive techniques, the retina has been one of the earliest target tissues for stem cell derived replacement strategies in regenerative medicine. There are several advances in differentiating hESCs or hiPSC into retinal cells, in order to transplant these cells into animal models of retinal degeneration or even into humans during clinical trials.

In first clinical trials, the successful sub-retinal implantation and functional integration of non-retinal, multipotent human central nervous system stem cells or hESC-derived RPE cells has been performed in patients, which suffer from AMD or Stargardt's macular dystrophy (see also ClinicalTrials.gov. numbers NCT01632527, NCT01674829, NCT01691261, NCT01344993, and NCT01469832). In a current clinical study in Kobe, Japan, the transplantation of hiPSC-derived retinal cells into the subretinal space of patients suffering from AMD will be conducted. The goal of this project is to differentiate patient skin fibroblasts into retinal cells, in order to transplant these cells into the subretinal space of the same patient. An advantage of iPSC-derived retinal cells is that these cells are non-immunogenic and provoke no response of the host immune system. Nevertheless, there are also several disadvantages in regard to iPSC-derived retinal cells, for example the risk of mutations and teratoma formation.

Key-terms

central nervous system, embryonic development, ESCs, extracellular matrix, iPSCs, Müller glia, multipotency, niche, personalized medicine, pluripotency, precursor cell, progenitor cell, reprogramming, retinogenesis, stem cells, TAPs, therapy, totipotency, transplantation

Questions

- Which cells are totipotent, pluripotent, multipotent?
- What are symmetric and asymmetric cell divisions?
- Which division mode leads to stem cell self-renewal (homeostasis)?
- Which factors can influence the division mode of a stem cell?
- Which components build the stem cell niche?
- Why is the research on human embryonic stem cells (hESCs) ethically problematic?
- What are teratomas?
- What is personalized medicine?
- What are the great advantages of personalized medicine?
- What are the disadvantages of personalized medicine?
- Give examples of adult tissue stem cells!
- What is the "Yamanaka cocktail"?
- What advantages has direct conversion of cells over induced pluripotency?
- What are the advantages of automated cell culture?
- Which cells are stem cells in the central nervous system?
- Which cells are stem cells in the eye?

Further readings

Gage, F. H. & Temple, S. 2013. Neural stem cells: generating and regenerating the brain. *Neuron* 80(3): 588–601.

Sanchez Alvarado, A. & Yamanaka, S. 2014. Rethinking Differentiation: Stem Cells, Regeneration, and Plasticity. *Cell* 157(1): 110–119.

Wohl, S. G., Schmeer, C. W. & Isenmann, S. 2012. Neurogenic potential of stem/progenitor-like cells in the adult mammalian eye. *Prog Retin Eye Res* 31(3): 213–242.

References

Belecky-Adams, T. L., Chernoff, E. C., Wilson, J. M., et al. 2013. Reactive Muller glia as potential retinal progenitors, Neural stem cells – new perspectives. Dr. Luca Bonfanti (Ed.), Publisher: InTech, DOI: 10.5772/55150.

Cepko, C. L., Austin, C. P., Yang, X., et al. 1996. Cell fate determination in the vertebrate retina. *Proc Natl Acad Sci USA* 93(2): 589–595.

Cook, T. 2003. Cell diversity in the retina: more than meets the eye. *Bioessays* 25(10): 921–925.

Doetsch, F. 2003. The glial identity of neural stem cells. *Nat Neurosci* 6(11): 1127–1134.

Doetsch, F., Garcia-Verdugo, J. M. & Alvarez-Buylla, A. 1997. Cellular composition and three-dimensional organization of the subventricular germinal zone in the adult mammalian brain. *J Neurosci* 17(13): 5046–5061.

Franze, K., Grosche, J., Skatchkov, S. N., et al. 2007. Muller cells are living optical fibers in the vertebrate retina. *Proc Natl Acad Sci USA* 104(20): 8287–8292.

Fischer, A. J. & Reh, T. A. 2003. Potential of Muller glia to become neurogenic retinal progenitor cells. *Glia* 43(1): 70–76.

Ladewig, J., Koch, P. & Brustle, O. 2013. Leveling Waddington: the emergence of direct programming and the loss of cell fate hierarchies. *Nat Rev Mol Cell Biol* 14(4): 225–236.

Locker, M., Borday, C. & Perron, M. 2009. Stemness or not stemness? Current status and perspectives of adult retinal stem cells. *Curr Stem Cell Res Ther* 4(2): 118–130.

Malatesta, P., Hartfuss, E. & Gotz, M. 2000. Isolation of radial glial cells by fluorescent-activated cell sorting reveals a neuronal lineage. *Development* 127(24): 5253–5263.

Moretti, A., Laugwitz, K. L., Dorn, T., et al., 2013. Pluripotent stem cell models of human heart disease. *Cold Spring Harb Perspect Biol* 5(11).

Pollak, J., Wilken, M. S., Ueki, Y., et al. 2013. ASCL1 reprograms mouse Muller glia into neurogenic retinal progenitors. *Development* 140(12): 2619–2631.

Sancho-Martinez, I., Baek, S. H. & Izpisua, J. C. 2012. Belmonte, Lineage conversion methodologies meet the reprogramming toolbox. *Nat Cell Biol* 14(9): 892–899.

Takahashi, K., Tanabe, K., Ohnuki, M., et al. 2007. Induction of pluripotent stem cells from adult human fibroblasts by defined factors. *Cell* 131(5): 861–872.

Takahashi, K. & Yamanaka, S. 2006. Induction of pluripotent stem cells from mouse embryonic and adult fibroblast cultures by defined factors. *Cell* 126(4): 663–676.

Vierbuchen, T., Ostermeier, A., Pang, Z. P., et al. 2010. Direct conversion of fibroblasts to functional neurons by defined factors. *Nature* 463(7284): 1035–1041.

Olivia A. Masseck, Katharina Spoida and Stefan Herlitze

12 Optogenetics

In 1979, the Nobel laureate, Francis Crick mentioned in his article in Scientific American that controlling one type of cell in the brain while leaving others unaltered would be the major challenge facing neuroscience. 35 years later, scientists are now able to control one type of cell in the brain with high temporal and spatial precision using optogenetic techniques. Optogenetics is a combination of genetics and optical stimulation to control cellular signaling processes *in vivo* and *in vitro* by using light activated proteins. The following chapter retraces the historical roots of the field of optogenetics and the development of the optogenetic toolbox. The last part of the chapter will provide an overview about recent investigations and therapeutical applications of optogenetics in neurological disorders.

Box 12.1: Optogenetics.

> Optogenetics has been highlighted as the method of the year in 2010 by Nature Magazines and represents a new technique with promising potential for therapeutic applications. It incorporates the fast spatial and temporal kinetics of light with genetic targeting to manipulate well-defined neuronal populations in intact brain circuits. Introducing naturally occurring light-sensitive proteins, so-called opsins into neurons, renders the cell light-responsive and allows precise control of neuronal activity in particular cell types. Molecular and engineering approaches created a substantial optogenetic toolbox, with light-sensitive proteins comprising ion-flow modulators, regulators of well-defined biochemical signaling pathways, and protein interaction for a broad range of applications.

12.1 Early developments

A major challenge for neuroscientists to control the excitation of a neuron or a neuronal population was to find a way to do this with high-temporal and spatial precision but without harming, injuring or killing the cell. Electrodes or glass pipettes, for example, which are commonly used to control and monitor potential changes of neurons, destroy the cell within a certain amount of time once penetrated and do not allow to control a defined neuronal population such as excitatory or inhibitory neurons. Light would be an ideal candidate to control the physiological state of a neuron, because a light beam can be positioned exactly onto one neuron or groups of neurons and will not be harmful to the cells when low energy light is used. The question, therefore, was how to make a neuron light responsive? This question was first asked by a group of scientists who looked more closely at the fly phototransduction mechanism as a possible concept. In comparison to our human eye (i.e. the vertebrate eye), flies and other invertebrates convert light into a depolarization response. The depolarization of a neuron is the necessity to fire an

Table 12.1

Name	Channel variant	Organism	Mechanism	Peak response λ	Ions	Ion flow	Characteristics
Depolarization							
Channelrhodopsin-1	VChR1	Volvox carteri (multicellular green algae)	cation channel	535 nm	Na⁺, H⁺, Ca²⁺, K⁺	inward	red-shifted
	ChR1	Chlamydomonas reinhaardtii (unicellular green algae)	proton channel	500 nm	H+	inward	
Channelrhodopsin-2	ChR2		cation channel	470 nm	Na⁺, H⁺, Ca²⁺, K⁺	inward	
	ChR2 (H134R)		cation channel	470 nm	Na⁺, H⁺, Ca²⁺, K⁺	inward	larger photocurrent
	ChR2 (T159C)		cation channel	470 nm	Na⁺, H⁺, Ca²⁺, K⁺	inward	larger photocurrent
	ChR2 (E123T)		cation channel	490 nm	Na⁺, H⁺, Ca²⁺, K⁺	inward	ChETA; faster deactivation kinetics
	ChR2 (L132C)		cation channel	474 nm	Na⁺, H⁺, Ca²⁺, K⁺	inward	CatCh; high light-sensitivity and Ca²⁺ permeability
	ChR2 (C128T/S/A)		cation channel	470 nm activation / 590 nm deactivation	Na⁺, H⁺, Ca²⁺, K⁺	inward	step function opsins (SFO); increased life time of the open state; deactivation with longer wavelengths

Table 12.1 (continued)

Name	Channel variant	Organism	Mechanism	Peak response λ	Ions	Ion flow	Characteristics
Hyperpolarization							
Halorhodopsin	NpHR2.0	*Natromonas pharaonis* (Haloarcheon)	chloride channel	580 nm	Cl^-	inward	contains an ER export motif from an inward rectifying K^+ channel to reduce accumulation in the ER
Halorhodopsin	NpHR3.0	*Natromonas pharaonis* (Haloarcheon)	chloride channel	580 nm	Cl^-	inward	proton pump; contains a trafficking signal at the end of the C-terminus from an K^+ channel (Kv2.1) for increased membrane targeting
Archaerhodopsin-3	Arch-3	*Halorobrum sodomense* (Archeon)	Proton pump	575 nm	H^+	outward	proton pump
Archaerhodopsin-3	ArchT	*Halorobrum sodomense* (Archeon)	Proton pump	575 nm	H^+	outward	> 3 fold higher light-sensitivity in comparison to Arch
Bacteriorhodopsin	BR	*Halobacterium salinarum*	Proton pump	540 nm	H^+	outward	proton pump
MAC	MAC	*Leptosphaeria maculans* (fungus)	Proton pump	470/580 nm	H^+	outward	proton pump

Table 12.1: (continued)

Name	Abbreviation	Origion	Organism	Mechanism	Peak response λ	cell response/ second messenger	characteristics
Modulation							
vertebrate Rhodopsin	vRh	bovine, rat	rods	Gi/o	475 nm	hyperpolarization	GIRK channel activation via Gi/o proteins; efflux of K+ ions
vertebrate short wavelength opsin	vSWO	mouse	cones	Gi/o	380 nm	hyperpolarization	GIRK channel activation via Gi/o proteins; efflux of K+ ions; bleaching resistant; increased light-sensitivity
vertebrate long wavelength opsin	vLWO	human	cones	Gi/o	560 nm	hyperpolarization	GIRK channel activation via Gi/o proteins; efflux of K+ ions; bleaching resistant; increased light-sensitivity
melanopsin	vMo	human	ipRGSCs	Gq/11	480 nm	PLC PLD Ca^{2+} depolarization	bistable pigment; activation of PLCβ; PLD mediated activation of transcription factors; activation of TRPC3 channels
OptoXR α1-adrenergic receptor	Opto-α1AR	bovine	rods	Gq/11	475 nm	PLC Ca^{2+} depolarization	control of α1-adrenergic receptor signaling pathway
OptoXR β2-adrenergic receptor	Opto-β2AR	bovine	rods	Gs	475 nm	cAMP	control of β2-adrenergic receptor signaling pathway
jellyfish opsin	JellyOp	Carybdea rastonii (jellyfish)	visual cells	Gs	500 nm	cAMP MAPK	bleaching-resistant

Table 12.1 (continued)

Name	Abbreviation	Organism	Origion	Chromophore	Mechanism	Peak response λ	cell response/ second messenger	characteristics
photo-activatable adenylyl cyclase (PAC)	euPAC	Euglena gracilis (flaggelate)	photorecep-tor cells		AC	480 nm	cAMP PKA CNG	light-induced increase of cAMP levels; constitutive dark activity
	bPAC/BlgC	Beggiatoa sp. (soil bacterium)			AC	480 nm	cAMP/cGMP PKA CNG	no constitutive dark activity; uses flavin as a chromophore

Protein family	Variant	Organism	Chromophore	Domain	Peak response λ	Interaction partner	Application
Phytochromes	Phytochrome B (PhyB)	Arabidopsis thaliana (flowering plant)	phycocyanobilin (= linear tetrapyrrole)	PAS, GAF, Phy		PIF3	Recruitment of proteins to the plasma membrane; Activation of Rho GTPase
Phototropins	Phototropin 1	Avena sativa (grain oat)	flavin nucleotides	LOV1/2		Jα	Activation of PA-Rac1 = member of the Rho family GTPases; control of cell mobility; recruitment of cytoplasmatic scaffolding proteins to subcellular locations; light dependent degradation of proteins; lumitoxins
				FKF1		GI (GIGANTEA)	Recruitment of RAc1; control of gene transcription
Cryptochromes	CRY2	Arabidopsis thaliana (flowering plant)	FAD	PHR (photolyase-homologous regions), CCE (Cryptochrome C-terminal Extension)		CIB1 (CRY-interacting basic helix loop helix (bHLH) protein)	light dependent control of reporter gene transcription; reconstitution of Cre-recombinase; LITEs (light-inducible transcriptional effectors); light induced recruitment of 5-ptase

action potential (AP). Therefore, why not express the visual cascade of the fly's eye, which should cause a depolarization, in a neuron? The visual cascade of the fly *Drosophila melanogaster*, one of the model organisms in biology research, is probably the best described and characterized signaling cascade in any animal kingdom. Gero Miesenböck's group, therefore, coexpressed in frog (*Xenopus laevis*) oocytes 10 different proteins of the fly's visual cascade to find the minimal components, which are necessary to elicit a light response in oocytes. The three necessary components, which the group called "chARGed", were rhodopsin (the light-sensitive G protein-coupled receptor (GPCR), G protein αq subunit (the intracellular proteins, which determines the signaling pathway), and arrestin-2 (an intracellular protein necessary for termination of the GPCR response). chARGed was then expressed in cultured hippocampal neurons and indeed, light-activation of chARGed expressing neurons induced AP firing. However, onset of AP firing and firing pattern and return to baseline activity was difficult to control.

In the next approach, the photochemical gating of ion channels was used to control the activity of neurons. The principle behind photochemical gating is to synthesize a photoactivatable ligand or channel blocker, which can be switched on/off or cleaved of by a specific wavelength of light. These molecules are so called caged compounds, since they are first inactive until released by light. Photoactivatable ligands have been synthesized for several non-selective cation channels such as TRPV1 (the hot chili pepper (capsaicin) receptor), TRPM8 (the menthol receptor), and P2X2 (an ATP receptor). Expression of these ion channels in neurons and perfusion of the neurons with the caged compounds made neurons light-responsive, i.e. absorption of photons in the near UV range (355 nm) of the ligand attached chromophore causes the relieve of the ligand, the activation of the ion channel, the influx of Na^+ and Ca^{2+} ions, and consequently the depolarization and AP firing of the neuron. The photochemical approach was further tested in flies. Expression and light-activation of P2X2 in neurons involved in escape behavior for example caused jumping, wing beating, and flying even in decapitated flies. This was the first demonstration that a behavior of an animal via activation of larger groups of neurons can be controlled by light.

In another approach, a photoisomerizalbe azobenzene (AZO) was conjugated to a K^+-channel blocker (QA), which could be tethered to a cysteine residue outside of the channel by a maleimide group (MAL). Certain types of K^+ channels are blocked by quarternary ammonioum (QA) ions, such as tetraethylammonioum, by binding to the outside pore region of the channel. Thus, tethering of the MAL-AZO-QA molecule to the outside of the channel allowed to change the orientation of the QA blocker by switching the *cis-trans* configuration of the AZO group using 380 and 500 nm light. The MAL-AZO-QA blocks the channel in the trans configuration (induced by 500 nm), but not in the *cis* configuration (induced by 380 nm). Since K^+ efflux hyperpolarizes neurons, the opening and closing of K^+ channel can be used to reduce or increase the firing of neurons which receive excitatory drive or are spontaneously active. Thus, expression of the engineered K⁺ channel in cultured hippocampal neurons and application of MAL-AZO-QA could be used to induce AP firing by 500 nm light (K⁺ channel block) and silence neurons by 380 nm light (K⁺ channel opening). The disadvantage of the photochemical gating for basic neurobiology research is that synthesized ligands have to be applied during the experiments, which is in particular difficult for *in vivo* experiments in vertebrates.

The breakthrough in optogenetics came with a light-activated, non-selective cation channel from the green algae *Chlamydomonas rheinhardii* called channelrhodopsin2 (ChR2). ChR2 was first cloned in 2003 and was for the first time used to control neuronal excitation in 2005 by three different groups in cultured hippocampal neurons, chicken spinal cord, and the nematode worm *Caenorhabditis elegans*. Since ChR2 does not need further application of a synthesized ligand *in vivo*, is a relatively small protein, which can be used in virus and genetic approaches and reliably works in various systems and cell-types to induce APs, it immediately became the optogenetic tool of choice to control the excitability of neurons with high temporal and spatial precision.

Box 12.2: Rhodopsins.

Rhodopsins are about 30–60 kDa membrane-bound light-sensitive proteins. Two main parts constitute the rhodopsin protein: The apoprotein, termed opsin, and the chromophore retinal. Retinal is an organic vitamin A-based co-factor and constitutes the prosthetic group of the opsin. It is covalently bound to a conserved lysine residue at helix 7 through a Schiff base linkage, and serves as the photon capturing unit. Retinal may exist in different isoforms such as all-*trans*, 11-*cis* or 13-*cis*. Opsin genes are divided into two distinct subfamilies: Microbial type I opsins are found in prokaryotes, algae, and funghi, whereas type II opsins are only employed by higher eukaryotes. Although both opsin families share a common architecture with 7 transmembrane α helices, they have almost no sequence homologies. Animal rhodopsins belong to the family-A GPCR superfamily and transduce light cues into chemical signals via activation of G proteins. Type II opsins prefer retinal in the 11-*cis* configuration. Upon absorption of a photon, 11-*cis* retinal is isomerized into all-*trans* retinal. The isomerization step initiates a structural reorganization of the protein and subsequent activation of downstream G protein signalling. All-*trans* retinal is released by the opsin and replaced by another 11-*cis* retinal molecule. When expressed in neurons, animal rhodopsins can be utilized to control an inhibitory signaling cascade. In contrast, microbial opsins form light-driven ion channels, which most preferentially bind all-*trans* retinal in the dark state. Upon photon capturing, the isomerization of all-*trans* retinal into 13-*cis* retinal induces conformational changes of the protein. As a result, charged particles (ions) translocate via the opsin protein. Channelrhodopsin is the most prominent microbial opsin used for optogenetic applications. It was first discovered in the unicellular algae *Chlamydomonas reinhardtii*, where it is utilized for visible light perception and phototaxis. When expressed in neurons, it integrates into the cell membrane. Upon illumination channelrhodopsin causes an influx of positively charged ions such as sodium and protons into the cell, thus depolarizing the cell. Microbial Opsins exclusively encode ion flow modulators, which are used to control the excitability of a neuron by directly affecting membrane potential.

Box 12.3: Photopharmacology.

Photochemical gating is an opto- or photopharmacology method. Photopharmacology means that a molecule, such as a drug, can be switched between two conformations ideally by two different wavelengths of light. For example, the active conformation is stabilized by shorter wavelengths of light, while the non-active ("silent") conformation is stabilized by the longer wavelength of light. Using this approach, ion channels, such as voltage gated K^+ channel, GPCRs such as metabotropic glutamate receptors, enzymes or any protein of choice can be activated or blocked depending on the wavelength of light, when a specific agonist or antagonist is available. These molecules can also be directly tethered to the protein via a cysteine bond. The cysteine mutation, which should be in close proximity to the binding site of the drug, can be introduced into the protein by genetic engineering. In this case, only genetically engineered proteins can be activated/ blocked by light. Photopharmacology in animal models has been successfully applied to restore visual responses and to reduce pain.

12.2 Fast on and off switches

12.2.1 Depolarization of neurons: Channelrhodopsin

In 2002, the first opsin related protein (Channelrhodopsin 1 (ChR1) from *Chlamydomonas rheinhardii* was cloned and expressed in *Xenopus* oocytes. The biophysical characterization revealed that ChR1 is a proton pump/channel that can be maximally activated by 500 nm light when all-*trans* retinal is applied to the media. Subsequently, in 2003, ChR2 was cloned and characterized. Native ChR1 and ChR2 are involved in generating photocurrents in *C. rheinhardii*. ChR2, which is also a proton channel, is capable of conducting Na^+ ions. The ion channel can be opened by blue light (475 nm) within a few milliseconds, closes rapidly once light is switched off and reveals an inward rectification, i.e. a large inward current and little outward current. As mentioned above, it was immediately recognized that light-activation of an inward Na^+ current could be used to trigger APs in neurons (Figure 12.1, 12.2a). Using the C-terminally truncated version of ChR2, AP firing was demonstrated in various neuronal systems. ChR2 used in optogenetic experiments is 315 aa long and needs retinal as a co-factor. Interestingly, retinal compounds are available in the brain and are sufficient to supply ChR2 and other opsin variants with the necessary co-factor. After the functional demonstration of ChR2 in neurons, ChR2 became optimized for different optogenetic applications and other microbial opsins were cloned from more than over 120 different algae and *Chlamydomonas* species including *Volvox carteri* and *Dunalelia salina*. However, most mutagenesis studies have been performed on the ChR2 from *C. rheinhardii*. Mutations were introduced into aa close to or involved in retinal binding, which affect the absorption, ion selectivity, and channel kinetics. The most commonly used ChR2 mutations are the E123T, C128T/S/A, L132C, H134R, and T159C. The E123T mutation is called ChETA and has much faster deactivation kinetics (i.e the current declines much faster to baseline levels once light is switched off). Therefore, this mutation is useful to trigger AP firing reliably with high frequency up to 200 Hz. Mutation at position C128 create the step function opsins. Mutations at this position increase the life time of the open state of the channel, which means that ion flux over the membrane is maintained for long times up to several minutes once light is switched off. The deactivation kinetics can be accelerated by switching to longer wavelength of light. These constructs can be used to induce long-time depolarization into neurons to either induce a so called depolarization block or to trigger AP firing over long time periods relying on the intrinsic AP firing rate. The L132C mutation CatCh was designed to create a channel with high-light sensitivity and Ca^{2+} permeability. Influx of Ca^{2+} into neurons also contributes to the depolarization of neurons, but is also involved in modulation of other signaling cascades and ion channels. Thus, controlling the influx of Ca^{2+} in neurons could be used for other purposes such as excitation-transcription (i.e. inducing gene transcription) and excitation-transmission coupling (i.e. triggering transmitter release directly at the synaptic terminal). The H134R and T159C mutations produce larger photocurrents and are therefore useful for driving APs at lower opsin expression levels.

One important aspect for controlling neuronal networks using light-activated proteins is that longer wavelengths of light are preferable for two reasons: 1. The shorter the wavelength of light the higher is the energy. High-energy light may cause photo-damage to the cell or the tissue as you experience when you get sunburn. 2. Longer wavelengths of light penetrate tissue better than shorter wavelengths of light. Therefore, red-shifted variants from different algae have been isolated and characterized. For example, ChR1 from *Volvox carteri*, a cation conducting channelrhodopsin variant, has a 70 nm red-shifted maximal light-activation peak in comparison to ChR2, which allows to drive APs with 589 nm light and might be used alongside with other ChR variants to control neuronal networks. Chrimson from *C. noctigama* is another red shifted version of ChR2 with a spectral peak at 590 nm. Chrimson can be coexpressed alongside the 45 nm blue shifted ChR variant Chronos to activate two separate neuronal populations with blue and red light.

12.2.2 Hyperpolarization of neurons: Halorhodopsin, Arch and MAC

Besides switching neuronal networks on, it is also desirable to switch neurons off. Neurons reduce the excitability by increasing anion influx into the cell or cation efflux out of the cell (Figure 12.1). Various anion (i.e. Cl−) and cation (i.e. H+) transporters are expressed in microbial opsins including halorhodopins, bacteriorhodopsin or archaerhodopsin. Cl^--transport is used by halophilic prokaryotes for osmotic balance and to maintain Cl^--dependent cellular processes, such as growth and motility. A proton (H^+) gradient over the membrane is commonly used in bacte-

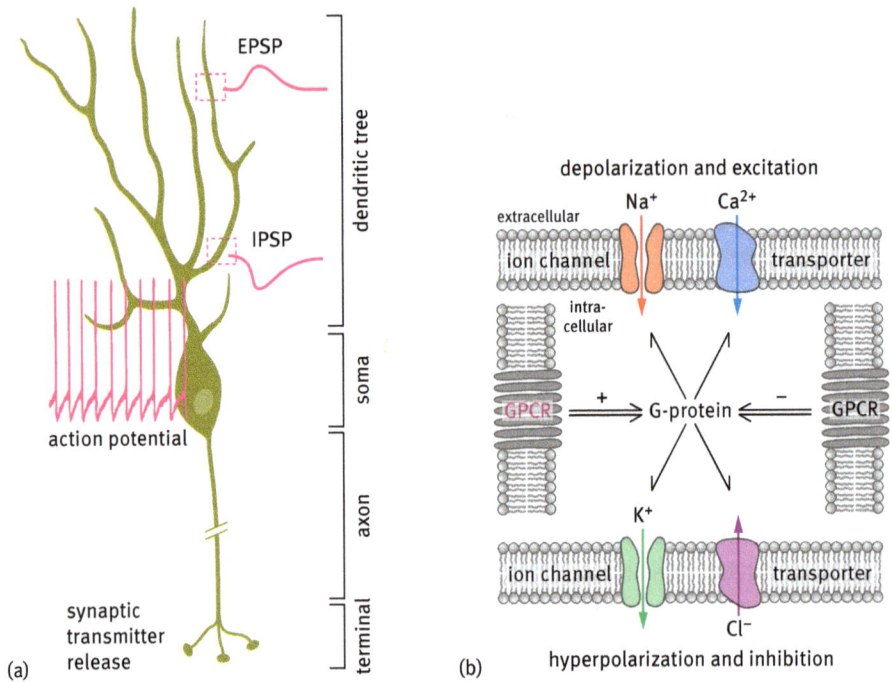

Fig. 12.1: (a) A neuron can be divided into four parts. The dendritic tree receives excitatory (i.e. excitatory postsynaptic potentials (EPSPs)) and inhibitory information (i.e. inhibitory postsynaptic potentials (IPSPs)) from other neurons. Excitatory and inhibitory information is compared at the soma. If excitatory input is larger than the inhibitory input and the depolarization exceeds a threshold value, action potentials are elicited. The information given by the strength and duration of the depolarization is encoded in the AP firing frequency. APs travel along the axon. At the synaptic terminal, the AP electrical signal is converted into a chemical signal, the neurotransmitter, which transfers neuronal information to the next neuron. The firing frequency of the APs determines the amount of transmitter which is released. (b) To depolarize or hyperpolarize neurons, or different ion channels or ion transporters are opened and regulated by modulatory mechanisms. Depolarization of the cell membrane is achieved by the opening or activation of Na^+ and Ca^{2+} channels, which leads to an influx of these ions along their concentration gradients into the neuron. In contrast, hyperpolarization is mediated either by opening of K^+ channels, which leads to an efflux of K^+ ions or opening/activating Cl^- channels/transporters, which leads to an influx of Cl^- into the cell. All channels and transporters can be increased (+) or decreased (−) in their activity by G protein coupled receptors (GPCRs), which activate different forms of G proteins. These GPCRs are activated by the neurotransmitter, and therefore represent a negative or positive feedback loop for the activity of a given neuron.

ria for energy (ATP) production. Halorhodospin from *Natromonas pharaonis* (NpHR) was first used to control the hyperpolarization of neurons using yellow light. NpHR has an excitation maxima around 580 nm and can therefore be used alongside with ChR2 to switch neurons on and off with two different wavelengths of light (Figure 12.2b). Two variants of NpHR have been developed to improve membrane trafficking and tolerability in neurons. NpHR2.0 contains an ER export motif from an inward rectifying K^+ channel to reduce accumulation of NpHR in the ER.

Box 12.4: Control and modulation of neuronal activity.

Neuronal communication relies on APs and the variation in AP firing frequencies. APs can be either elicited or suppressed or the AP firing frequency can be modulated to encode for example sensory information in the brain (Figure 12.1b). In order to elicit, modulate or suppress an AP, neurons contain ion channels and GPCRs in their excitable membrane. In order to excite or silence a neuron, the ion flow over the membrane has to be regulated. Several ion gradients exist over the membrane, which are maintained by transporters/ion pumps. In the adult mammalian brain, neurons have a high concentrations of Na^+, Ca^{2+}, and Cl^- extracellular and a high concentrations of K^+ intracellular. To elicit an AP, the membrane has to be first depolarized and then re-/hyperpolarized. Depolarization of a cell membrane occurs when Na^+ or Ca^{2+} ions flow into the cell, while hyperpolarization occurs when K^+ ions flow out of the cell or Cl^- ion flow into the cell.

At a certain depolarization threshold, an AP is elicited, which represents the fast influx of Na^+ into the cell followed by the delayed efflux of K^+ out of the cell. Thus, opening Na^+ or Ca^{2+} channels will depolarize a cell. Once a certain threshold is reached, an AP will be fired. In contrast, opening of Cl^- or K^+ channels will hyperpolarize the cell and will shunt AP firing. The maximal firing frequency of APs in the brain is about 1000 Hz. GPCRs can now modulate the opening or closing of ion channels. They can also regulate how many channels should be transported to the plasma membrane. However, GPCRs act on a slower time scale than ion channels. Ion channels are opened within μsec, while GPCRs act on a millisecond, second to minute time scale. Therefore, also GPCRs modulate the firing of neurons.

Box 12.5: Neurons: Fundamental units of information processing.

Neurons can be functionally and morphologically divided into three parts: The dendritic tree, which mainly receives information from other neurons, the soma with the axon hillock, where the AP originates and the axon with its presynaptic terminal, where the AP is running along to communicate the signal to the next neuron (Figure 12.1a). The presynaptic terminal contains the synaptic transmitter release machinery. At the presynaptic terminal, the AP (i.e. an electrical signal) is transformed into a chemical signal, i.e. the transmitter. The amount of transmitter release is given by the AP firing frequency. This transmitter release can be modulated again by GPCRs. Most GPCRs at the presynaptic terminals such as the cannabinoid receptors, which are activated also by THC produced by the cannabis plant, inhibit transmitter release via inhibition of presynaptic Ca^{2+} channels.

The transmitter binds postsynaptically onto ion channels and/or GPCRs to hyperpolarize or depolarize the dendritic tree of the connected neuron depending whether Na^+/Ca^{2+} or K^+/Cl^- channels are opened. GPCRs again modulate these ion channels to increase, decrease, shorten or prolong the influx of ions over the membrane. Neurons can receive many thousands of excitatory (depolarization) or inhibitory (hyperpolarization) inputs from other neurons. The excitatory and inhibitory information runs along the dendritic tree to the soma. At the soma, all the different inputs are compared and integrated to decide if an AP should be fired.

(a) depolarization and excitation

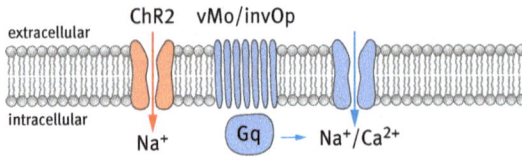

(b) hyperpolarization and inhibition

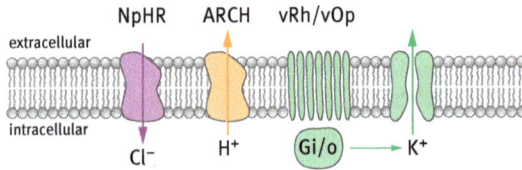

(c) modulation of intracellular signaling pathways

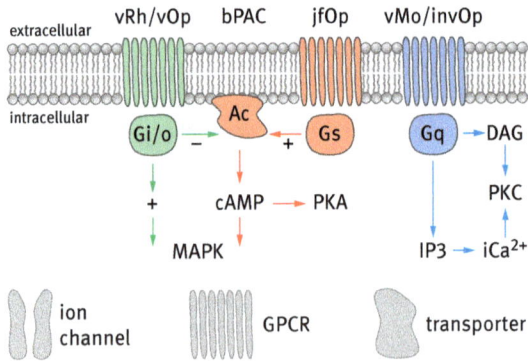

In addition, NpHR3.0 contains at its C-terminus the trafficking signal of another K$^+$ channel (Kv2.1), which increased membrane targeting and light-induced hyperpolarization in neurons.

In addition to H$^+$-pumps, proton pumps from the archaeon *Halorubrum sodomense* (archaerhodopsin-3, Arch) and from the fungus *Leptosphaeria maculans* (Mac) can also be used to hyperpolarize neurons (Figure 12.2b). Arch has an excitation peak around 575 nm and Mac around 540 nm. Mac is also capable of hyperpolarizing neurons at shorter wavelengths around 470 nm. Another archaerhodopsin from *Halorubrum* strain, TP009 has been isolated and is called ArchT. ArchT has a >3 fold higher light-sensitivity in comparison to Arch and has been tested for function in the cortex of awake non-primate monkeys.

◀ Fig. 12.2: Optogenetic tools to control neuronal activity can be divided into three main groups, depending on whether neuronal activity should be induced, blocked or modulated. (a) To control action, potential firing neurons have to be depolarized. Two optogenetic tools are commonly used. ChR2 and its variants are non-selective cation channels capable of conducting Na^+ ions to depolarize neurons. Light-induced depolarization and recovery from depolarization once light is switched off is very fast and occurs within milliseconds. Melanopsin and invertebrate (fly) rhodopsins are GPCRs, which couple to the Gq pathway. Melanopsin and fly rhodopsin most likely activate Na^+ and Ca^{2+} conductances in the cell to induce neuronal activity. Control of neuronal activity is slower in comparison to ChR2 and occurs within a second. (b) For fast inhibition of neuronal activity, two ion transporters are used. Halorhodopsin (NpHR) variants hyperpolarize neurons via transport of Cl^- ions from the outside to the inside, while archaerhodopin (ARCH) uses the transport of protons from the inside to the outside of the neuronal membrane. Hyperpolarization occurs within milliseconds. Vertebrate rhodospins/opsins are GPCRs, which couple to the Gi/o pathway. Activation of the Gi/o pathway can lead to the opening of K^+ channels at the soma and the closing of Ca^{2+} channels at the synaptic terminal. The modulation of the channels occurs within milliseconds to seconds. Opening of K^+ channels inhibits AP firing, while closing of presynaptic Ca^{2+} channels reduces and/or inhibits synaptic transmitter release. Thus, the light-activated GPCRs coupling to the Gi/o pathway can modulate neuronal activity at different subcellular domains, the electrical signal at the soma and the chemical signal at the presynaptic terminal. (c) To modulate neuronal activity and neuronal fate such as synapse formation and dendritic growth GPCR pathways and intracellular signaling pathways have to be activated. These signaling pathways act on a second to minute/hour time range. Three main GPCR pathways can be distinguished, i.e. Gi/o, Gs and Gq pathways. Gi/o pathways can be controlled by vertebrate rhodopsin/ospins (vRh/vOp), which inhibit membrane associated adenylylcylases to reduce the second messenger molecule cAMP. To increase cAMP levels in the cell either photo-activated adynylylcylases such as bPAC or Gs coupled light-activated GPCRs, such as jellyfish opsin (JellyOp), can be used. The Gi/o and Gs pathway both also activate the MAP-kinase pathway, which also is involved in regulating gene transcription. The Gq pathway, which can be controlled by vMo and invOp results in the activation of another protein kinase called protein kinase C (PKC). It involves the activation of phospholipase C in the plasmamembrane, which breaks down the membrane phospholipid phosphatidylinositol 4,5-bisphosphate (PIP_2) into inositol 1,4,5-trisphosphate (IP_3) and diacylglycerol (DAG). IP_3 leads to the release of intracellular Ca^{2+} from internal stores, such as the endoplasmatic reticulum. Ca^{2+} together with DAG activates PKC to phosphorylate and modulate the activity of target proteins.

12.3 Modulation of intracellular G-protein signals

G protein-coupled receptors (GPCRs) represent a large family of transmembrane proteins, which are involved in modulation of every aspect of animal physiology including vision, olfaction, taste, sensation, attention, reward, and fear. Consequentially, malfunctioning of GPCRs is associated with diseases such as obesity, anxiety, depression, migraine, and diabetes. Because of the crucial role of GPCRs in all aspects of our daily life, understanding and precisely controlling these pathways in health and disease is important, in particular for pharmaceutical and clinical interventions. Unfortunately, most pharmaceutical compounds such as GPCR agonists and antagonists are unspecific, activating more than one receptor type.

Therefore, new techniques had to be developed to decipher the specific physiological function of GPCRs. GPCRs are 7 transmembrane receptors, which can be activated by external physical and chemical stimuli, such as light, odors or neurotransmitters. The external stimulus is converted into intracellular signals via the activation of a GTP binding protein (G protein). The activated G protein, which consists of a G protein α and $\beta\gamma$ subunit can then activate diverse signaling cascades. There are four main pathways which are commonly activated by the four main families of G proteins, i.e. the $G_{i/o}$, G_q, $G_{12/13}$, and G_s. G_q coupled GPCRs often activate the G_q and the $G_{12/13}$ signaling pathways. Activation of the GPCR pathway can mediate relatively fast effects in the time range of milliseconds to seconds or slower effects in the time range of seconds, minutes, and hours. The fast modulation is normally mediated via the direct interaction between the G protein and ion channel, which can be activated or inactivated.

12.3.1 The $G_{i/o}$ pathway

Activation of the $G_{i/o}$ pathway in neurons normally leads to the fast modulation of ion conductances (Figure 12.2b). At the soma and dendrites of neurons, activation of the GPCRs causes the activation of G protein and binding of G protein $\beta\gamma$ to a K^+ channel, called G protein inwardly rectifying K^+ (GIRK) channel, which causes the efflux of K^+ and the hyperpolarization of the cell membrane. The hyperpolarization is induced within a second and recovers within a few seconds. This process also underlies the slowing of the heart rate mediated by the release of acetylcholine by the vagus nerve. At the presynaptic terminal, activation of the $G_{i/o}$ pathway causes the fast inhibition of presynaptic Ca^{2+} channels which are responsible for triggering synaptic transmitter release. Inhibition of the presynaptic Ca^{2+} channels inhibits and/or reduces the release of the transmitter. Thus, activation of the $G_{i/o}$ pathway in neurons causes a reduction of neuronal excitability.

The visual cascade of vertebrates uses light-activated GPCRs, which couple on the intracellular site to the G protein transducin (Figure 12.2c and Figure 12.3). Transducin belongs to the $G_{i/o}$ protein family. Vertebrate rhodopins (vRh), the photosensitive GPCR from the vertebrate rods, was first expressed in neurons to control the pertussis toxin sensitive $G_{i/o}$ pathway. Pertussis toxin is produced by the bacterium *Bordatella pertussis*, which causes the whooping cough. Activation of vRh at the somatodendritic site of neurons causes the activation of the GIRK channels and the reduction in AP firing (Figure 12.2b). Activation of vRh at the presynaptic terminal inhibits presynaptic Ca^{2+} channels and reduces synaptic transmitter release. The signal transduction in rods underlying vision in dim light is relatively slow in comparison to daylight vision using cones. This is also reflected at the GPCR level. GIRK channel modulation by cone opsins is much faster in comparison to rod rhodopsin and modulation does not decline in amplitude during repetitive stimula-

(a)

retina

light

optic nerve

(b)

retina

cones

rods

bipolar cells

ganglion cells

light

(c) visual cascade in the vertebrate eye

light

light

vRh/vOp

open → closed

extracellular

intracellular

cGMP

Gt → PDE → Na+

GMP

(d)

11-*cis*-retinal

Opsin

all-*trans*-retinal

Opsin

N

Opsin

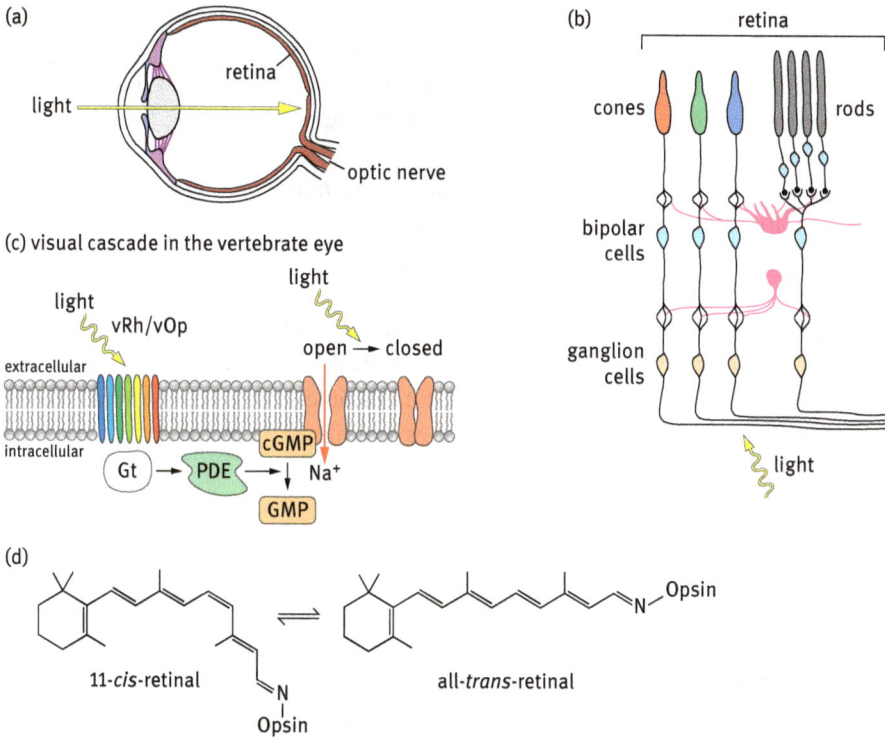

Fig. 12.3: (a) In vertebrates, such as humans, light enters the eye through the pupil and is direct-ly projected onto the retina, a relatively complex structure, which contains the photoreceptor cells and various other cell types for light processing. (b) There are two types of photoreceptor cells. Cones are for daylight and color vision, while rods are used during dim/low light condi-tions. Both cones and rods project directly onto biopolar cells and biopolar cells directly onto gan-glion cells. Light information processing also involves indirect input from interneurons, such as horizontal cells and amacrine cells (shown in light red). At the ganglion cell level, the processed light information is encoded into AP firing frequencies and sent into higher order brain areas via the optic nerve. (c) The visual response in the photoreceptor cells involves a GPCR signaling cas-cade. Vertebrate rhodopsin or vertebrate cone opsins activate the G protein transducin. This G protein activates a phosphodiesterase (PDE), which hydrolyzes cGMP to GMP. cGMP opens a non-selective cation channel, which causes the depolarization of the cell membrane. Therefore, in the dark photoreceptors are depolarized. In the light, the cell membrane is hyperpolarized, since in-tracellular cGMP concentration is decreased. (D) The light sensitive molecule in vertebrate opsins is a derivative from vitamin A, 11-cis retinal, which is covalently attached to the opsin protein via a Schiff base linkage. Photo-excitation leads to the isomerization of 11-cis to all-trans retinal. In the vertebrate retina, all-trans retinal is converted back to 11-cis retinal via numerous biosynthet-ic steps involving different enzymes and cell-types. This process is called photocycle. In some op-sins, designated bistable opsins, such as melanopsin, all-trans retinal can be photo-reconverted by absorption of a second photon.

tion. Cone opsins can be further distinguished according to their wavelength specificity into short-(SWO), mid-(MWO) and long-(LWO) wavelength opsins for blue, green, and red color vision. Co-expression of SWO from mice and LWO from human can be used to control two GPCR pathways simultaneously using blue (400 nm) and orange (590 nm) light as has been demonstrated in hind/midbrain regions of mice (Masseck et al. 2014). One interesting feature of the vertebrate opsins is that they are 100–1000 times more light-sensitive in comparison to the above mentioned microbial opsins. The light-sensitivity is important for activating signaling cascades in deep brain or tissue structure and the vertebrate opsins may therefore have advantages for neuronal silencing in comparison to NpHR, Arch or Mac.

The cone opsins have also been used to control slower, second messenger mediated $G_{i/o}$ pathways (Figure 12.2c). SWO, for example, was used to control growth cones and neurite outgrowth in cell cultures and hippocampal neurons.

12.3.1 The G_q pathway

Activation of the G_q pathway involves the activation of phospholipase Cβ (PLCβ) . PLCβ cleaves the membrane-associated phospholipid phosphatidylinositol 4,5-bisphosphate (PIP$_2$) into the intracellular messenger inositol 1,4,5-trisphosphate (IP$_3$) and the membrane-associated diacylglycerol (DAG), which recruits and activates PKC. IP$_3$ binds to IP$_3$ receptors located on the endoplasmatic reticulum (ER), which results in the release of Ca^{2+} from the ER into the cytoplasm. The intracellular increase in Ca^{2+} can activate a store operated Ca^{2+} channel at the plasma membrane, which leads to the influx of extracellular Ca^{2+}. The $G_{q/11}$ pathway can also activate phospholipase D and the NFκB a protein complex that controls transcription.

Melanopsin is a $G_{q/11}$ coupled GPCR expressed in a subset of intrinsically photosensitive retinal ganglion cells (ipRGCs). ipRGCs convert information from photoreceptors into APs. Melanopsin is part of the third photoreceptor system in the mammalian retina involved in photoentrainment. Photoentrainment is the detection of changes in the quantity and quality of light during the 24 h day/night cycle. The maximal excitation of melanopsin is around 480 nm. Melanopsin is most likely a bistable photopigment. Similar to invertebrate photoreceptors melanopsin regenerates the light-sensitive chromophore by photoconverting all-*trans* retinal into 11-*cis* retinal within the receptor protein. Expression of melanopsin in heterologous expression systems, such as oocytes from the frog *Xenopus laevis*, human embryonic kidney (HEK293) cells and neuroblastoma cell lines leads to the increase in intracellular Ca^{2+} via activation of $G_{q/11}$ (Figure 12.2c), PLC and the activation of a transient receptor potential channel (TRPC3) (Figure 12.2b), which might be involved in the light response in ipRGCs. Heterologous expression systems are used to characterize the biophysical properties of ion channels, GPCRs and transporters in a cellular environment, where the exogenously expressed proteins of interest do not exist. Since light-induced activation of melanopsin activated the G_q pathway, mela-nopsin was tested in various systems for its ability to control neuronal activity. First expression of melanopsin in blind mice in RGCs make RGCs, which normally do not respond to light, light responsive. Most importantly, blind animals also now show light responsiveness in certain behavioral test (see below). Expression of Mel-anopsin in neurons involved in modulating sleep and wakefulness induces long-lasting activation of neuronal activity in behaving mice. Expression of melanopsin in inhibitory GABAergic neurons can be used to control neuronal excitability repet-itively on a fast time scale in hindbrain regions to relieve anxiety behavior. In addition, expression of melanopsin in the nematode *Caenorhabditis elegans* in spe-cific neurons underlying locomotion, enhances movements in a PLCβ dependent manner. Recently, melanopsin was also used to control the contraction of cardio-myocytes, the cells that constitute the cardiac muscle.

As mentioned above, activation of the $G_{q/11}$ pathway also activates transcrip-tion factors in a Ca^{2+} dependent manner, suggesting that activation of melanopsin can be used to control the transcription of genes of interest. Indeed, expression of melanopsin in heterologous expression systems revealed that light-activation of melanopsin induces the NFAT (transcription factor)-dependent expression of a transgene, i.e. glucagon-like peptide-1. This system was then used in diabetic mice to control their insulin levels.

12.3.3 The G_s pathway

Activation of the G_s pathway leads to the activation of adenylyl cylases, which produce the intracellular second messenger cAMP. cAMP can activate various effec-tor proteins including ion channels (e.g. L-type Ca^{2+} channels), protein kinase A (PKA) or small GTPases such as Rap, which is part of the mitogen-activated protein kinase (MAPK) pathway. One important aspect of G_s signaling is the regulation of gene transcription by CREB, the cAMP response element binding protein, a tran-scription factor. CREB activates genes, which contain the specific DNA recognition sequence CRE, the cAMP response element.

The most basic eyes in animal kingdom can be found in jellyfish. Jellyfish have ciliary visual cells, which respond to blue and green light. The visual cascade of jellyfish uses the jellyfish Opsin (JellyOp), which activates the G_s pathway leading to cAMP production (Figure 12.2c). Expression of jellyfish opsin in heterologous expression systems revealed that JellyOp is maximally sensitive to 500 nm light, can repetitively increase intracellular cAMP levels and on a slower time scale can activate the MAPK pathway. Based on these properties, the jellyfish G_s coupled opsin is probably more suitable for *in vivo* studies than the first generation of Op-toXR controlling the G_s pathway. The OptoXRs are chimeras between the vertebrate rhodopsin and the adrenergic receptors, where the intracellular loops of the GPCRs where exchanged to transfer G_s and G_q signaling specificity. OptoXR stands for a light-activated GPCR, which controls a certain signaling pathway in relation to a specific receptor type such as the α1-adrenergic or β2-adrenergic receptor.

The second possibility to control the G_s pathways is to control directly cAMP levels using photo-activatable adenylyl cyclases (PAC) (Figure 12.2c). PAC from the flaggelate *Euglena gracilis* (euPAC) can be used to control intracellular cAMP levels in heterologous expression systems and in flies. euPAC consists of the flavoproteins euPACα and euPACβ, is maximally activated around 480 nm and increases cAMP levels around 100 fold. Light-induced increase in cAMP levels is sufficient to regulate cAMP targets such as PKA and cyclic nucleotide-gated (CNG) channels. Neuronal expression of euPACα in flies was sufficient to alter grooming behavior. For optogenetic applications, euPACα has two disadvantages. First, it is a relatively large protein made up of 1000 amino acids (aa), and has a constitutive activity in the absence of light (dark activity). The bacterial light-activated adenylyl cyclases (BlaC or bPAC) from *Beggiatoa* contain a small blue light sensitive domain, called BLUF domain (Figure 12.4a). These BLUF (sensor of blue light using FAD) domains contain flavin as a chromophore, which is available in pro- and eukaryotic cells. Therefore, these domains are ideal tools for optogenetic application *in vivo*, because no external ligands have to be applied. BlaC/bPAC reveals no detectable dark activity and increases cAMP levels when activated by blue light. Introducing a triple mutation into BlaC/bPAC converts the adenylyl cyclase activity into a guanylate cyclase activity (BlgC), given the possibility to not only control cAMP but also cGMP levels in cells.

cGMP is also a second messenger and regulates cGMP-gated channels in the vertebrate eye, activates protein kinase G (PKG), and is, for example, involved in relaxation of smooth muscles, i.e. blood vessels, glycogenolysis, and apoptosis. BlaC/bPAC was used to control cyclic nucleotide gated channels in heterologous expression systems and hippocampal neurons and was sufficient to impair grooming behavior in flies.

Box 12.6: Tissue, cell-type and Circuit Specificity.

One important aspect for the investigation of the function of certain signaling processes involved in diseases and the development of possible future therapeutic treatments is that different cell-types with different physiological function within the brain, heart or pancreas can be controlled specifically. For example, in the brain, various cell-types such as astrocytes, microglia or neurons have different functions. In addition, many different types of neurons can be found in the CNS. These neuron types, for example, use different transmitters such as the excitatory transmitter glutamate, the inhibitory transmitter GABA or modulatory transmitters such as serotonin or dopamine. In order to control signaling pathways with cell-type specificity, various systems have been developed. The most common system is the so-called Cre-lox system. This system uses the bacterial Cre recombinase, which is expressed in the cell-type of choice such as serotonergic neurons or GABAergic neurons. The specific expression of the Cre recombinase is given by a specific promoter or enhancer sequence of a gene, which is only expressed in the cell-type of choice. For example, tyrosine hydroxylase is the rate-limiting enzyme for the synthesis of the catecholamines, dopamine, noradrenaline, and adrenaline and is only expressed in catecholaminergic neurons. The information for the specific expression is localized in the promoter/enhancer region of the gene. Promoter/enhancer regions are regulatory DNA sequences, which can be found 5′ of the start codon of the gene. Therefore, expression of Cre-recombinase under the promoter/enhancer sequence of the tyrosine hydroxylase gene leads to specific expression of Cre in catecholaminergic neurons. Cre recombinases recognize specific DNA sequences, so called loxP sites, to mediate specific DNA recombination events. DNA sequences between two loxP sites are called floxed DNA. Important for the recombination event is the orientation of the flanking loxP, i.e. DNA sequences between loxP sites oriented in the same direction will be excised, while DNA sequences between loxP sites oriented in the opposing direction will be inverted. Thus, expression of, for example, an inverted or inactive gene such as ChR2, due to a DNA insertion in the brain, can be activated specifically and only in Cre-expressing neurons, when the ChR2 gene is floxed. Thus, by use of cell-type specific promoters for expression of Cre recombinases and a floxed gene, a specific cell-type and signaling pathway can be switched on or off.

Box 12.7: RASSLs and DREADDs.

One approach, which has been developed in recent years, is to modify GPCRs in a way that they are not activated by their native, endogenous ligand, but are activated by a synthetic drug, which can be orally or intravenously supplied. RASSLs and DREADDs represent such GPCRs.

RASSLs are receptors activated solely by synthetic ligands. Various RASSLs have been developed. The D113S mutation in the $G_{i/o}$ coupled β_2 adrenergic receptor (β_2AR) is insensitive to endogenous catecholamines, but can be activated by ketone ligands and catechol ester, which do not have affinity to the wild-type β_2AR. β_2AR RASSLs were further developed by reengineering the ligand binding site, altering the agonist-induced down regulation and the phosphorylation of the GPCR by kinases and fusion of the modified GPCR to $G_{\alpha s}$. This receptor is also insensitive to β-adrenergic agonists, but can be activated by a non-biogenic agonist, L156870. β-adrenergic receptors are important modulators of, for example, smooth muscles, blood vessels, heart rate, and renin secretion from kidney and are, therefore, important drug targets for asthma and heart failure. $G_{i/o}$-coupled Ro1 and Ro2 receptors are derived from the human κ–opiod receptors and can be activated by nanomolar concentrations of the synthetic ligand spiradoline. The modified opiod receptors have decreased affinity for endogenous opioid peptides. Opiod receptors bind opiod-like compounds and are involved in pain perception, mood, and consciousness. Other RASSLs are derived from 5-HT_4 receptors, which is a G_s coupled GPCR. The single point mutation D100A in the transmembrane region TM-3 makes the GPCR insensitive to serotonin, but the receptor can still be activated by synthetic ligands. Replacing the C-terminus of the RASSL-5-HT_4 with the C-terminus of the G_q-coupled 5-HT_{2c} receptor increased G_q signaling, while replacement of the intracellular loop 3 within RASSL-5-HT_4 by the corresponding loop of the $G_{i/o}$ coupled 5-HT_{1A} receptors converts the GPCR to a $G_{i/o}$ coupled receptor. The 5-HT_4 receptor is expressed in the peripheral and central nervous system and has been suggested to play important roles in learning and memory, mood control, and gastronintestinal transit. The mutation F435A in TM6 in the human histamine H1 receptor, a $G_{q/11}$ coupled receptor, decreases the affinity for histamine and increases affinity for synthetic H1 agonists. This receptor is expressed in the central nervous system, the heart, smooth muscles, and endothelial cells. Drugs targeting this receptor are used as anti-allergy drugs.

DREADDs are designer receptors exclusively activated by designer drugs and represent the next generation of RASSLs. The clear advantages of DREADDs are that they are activated solely by synthetic ligands, have low constitutive activity, and cannot be activated by any known endogenous ligand. DREADDs have been developed by a random mutagenesis approach using the muscarinic acetylcholine receptor 3 (mAChR3). By introducing two point mutations (Y149C/A239G) into the human mAChR3, the G_q coupled GPCR can be activated by the pharmacologically inert ligand clozapine N-oxide (CNO) and is insensitive to Ach. Based on this approach, various DREADDs have been developed to create receptors with coupling preference for the $G_{q/11}$, $G_{i/o}$, G_s and G_s/G_q pathway (Wess 2013). In addition, a DREADD has been designed by introducing the R3.50L into the G_q-DREAAD (rM3Darr) to specifically activate arrestin signaling. Arrestins are involved in termination of the GPCR response. They bind to the activated GPCR after the phosphorylation by G protein-coupled receptor kinases (GRKs) and block further activation of the G protein signal via internalization of the GPCR. They are also involved in activation of G protein independent signal termination. The rM3Darr does not activate the G_q pathway anymore, but induced arrestin-2,3 dependent ERK1/2 phosphorylation. DREADDs have now been applied in many animal studies in particular in the brain to control GPCR pathways involved in seizure activity, amphetamine sensitivity, sleep and wakefulness, food intake, respiratory reflexes, fear conditioning, and depression. In addition, insulin release could be increased to improve glucose tolerance in obese mice.

12.4 Regulation of Protein-Protein interactions

Bacteria, fungi, plants, and animals do not only use light for photosynthesis and light sensing, but also to regulate physiological processes such as morphogenesis and circadian rhythms. Various light sensing systems in prokaryotes and eukaryotes have been developed based on sensing red light (i.e. phytochromes and bacteriophytochromes) and blue light (i.e. phototropins and cryptochromes).

12.4.1 Phytochromes

Phytochromes are activated by red and far red light (Figure 12.4b). The light-sensing molecule is a covalently bound linear tetrapyrrole, bilin chromophore. The chromophore is bound within GAF domain via covalent linkage to a cysteine residue. Together with the PAS and PH4 domain, the PAS, GAF, PH4 constitute the photosensory core which is sufficient for photoisomerization of the chromophore and subsequently the protein. Photoisomerization triggers the transition of the phytochrome between two conformational states. The Pr state absorbs red light (630 nm), while the Pfr state absorbs far red light (750 nm). Therefore, switching between red (630 nm) and far red light (750 nm) changes the conformation of the chromophore which leads to protein conformational changes. In the small flowering plant *Arabidopsis thaliana*, for example, the photoisomarization of a phytochrome (phytochrome (PhyB)) is used to control the interaction with a transcription factor (phytochrome interaction factor 3 (PIF3)). PhyB and PIF3 only interact during exposure to red light, but not in the dark or infrared light. The interaction

(a) BLUF

blue light

extracellular

intracellular

FAD
BLUF — Ac   450 nm
FAD   ⇄   cAMP
BLUF — Ac   dark

(b) phytochromes

red light
650 nm   750 nm

extracellular

intracellular

PCB 650 nm PCB
PhyB   ⇄   PhyB — PIF3
Pr state 750 nm Prf state

PIF3

(c) LOV

blue light

extracellular

intracellular

450–
FMN 470 nm   FMN
LOV2   ⇄   Jα — LOV2
Jα   dark

(d) CRY

blue light

extracellular

intracellular

FAD 488 nm FAD
CRY2   ⇄   CRY2 — CIB1
dark

CIB1

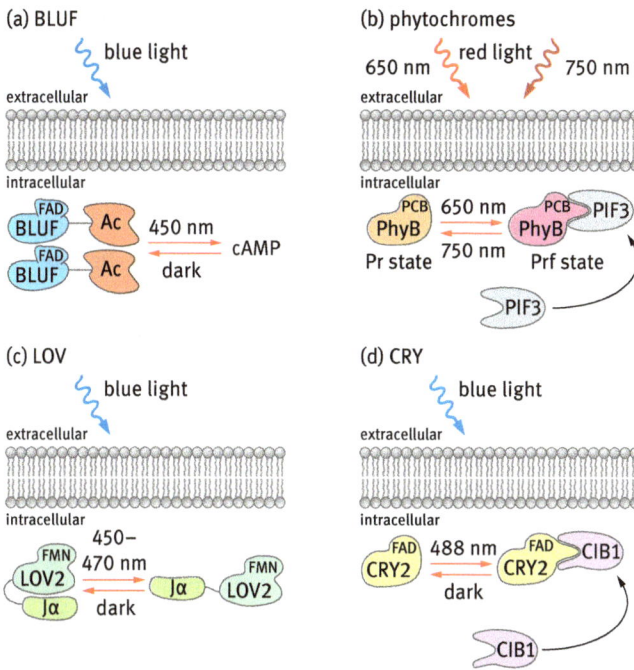

Fig. 12.4: Various light sensing systems in plants, fungi, and bacteria have been developed, which can be used to control protein-protein interactions and the activity of proteins. (a) BLUF (blue light sensor using FAD) domains use flavin adenine dinucleotide (FAD) to detect blue light. BLUF domains are usually dimeric and can be covalently linked to effector domains. For optogenetic applications, BLUF domains coupled to catalytic, enzymatic like, protein domains involved in for example cyclic nucleotide metabolism (i.e. adynylyl and guanylyl cyclases). For example, bPAC from *Beggiatoa* is a 350 aa long protein with a single BLUF domain, which is C-terminally linked to a Type III adenylyl cyclase. In contrast, euPACa is 1019 aa long and contains 2 BLUF and two adenylyl cyclase domains. (b) Phytochromes are activated by red and far red light. The photosensory core domain of phytochromes binds covalently a light-sensing molecule, a linear tetrapyrrole chromophore, which can switch the phytochrome between two conformational states (Pr state and Pfr state) depending whether red or far red light is used. In *Arabidopsis thaliana,* the photoisomerization of phytochrome PhyB is used to control the interaction with the transcription factor PIF3. For optogenetic application, PhyB and PIF3 can be linked to different proteins or different cellular compartments to recruit proteins to the plasma membrane or to control protein-protein interactions. (c) LOV (light-oxygen-voltage) domains use flavin mononucleotides (FMN) to sense blue light. LOV domains are found in phototropins, which are plant photoreceptors. In *Avena sativa,* light-activation of phototropin 1 induces the conformational rearrangements between the LOV2 domain and the C-terminally located Jα helix domain. The unfolding mechanism between LOV2 and Jα has been used in various applications to control gene transcription, toxin block of ion channel and activity of small G proteins. (d) CRY (cryptochromes) use flavin adenine dinucleotide (FAD) to detect blue light. CRY2 from *Arabidopsis thaliana* interacts with the transcription factor CIB1 when blue light is switched on. In optogenetic applications, CRY2 and CIB can be fused to proteins of choice to control protein-protein interactions involved in for example gene transcription and membrane recruitment of proteins.

involves the 650 aa long photosensory core domain of PhyB and the 100 aa long binding domain of PIF3. Therefore, the system can be used to control the interaction of two proteins by red and far-red light, when fused to the PhyB and PIF3 domains and addition of the chromophore phycocyanobilin. Indeed, the system was used to recruit proteins to the plasma membrane. Localization of the PhyB domain at the plasma membrane could control the cellular trafficking of a PIF3 domains-YFP fusion construct between the cytoplasm and the membrane by switching between 650 nm and 750 nm light. The system could also be used to recruit and activate Rho GTPAse at the plasma membrane and controlling the actin skeletal dynamics in living cells. Rho GTPases are small G proteins and belong to the Ras superfamily. Rho GTPases are important for regulating cytoskeletal dynamics, cellular movement, mitosis, and phagocytosis.

12.4.2 Phototropins

Phototropins are plant photoreceptors that contain light-oxygen-voltage (LOV) domains to sense blue light using flavin nucleotides as chromophores (Figure 12.4c). Phototropins are light-regulated serine/threonine kinases, which comprise of two LOV domains (LOV1 and LOV2) and the serine/threonine kinase domain. The kinase activity is induced by blue light-absorption of the LOV2 domain. In the cereal grain oat *Avena sativa*, light-activation of phototropin 1 causes conformational rearrangements between the LOV2 domain and the C-terminus, which contains a α helix domain designated Jα helix. LOV2 and Jα helix interact in the dark and dissociate during blue light application. Thus, fusion of LOV2 and Jα domains to protein domains of interest can regulate protein interaction by blue light. This approach has been used to photoactivate a mutant Rac1 (PA-Rac1). Mutant Rac1 was only active when LOV2-Jα were not interacting. Rac1 also belongs to the Rho family of GTPases and regulates various cellular processes such as cell-cycle, cell adhesion and cell motility. Rac1 is involved in the development of various forms of cancer. The light-controlled PA-Rac1 has been used to control cell mobility in cell culture systems and to block the cocaine-induced plasticity changes in neurons of the nucleus accumbens in mice, a brain region associated with drug addiction and pleasure.

The light-controlled interaction between FKF1 and nuclear protein GIGANTEA (GI) regulates the day-length dependent photoperiodic flowering in *Arabidopsis thaliana*. FKF1 contains the blue light-detecting LOV domain. 450 nm light induces the covalent binding of flavin mononucleotide (FMN) to a cysteine residue in FKF1, which then interacts with GI within minutes. Dissociation of the FKF1-GI complex is caused by the hydrolyzation of the cysteinyl-flavin bond. This process is very slow and can take several hours at least in heterologous expression systems. Mutated forms of FKF1 and GI, which localize to the cytoplasm were used to control the membrane recruitments of Rac1 and lamellipodia formation. In addition, this system was used to control gene transcription by controlling the interaction between the DNA binding domain (Gal4) and the activation domain (VP16) by fusion of GI to Gal4 and FKF1 to VP16. 450 nm light induced the complex formation of FKF1-VP16/Gl-Gal4 and transcription of the reporter gene luciferase.

The phototropin 1 derived LOV2 domain was also used to create TULIPs (tunable, light-controlled interacting proteins tags). TULIPs can be used to target and localize specific proteins to a subcellular location and to control specific signaling pathways. The TULIP approach is based on the idea that PDZ domains will recruit interacting peptides to a specific subcellular structure. PDZ domain containing proteins are scaffolding proteins and are involved in anchoring plasmamembrane proteins to the cytoskeleton and to organize signaling complexes. For example, PSD95, a neuron specific protein involved in organization of the postsynaptic, glutamatergic synapses, while InaD clusters the visual signaling cascade in the *Drosophila* eye. PDZ domains are small globular domains, which bind C-terminal peptides of proteins in subcellular structures. The Erbin PDZ domain has been used to engineer chimeric, clam-shell-like ePDZ versions, which bind peptides within the clam-shell. Thus, fusion of the ePDZ binding peptide to the C-terminus of LOV2-Jα (LOVp) allows the release (uncaging) of the peptide by light to interact with the ePDZ clam-shell. The approach was used to recruit the cytoplasmic ePDZ to various subcellular locations in yeast (i.e. plasma membrane, eisosomes, and bud neck) and mammalian cells (i.e. plasma membrane and mitochondrial outer membrane). By fusion ePDZ to the scaffolding protein Ste5, which orchestrates the activation of the MAPK pathway in yeast and anchoring the LOVpep to the plasma membrane allows plasmamembrane recruitment of Ste5, MAPK, and Cdc42 pathway activation. Both pathways are recruited to the plamamembrane during activation of a GPCR by a yeast pheromone, which initiates mating. Thus, light-mediated recruitment of Ste5 to the plasmamembrane, activated MAPK, and Cdc42 pathways and mating related phenotypes.

Another approach using LOV2 domains was to control protein stability with light. Here, the LOV2-Jα domain was fused to a protein degradation-inducing sequence called degron. When fused to the C-terminus of a protein of choice, the 37 aa long degron peptide of the ornithin decarboxylase from mice induces proteasomal degradation. The degradation mechanism is conserved in fungi, plants, and vertebrates. Fusion of a truncated 23 aa long degron peptide to the C-terminus of LOV2-Jα N-terminally tagged with the red fluorescent protein (RFP) was sufficient to degrade itself. The construct is called (RFP)-PSD for photosensitive degron and is sufficient to degrade proteins within 4 h. The N-terminal fusion of the PSD module to the protein of choice induced light-dependent degradation of proteins in different assays and proteins. Fusion of PSD to different yeast proteins allowed control of cell cycle and cell growth in yeast.

LOV2-Jα has also been used to control the activation of a Ca^{2+} channel Orai1, which plays an important role in activation of T-lymphocytes. Orai1 is activated by

Stim1 via relocation to the plasma membrane. LOVS1K is a fusion product of the C-terminal fragment of Stim1 (233–450). LOVS1K reversibly translocates to the plasma membrane to activate Orai1 by blue light.

In another approach, the LOV2-Jα domain was expressed extracellular via anchoring the light-activated domain to the transmembrane domain of the mCherry tagged PDF-receptor. The LOV2-Jα domain was coupled with different peptide toxins to block ion channels. With this Lumitoxin approach, fusion of different versions of the mamba snake toxin α-dendrotoxin and the toxin Conkunitzin-S1 from the cone snail *Conus striatus* allowed toxin block for the voltage gated K^+ channels Kv1.1, Kv1.1 and Shaker with toxin specificity.

12.4.3 Cryptochromes

Cryptochromes (CRY) are photolyase-like photoreceptors involved in circadian rhythms in animals and plants, growth and development in plants, and magnetoreception in birds. Cryptochromes are flavoproteins and are sensitive to blue light (Figure 12.4d). The chromophore of cryptochromes is FAD. Absorption of blue light by the oxidized FAD leads to formation of a radical, the activated signaling state, followed by the reduction to $FADH^-$ during green light exposure. $FADH^-$ reoxidizes to FAD in the dark. CRY2 from *Arabidopsis thaliana* comprises of two domains, the approximately 500 aa long, N-terminal PHR (photolyase-homologous regions) domain, which non-covalently binds FAD and another pterin chromophore and the 110 aa long C-terminal CCE domain (Cryptochrome C-terminal Extension). CRY2 interacts in a blue light-dependent manner with the transcription factor CIB1 (CRY-interacting basic helix loop helix (bHLH) protein). The interaction between CRY2 and CIB1 is induced within seconds and reverses within minutes. Mutations of predicted nuclear localization signals in CIB1 and CRY2 leads to cytoplasmic localization of both proteins in mammalian heterologous expression systems. Recruitment of cytoplasmic CRY2 to the plasma membrane could then be repetitively induced by a blue (488 nm, 25 μW) light pulse within seconds when the C-terminally truncated, plasma membrane attached version of CIB1 (CIBN) was coexpressed in mammalian cells. Dissociation from the plasma membrane occurred within 12 min. This system was further used to control the transcription of reporter genes using Gal4 system (see above) and a split Cre-recombinase, which could be recombined by light. Here, the N-terminal part of Cre-recombinase was fused to CRY2, while the C-terminal part was fused to CIBN. Blue light induced the functional reconstitution of Cre and the expression of a floxed GFP. The Gal4-based system could also be established in zebra fish.

The light-controllable interaction between CRY2 and CIB has also been used to engineer LITEs (light-inducible transcriptional effectors). LITE uses TALEs (transcription activator-like effectors) from the bacterium *Xanthomonas* fused to CRY2 and a transcription effector domain (TED) fused to CIB1. TALE-CRY2 will bind to a specific promoter region of a gene of interest and will recruit TED-CIB1 to the transcriptional start site of the promoter once light is switched on to modulate and/ or induce expression of the gene. The system promises to be designed for every gene of interest and can be combined with activating or repressing effectors to increase or suppress transcription *in vivo*. Using VP64 as a TED, various genes could be upregulates such as Neurog2 and Grm2 in cell culture systems, cortical neurons in culture, and *in vivo*.

In another approach, the inositol 5-phosphatase domain of OCRL (5-ptase; also known as the Lowe oculocerebrorenal syndrome protein) was fused to CRY2 to control the lipid components of the cell membrane phosphoinosides $PI(4,5)P_2$ and $PI(3,4,5)P_3$ by light. CIB localized to the plasma membrane by fusion to a targeting motif. Blue light induces the fast (within 10 sec) recruitment of 5-ptase to the plasmamembrane. Membrane localization reversed within 10 min after light application. Light induced recruitment of 5-ptase reduced $PI(4,5)P_2$ at the plasmamembrane with similar kinetics and as a functional consequence reduced KCNQ2/3 currents in heterologous expression systems. The KCNQ2/3 channel, also called M channel, is an important modulator of neuronal excitability. The activity is regulated via GPCR induced depletion of $PI(4,5)P_2.$

12.5 Therapeutic applications

Optogenetic technology is currently used extensively to investigate signaling cascades and neuronal circuits to understand basic biological principles in health and disease (Figure 12.5). However, the technology also provides the opportunity to develop new therapeutic strategies to replace and improve currently applied techniques such as deep brain stimulations and to cure diseases such as blindness.

The first problem in optogenetic therapy is how to deliver optogenetic tools to the organs, tissues or cell-types, which shall be controlled and cured. Currently, two approaches are being considered, i.e. a virus and a cellular approach.

12.5.1 Delivery of optogentic tools to intact systems

Adeno-associated viruses (AAV) are currently developed for gene therapy for several reasons: AAVs have not been associated with human disease, have a small genome, which can easily be genetically modified, infect many different dividing and non-dividing cells and cell-types, and mediate long-term transgene expression, without integration into the genome. AAVs need for efficient viral replication and propagation a helper virus such as adenovirus or herpes virus. Therefore, for production of infectious human AAV virions, helper free AAV production systems have

1. choose cDNA opsin variant

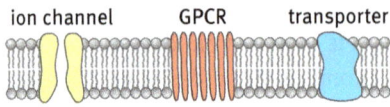

ion channel   GPCR   transporter

2. create AAV floxed vector construct

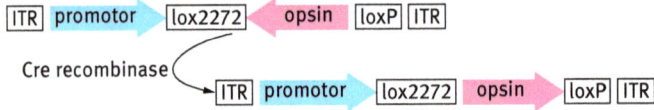

| ITR | promotor | | lox2272 | | opsin | loxP | ITR |

Cre recombinase

| ITR | promotor | | lox2272 | opsin | | loxP | ITR |

3. produce AAV

pHelper   pAAV-opsin   pAAV-RC

HEK$_{293}$ (E1) cells

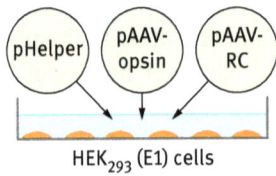

4. stereotactically inject virus into brain or other tissue and implant light-guide

AAV

Cre

fiber optic

Cre

5. control the cell-type and signaling pathways of interest by light

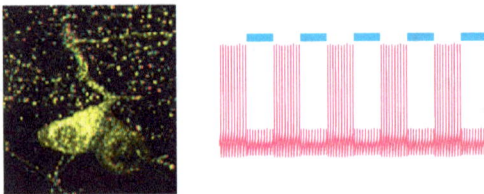

6. monitor and analyze physiological and behavioral read-out

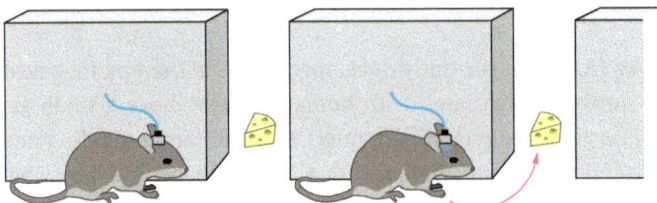

◄ Fig. 12.5: To perform an optogenetic experiment, six steps have to be considered. 1. The experimenter has to choose an optogenetic probe, depending on whether a neuron should be switch on or off or modulated over long time periods. 2. Once a certain opsin has been selected, the cDNA encoding the opsin has to be genetically engineered into an adeno-associated virus vector. For cell-type specificity, the opsin will be flanked with loxP sites, such as loxP or lox2272 to inactivate the opsin gene. The loxP sites are recognized by Cre recombinases. The Cre recombinase can be expressed in the cell-type of choice such as a certain neuronal circuit, and will recombine (in the example shown here invert) the cDNA of the opsin to create a functional gene, which is then transcribed and translated into the light-activated protein. For robust initiation of the transcription, a promotor is used. ITRs are AAV terminal repeats. These are cis elements required for AAV replication, packaging, integration, and rescue. 3. Once the floxed AAV opsin carrying vector has been engineered, AAV virus is produced. The current procedure for creating AAV is to separate the information necessary to produce a complete virus onto different expression vectors (plasmids) and the host cell. This guarantees the safe production of the virus, which can only be replicated in the specifically engineered HEK 293 cell line. 4. The purified opsin carrying AAV is stereotactically injected into a brain area of choice and a fiber optic for light-delivery is implanted. 5. 7–14 days after injection the Cre mediated induction of opsin expression can be visualized (left). In the example shown on the right, light activation of the neuronal circuit leads to inhibition of the action potential firing of a recorded neuron in the brain. 6. After verification of the functional expression of the light-activated protein, physiological and behavioral assays can be performed to investigate for example the role of a certain signaling pathway or cell-type in anxiety behavior.

been developed, where the necessary genes normally provided by the adeno- or herpes virus are supplied by a helper DNA plasmid and by the host cells. These recombinant, engineered AAVs (rAAV) are not able to replicate in the infected tissues/cells and have, therefore, a good safety profile.

Various AAV serotypes (AAV1–5 and AAV7–9) and over 100 variants have been isolated from adenovirus stocks and human and primate tissues. The different serotypes differ in their transduction efficiency of different cell-types and different tissues. For example, transduction efficiency in kidney is optimal for AAV2, in heart and pancreas for AAV8 and in the eye for AAV4 and AAV5 (Grieger et al. 2006). Therefore, depending on the application/disease and cell-type/tissue specific location of the disease, a certain serotype can be selected. Among the different serotypes AAV2 is the best characterized, because it was the first AAV, which could be cloned into a vector DNA. Recombinant AAV2 have been tested in various preclinical trials involving cystic fibrosis and rheumatoid arthritis patients.

A second strategy to deliver foreign DNA into human tissue and cell-types is called cell therapy. In cell therapy, a molecular engineered human cell line such as HEK293 cells or a stem cells can be used to express gene and gene cascades of interest. Such an approach has been used to engineer a light-induced, melanopsin dependent gene transcription device in cells, which were implanted in type II diabetic mice to control glucose levels by light.

12.5.2 Cell-type specificity

One important aspect of using optogenetic approaches for gene therapy is that only a limited number of cells and a specific type of cell should be manipulated and infected by AAV to reduce possible side effects of unspecific expression of the opsins. Cell-type specificity is achieved via specific promoters, which drive the expression of a certain gene specifically in a subpopulation of cells (see Box). These promoters have to be small (< 2 kb of size), because of the limited capacity of the AAV virus. Various promoters have been used so far for expression of opsins in the brain. To achieve expression in every cell-type the most frequently used promoters are those of CMV (cytomegalovirus) and EF1α (Elongation factor 1-α), a protein involved in translation. These promoters are used to induce robust expression in almost every cell-type and are often combined with the Cre/lox system (see above). Unfortunately, only a limited number of cell-type specific promoters have been developed for virus approaches. Syn1 (synapsin 1, a protein involved in the regulation of neurotransmitter release) and CamKII (Ca^{2+}/calmodulin-dependent protein kinase II, a protein famous for its role in plasticity in the brain) promoter are commonly used for specific expression in neurons and some promoters such as the cone arrestin 3 promoter have been developed for cone photoreceptors in the eye. It will be important for the application of gene therapy that more specific promoters will be established or alternative strategies for cell-type specific gene delivery will be developed.

12.5.3 The cure of blindness: Retinitis pigmentosa

The most obvious application of light-activated proteins in disease is blindness. Blindness is often associated with malfunction and degeneration of photoreceptor cells. Retinitis pigmentosa is the most common disease with retinal degeneration in humans, which affects about 3 million people world-wide. It is an inherited disease with heterogeneous genetic origin and can affect mutations in any of more than 60 genes. The disease is associated with the degeneration of the retina, starting with photoreceptor cells. The disease and disease phenotypes progress slowly over time starting in childhood or early adolescence. The first signs of the disease are night blindness, which progresses to complete loss of vision. A therapy is currently not available, but mouse models have been developed, the rd mice, which mimic the human disease to explore possible treatments.

The retina of humans is the imaging processor of the eye. Light is sensed by two types of photoreceptor cells. Rod photoreceptors respond to light at low light intensities and are responsible for dim light vision, while cone photoreceptors respond to light at higher light intensities and are responsible for color vision. In photoreceptor cells, changes in light is transformed into modulation of transmitter release onto two types of bipolar cells (ON and OFF), which respond to light either with an increase (ON) or decrease (OFF) in synaptic transmitter release. Bipolar cells are coupled to ON and OFF ganglion cells, respectively, which encode the information received from bipolar cells into AP firing frequency. This information is sent to higher brain centers, i.e. the visual cortex. It is important to note that the information processing within the retina is much more complex than described above, since more than 60 cell-types are involved for parallel processing of visual images.

Degeneration in retinitis pigmentosa affects first rods and than cones. Therefore, in order to restore vision non-degenerated cones, bipolar cells or ganglion cells could be made light sensitive to restore vision. Various strategies and approaches have been applied to achieve this goal in animal models using ChR2, NpHR2, and melanopsin. These opsin variants are all bistable opsins, which regenerate the light-activated chromophore within the receptor protein. Therefore, supply of all-*trans* or 11-*cis* retinal in the eye should be sufficient for continuous light responses. In a chemical-optogentic approach LiGluR in combination with its organic light-sensitive ligand has been used. LiGluR is a light-gated engineered glutamtate receptor (GluR), where a synthetic caged glutamate compound is tethered to the outside of the channels and can be reversibly released by light. Activation of LiGluR leads to the depolarization of the cell.

Expression of opsins in cones: In various blind rd patients, cone photoreceptor are not completely degenerated. A light response of cones results in the hyperpolarization of the photoreceptor cell. Therefore, expression of the Cl^--pump NpHR should result in hyperpolarization of cones and induce cone mediated light responses. In fact, expression of NpHR in cone photoreceptors in rd mice restores light sensitivity in the cone pathway (Busskamp et al. 2010). Photoresponses in cones are faster than in wt mice, because the naturally occurring cone opsins are GPCRs activating a second messenger pathway, while the transport of Cl^- via NpHR is much faster and immediately changes the resting membrane potential. Light-activation of NpHR expressing cones results in ON and OFF responses and lateral inhibition, which is important for spatial contrast of RGCs comparable to responses in wt mice. These results indicate that the visual cone pathway is intact and can be recruited by light-activation of NpHR, which indeed results in corresponding light response activity in the visual cortex and visual guided behavior. Rd mice expressing NpHR in cones but not blind mice avoid bright areas and can follow moving visual cues. NpHR was also tested for possible human therapeutic application. Expression of NpHR in human *ex vivo* retinas renders light sensitivity specifically to cones, when the promoter fragment of cone specific arrestin-3 was used (Busskamp et al. 2010). Achieving cell-type specificity in human gene therapy is important to exclude and minimize side effects of the transgene expressed in other cell-types.

Expression of opsins in bipolar cells: On-bipolar cells respond to light with a depolarization. ChR2 induces immediate depolarization of neurons and can, therefore, be used to mimic activation of on-bipolar cells by light. Indeed, expression of ChR2 specifically in on-bipolar cells induced light-evoked responses in RGCs and in the visual cortex. Animals are also able to increase the locomotor activity during light application and to detect visual cues.

Expression of opsins in ganglion cells: Ganglion cells in the retina integrate the visual stimuli from cones and rod and convert the information into changes in AP firing. ON ganglion cells increase their activity during light, while OFF ganglion cells decrease their activity during light. Thus, expression of optogenetic tools such as ChR2, LiGluR or melanopsin in ON ganglion cells should be used to depolarize the neuron, while expression of NpHR in OFF ganglion cells should be used to inhibit these cells. Indeed, various approaches have been tested in RGCs. Expression of ChR2, Melanopsin, and LiGluR in ON and OFF RGC restores light responses in RGCs and the visual cortex. In addition, expression of the different light-activated proteins restores/improves the pupillary reflex in the blind mice. In different behavioral assays, mice expressing light-activated proteins in RGCs could discriminate a light stimulus form the dark, were able to avoid light in the open field test and could detect visual cues. Interestingly, coexpression of ChR2 and NpHR2 in RGCs restored the ON and OFF responses of ganglion cells, i.e. ChR2 expression RGCs are activated by light, while NpHR2 expressing RGCs are inhibited by light. ChR2 expression in RGCs in blind mice was also used in combination with a prosthetic system that uses light pulses to mimic the retinal neuronal code when a picture is presented. By using this prosthetic device, RGC firing pattern are similar to firing patterns from wild-type mice. In addition, this method restores the tracking of objects in blind mice, which is not the case in ChR2 expressing blind mice without the prosthetic system. The study suggest that a combination of a prosthetic device (e.g., engineered goggles) encoding the optogenetic control of ON and OFF responses in blind patients might be currently best suitable to regain perfect vision.

12.5.4 Application to other diseases

Various studies have been performed in animal models to control neuronal circuits involved in human diseases such as Parkinson's, schizophrenia, addiction, obesity, sleep disorders, epilepsy, depression, and anxiety. The basic strategy in animal models is currently to activate, silence or modulate specific neuronal pathways, which cause alterations in disease associated behaviors. This strategy should eventually identify the neuronal cell-type involved in the disease process. In the reverse approach, the cell-type or circuit is then controlled to improve or reverse the disease phenotype. Once the disease circuit and involved cell-type have been identified, they can then be targeted with therapeutic approaches such as drugs, electrical stimulation (e.g., deep brain stimulation (DBS)) or viral expression of pharmacogenetic and/or optogenetic tools. There are indeed main advantages of using optogenetic approaches to control disease states. Optogenetic approaches allow for precise temporal and spatial control of neuronal signaling events in the millisecond to second time range, while drugs act on a very long time scale, mostly with low specificity and throughout the body. DBS can target a specific area in the brain with high temporal resolution, but it is impossible to control only one specific cell-type by this approach. In the following section, we will give some examples of the exciting work, which has been performed to understand, improve or reverse disease phenotypes.

Anxiety disorders represent the most common class of psychiatric diseases. Various small neuronal circuits have now been identified using optogenetic techniques in the dorsal raphe nuclei (DRN) and in the amygdala, which control anxious states. For example, control of GABAergic neurons in the DRN by melanopsin to activate $5\text{-HT}_{2c}/G_q$ coupled signals relieves anxiety in mice (Spoida et al. 2014). The experiments suggest that controlling only a small number of 5-HT_{2c} receptors by light in defined subpopulation of neurons in the midbrain is sufficient to control anxiety.

Parkinson's disease (PD) is the second most common neurodegenerative disorder. PD is associated with the degeneration of dopaminergic neurons originating in the substantia nigra, a part of the basal ganglia located in the midbrain. Common treatments of the disease are to increase dopamine levels in the brain or to stimulate basal ganglia circuits, such as the subthalamic nucleus (STN) using deep brain stimulations (DBS). In an optogenetic approach, it could be demonstrated that the light-control of layer 5 cortical neurons in the primary motorcortex projecting to the STN were sufficient to improve parkinsonian's symptoms in a PD rat model. The results suggest a new treatment avenue for improving parkinsonian's symptoms by controlling cortical neurons, located close to the surface of the brain, rather than controlling deep brain structures.

Epilepsy is characterized by epileptic seizures. There are various forms of epilepsy, which affects about 1% of the world population. The occurrence of epileptic seizures is related to an imbalance between excitatory and inhibitory circuits in the brain resulting in hyperexcitablity. Therefore, optogenetic tools can be used to control and/or counteract the imbalance of excitatory and inhibitory circuits to prevent or stop seizures. In fact, optogenetic silencing using NpHR of excitatory hippocampal neurons is sufficient to delay and reduce seizure activity in rodent models of epilepsy (Kokaia et al. 2013). Seizure activity occurs in many different brain areas, such as the thalamus and cortex, and therefore, optogenetic applications in animal models will dissect the cell-types and circuits, which are sufficient to prevent seizure activity. The identification of these cell-types in certain brain areas will be highly beneficial for future therapeutic treatment.

Spinal cord injury is associated with the malfunction of various important neuronal circuits involved in regulating breathing, bladder control, gut movement, and

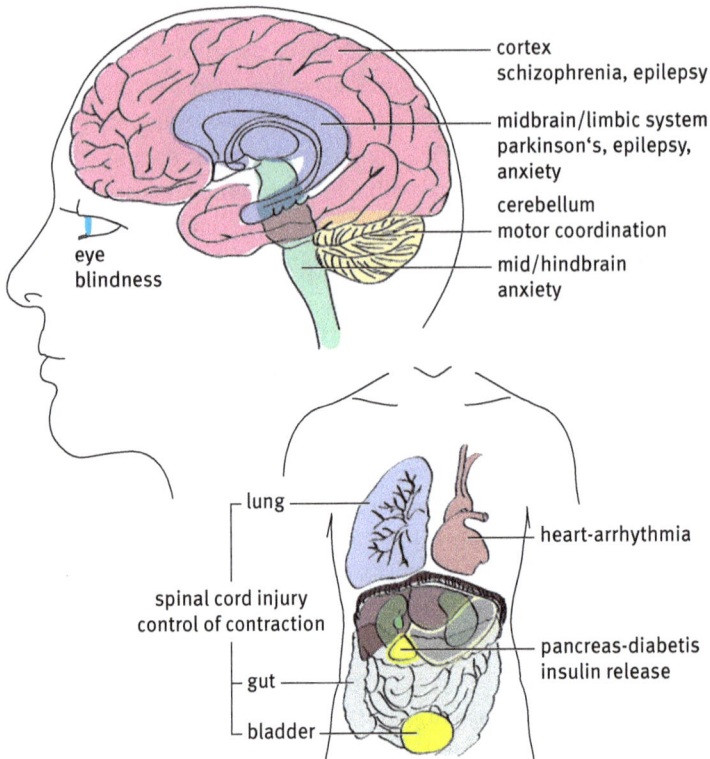

Fig. 12.6: In mice, rats, and non-human primates, optogenetic tools have been used in different brain regions and tissues to understand and control neuronal circuits and cell-types underlying disease. For example, controlling neuronal circuits in the cortex may help to alleviate symptoms in schizophrenia and to block epileptic seizures. Expressing of light-activated probes in the midbrain/hindbrain and limbic system have been used to understand Parkinson's disease and to relieve anxiety. Expression of light-activated proteins in the cerebellum is used to influence motor control and motor learning. The most promising application of light-activated proteins is the eye and the cure of blindness. Outside the brain light-activated proteins have a great potential to be used in spinal cord injury to control the contraction of muscles involved in breathing, gut, and bladder control. In addition, light-activated proteins have been used to control the heart contraction and the insulin release in fish and mice.

sexual organ function. One of the main leading causes of death in spinal cord injury is the loss of respiration. Loss of respiratory function occurs when the spinal cord is injured at the cervical level, which removes the respiratory drive into the sphrenic motor neuron pool. The sphrenic motor neurons control directly the contraction of the diaphragm, which draws air into the lung. In a rat model of spinal cord injury, which underwent a cervical spinal cord lesion, expression of ChR2 in the sphrenic motor neurons pool was sufficient to restore contraction of the diaphragm over long time periods (Alilain et al. 2009). The application of optogenetic tools for spinal cord injury has a great therapeutic potential. The neuronal circuits underlying for example breathing and urination are relatively easy to control. AAV carrying optogenetic probes can be precisely delivered to the corresponding spine segment and light sources can be implanted without major surgery.

Optogenetic approaches have not only been applied in brain neuronal circuits, but have also been used in other tissues such as the heart and to control pain. In principal, optogenetics can be applied in every tissue and organ to control for example insulin release, tumor growth, gut movement, kidney and liver function. The therapist just has to select a signaling pathway, which should be controlled by light. The possibilities are endless. However, for successful application of optogenetic gene therapy several issues have to be addressed: First, it has to be shown that in particular microbial opsins do not cause any immune responses or secondary effects in the host cell, which should be less problematic using human opsins. Second, for optogenetic approaches particularly in the brain, devices for light delivery have to be developed. In particular, wireless and multisite stimulation approaches to control LEDs or optical fibers would be beneficial for many applications. While the road to optogenetic gene therapy in humans is long, the optogenetic control of specific cell-types and signaling pathways involved in disease has the great potential to improve health and cure diseases, considering the fact that the GPCR superfamily is one of the main targets in drug discovery (Figure 12.6).

Key-terms

Adeno-associated virus, adenylyl cyclase, ARCH, blindness, BLUF domain, channelrhodopsin, Cre/loxP, cryptochromes, DREADDs, G protein-coupled receptor, gene therapy, halorhodopsin, LOV domain, opsin, melanopsin, optogenetics, pharmacogenetics, phytochromes, RASSL, rhodopsin.

Questions

- What are the main differences between optogenetics and photopharmacology?
- What are RASSLs and DREADDs?
- Which microbial opsins can be used to activate or silence neurons?
- Which membrane ion fluxes cause hyperpolarization and which depolarizations?
- Which are the main G protein coupled receptor pathways?
- What are the differences between the invertebrate and vertebrate visual signaling cascades?
- What are possibilities to control adenylyl cyclases?
- What are the current limitations for therapeutic applications of optogenetic tools?

- How is cell-type specific expression of optogenetic tools achieved?
- What is AAV?
- What are phytochromes and cryptochromes?
- How are LOV domains used to control protein-protein interactions?

Further readings

Kramer R. H., Mourot, A. & Adesnik, H. 2013. Optogenetic pharmacology for control of native neuronal signaling proteins. *Nat Neurosci* 16: 816–823.

Moglich, A., Yang, X., Ayers, R. A., et al. 2010. Structure and function of plant photoreceptors. *Annu Rev Plant Biol* 61: 21–47.

Neves, S. R., Ram, P. T. & Iyengar, R. G 2002. protein pathways. *Science* 296: 1636–1639.

Sahel, J. A. & Roska, B. 2013. Gene therapy for blindness. *Annu Rev Neurosci* 36: 467–488.

Tye, K. M. & Deisseroth, K. 2012. Optogenetic investigation of neural circuits underlying brain disease in animal models. *Nat Rev Neurosci* 13: 251–266.

http://www.openoptogenetics.org for an update on optogenetic constructs.

References

Alilain, W. J., & Silver, J. 2009. Shedding light on restoring respiratory function after spinal cord injury. *Front Mol Neurosci* 2: 18.

Busskamp, V., Duebel, J., Balya, D., et al. 2010. Genetic reactivation of cone photoreceptors restores visual responses in retinitis pigmentosa. *Science* 329: 413–417.

Grieger, J. C., Choi, V. W. & Samulski, R. J. 2006. Production and characterization of adeno-associated viral vectors. *Nat Protoc* 1: 1412–1428.

Kokaia, M., Andersson, M. & Ledri, M. 2013. An optogenetic approach in epilepsy. *Neuro-pharmacology* 69: 89–95.

Masseck, O. A., Spoida, K., Dalkara, D., et al. 2014. Vertebrate Cone Opsins Enable Sustained and Highly Sensitive Rapid Control of Gi/o Signaling in Anxiety Circuitry. *Neuron* 81: 2014. 1263–1273.

Spoida, K., Masseck, O. A., Deneris, E. S., et al. 2014. Gq/5-HT2c receptor signals activate a local GABAergic inhibitory feedback circuit to modulate serotonergic firing and anxiety in mice. *Proc Natl Acad Sci USA.* 111(17): 6479–6484.

Wess, J., Nakajima, K. & Jain, S. 2013. Novel designer receptors to probe GPCR signaling and physiology. *Trends Pharmacol Sci* 34: 385–392.

Hermann Lübbert

13 Drug development and registration

In the fields of medicine, biotechnology, and pharmacology, drug discovery is the process by which new candidate medicinal products (drugs) are discovered. Drug development demonstrates the efficacy and safety of new drugs. Although the term is mostly applied in relationship to clinical and toxicological research, it also includes chemical development through which a process is established which renders manufacturing at large quantities of the drug substance (the active pharmaceutical ingredient) and the final drug product possible at the highest standards of quality and purity.

The drug approval is the authorization required to offer, distribute or deliver an industrially manufactured medicinal product. The drug approval process requires the submission of a dossier to the agency responsible for the territory for which the marketing authorization is desired, describing all required information and every known aspect of the drug and its production in an amount of detail that is greatly above that required for scientific publications. Many such agencies exist around the world, and each one of them follows its own rules, which may be greatly different from one another. This forces the pharmaceutical industry to create all data required to satisfy the relevant agencies around the world, an immensely costly endeavor. Once an approval has been obtained, the approval dossier needs to be updated regularly according to new data or just to follow the constant elevation of administrative requirements. Any such amendment of a dossier is costly and work-intensive.

The exact medical indication, the dosage, and the mode of application (posology) are described in the product information, called "United States Prescribing Information (USPI)" in the US and "Summary of Product Characteristics (SPC)" in Europe. Every statement in the SPC must be clearly supported by the data presented in the approval dossier. The package insert (Patient Information Leaflet or PIL) is a summary of the SPC, written for the patients, and therefore with reduced scientific depth. The use of drugs outside its "label" (the indications and posology described in the SPC of USPI) is called "off-label use". Even though it is strictly prohibited for drug companies to promote off-label use, in most countries it falls under the medical freedom and personal responsibility of physicians. However, if applied it may cause problems with reimbursement systems or litigation risks for the medical profession.

Since the early and "easy" starting points for drug discovery have been extensively exploited, drug discovery has become an extremely lengthy and expensive process which, despite all advances in technology and understanding of biological systems, only very rarely leads to a new, approved therapeutic discovery. In 2010, the research and development cost of each new molecular entity (NME) was approximately US$1.8 billion (Paul et al. 2010). Some authors assume even much higher costs for each new drug. Following the initial discovery, new drugs require very expensive Phase I, II and III clinical trials, and most of them fail. It is the role of research management to create successful research and development departments, select the medically and commercially most promising drugs based on the evaluation of data available at the discovery stage and on realistic market forecasts, and then allow the immense spending for their development.

The highly regulated drug approval process combines expertise in multiple aspects of preclinical science, toxicology, chemistry, medicine, legislation, pharmacoeconomy, marketing, and corporate strategy. Discovering and developing new drugs with a good risk/benefit relationship for diseases with a high medical need and manufacturing and distributing them at the optimal price/reimbursement levels is without any doubt among the most diverse, difficult, and legally most complex tasks in the entire commercial world. With constantly growing scientific and legal requirements for drug development, the pharmaceutical industry has become largely unsuccessful in developing and approving the number of new drugs required to warrant the future of large international companies. Thus, new strategies are discussed in pharmaceutical industry, governments, and academia on the interaction of these institutions in drug discovery and development.

In the context of all this, many and highly diverse jobs exist and are constantly created for well-trained, intellectually flexible young scientists. Listing just some areas where job opportunities arise may cover preclinical research, clinical research, quality management, regulatory affairs, medical science, pharmacoeconomy, reimbursement, pharmacovigilance, patent law, stock analysis in the financial world, advertisement, marketing, and sales. All these jobs, when created in the pharmaceutical and/or biotech sector, are mirrored by corresponding experts working in the government agencies. The usual academic education of biologists merely provides training for some aspects of preclinical research. Even this is approached differently in the pharmaceutical industry which is forced to take on increasingly interdisciplinary approaches. All additional expertise must be gained in specific courses or by experience in the job. Because of this, many aspects of drug discovery, development, and approval are covered by various articles in books or even Wikipedia, and this chapter is an attempt to pull many of the aspects together to provide a combined understanding of the requirements.

Thus, this chapter aims at describing the responsibilities and professional interactions of many scientists working in the pharmaceutical industry, providing orientation for further reading and aiding young biologists in choosing the professional direction that is best suited to their personal talents.

13.1 Historical background

The history of pharmaceuticals was driven by several discoveries with historical dimensions. Initially, up to the 19th century, 311 preparations, mostly plant extracts and inorganic molecules were listed in the *British Pharmacopoeia* of 1864. Drugs were discovered by serendipitous discovery. By the end of the 19th century, developments in biomedicine, chemistry, and the traditional apothecaries' trade came together, forming the start of modern pharmacology and the pharmaceutical industry. Initially, few apothecaries worked along the idea that the medical activity of plant extracts is related to individual molecules, a much ridiculed concept at the time. Convinced of their hypothesis, they started to identify the active ingredient from traditional remedies. Examples of drug compounds isolated from crude preparations are morphine, the active agent in opium, and digoxin, a heart stimulant originating from Digitalis lanata. In parallel, advances in dye chemistry provided new chemical structures which were soon tried as pharmaceuticals. Organic chemistry also led to the synthesis of many of the natural products isolated from biological sources. Chemical modifications of natural products led to major discoveries, such as acetylsalicylic acid (Aspirin), heroin (a chemical derivative of morphine), or the various derivatives of the ergot alkaloids.

Chemistry was thus driving pharmaceutical industry up to the 1970s. It was only in the course of the 20th century when pharmaceutical principles were developed in an attempt to explain the action of drugs. Only once an active substance had been identified was an effort made to identify its molecular target. This approach is known as "classical pharmacology", "forward pharmacology", or "phenotypic drug discovery".

Alongside his discovery of *Salvarsan* in 1910, Paul Ehrlich and others developed the principle of the physical interaction of a "drug molecule" with a "target molecule". When George Hitchings and Gertrude Elion began working together in 1944, their discovery of dihydrofolate reductase, an enzyme required for the synthesis of nucleic acids, was the starting point in their search for molecules that address a specific biochemical pathway. The approach was immensely successful in finding antibacterial, immunosuppressant and anticancer drugs. Subsequently, small molecules were synthesized to specifically target a known physiological/ pathological pathway, rather than adopt the screening of stored compounds in animals and humans. In parallel, the concept of receptors binding specific signaling molecules was developed and successfully integrated into drug discovery by James Black in his discovery of the beta-adrenergic antagonists *pronethalol* in 1960 and *propranolol* in 1964.

Later chemical libraries of synthetic small molecules, isolated natural products or extracts from living organisms were screened in intact cells or whole organisms to identify substances that have a desirable therapeutic effect. The availability of the sequence of the human genome and other technological advances allowed rapid cloning and synthesis of large quantities of purified proteins. Cloning of human proteins enabled the screening of large libraries of compounds against specific targets believed to be relevant for specific diseases. This approach is known as "reverse pharmacology" and is the most frequently used approach today. It has become common practice to use high throughput screening of large compound libraries against isolated biological targets hypothesized to be disease modifying. Hits from these screens are then verified in cellular assays and, if confirmed, in appropriate animal model. Following the identification of screening hits, medicinal chemistry will optimize the hits to increase the target affinity, selectivity (to reduce potential side effects), efficacy, metabolic stability (to increase the biological half-life in the body), and bioavailability with respect to the desired application. Once a compound fulfills all of these requirements, drug development may be started. Even at this stage, however, a new molecule is several years away from its first application to humans.

The use of biologically active compounds in humans poses inherent safety risks, and already in the mid-19th century restrictions on the sale of poisonous substances were imposed in the USA and UK. Several tragic accidents triggered the development of an increasingly restrictive regulatory environment, starting with the concept of "prescription only" drugs in the early 20th century. Already at that time, some drugs, for example heroin as the active component in juice aimed at reducing coughing in children, were prohibited in the USA. The use of diethylene glycol as solvent for sulfonamide preparations caused more than 100 deaths in the USA in 1937. Europe, which in the 20th century was previously preoccupied by the two world wars, was aroused by the disaster around thalidomide (marketed as Contergan) and initiated the development of strict regulations around drug discovery, development, approval, and manufacturing. Since then, all aspects of the pharmaceutical industry are regulated to an extent that only very few small companies have managed to cope with and that has pushed many drug companies out of business. Particularly in Europe, where each national regulatory body can add local rules on top of the European rules (thereby saving and creating jobs in local agencies) this can be extremely tiresome and hard to cope with. Nevertheless, the immense control that surrounds the drug business in every aspect is the consequence of several historical disasters, and reminds us that the pharmaceutical industry is working with biologically active, potentially dangerous compounds. Any drug with desired biological effects will also display unwanted effects and pharmaceutical companies must work together with medical agencies in defining acceptable risk/benefit ratios.

13.2 Drug discovery

13.2.1 Drug targets

Drug Targets are naturally occurring cellular or molecular structures that new drugs may bind to in order to modify the disease or its symptoms. The criteria a new drug target has to fulfil in order to be selected for a new drug discovery program is heavily debated within the pharmaceutical industry (Rask-Andersen et al. 2011). If a particular company sets rules for target selection that are too strict, it may miss important developments. Alternatively, its research management may waste financial resources by working on targets failing later in the development process if criteria are too loose. However, a full understanding of just how a target is related to a disease is not necessarily required to select "new" or "established" targets. In 2011, it was estimated that only 435 human proteins served as therapeutic drug targets for all FDA-approved drugs.

"Established targets" are scientifically well understood and their relevance is supported by a lengthy publication history or by other, perhaps less selective drugs working on the target. This does not necessarily imply that the mechanism of action of the other drugs acting through the target is fully understood. "Target validation" is the process of gathering functional information about a potential target. All previously available information reduces the cost of the development of new drugs on the target. Important is not only the relevance of a target for a disease, but also the "drugability" of the target, i.e. the likelihood of finding chemicals selectively binding to the targets. Most currently available drugs are directed towards human targets, about 17 % of all drugs target infectious organisms. In fact, the vast majority of human drug targets are receptors, enzymes, transporters, and ion channels. In essence, these are proteins that as part of their natural function bind ligands, substrates or naturally occurring toxins (Dollery 1999). Evolution has thus formed them such that small molecules can interact with reasonable affinity and selectivity (Box 13.1).

Box 13.1: Human drug targets.

A drug target is the molecular structure, in most cases a protein, to which a drug binds with a certain affinity. Targets of current drugs are (Dollery 1999):
36 % G-protein coupled receptors
9 % ionotropic receptors
9 % kinase-linked receptors
9 % nuclear receptors
38 % enzymes
11 % transporters
5 % ion channels
8 % miscellaneous

13.2.2 Screening and design

The next step in drug discovery, following target selection, encompasses the identification of "lead structures". These are most frequently found by high-throughput, fully automated screening (HTS) of compound libraries on assays with the chosen target. The assay design, often based on functional readouts from response systems linked to the target expressed in cell culture, is crucial for getting reliable hits with few false positive or negative responses. The compounds to be screened may include collections of chemical entities from within the company or purchased from commercial sources, derived from combinatorial chemistry approaches, or natural compounds. From 1981 to 2006, 63 % of the 974 new small molecules in drug development were directly derived from nature or semisynthetic derivatives of natural products (Warren 2011). For certain therapy areas, such as antimicrobials, antineoplastics, antihypertensive, and anti-inflammatory drugs, the numbers were even higher and, in many cases, these products had been used traditionally for many years. The high success rate of natural products is due to the fact that evolution has shaped them for specific binding to and functional interaction with a particular protein domain. In evolution, a limited number of protein domains has, through gene rearrangements, been used repeatedly in entirely different frameworks. Thus, secondary metabolites synthesized in plants, bacteria, or fungi for a particular biological reason may display totally different functions upon interaction with similar domains in human proteins (Newman et al. 2007).

Despite the implied potential, only a minute fraction of Earth's living species has been tested for bioactivity and it may seem surprising that most pharmaceutical giants resigned from working with natural compounds. Reasons are the difficulty of screening extracts from whole organisms which frequently generate false positives or mixed responses that disappear during compound isolation, and the difficulty of manufacturing the compounds later. Many natural compounds display complicated structures which can hardly be synthesized in the required amounts of tons, and growing their natural hosts may be tedious or impossible. Just imagine the isolation of a ton of a secondary metabolite from Mediterranean sponges! Thus, in spite of the enormous historic success of natural compounds and the fact that they cover a chemical space unreachable by synthetic compounds, research in them has been decreasing drastically in the last two decades.

Which criteria a lead structure has to fulfil in addition to binding to the chosen target is a matter of constant debate and the term lead is, therefore, defined differently in different companies (Leeson et al. 2007; Feher 2003). It is, however, common practice to call a compound "lead" only once it fulfils limited criteria that may, in addition to the affinity towards the target, be directed at selectivity (i.e. affinity to proteins similar to the target), bioavailability (i.e. the uptake of the drug into the body with the desired application), pharmacokinetics (i.e. the distribution and stability in an organism), and some chemical criteria. Such criteria are called ADME properties, an abbreviation often used in pharmacology for "absorption, distribution, metabolism, and excretion". Together, they describe the disposition of a pharmaceutical compound within an organism. The four criteria all influence the drug levels and kinetics of drug exposure to the tissues, and hence influence the performance of the compound as a drug. An abbreviation also used frequently is ADMET, in which toxicity is added to the four other criteria.

It is very unlikely that a perfect drug candidate will emerge from early screening runs. It happens more often that several positive hits are identified, allowing medicinal chemists to look for structural similarities or common pharmacophores. The chemists will then modify the compounds aiming at

- increased activity against the chosen target;
- reduced activity against unrelated targets;
- improved ADMET;
- penetration through polarized cells (e.g. Caco-2) to predict uptake in the gut;
- improved or decreased blood-brain barrier penetration (in cases where the drug target is inside or outside the central nervous system);
- initial safety tests such as Ames test (for mutagenicity) or hERG activity (the product of the "human Ether-à-go-go-Related Gene", a potassium channel predictive of irregularities in heart beat);
- behavioral tests in animals;
- efficacy in animal disease models.

Following several iterations of derivatisation, optimized compounds may be selected for further *in vitro* and *in vivo* screening. Eventually, one or several compounds may be selected for formal preclinical development.

While HTS is a commonly used method for novel drug discovery, it is not the only method. Frequently, molecules with the desired properties may be available from a natural product or as a marketed drug where the molecule can be approved ("me too" drug). Other methods, such as virtual screening of known compounds using computer-generated models of the interaction of a ligand with its binding partner may also be successful.

Drug design describes an approach whereby the biological and physical properties of the target are modelled and virtual compounds are designed in the computer and fitted to the model. Not many successes have been described based on molecular modelling alone, but modeling is routinely used when the potency and properties of lead structures are optimized.

13.3 Drug development

Once a new drug candidate has been discovered and selected for development, it will undergo stringent preclinical and clinical testing before it can be brought to market. This process is called drug development and, altogether, takes roughly 10 years or longer. The end of the drug development process should be a collection of data that can be assembled in an approval dossier and submitted to the federal agencies for drug approval.

13.3.1 Pre-clinical phase

While new drug candidates were selected for development mostly with respect to their promising activity against the chosen biological target, knowledge about the toxicity, safety, pharmacokinetics, pharmacodynamics, secretion, and metabolism of the New Chemical Entity (NCE) in humans is still very restricted. Pre-clinical development assesses all these parameters prior to human clinical trials. A recommendation has to be established based on rigorous data for a First Human Dose (FHD) that in all likelihood is safe while enabling sufficient drug concentrations at the site of the interaction with its target in the human body.

In parallel, chemical development is required to establish the physicochemical properties, stability, and solubility of the NCE. The synthetic pathway, rather irrelevant for the bench scale required in the drug discovery phase has to be optimized to avoid toxic solvents or byproducts and reduce cost for manufacturing at the scale of tons. A galenic form needs to be developed in which the NCE will be applied and stabilities of the NCE and the final medicine have to be determined under NCI (see below) conditions. Options for galenic forms may include capsules, tablets, aerosol, intramuscular injectable, subcutaneous injectable, or intravenous formulations, tansdermal formulations, topical formulations or nasal formulations. The galenic form determines how fast the drug is released in the body (e.g. slow-release formulations) or whether the NCE is released in the stomach or the gut. The latter may be advisable if the NCE causes stomach toxicity problems. In drug development, these processes are jointly called CMC (Chemistry, Manufacturing, and Control). Any new NCE and new medicine entering the human body within a clinical trial or later during the marketing phase must be manufactured under the protocols of Good Manufacturing Practice (GMP). A GMP procedure for manufacturing an Active Pharmaceutical Ingredient (API) of a new medicinal product must be established and approved by the local authorities. GMP regulates each and every detail of drug manufacturing and analytics, rendering the final product highly reproducible with very low batch variability. All analytical methods need to be validated in rigorous procedures. Since APIs are biologically active molecules, variation in production quality is considered a major safety risk, and therefore reduced in every possible way. Any batch that is to be applied to humans, either in clinical trials or as marketed product, has to meet the Specifications (analytical limits of the API and the final drug product agreed upon with the authorities) as confirmed by the Head of Quality Control (QC) and has to be "released" by the Qualified Person (QP), somebody with the necessary education (either chemist or pharmacist) and sufficient professional experience. Batch release not only requires the coherence of the results of chemical or microbiological analyses with the specifications, but also the confirmation that transport was performed by a specified drug transporter according to the specific requirements of the drug and the rules of Good Distribution Practice (GDP). For several positions within the pharmaceutical industry, including the QP, the QPPV, the head of QM, the Information Officer (responsible for the scientific information distributed about the medicinal products of the company) or head of QC, the people in charge must be personally registered with the authorities and they carry personal legal responsibility and are, in this function, independent from senior management. Every company qualifying as a drug company, irrespective of its size, has to fill all these positions with qualified employees or contractors before any drug approval will be granted.

Drug development generally aims at satisfying the regulatory requirements of drug approval authorities. These agencies not only request certain tests, but also require the company to perform safety-related preclinical tests under Good Laboratory Practice (GLP). Even though GLP requirements are not as stringent as GMP requirements, the amount of testing and documentation is way above that generally present in academic laboratories. In consequence, GLP can only be applied to standard tests, routinely employed to varying test samples in specialized laboratories. Preclinical testing generally includes a number of assays designed to determine the toxicity of novel APIs and medicines prior to first use in man. It is a legal requirement that an assessment of major organ toxicity be performed (effects on the heart and lungs, brain, kidney, liver, and digestive system) with escalating drug doses and over increasing time periods, as well as effects on other parts of the body that might be affected by the drug (e.g. the immune system or the brain if the NCE penetrates the blood-brain barrier or the skin if the new drug is to be delivered through the skin). It is required for most of these tests to perform them in one or two animal species, quite frequently rats and dogs. Dogs are used since they react very sensitively to slightly toxic agents, and toxicity is most easily detected. Toxicity is not only required for the NCE, but also for all metabolites of the NCE that occur in humans. These metabolites may not all be known prior to the first application of the drug to humans, but need to be determined very early on in the clinical phase. Some metabolites may already be formed when incubating the NCE with extracts from human liver cells. In many instances, such metabolites are also formed by the animal species used for toxicology testing. If so, they are automatically tested alongside the NCE. If, however, this is not the case, human metabolites need to be synthesized and separate toxicology testing performed.

Even after the clinical phase has already started, preclinical development continues. In addition to the tests required for the first application to humans, long-term or chronic toxicities are determined, as well as effects on systems not previously monitored (fertility, reproduction, immune system, neurological effects etc.). The compound will also be tested for its capability to cause cancer (carcinogenicity testing).

13.3.2 Clinical phase

Clinical trials, which aim at the demonstration of a drug's safety and efficacy, are immensely costly, and planning a good study is an art in itself. The high failure rate of investigational drugs even at late stages of costly clinical development programs (Hay et al. 2014) render the design of the trials one of the most critical aspects of the entire drug discovery and development process.

Clinical trials need approval by the drug authorities of all countries in which patients are included in the trial and, in addition, by the ethical committees responsible for the clinical trial centers.

This requires a dossier describing the planned clinical study and enabling the clinical investigator do perform the study (the Investigator's Brochure or IB) and a detailed description of all preclinical data and CMC information (the Investigational Medicinal Product Dossier, IMPD) including the Drug Master File (DMF) describing the GMP process for manufacturing the API, alongside with all analytical methods. In Europe, the DMF is called Active Substance Master File (ASMF) since European agencies generally try to avoid the word "drug".

The Clinical Trial Protocol is a document used to define and manage the trial. It describes the scientific rationale, objective(s), design, methodology, statistical considerations, and organization of the planned trial. All of its practical aspects are, in greater detail, reflected in the IB. The IB constitutes a precise study plan to assure safety and health of the trial subjects and to provide an exact template for trial conduct by investigators. All study investigators are expected to strictly observe the protocol and, for each patient, fill out the Case Report Form (CRF) provided by the sponsor of the study, allowing data to be combined across all investigators/sites. Study monitors control the performance and the clinical documentation of every single patient to reduce the number of protocol violations which may easily happen in the clinical workload.

In Europe, the approval for clinical trials is provided by the national drug agencies rather than the European Medicines Agency (EMA). Prior to initiating a clinical trial in the US, an Investigational New Drug application (IND) needs to be submitted to and approved by the Food and Drug Administration (FDA), in most instances, its Center for Drug Evaluation and Research (CDER).

Clinical trials for drug development involve three steps:

- Phase I trials are usually performed in healthy volunteers. The goal is the determination of safety and dosing. Several phase I trials are required to cover aspects such as dose escalation or longer term (2- to 4-week) studies. Alongside such trials, the ADME properties of the new drug are tested in humans. During dosing periods, study subjects typically remain under supervision in specialized phase I units for 1 to 40 days and nights. About one third of the new drug candidates fail in phase I, in spite of the rigorous preclinical testing.
- Phase II trials usually involve the first treatment of patients. The goal is to explore safety in small numbers of patients and to get an initial indication of efficacy. Phase II trials include at least one trial for dose finding. Usually, they involve 50 to 150 patients. About two thirds of all new drugs that successfully passed phase I trials fail in phase II.

– Phase III trials are large, pivotal trials to determine the drug's safety and efficacy in sufficiently large numbers of patients. The size of the trial must be sufficient to determine the risk/benefit relationship. Thus, the lower the acceptable risk for the disease and the lower the expected benefit, the larger the trial. Phase III trials may include between 100 and several thousand patients. About one third of the drugs that were still successful in phase II fail in phase III, mostly due to problems with clinical safety or lack of efficacy.

The term phase 0 trial is occasionally used for first-in-human trials with single, subtherapeutic doses of the study drug. A sensitive assay for the API is required since phase 0 trials are often performed prior to extensive toxicology, and therefore, only minute concentrations of the drug can be applied. Most often the drug is labelled with a short-living radioisotope which also allows the detection of metabolites of the API. With a small number of subjects (10 to 15), phase 0 trials may gather preliminary information on pharmacodynamics (what the drug does to the body) and pharmacokinetics (what the body does to the drug). Drug companies frequently also perform phase IV trials, which can however only be performed with already approved drugs. Phase IV trials are performed to collect additional safety data and often requested by the agencies as post-approval obligations. In the trial, the drug has to be applied according to the posology and in the indication for which approval is granted.

All human trials have to be performed according to the "Declaration of Helsinki", developed by the World Medical Association (WMA) as a statement of ethical principles for medical research involving human subjects, identifiable human material, and data stemming from individuals. In addition, studies have to be performed according to "Good Clinical Practice" (GCP) guidance issued by the International Conference on Harmonization of Technical Requirements for Registration of Pharmaceuticals for Human Use (ICH). All study centers involved must provide evidence for recent training in GCP according to the most recent rules. The ethical committee responsible for the center (in the US called Institutional Review Board, IRB) will control each center's ability to work according to GCP.

While phase 0 and phase I studies are exclusively aiming at the determination of the properties of a new drug, phase II and phase III trials are designed to characterize a new drug's safety and efficacy. In medical jargon, effectiveness is how well a treatment works in practice and efficacy is how well it works in a clinical trial. Volunteers in phase I receive financial incentives, which is not allowed for patients in phase II and III trials.

The goal of phase II trials is to define the dose, exact patient population and the posology which will then be applied in phase III trials. The sponsor designs the trial in coordination with a panel of expert clinical investigators. The precise definition of the disease state and, accordingly, the selection of patients can be crucial for the success of a clinical trial. Later on, the drug approval will only be awarded for the exact patient population that has been tested in clinical trials. A reasonable balance is thus required between, on the one side, selecting the patients carefully and, on the other side, getting an approval that is restricted to the selected patient group with predetermined characteristics. In addition, the biggest barrier to completing clinical studies is the shortage of people who take part. All drug and many device trials target a subset of the population, meaning not everyone can participate. It is a challenge to find the appropriate patients and obtain their consent, especially when they may receive no direct benefit (because they are not paid, they have to visit the doctor's office more often than required for routine treatment, the study drug is not yet proven to work, or the patient may receive a placebo).

The rationale behind performing clinical trials on a very homogenous patient group is that this may reduce statistical variation, thereby providing higher statistical power. Other medical conditions or interactions with other drugs may interfere with the action of the study drug, and thus be excluded in pivotal trials (pivotal trials aim at an approval and are thus designed with the highest possible quality and control). Immunosuppressed patients may react differently and cause variability in the study outcome. Women with childbearing potential cause an extra safety concern and are therefore usually excluded. Elderly people must be included in the trial if it cannot be excluded that the drug is provided to them later. Clinical testing in children has recently been made obligatory, unless the sponsor gets a "pediatric waiver" from the agency since there is no risk of drug application to children. Trials in children are difficult to perform and few parents feel comfortable with their child participating in a clinical trial.

Most phase II and III trials are performed at several independent clinical centers, and differences between the results of the centers are monitored. All patients have to consciously sign an "Informed Consent" form in which they confirm that they were fully informed about the trial including any potential safety risks and agree to participate. An unconscious patient cannot be informed, and therefore not included in a clinical trial. During the trial, independent investigators recruit patients, administer the treatment(s) and collect data on the outcome of the treatment and the patients' health for a defined time period. Safety data include measurements such as vital signs, concentration of the study drug in the blood and/or tissues, changes to symptoms, and general health outcomes. Every adverse event (AE) is recorded, regardless of its relationship to the study drug. Serious, potentially fatal, adverse events (SAE) require instant reaction, potential hospitalization and disposing the study drug. Clinical centers have, for each patient, a sealed safety envelop which may, decided by the physicians, be opened in the case of an SAE to check which study drug the patient was given. Most SAEs and AEs are not related to the study drug, they include accidents or other health problems in elderly. This may render it difficult to distinguish a "related" from an "unrelated" event, and many events are ranked as "potentially related". All related and potentially related

AEs and particularly SAEs must be discussed by the sponsor and further experiments in animals or in humans may be required to avoid strong safety warnings or even a rejection of the drug approval. Unless a relationship between the event and the drug can convincingly be excluded, a warning will be contained in the product and patient information later. This is the reason for the long list of potential side effects we all experience particularly in newer drugs which have generally been more rigorously tested than older drugs.

Currently, some Phase II and most Phase III drug trials are designed as randomized, double-blind, and placebo-controlled studies. Randomized trials are those in which each study subject is randomly assigned to receive either the study medication or a comparator such as placebo or another drug. Blind refers to the knowledge of the patient or, if double-blind, patient and medical personnel about the treatment a subject receives. The intent is to prevent clinical researchers from treating the two groups differently. In a placebo-controlled trial, a group of patients receives a placebo (a fake treatment which cannot without chemical analysis be distinguished from the samples containing the study drug). This allows isolating the effect of the new drug from the placebo effect. The comparison to a placebo shows whether the new drug is better than giving nothing to the patient (i.e. the placebo drug). Placebo effects occur in every clinical trial and are thought to be the consequence of the release of endorphins from the nucleus accumbens (Lidsotone 2005). They are in all likelihood the same effects that constitute the successes of alternative medicine.

Testing the drug against the standard-of-care therapy is more stringent than the placebo-control. The new drug has to be at least as good (non-inferior) or better (superior) than the standard-of-care therapy. Such "active comparator" (also known as "active control") trials are a requirement for getting approval of the EMA which intends to restrict new approvals to medical improvements. The FDA does not yet require active comparator trials but may be more stringent on safety and CMC aspects. Wherever ethically possible (not treating a patient for the duration of the study may be harmful), a placebo group should also be included in a comparator trial providing additional statistical power since the efficacy of the two other study arms can be documented in the first place.

A full series of clinical trials may cost hundreds of millions of Euros. The money is usually borne by the sponsor, in most cases a pharmaceutical or biotechnology company. Programs exist in the US which provide funding for clinical trials if a public interest in the new medication exits. Since most organizations lack the necessary capacity and knowledge to perform all aspects of clinical trials on their own, most trials are managed alongside with specialized contract research organizations (CROs) or an academic clinical trial unit. Experience shows, however, that the sponsor's insight into the drug and the disease is a necessary prerequisite in designing and performing a successful clinical trial. Just contracting a CRO for study administration is not sufficient and may result in major mistakes devaluing the entire study.

13.4 Statistical analysis

13.4.1 Statistical power

The number of subjects impacts the "power" of a study which describes the likelihood to reliably detect and measure effects of the treatment. Larger numbers of participants are related to greater statistical power (and higher cost). The statistical power estimates the ability of a trial to detect a difference of a particular size (or larger) between treatment and control groups. The calculation of statistical power resides on expectations on the outcome of the study. These expectations are generally based on literature data or, for phase III studies, on the results of phase II. Power calculations must be performed as part of the study application, since studies that are not promising relevant results or are heavily overpowered are considered unethical. Thus, regulatory bodies are in charge of controlling power estimations for clinical trials prior to approval of the study.

13.4.2 Statistical and epidemiological analysis plan

Once a study protocol is final, the creation of the Statistical Analysis Plan (SAP) can commence. The "International Conference on Harmonisation of Technical Requirements for Registration of Pharmaceuticals for Human Use (ICH), ICH E9 'Statistical Principles for Clinical Trials'" provided guidance on the content of SAPs. The SAP must be finalized prior to completion of the clinical part of the study. It describes all details of how the study is evaluated.

Statistical analyses will be performed separately for the treatment period and the follow-up period. The statistical analysis of the treatment period will be performed after database lock and final unblinding.

A clinical trial usually has one primary endpoint and several secondary endpoints. The primary endpoint in pivotal trials usually describes the overall efficacy of the trial drug. The parameter by which improvement of the disease status is measured must be discussed with and approved by the regulatory agency. Secondary endpoints can include patient subgroups or other disease parameters. Safety is an important aspect of every clinical trial and must be assessed in the SAP as well. The statistical analysis plan must describe the detailed statistical analysis. Once the data base is locked, the SAP cannot be modified any more. It should describe analyses unambiguously and in sufficient detail for another statistician to be able to repeat them.

The SAP must integrate the complex hypotheses underlying all aspects of a clinical trial in a way that controls the chances of drawing incorrect conclusions. The SAP details the mathematical transformations that will be performed on the observed data in the study and the patterns of results that will be interpreted as supporting alternative answers to the questions. It will also explain the rationale behind this decision making process and the way that this rationale has influenced the study design. An important part of the statistical analysis plan will explain how problems in the data will be handled in such calculations, for example missing or partial data. Thoughtful specification of the way missing values will be handled need to be outlined to overcome such problems.

The SAP should be sufficiently detailed so that it can be followed in the exact same way by any competent analyst. Thus, it should provide clear and complete templates for each analysis. Confirmatory hypothesis tests will be performed for the primary and secondary efficacy endpoints. Confidence intervals must be predefined for any variable and will be presented for the primary and secondary endpoints. A trial which just misses the predefined confidence interval in the primary endpoint is useless and all the cost of the trial wasted.

All other data are mostly analyzed descriptively and in an exploratory way. Continuous data are, in most cases, summarized by means of descriptive statistics, which may include number of patients, mean, standard deviation (SD), median, quartiles, and range (minimum and maximum), but lacks a determination of confidence intervals.

A feature common to most studies is that some not pre-specified analyses will be performed in response to chance observations in the data. It is important to distinguish between such data-driven analyses and the pre-specified findings. In most cases, pre-specified statistical calculations including confidence intervals will be kept to a minimum since statistical power is reduced alongside with larger numbers of pre-specified analyses.

In any case, the SAP should include a brief description of the trial design, and the statistical methods section of the protocol. Analysis populations such as the Intent to Treat (ITT) or Per Protocol (PP) populations need to be defined in an unambiguous fashion. In general, the ITT population includes all patients of a treatment group that were randomized (and usually got at least one treatment), the PP population includes all patients treated exactly according to the study protocol without any relevant protocol violation. Thus, the PP population reflects the potential of a drug, the ITT population may be interpreted as the result in real life. The SAP should state overall level of significance, and any other relevant information that will be used in the majority of analyses (e.g. treatment of missing data, adjustment for covariates, adjustment for multiplicity, adjustment of p-value due to interim analyses, if any). A list of analyses should describe each test group precisely, the analysis method to be used, any sensitivity analyses, any deviations from the methods listed in 'Overall Statistical Principles', and any subgroup analyses.

13.5 Pharmacovigilance

Pharmacovigilance (PV or PhV), also known as Drug Safety, is the pharmacological science relating to the collection, detection, assessment, monitoring, and prevention of adverse effects with pharmaceutical products. It is concerned with identifying the hazards associated with pharmaceutical products and with minimizing the risk of any harm that may come to patients.

It focuses on adverse drug reactions (ADRs), which are defined as any response to a drug which is noxious and unintended, including lack of efficacy. Overdosing, misusing and abusing a drug or drug exposure during pregnancy and breastfeeding or in patient populations that are excluded in the drug approval are also of interest (even without adverse event itself), because they may provide additional safety information or result in an ADR.

Any company performing a clinical trial, applying for drug approval or marketing a drug is required to have a responsible person with the required professional experience, the Qualified Person for Pharmacovigilance (QPPV). The QPPV is responsible for assessing all information related to adverse events, any complaints of physicians or patients, and any information regarding non-authorized usage of the drug. The data must be stored in a "qualified" database to which the responsible regulatory authority has constant access. Information received from doctors, patients, members of the own sales force or healthcare providers via pharmacovigilance agreements (PVAs), as well as other sources such as the medical literature, plays a critical role in providing the data necessary for pharmacovigilance to take place. Regular literature screening with respect to any use of the drug in all countries where the product is marketed is a legal requirement and the company has to document the process and the results of the literature screening.

13.6 The quality management system

The quality management system (QMS) describes relevant business processes in Standard Operating Procedures (SOPs). SOPs are controlled documents that can only be adapted according to rules described in a more fundamental SOP. The goal is to secure the quality and the reliability of the company's processes and products. The QM-handbood describes the structure of the QM system and the interaction between the QM-system, the QM managers, the company management, other relevant functions within the company and the employees. The Quality Management Team reports directly to senior management and its work is periodically controlled by governmental and/or notified bodies (Box 13.2).

The idea of QMS was borne in the last decades of the 20th century, and has increasingly been legally required, and therefore, implemented in the 21st century. Of all QMS regimes, the ISO 9000 family of standards is probably the most widely implemented worldwide – the ISO 19011 audit regime deals with quality and sustainability and their integration.

It is left to the companies to determine the necessity for, or extent of, some quality elements and to develop and implement procedures tailored to their particular processes and devices. Nevertheless, the governmental bodies will look into the extent of the QM system and how it is applied in daily life during their regular audits. ISO9001 requires that the performance of QM processes must be measured, analyzed, and continually improved. Potential problems in the workflow are identified and Corrective and Preventive Actions (CAPAs) are integrated. The results of this form an input into the annual "management review process".

A QM system is required and may be inspected by the authorities at the latest by the time a company initiates clinical trials. All companies working together or

Box 13.2: The quality management (QM) system.

The QM system describes a company's processes and how such processes are controlled and how deviations from regulated processes are documented and mistakes corrected. A QM system is mandatory for any company active in pharmaceuticals or medical devices and is regularly inspected by the responsible regulatory bodies. It includes aspects of:

- personnel training and qualification
- product design
- product documentation
- purchasing
- product identification and traceability at all stages of production
- controlling and defining production and process
- defining and controlling inspection, measuring and test equipment
- validating processes
- product acceptance
- controlling nonconforming product
- instituting corrective and preventive action when errors occur
- labeling and packaging controls
- handling, storage, distribution, and installation
- archiving
- servicing
- statistical techniques
- pharmacovigilance procedures
- order processing
- production planning
- calibration
- internal audits
- corrective action
- preventive actions
- identification, labeling, and control of non-conforming product to preclude its inadvertent use, delivery or processing
- product call-back
- purchasing and related processes such as supplier selection and monitoring

using one another's services are required to audit the other company's processes and QM system. Audit Reports are part of the drug approval documentation and thus official documents that will be made available to the federal drug agencies.

13.7 The drug approval process

If a compound finally emerges from all these tests with an acceptable toxicity and safety profile, and it can further be demonstrated to have the desired effect in clinical trials, then it has to be submitted for marketing approval in the various countries before it can be sold.

The process of reviewing and assessing the dossier to support a medicinal product in view of its marketing (also called licensing, registration, approval, etc.) is performed within a legislative framework which defines the requirements necessary for application to the concerned (competent) regulatory authority. The application is filed with the competent drug regulatory authority in the concerned country, which can be either an independent regulatory body or a specialized department in the ministry of health. Authorization processes follow either a purely national procedure, with rules and requirements as per current national legislation, as it occurs in most of countries worldwide. In the European Union, it may either follow a centralized approval or a mutual recognition or decentralized procedure.

The application dossier for marketing authorization is called New Drug Application (NDA) in the USA or Marketing Authorization Application (MAA) in the European Union and other countries, or simply registration dossier. Basically, this consists of a dossier with data proving that the drug has the quality, efficacy, and safety properties suitable for the intended use and additional administrative documents. The content and format of the dossier must follow rules as defined by the competent authorities. For example, since 2003, the authorities in the United States, the European Union, and Japan ask for the Common Technical Document (CTD) format, and more recently, its electronic version – the electronic Common Technical Document (eCTD).

The European Medicines Agency (EMA) and the US Food and Drug Administraion (FDA) share similar objectives, including "promoting and protecting public health, evaluating the safety and efficacy of therapeutic products, working collaboratively with outside experts, reducing the regulatory burden through international harmonization, providing regulatory and health information, and enhancing product development." However, these agencies differ in structure and benefit-risk assessment.

The FDA is a centralized agency that oversees the drug development process in a single country (the USA), whereas the EMA is a reviewing body that manages the process in many European nations. Within the FDA, drug evaluation applications and the drug development process are monitored by the FDA's own staff. In the EMA, the assessment is conducted by the national agencies of the member states. According to the EMA's Web site, the agency brings together the scientific resources of more than 40 national competent authorities in 30 European Union (EU) and European Economic Area-European Free Trade Association countries in a network of more than 4,500 European experts. Once the EMA renders an opinion, approval is granted or denied by the European Commission which in all likelihood will follow the recommendation of the EMA.

In the EU, an application for a marketing authorization license is filed with the EMA, which is valid in all EU member states, plus the European Economic Area-European Free Trade Association countries of Iceland, Liechtenstein, and Norway. This centralized authorization procedure is mandatory "for all medicinal products developed by biotechnologic process; for new active substances indicated for the treatment of acquired immune deficiency syndrome, cancer, neurodegenerative disorder, or diabetes; and also for designated orphan medicinal products." For drugs that do not fall under these categories, companies may apply for a centralized marketing authorization if the drug constitutes a significant therapeutic, scientific, or technical innovation. Other authorization procedures – the national procedure, decentralized procedure, and mutual-recognition procedure – exist for drugs that do not fall within the scope of the centralized procedure.

Although the FDA and the EMA have similar evaluative processes, the final outcome of the benefit-risk assessment is not necessarily the same in all cases. Clinical investigations of new drugs in the United States compare the drug with a placebo. In the EU, the benefit-risk assessment has become increasingly based on (more difficult) comparisons between the new and existing drugs. A three-armed study using placebo and an active treatment as controls is considered the preferred method of benefit-risk assessment in the EU, when possible.

Despite the differences between these bodies, the FDA and EMA recently standardized the orphan medicines designation process. Orphan drugs are those developed for rare diseases with prevalences below a certain threshold that varies between different countries, affecting fewer than five in 10,000 people in the EU and fewer than 200,000 people in the United States. An application for an orphan drug is facilitated in various procedural and financial aspects. In an effort to simplify part of the orphan medicines designation process, in November 2007, the EMA and the FDA adopted a common application form for drugs for rare diseases in both jurisdictions. This common application format allows sponsors to apply to both jurisdictions at the same time with one application.

The type of application may vary according to status of the active ingredient. Thus, if the application concerns a new active ingredient (new active substance, new chemical entity, new molecular entity), one talks about a full application.

Once an active ingredient is authorized, any additional strengths, pharmaceutical forms, administration routes, presentations, as well as any amendments (changes to the existing marketing authorization), and extensions shall also be granted an authorization or be included in the initial marketing authorization, subject of a shorter application.

Special consideration is to be given to application for authorization of biological products and biotechnology products, homeopathic products, herbal drugs, radionuclide generators, kits, radionuclide precursor radiopharmaceuticals, and industrially prepared radiopharmaceuticals. For such products, requirements are specific and may be more or less elaborate reflecting the nature of the active ingredient.

In most countries, a marketing authorization is valid for a period of 5 years. In regular intervals during this time and later, reports on new safety information (Periodic Benefit-Risk Evaluation Report, PBRER, formerly called Periodic Safety Update Report, PSUR) on all drugs need to be provided to the regulatory agencies. After the initial 5-year period of approval, the marketing authorization can be renewed upon providing minimal data demonstrating that quality, efficacy, and safety characteristics are maintained and the risk-benefit ratio of the medicinal product is still favorable. In the European Union, after one renewal, the marketing authorization shall remain valid for an unlimited period, unless the competent regulatory authority decides otherwise.

If the marketing authorization is not renewed in due time as requested by the local legislation, in order to maintain the pharmaceutical product on the market, one can apply for re-authorization (re-registration). In such situations, the applicant may be requested to submit all items required for a full application.

Marketing authorization may be withdrawn, suspended, revoked or varied by regulatory authorities if under normal conditions of use, the benefit over risk ratio is no more favorable, the product is harmful, or lacks therapeutic efficacy; also, one of the above actions can be taken if the qualitative and quantitative composition or other qualitative aspects (control) are not according to the specification.

Marketing authorization may be also withdrawn, suspended or revoked if the marketing authorization holder or its representative does not fulfill other legal or regulatory obligations necessary to maintaining of product on the market, as per the legislation in force.

Also, the marketing authorization is withdrawn in the EU if the product is not placed on the market within 3 years after granting of authorization or if it was not any more marketed for 3 consecutive years (so-called "sunset clause"). While the sunset clause applies in each single nation with national approvals, for centralized EU approvals marketing in at least one EU member state is sufficient to obey to the rule.

13.8 The common technical document

The Common Technical Document (CTD, Figure 13.1) specifies the content of the application dossier for the registration of medicines across Europe, Japan, and the

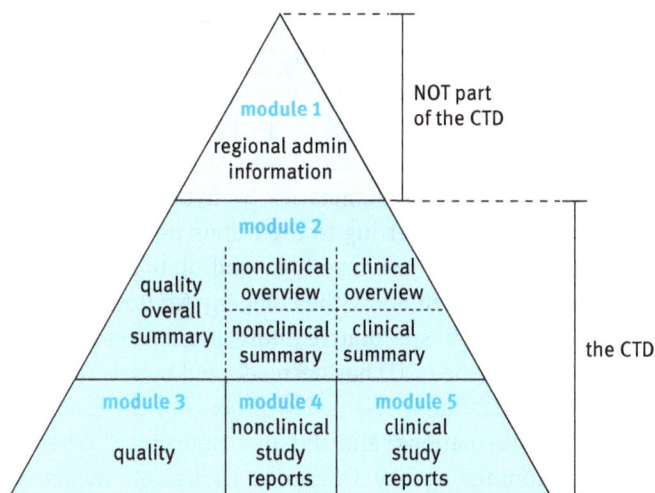

Fig. 13.1: The Common Technical Document is divided into five "modules", presented in the dossier at increasing detail (www.fda.gov/downloads/forindustry/userfees/prescriptiondruguserfee/ucm272444.pdf). It consists of the following modules:
1. Administrative and prescribing information
2. Overview and summary of modules 3 to 5
3. Quality (pharmaceutical documentation)
4. Preclinical (Pharmacology/Toxicology)
5. Clinical – efficacy (Clinical Trials)

United States. It is an internationally agreed format for the preparation of applications regarding new drugs intended to be submitted to regional regulatory authorities in participating countries. It was developed by the European Medicines Agency (EMA, Europe), the Food and Drug Administration (FDA, US), and the Ministry of Health, Labour and Welfare (Japan). The CTD is maintained by the International Conference on Harmonisation of Technical Requirements for Registration of Pharmaceuticals for Human Use (ICH). For most nations, the CTD has to be submitted in electronic form (eCTD, Figure 13.1).

The application of the eCTD to the agencies is called New Drug Application (NDA) in the US or Marketing Authorization Application (MAA) in Europe. Detailed subheadings for each module are specified for all jurisdictions. The contents of Module 1 and certain subheadings of other modules will differ, based on national requirements. The eCTD is organized as a series of increasingly detailed summaries covering CMC, nonclinical and clinical data. The most comprehensive summary at the top of the pyramid is the product information made available to medical personnel and patients, in the EU called Summary of Product Characteristics (SPC) and Patient Information Leaflet (PIL) as well as the layout and the labelling of the inner and outer packages. For a centralized EU approval, these have to be provided in all EU languages, while the major parts of the eCTD will be presented in English.

This comprehensive summary is the definitive description of the product in terms of its chemical, pharmacological and pharmaceutical etc. properties, and its clinical use. Down the eCTD pyramid, the summaries and descriptions of the various aspects become more elaborate, and at the bottom of the pyramid all primary data including laboratory results and every individual CRF from clinical trials are presented. Such data must be connected to the summaries by hyperlinks to allow reviewers to rapidly find all original data referring to the claims presented in the summaries. An eCTD may contain up to 30,000 hyperlinks and up to several hundred thousand pages. Since the eCTD is written by the company for licensing purposes only, by its nature it will not contain speculative information. After the United States, European Union, and Japan, the eCTD has been adopted by several other countries including Canada and Switzerland. The review process of the eCTD follows specific timelines in the various nations, and the applicant knows when to expect feedback and questions from the agency. Questions are usually numerous and include requests for additional tests or even clinical trials. In Europe, the agency provides long lists of questions at specific times, while constant interaction with the FDA is required during the review process in the US. All questions need to be answered by the applicant before the new medicinal product will be approved. If the available time is insufficient to provide the answers or adhere to the proposals, the agency may agree to a preliminary drug approval and request certain tests, modifications in manufacturing or safety studies within a specified time post approval (post approval obligations). Regulatory Affairs (RA) managers are informed about the expectation of the agencies through a vast body of legal rules and information and through meetings with the agencies in which relevant topics can be discussed. Rules and information by the agencies can be legally binding acts, expressed in "Regulations", "Directives" or "Decisions", or soft law in the form of "Resolutions", "Communications", "Guidelines" or "Notes to applicants". Whether legally binding or soft law, it is advisable to strictly follow those rules and watch out for any changes. For agency meetings, the applicant provides the necessary information beforehand and formulates its questions and proposed answers. The agency will then agree with the answer or disagree, but in most instances not provide clear guidance otherwise. Therefore, the precise wording for questions and proposed answers is of utter importance. Every national agency or the EMA is available for such meetings at quite variable costs. Working together with the agency in the process of drug approval is most formalized in the US, where pre-IND or pre-NDA meetings are routinely integrated into the development and approval cycle.

Unfortunately, while respecting the common eCTD format, the FDA has adopted some additional specific rules. It does f.i. since a few years require the applicant to transform all clinical data to a specific format developed by the "Clinical Data Interchange Standards Consortium (CDISC)", a non-profit organization in Austin, Texas. Since FDA reviewers tend to repeat statistical testing on their own, this data format is intended to facilitate working with data from multiple pharmaceutical companies or CROs. Using this format, the company must combine the data of all clinical trials in one "integrated analysis". Both efficacy and safety results of the integrated analysis are then summarized in the Integrated Summary of Efficacy (ISE) and the Integrated Summary of Safety (ISS). Also, as part of the approval process, the FDA locally inspects all manufacturing sites and controls their processes and QM systems. European agencies do the same, but quite often later during the marketing phase of a drug.

The use of the eCTD does not come to an end with drug approval. It is a controlled document (i.e. a document that cannot be modified without an electronic history of changes) and any modification must be agreed upon with the agency in the form of a (can be very costly) Amendment. The eCTD thus accompanies a drug through its entire commercial life and a summary as well as the agency's assessment, in the EU called the European public assessment report (EPAR), are publicly available on the agency's webpage.

Key-terms

Drug Approval Process, Drug Development, Drug discovery, Drug targets, lead structures, Pharmacovigilance, Pre-clinical phase, Clinical phase, Quality Management, Statistical and epidemiological analysis plan

Abbreviations

ADR	Adverse Drug Reaction
AE	Adverse Event
API	Active Pharmaceutical Ingredient
ASMF	Active Substance Master File
CAPA	Corrective And Preventive Action
CDER	Center for Drug Evaluation and Research
CDISC	Clinical Data Interchange Standards Consortium
CMC	Chemistry, Manufacturing and Control
CRF	Case Report Form
CTD	Common Technical Document
DMF	Drug Master File
eCTD	electronic Common Technical Document
EMA	European Medicines Agency
EPAR	European public assessment report
FDA	Food and Drug Administration
FHD	First Human Dose
GCP	Good Clinical Practice
GDP	Good Distribution Practice
GLP	Good Laboratory Practice
GMP	Good Manufacturing Practice
HTS	High-Throughput Screening

IB	Investigator's Brochure
ICH	International Conference on Harmonization of Technical Requirements for Registration of Pharmaceuticals for Human Use
IMPD	Investigational Medicinal Product Dossier
IND	Investigational New Drug
IRB	Institutional Review Board
ISE	Integrated Summary of Efficacy
ISS	Integrated Summary of Safety
ITT	Intent To Treat
MAA	Marketing Authorization Application
NCE	New Chemical Entity
NDA	New Drug Application
PBRER	Periodic Benefit-Risk Evaluation Report
PIL	Patient Information Leaflet
PP	Per Protocol
PSUR	Periodic Safety Update Report
PV or PhV	Pharmacovigilance
PVA	PharmacoVigilance Agreement
QMS	Quality Management System
QP	Qualified Person
QPPV	Qualified Person for Pharmacovigilance
RA	Regulatory Affairs
SAE	Serious Adverse Event
SAP	Statistical Analysis Plan
SOP	Standard Operating Procedure
SPC	Summary of Product Characteristics
USPI	United States Prescribing Information
WMA	World Medical Association

Questions

- What is described in a drug „label" and how is the "label" officially called in the US and in Europe?
- What is off-label use of drugs?
- Give two examples of drugs derived from crude plant extracts.
- Give at least one example of a drug that is a chemical derivative of a natural compound.
- What is "phenotypic drug discovery"?
- Why was dihydrofolate reductase so important for the development of drug discovery concepts?
- What is "reverse pharmacology"?
- What is meant by "drugability" of a target and what influences drugability?
- What is high-throughput drug screening?
- What are the ADMET properties of a drug?
- What needs to be tested in lead optimization?
- What are the difficulties with natural products in drug discovery?

- After which stage do drugs in development need to be produced according to GMP?
- What is then function of the "Qualified Person"?
- Is all research in pharmaceutical companies performed according to GLP?
- What is the role of ethical committees in clinical research?
- What is the difference between the IB and the IMPD?
- What is the role of study monitors?
- What is the US equivalent of the European Medicines Agency?
- Who authorizes clinical trials in Europe and in the USA?
- What is the difference between phase II and phase III trials?
- What are bioavailability and pharmacokinetics?
- What is the statistical analysis plan and when must it be finalized?
- What is the difference between the PP and the ITT population in a clinical trial?
- What is pharmacovigilance and when is it required?
- What is an eCTD?
- What is a generic approval?

Further readings

Gad, S. C. 2005. Drug discovery handbook. *Wiley-Interscience/J. Wiley.*
Pocock, S. J. 2004. Clinical Trials: A Practical Approach. *John Wiley & Sons.*
Rang, H. P. 2007. Drug Discovery and Development: Technology in Transition. *Churchill Livingstone.*

References

Dollery, C. T. 1999. *Therapeutic drugs.* Edinburgh: Churchill Livingstone.
Feher M. & Schmidt, J. M. 2003. Property distributions: differences between drugs, natural products, and molecules from combinatorial chemistry. *J Chem Inf Comput Sci* 43: 218–227.
Hay, M., Thomas, D. W., Craighead, J. L., et al. 2014. Clinical development success rates for investigational drugs. *Nat Biotechnol* 32: 40–51.
Leeson, P. D & Springthorpe, B. 2007. The influence of drug-like concepts on decision-making in medicinal chemistry. *Nat Rev Drug Discov* 6: 881–890.
Lidsotone, S. C., Fuente-Fernandez, R., Stoessl, A. J. 2005. The placebo response as a reward mechanism. *Semin Pain Med* 3: 37–42.
Newman, D. J. & Cragg, G. M. 2007. Natural products as sources of new drugs over the last 25 years. *J Nat Prod* 70: 461–477.
Paul, S. M., Mytelka, D. S., Dunwiddie, C. T., et al. 2010. How to improve R&D productivity: the pharmaceutical industry's grand challenge. *Nat Rev Drug Discov* 9: 203–214.
Rask-Andersen, M., Almén, M. S. & Schiöth, H. B. 2011. Trends in the exploitation of novel drug targets. *Nat Rev Drug Discov* 8: 549–590.
Warren, J. B. 2011. Drug discovery: lessons from evolution. *Br J Clin Pharmacol* 71: 497–503.

Axel Mosig

14 Bioinformatics

In the past 50 years, developments in biology and biotechnology yielded ever-increasing amounts of digitizable observations, which has naturally led to the utilization of computational approaches to obtain insights from biological data. This development was certainly initiated by the availability of peptide sequences of proteins, which led to the first computational methods for sequence comparison in the early 1970s. This development was continued at faster pace by the ability to sequence DNA in ever-faster rates at ever-lower costs, finally enabling the sequencing of whole genomes. In fact, bioinformatics took center stage in the race to sequence the human genome, whose sequence was first published in 2001. Computational means to compare protein and DNA sequences also enabled to trace evolution directly at the sequence level, which led to important contributions from bioinformatics to reconstruct evolutionary relationships between genes and organisms, and to understand mutation patterns at different time scales. A certainly fascinating aspect of bioinformatics is that the field helps to unveil the intertwined relationships between sequence, structure, function and evolution of genes.

As defined by the National Institute of Bioinformation technology (NCBI), Bioinformatics is the field of science in which biology, computer science, and information technology merge into a single discipline. In a broader sense, mathematics and statistics are also an important ingredient of many bioinformatics tools and techniques. There are three important sub-disciplines within bioinformatics: the development of new algorithms and statistics with which to assess relationships among members of large data sets; the analysis and interpretation of various types of data including nucleotide and amino acid sequences, protein domains, and protein structures; and the development and implementation of tools that enable efficient access and management of different types of information. This chapter is focused on introducing bioinformatics approaches on the basis of *what* problem they address and where they can be used to answer biological questions. Understanding *how* these methods work requires some background in the underlying disciplines of computer science and statistics and is thus not a subject of this chapter. The questions of what bioinformatics approaches do and how they work are yet closely connected, and students interested in a deeper understanding are encouraged to study this in the corresponding textbooks, for example the one Durbin et al. (1998).

After a brief introduction on the history and scope of bioinformatics, this chapter introduces approaches to sequencing and annotating whole genomes in Section 14.2. The subsequent Section 14.3 deals with comparing sequences across different species, where we deal with the essential concepts of sequence alignments and homology search, as well as the comparison of complete genomes. Section 14.3 introduces bioinformatics approaches for handling RNAs that are not being translated into proteins – so-called non-coding RNAs – as an important class of cellular molecules. Characterizing and comparing the overall repertoire of what is being transcribed in different cell or tissue types is subject of section 14.4. These sections cover essential core topics for biological sequence analysis. Yet, sequence analysis represents only a fraction of the scope of modern bioinformatics. Correspondingly, Section 14.5 sketches the key questions behind other subfields of bioinformatics such as proteome analysis or computational bioimaging.

Throughout this chapter, we will use different components of the telomere complex to illustrate the application of bioinformatics approaches. The telomere complex, briefly reviewed in Box 14.1, is a formidable system to exemplify a broad range of bioinformatics approaches, as it involves protein coding genes, non-coding RNA genes, and also genomic repeat elements.

14.1 Introduction

The root of the development of Bioinformatics certainly lies in the development of techniques for sequencing DNA and proteins, and the scientific questions resulting from the availability of corresponding sequencing data. Yet, the rise of bioinformatics as an important part of the life sciences was not limited to the analysis of protein and DNA sequences, as also other biotechnological developments brought forth digitizable observations. At the level of transcribed RNA, microarray technology allowed to determine expression levels of messenger RNAs in a genome-wide manner, which produces high-dimensional data whose interpretation relies on a chain of computational analysis steps. For the level of proteins and metabolites, different types of mass spectrometry provide precise snapshots of the biochemical status of cellular or tissue samples whose interpretation again requires substantial computational effort. Recent developments in fluorescence microscopy, as well as other imaging techniques, have established microscopy as location proteomics, allowing to study location patterns of biomolecules and their change under different conditions. For the analysis of location patterns in microscopic images, image processing entered the list of computational techniques relevant for life science research. The mentioned techniques of transcriptomics, proteomics or microscopy are rather few examples among a long list of technological developments in biology that led to new questions and approaches in bioinformatics, and this list will certainly expand further with new biotechnological developments.

With the digitalization of observations – both at the level of sequences and further downstream at phenotypical level – and the consequential utilization of computational analysis, biology is confronted with a new situation. On the one hand, the biological systems under consideration are inherently heterogeneous and exhibit appearingly ambiguous traits. Computational methods, on the other hand, inherently are subject to the unambiguity of mathematics, statistics and computer science. Closing this gap between biological systems and computational methods obviously requires a great degree of abstraction. Achieving a good match between the traits of a biological system and the explicit or implicit assumptions behind a bioinformatics algorithm certainly is the key behind many of the successful algorithms and corresponding software tools that are nowadays naturally part of biology and biotechnology.

Box 14.1: Telomerase and the Telomere Complex as a Showcase System for Bioinformatics.

Throughout this chapter, we will use a specific molecular mechanism, the so-called telomerase, as an accompanying example of how bioinformatics can be used to study sequence, structure, function, and evolution of such a molecular mechanism. Recall that during DNA replication, there is an asymmetry between the two strands of double stranded DNA: Termination of chromosomal replication of the leading strand precedes replication of the lagging strand, leaving an inevitable gap of nucleotides in the leading strand that are not complemented in the lagging strand. If these gaps were not further complemented after termination, this would lead to a shortening of the chromosomes after each replication. Almost all eukaryotes resolve this termination problem through the so-called telomerase as a mechanism to compensate the uncomplemented nucleotides at the chromosomal ends. The DNA sequence at the chromosomal ends consists of a short motive (e.g. TTAGGG in vertebrates), the so-called telomeric repeat, that is repeated many times at the end of the chromosome. The telomerase now complex contains a reverse transcriptase that reversely transcribes the telomeric repeat from an RNA molecule, and uses the reversely transcribed single-stranded DNA to compensate the unpaired nucleotides in the lagging strand.

Beside the telomeric repeat, the two main functional players involved in telomerase are the telomerase reverse transcriptase (TERT) protein and the telomerase RNA component (TERC), which is a single-stranded RNA transcript containing the reverse complement of the telomeric repeat as a template for reverse transcription.

Specific showcases illustrating the use of bioinformatics in the context of telomerase and the telomere complex are scattered throughout this chapter:
- The telomeric repeat as an example for genomic repeats in Section 14.1
- Alignments of the TERT protein in Section 14.3.1
- Homology search for TERT in Section 14.3.2
- TERC as a pseudoknotted non-coding RNA in Section 14.4.1
- TERC illustrating the difficulty of non-coding RNA homology search in Section 14.4.2.

The discipline of bioinformatics lies at the junction of two developments that took their roots in the 1950s. In the life sciences, the discovery of DNA as the carrier of genetic information was a major trigger for the "molecularization" of biology in subsequent decades. On the other hand, the introduction of the concept of universal computation by Alan Turing (1936) and its physical realization by Claude Shannon and John von Neumann (1957) – both deeply rooted in the underlying mathematical foundations – paved the way for computer science as a new discipline, which has accompanied the digitalization of not just biology and other sciences that took place in the past 50 years. The interplay between the outgrowths of these initially unrelated roots is certainly what makes up a large part of the fascination for bioinformatics as an interdisciplinary field.

14.2 Genome sequencing

The availability of a fully sequenced genome for an organism is an invaluable, in some cases indispensable resource when studying traits of a specific organism, or comparing traits among different organisms. As an example of how the availability of a genome sequence yields immediate functional insights, consider the mechanism of chromosome maintenance in fruitfly. Most higher organisms avoid the shortening of chromosomes after duplication by the telomere complex (see Box 14.1). Taking a closer look at the chromosomal ends of the fruitfly genome, it is immediately noticeable that the fruitfly genome unlike the genome of most other eukaryotic organisms lacks the telomeric repat – an obvious indicator that flies must utilize a different mechanism to maintain their chromosome ends after replication. Further evidence of this well-known fact is obtained when performing homology search for the *TERT* gene, which cannot be found in the genome of *D. melanogaster* as we will see in Section 14.3, providing further evidence that indeed fruitfly does not possess a telomere complex and maintains its chromosomal ends using a different molecular mechanism.

14.2.1 Shotgun sequencing

Approaches to sequencing DNA have progressed tremendously since the first approaches developed in the 1970s, now allowing to sequence billions of nucleotides per day on a state-of-the-art sequencing device. However, even the most advanced sequencing techniques allow only relatively short stretches of DNA to be sequenced. In fact, the maximal so-called read length, i.e., the length of a DNA fragment that can be sequenced, ranges between roughly 1000 consecutive nucleotides in the case of conventional gel-based Sanger sequencing, and 30 to 500 nucleotides for next-generation sequencing devices. Such read lengths are hardly sufficient to sequence reversely transcribed messenger RNA, and way too short to sequence complete chromosomes, even of most prokaryotic organisms with relatively short genomes.

In order to sequence complete genomes under the constraint of limited read length, the approach of shotgun sequencing was developed already in the late 1970s in order to sequence prokaryotic genomes. The idea behind shotgun sequencing is to fragment chromosomal DNA into shorter pieces using ultrasound, so that each fragment is shorter than the read-length limitation of the respective sequencing technology to be used. The DNA sample material will contain multiple copies of the chromosome to be sequenced, and the application of ultrasound breaks the chromosome into fragments in a random fashion. As illustrated in Figure 14.1, one can utilize overlap information between fragments to infer longer stretches of DNA, and eventually a complete chromosome. The computational task in shotgun sequencing is thus to identify sufficiently long overlapping areas: whenever the 5′

```
CCGAGTTCGCCCCCTCGGGATGACGTTACA
        CCCCCCTCGGAATGACGTCACA..................................................CCACCCATACAGAGCCCAGCAAGCT
            GGAATGAAGTCACATGCAATGG      ACACTAACCACGTCACTTCA              GGCGGGTCCCACCCATACAGAGCCCA
GCACCGACTTCGCCCCCCCCGGAATGACG          AATGGCGTCACACACTAACCA        AGTGGTCTGGCTCGGCGGGTCCCAC       AGAGCCCAGCAAGCTAAGATCCACT
```
```
GCACCGAGTTCGCCCCCCTCGGAATGACGTCACATGCAATGGCGTCACACACTAACCACGTCACTTCA     AGTGGTCTGGCTCGGCGGGTCCCACCCATACAGAGCCCAGCAAGCTAAGATCCACT
```
  contig 1   contig 2

Fig 14.1: Illustration of shotgun sequencing. The two contigs can be assembled from conventionally sequenced reads. The paired end read (highlighted in red) yields the relative position of the two contigs to each other in a scaffold, as the length of the paired-end-read is known with high precision. The sequence of the complete scaffold (which ideally will be a complete chromosome) can be obtained when a sufficient number of contigs and paired-end reads is available.

Table 14.1: Genome Sizes and gene numbers of selected organisms.

Species	# of chromosomes	# of ORFs	# of nucleotides
E. coli (K-12 strain)	1	4,288	4.6×10^9
C. elegans	11	20,541	103×10^9
D. menangogaster	8	13,937	168×10^6
M. musculus	21	23,148	2.8×10^9
H. sapiens	23	20,687	3×10^9
A. thaliana	5	27,206	157×10^6
S. cerevisiae	16	5,097	12×10^6

Table 14.2: Repeat types in eukaryotic genomes.

Simple Repeats	Duplications of simple sets of DNA bases, typically 1–5 nucleotides long
Tandem Repeats	Duplications of more complex 100–200 base sequences typically found at the centromeres and telomeres of chromosomes
Segmental Duplications	Large blocks of 10–300 kilobases which are that have been copied to another region of the genome
Interspersed Repeats	Different types of transposable elements

end of one read is identical to the 3′ end of another read, they can be assembled into a longer read. State-of-the-art approaches to shotgun assembly perform this in a more advanced fashion by establishing a combinatorial graph structure from the overlap information (which may be perturbed by sequencing errors), and then inferring the sequence using this graph structure.

While shotgun sequencing proved viable for sequencing prokaryotic genomes as early as 1979 through the work of Roger Staden (1982), its application to larger eukaryotic genomes was inhibited mainly by the phenomenon of so-called repeat regions, i.e., sequences of nucleotides that occur more than once within the same chromosome or genome. Several types of such repeat regions are known, as summarized in Table 14.2. Note that we already encountered one particular type of repeat, namely the telomeric repeat that caps the chromosomal ends for processing by the telomere complex. While the telomeric repeat is functionally well understood, most other repeat regions, which e.g. in human cover more than 50 % of the genome, are functionally not understood. It is easily conceived that the quite abundant presence of repeat regions in eukaryotic genomes leads to a large number of ambiguous read overlaps: if for instance the 5' end of one read overlaps the 3' end of two other reads, which however have been obtained from different genomic loci, it becomes impossible to decide which of the two possible overlaps leads to the correct assembly. It is mainly due to the presence of repeats that it was initially believed that shotgun sequencing is principally infeasible for sequencing large eukaryotic genomes.

The applicability of shotgun sequencing to sequence eukaryotic genomes was finally made possible by joint progress in experimental procedures and computational approaches. On the experimental side, pairwise end sequencing was established. Here, chromosomal DNA is fragmented into longer fragments of different sizes, usually 2, 10, 50, and 150 kilo bases (kb). For each fragment, both the 5' end and the 3' end are being sequenced. Naturally the length limitations of either Sanger or next generation sequencing apply for sequencing the two ends. Yet, along with the two sequences at both the 3' and the 5' end – the so-called mate-pair sequences – their distance is known within relatively precise boundaries. Obviously, knowing this distance is very helpful to disambiguate non-uniquely overlapping reads. On the computational side, James Weber and Gene Myers demonstrated in a seminal paper in 1998 (Weber et al. 1997) that paired end sequencing can be utilized in a systematic manner in order to sequence eukaryotic genomes, in particular the human genome, in a very efficient manner. While the approach was initially rejected in particular by the then already started initiative to sequence the human genome using a much less efficient, so-called BAC-to-BAC approach, it was finally employed for sequencing the human genome as well as dozens of other eukaryotic genomes since.

Technically, whole-genome shotgun sequencing works in several stages. First, unambiguously overlapping reads are assembled into longer sequence stretches referred to as contigs, which are typically few thousand nucleotides in length. In the next round, information from long-range mating pairs is used to arrange contigs in the correct order and distance to each other, producing so-called scaffolds. While in an ideal assembly, scaffolds turn out to be complete chromosomes, this is not necessarily always the case, e.g., due to the presence of complex repeat patterns or too few reads available for obtaining a unique assembly. An illustration of whole-genome shotgun sequencing is illustrated in Figure 14.1.

While nowadays genome sequencing is commonly pursued as a routine task in numerous laboratories, it still comes with some caveats resulting from different potential sources of errors. Any sequencing technology produces sequencing errors at a generally low, but always noticeable level. A general problem in genome assembly is to distinguish sequencing errors from biological genomic variability, which can be addressed by sequencing more reads so that each genomic location is covered by several reads. Thus, an important parameter in any genome sequence is the coverage, i.e., the ratio between the summed up lengths of all reads used for a genome assembly versus the length of the genome. Obviously, higher coverage means higher quality of the genome assembly, but also makes sequencing and computation for assembly more expensive. While the human genome is currently available at 12X coverage, some genome projects may involve coverages as low as 3X or even 2X, yet with correspondingly low confidence that may limit some further analyses of the obtained genome sequence, and also decreasing chances to obtain complete chromosomes from scaffolds. Genomes sequenced at 2X coverage are certainly to be taken carefully, as they may contain larger types of errors such as misplaced or wrongly oriented scaffolds within chromosomes.

Whole genome shotgun sequencing has originally been conceived for conventional Sanger sequencing, where typical read lengths in paired end sequencing range between 500 and 1000 nucleotides. It has yet been adapted to next-generation sequencing approaches, where read lengths are generally shorter, for some technology platforms as low as 100 nucleotides, yet at orders of magnitude lower sequencing costs per nucleotide. Different protocols have been developed to deal with such short read lengths, eventually by combining NGS with a certain amount of conventional Sanger sequencing to obtain high quality genome assemblies. As sequencing technology evolves rapidly, it is likely that shotgun approaches will remain the method of choice also for new approaches to sequencing, even if their characteristics may change e.g., towards longer, but lower quality read lengths. As sequencing costs drop exponentially over time, many side effects of new sequencing techniques can be cured by higher coverage.

14.2.2 Genome annotation

Once a genome has been sequenced, it is naturally a challenge to annotate functionally relevant regions in the genome. The most prominent regions to be annotated are arguably protein coding genes, which yet cover only a relatively small part of a typical eukaryote genome – in the human genome, in fact, only 2 % of the 3 billion nucleotides code for protein. A large fraction of the human as well as many other eukaryotic genomes is repetitive. In human, roughly 50 % of the genome are recognized as repetitive elements. Among the remaining non-repetitive part, some estimates exceed the number of 80 % of the human genome being transcribed in some cellular context. While being transcribed in some context is not a sufficient criterion to identify a genomic locus as functional – there may be randomly transcribed "transcriptional noise" – such findings suggest that many non-coding regions in the genome function as transcripts. In fact, there is a large and growing repertoire of so-called non-coding RNA genes, some of which belong to evolutionarily ancient and well-conserved families. Many such non-coding gene families exist beyond the well-known textbook examples of transfer RNAs, ribosomal RNAs or RNAse P genes.

Genome annotation generally goes farther than just annotating what is being transcribed. Annotations of the human genome or model organisms contain numerous annotations of regulatory regions such as transcription factor binding sites or so-called CpG islands, i.e., genomic sequence stretches whose length exceeds 200 nucleotides where the percentage of GC nucleotides is significantly larger than expected in a random genomic region of same length. Such CpG islands are commonly found in promoter sequences upstream to the transcription start sites of protein coding regions. Also, sites of DNA methylation, which are strong regulators of gene expression, are a typical component of genome annotations.

Genome annotations are obtained from a combination of experimental data and computational predictions, and typically made available through so-called genome browsers. These genome browsers make all predictions and genome-wide experimental data accessible in so-called tracks. E.g., there are annotation tracks for computationally predicted mRNAs, along with tracks linking to gene expression data for mRNAs. Other annotation tracks relate to non-coding RNA genes, either computationally predicted or experimentally validated. Correspondingly, other annotation tracks are available for repetitory and regulatory elements, genomic variation, and indicators for conservation across related species.

While the involvement of experimental data for genome annotation is beyond the scope of this chapter, there is a broad range of computational approaches for genome annotation, even for the restricted task of annotating protein coding genes. Annotating mRNAs in a genome can generally be achieved based on two strategies, namely a *de-novo* approach focusing exclusively on statistical signature of protein coding genes within the genome under consideration. Alternatively or supplementary to a *de-novo* annotation, one can use a comparative annotation by identifying regions that are homologous to genes from a closely related species. Statistical features used for *de-novo* annotations are obtained from genomic signatures indicative of open reading frames, which is relatively straight forward in prokaryotic genomes. Here, the genome can be scanned for start codons followed by a sequence of coding triplets and finally a stop codon. If the number of amino acids coded for by the putative coding gene lies within a range typical for protein coding genes, there is already a significant evidence for a protein coding gene. Yet, such procedure would report protein coding genes even in a randomly generated "genomic" sequence, where a corresponding constellation of start codon, coding triplets, and stop codon is somewhat likely to occur if the length of the random sequence is in the range of typical genome sizes, which would lead to false positive annotations of genes. To eliminate false positive predictions, one can utilize further features exhibited by typical proteins. For example, the repertoire of 20 amino acids is not evenly distributed among known proteins. Among human proteins, for instance, rather than each amino acid occurring at a level of 5 %, tryptophan and cysteine are relatively rare peptides, whereas glutamine and serine are much more frequent than other amino acids. Knowing this distribution can be utilized in a

C.elegans TERT
chrl:8,768,514-8,771,171

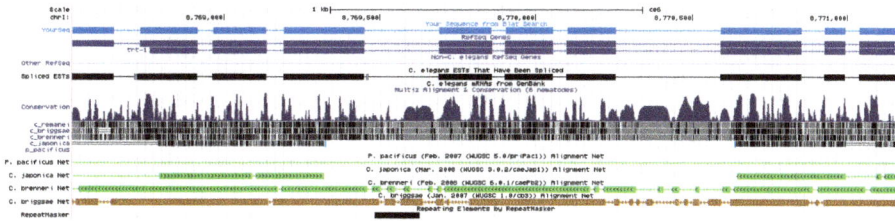

genome.ucsc.edu

C. elegans TERT protein sequence	MAPTIKSSLTTLHEEIRKYDKRRHWNTIIRKSAHLSLVRNIFKKKRRQKLNYRHMSTFLASSENISNIRDTVVLLEINKLSQSILSSRNIFELRKLKPTRKM VEIEHCFVSKNDTFVACSKLISMFHLEKYIGSTNCANILEWITKMLANNGGYBKMKPFVYKEAPFTRAFRASKVMKNVDPNVKRNIQTSIVNNLKSGVHWWALMALRQVMIPIVIKE ERVLLWRDGYLNILTKEIKDFKQRYIVQKAPHFIRPNVATFKLSISRQKLRPLFRRKAIDKKTETMQWKKLNSMLSWCLEKSGVYRHTIRDSCKKVSDFLKKNSQNSKIIGYTADV SKCFSTVNHDVLISIIDRLFSQEHDIYTVCGKGRNHGGFHKLLFCSAGTELNAHEALRRKMELKGVFNFEVCYRHEMSSSTTLYSVIRTTLSTYYYKRGPTSWRITKGVPQGHPISS NLAHMYLNNFEQKYWSNEKEDSRIVFCRYEDDFIPITTENSLFEKMMKPLSTGNNTHFLTANPKKFKKSERCGASQVLQWCGVKLDFQSGNCFIRRRCKDGVARQFLIKLQ

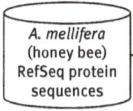

blast.ncbi.nlm.nih.gov

Range 1: 545 to 606 GenPept Graphics

Score	Expect	Method	Identities	Positives	Gaps
32.3 bits(72)	0.84	Compositional matrix adjust.	19/64(30%)	32/64(50%)	2/64(3%)

```
Query  440  KGVPQGHPISSNLAHMYLNNFEQKYWSNEKEDSRIVFCRYEDDFIFITTENSLFEKMMKP  499
            KG+ QG  +S  L+ +Y N     K  S    I+  +Y DD ++IT   +L E+ ++
Sbjct  545  KGIVQGSMLSDIYYNYILNKEMSTYLKTGEII--KYMDDILYITENKTLABQFLEL      602

Query  500  LSTG  503
            G
Sbjct  603  TKKG  606
```

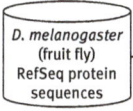

A. mellifera (honey bee) RefSeq protein sequences → **blastp** →

blast.ncbi.nlm.nih.gov

D. melanogaster (fruit fly) RefSeq protein sequences → **blastp** →

Fig 14.2: Example of *C. elegans* TERT (shown within its genomic context in the UCSC genome browser) as a blast query against the protein sequences of *D. melanogaster* and *A. mellifera*. Blast reporting no similarities found indicates that indeed *D. melanogaster* does not possess a telomerase.

statistical manner for reliable identification of real protein coding sequences; for example, if an open reading frame contains many tryptophan residues, it is unlikely to code for a protein. Computational gene prediction approaches utilize such statistics on known proteins to make gene annotations more reliable.

Beside the human genome, the genome of all essential model organisms such as *M. musculus* (mouse), *D. melanogaster* (fruit fly), *C. elegans* (worm), *S. cerevisiae* (baker's yeast) or *A. thaliana* (thale cress) have been sequenced and annotated. For model organisms and genomes of other frequently investigated species, the *RefSeq* database has become the predominant and well curated database for nucleotide sequences and their protein products, linked to their genomic coordinate. Popular genome browsers are thus often built on the basis of RefSeq annotations, and also provide links to RefSeq entries; refer to the coding region of *C. elegans* TERT displayed in the UCSC genome browser in Figure 14.2 as an example.

While identifying open reading frames in prokaryotes is relatively easy, things are more involved in eukaryotes, where mRNA transcripts are usually spliced from

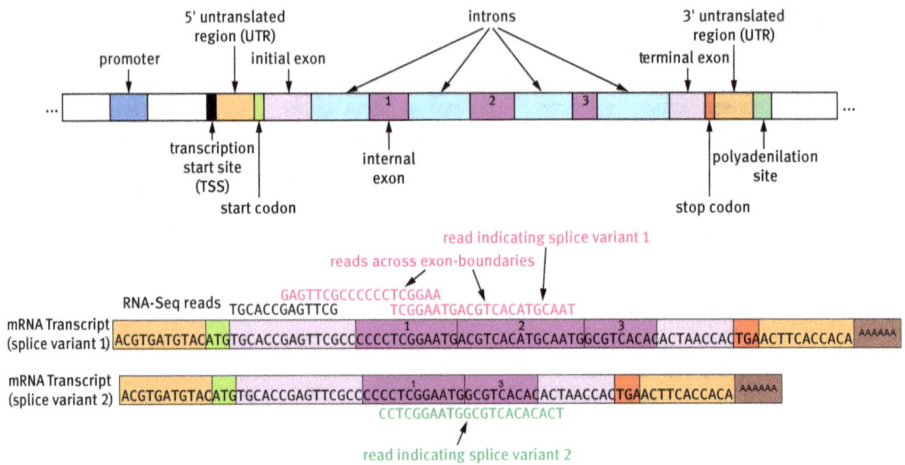

Fig 14.3: Architecture of a eukaryotic gene, where mapping reads needs to take into account exon boundaries. Reads crossing exon boundaries are indicative of splice variants.

several exons (see Figure 14.3). More specifically, open reading frames in the human genome stretch across 8.8 exons on average, where 80 % of all exons are shorter than 200 bases. Intron length varies between less than 20 up to more than 10,000 nucleotides in some cases; in a typical mRNA, not all exons are spliced, but some are skipped. Which exons are spliced indeed depends on cellular context, and as indicated in Figure 14.3, RNA-Seq may identify splice variants. Due to these difficulties, gene annotation in eukaryotic genomes relies on a combination of comparative approaches with the inclusion of high-throughput experimental data, such as ESTs and RNA-Seq data discussed later on in Section 14.4.3.

14.2.3 Partial genome sequencing and resequencing

The dramatic drop in sequencing costs made it possible to not just sequence one reference genome of a species, but also sequence the genomes of several – in some cases hundreds or thousands – of individuals of one species. This allows to unveil the genomic variability within species or populations, and associate this variability with correlated phonotypical variability. One such genotype-phenotype association that is obviously of high relevance is the correlation between variable genomic loci and susceptibility to disease.

Genomic variability between individuals occurs in different types. The arguably most simple type of variation is a single nucleotide polymorphism (SNP), which is a single nucleotide that is variable among individuals and flanked by nonvariable nucleotides. SNPs generally do not occur randomly, but are restricted to certain genomic loci. In the human genome, at the time of writing roughly 50 million of the 3 billion genomic nucleotides constitute validated SNPs. Another type of genomic variability is constituted by so-called copy number variations (CNVs). CNVs affect longer contiguous genomic regions, whose size ranges from few thousands of bases up to several millions. A CNV is given if such genomic region is either duplicated or deleted. Recent studies identified in the order of thousands of CNVs in the human genome using microarray technology; in more recent studies such as the 1000 genomes project, where the genomes of 1000 individuals are fully (re-)sequenced, promise to provide a more accurate and complete snapshot in the near future.

Whenever a reference genome for a species is available, the genome of an individual – the so-called re-sequencing of a genome – can be obtained using much less effort than a complete whole genome shotgun assembly. SNPs can be identified by mapping reads onto the organism's reference genome, i.e., determining a locus on the reference genome where the sequence is identical to the read up to a small number of mismatches. As mismatching positions during read mapping may result from either sequencing errors or SNPs, this process of SNP calling requires a higher coverage than whole genome shotgun sequencing, typically in the order of 20X or higher, depending on whether sample material from one individual is used or genomic DNA from several individuals is being pooled. Yet, mapping reads against a reference genome (also refer to Section 14.4.3) is a computationally much easier task than a shotgun assembly. For investigating genomic variability of larger cohorts, resequencing can be prohibitively expensive despite the ever decreasing costs for sequencing DNA. For such scenarios, partial resequencing approaches are available where sophisticated sample preparation protocols allow for instance resequencing of the exome, i.e., the protein coding part of the genome.

Software Tools and Implementations

Table 14.3: Software tools and resources for genome sequencing.

Celera assembler	Whole-genome shotgun assembly for Sanger sequencing	sourceforge.net/projects/wgs-assembler
Velvet	Whole-genome shotgun assembly, in particular for short read lengths	www.ebi.ac.uk/~zerbino/velvet/
ensEMBL	Sequences and annotations of eucaryotic genomes	www.ensembl.org
UCSC genome browser	Genome browser	genome.ucsc.edu
Phytozome	Sequences and annotations of plant genomes	www.phytozome.net
ENCODE	Study of functional elements in the huaman genome	www.encodeproject.org
RepeatMasker	Masking repeats in genomes	www.repeatmasker.org

14.3 Comparative genomics

As soon as experimental means were available to obtain the amino acid sequence of a protein or the nucleotide sequence of DNA, the question of how to compare evolutionarily or functionally related DNA or protein sequences raised more or less automatically. Any algorithmic approach to comparing sequences naturally involves an implicit or explicit assumption how the differences may have evolved, so that most sequence comparison approaches involve some concept of sequence evolution. As the right concept may vary in different settings, numerous approaches for different tasks related to the comparison of sequences have been proposed and implemented.

14.3.1 Sequence alignments

The first approaches to compare sequences, in the late 1960s typically protein sequences, were based on dot plots: Two sequences to be compared were drawn in a matrix, one sequence horizontally and the other sequences vertically. Whenever the peptide at position i in the first sequence is identical to the peptide at position j in the second sequence, at dot is printed at position (i,j) in the matrix. A resulting dot plot for the human vs. chicken TERT gene is shown in Figure 14.4. An obvious pattern is that the TERT peptide sequence contains sub regions that are fully identical between human and chimp. Yet, other sub regions exhibit little or no sequence identity at all. These divergent regions result from insertions, deletions, or substitutions on the level of DNA during the course of evolution.

While providing an intuitive understanding of sequence conservation, analyzing dot plots hardly leads to an objective measurement of sequence similarity, and becomes cumbersome as soon as more than two sequences will be compared. As sequences of many genes from numerous species became available, computational approaches to sequence comparison gained importance. The seminal work by Needleman and Wunsch (1970) captured the pattern of insertions, deletions, and

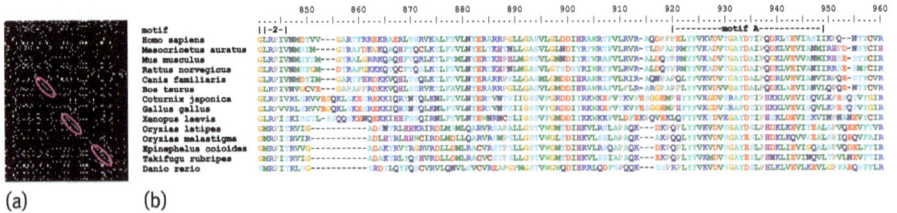

(a)   (b)

Fig 14.4: *Left:* Dot plot between subsequences of human and chicken TERT. The homologous peptides occur around the diagonal (highlighted in red). *Right:* Excerpt of 120 out of 1353 columns of a a multiple sequence alignment of vertebrate TERC (sub-)sequences.

substitutions in an algorithm, and proposed a way to find the maximal number of identical positions in two sequences when accounting for these three modifications. The Needleman-Wunsch algorithm essentially identifies diagonal elements in the dot-plot representation, which also made it possible to write aligned sequences as shown in Figure 14.4. Here, evolutionarily inserted or deleted sequences are indicated by gaps in the respective sequence, and identical positions are written within the same column. This representation of aligned sequences is particularly useful when dealing with more than two sequences.

A practical problem occurs when aligning more than two sequences. In general, one can generalize the Needleman-Wunsch approach to more than two sequences. Yet, rather than working on a dot-plot matrix, the structure to work on is a "dot-box" for three sequences, and for more than three sequences a "high-dimensional dot-hyper-matrix". While this is technically possible to implement and would lead to useful sequence alignments, it leads to infeasible running times in practice even when dealing with a handful of sequences only. To circumvent this problem, several approaches have been developed to make multiple sequence alignments feasible for practical purposes. The most common strategy is the so-called agglomerative strategy, which first infers a guide tree that displays similarity patterns of the sequences to be aligned. In this tree, similar sequences occur as close neighbors in the tree, while divergent sequences occur far apart, eventually only linked through the root of the tree. The alignment algorithm can now perform pairwise sequence alignments in a "bottom-up" fashion. For vertices that have two leaves as descendants, one can perform a conventional pairwise sequence alignment. For vertices closer to the root, the pairwise alignment for this vertex needs to align two previously computed alignments, which can also be done in an Needleman-Wunsch-like fashion. Finally, at the root of the tree, the complete multiple sequence alignment of all sequences is obtained.

A potential problem of the agglomerative approach is that if the topology of the guide tree accidentally groups two only distantly related sequences together, an erroneous partial alignment is produced at a low-level vertex. There is no chance for this error to be corrected as more and more sequences are aligned towards the root of the guide tree, thus producing an overall erroneous alignment. The use of a guide tree may also be problematic in a different sense. Multiple sequence alignments are the most common starting point for reconstructing phylogenetic trees, where each column in the alignment conveniently yields one phylogenetic character. The general point to be aware of is that if using a agglomerative alignment strategy for an alignment to infer a phylogenetic tree, a tree-reconstruction method has already been used to obtain the guide tree, constituting a circularity in the methodology. It has thus been carefully established for common practices in molecular evolution that the – potentially evolutionarily wrong – guide tree will not affect the trees inferred from agglomerative sequence alignments, which over more than two decades have remained the state-of-the-art approach for multiple alignments.

A specific problem to deal with when aligning two protein sequences is to appropriately take into account the biochemical similarity and dissimilarity among the 20 amino acids. For example, it will generally be more reasonable to align a hydrophobic residue in one protein with a hydrophobic residue in another protein, while aligning it with a hydrophilic one may be rather questionable. In order to reflect this in sequence alignments, so-called substitution matrices have been developed that assign a score to each potential substitution of amino acid x by amino acid y, thus producing a 20×20 scoring matrix. Rather than starting form biochemical properties of amino acides, both the PAM (Point accepted mutation) matrices developed by Dayhoff in 1978 and the BLOSUM (Blocks Substitution Matrix) matrices by Henikoff and Henikoff in 1992 start from an evolutionary perspective. In their basic form, the PAM matrix is obtained from substitutions observed in pairs of very closely related proteins, which exhibit a sequence similarity of at least 85 %. The resulting scoring matrix will reasonably represent substitution patterns over short evolutionary times, but may not properly represent substitutions over longer time periods, as there may be "transitive" substitutions, e.g., of Asp mutating into Glu, and then Glu into Leu, although no direct Asp-Leu mutation has been observed in the alignments to produce the basic PAM matrix. In fact, with increasing evolutionary times, biochemically more unlikely substitutions become more likely. To take this into account, PAM matrices are equipped with a time scale parameter. E.g., the PAM250 matrices is obtained by multiplying the basic PAM matrix 250 times with itself, which biologically corresponds an evolutionary time span of 250 times the average time span between the sequences that were used for obtaining the basic PAM matrix.

14.3.2 Homology search

The invention and breathtaking development of DNA sequencing technology along with the resulting huge repositories of genomic sequences has created homology search as another challenge for bioinformatics. For example, for a newly sequences eukaryotic genome one may ask whether the genome contains a copy of TERT gene. Correspondingly, one would search the genome for an open reading frame that translates to a protein with high sequence similarity to the TERT protein from a closely related species. More specifically, performing this search using the *C. elegans* TERT protein as a query against the ORFs annotated in the genome of the honey bee (*Apis mellifera*) as displayed in Figure 14.2, one obtains a unique hit exhibiting 30 % sequence identity. Note that performing the same search for *C. elegans* TERT in the fruit fly (*D. melanogaster*) genome fails to report any close homolog – indeed indicating the well-known fact that *D. melanogaster* as well as other drosophilids possesses an alternative, namely retrotransposon-based, mechanism to maintain chromosomal ends, which is somewhat exceptional within the eukarya. This example yet shows how homology search may provide immediate insights into the evolution of molecular mechanisms.

Note that homology search comes in different variants. In some cases, one may search a protein sequence within a database of protein sequences. In other cases, a protein sequence may be queried against a genomic DNA sequence, or both the query and the database may be DNA sequences. If either the query or the database sequence are DNA, it needs to be taken into account whether each sequence codes for proteins, so that similarity scoring may be performed using PAM or BLOSUM matrices in translated amino acids. In any case, the general terminology in homology search is that a typically short query sequence is given (typically the protein or DNA sequence of an individual gene), which is searched against a database. The database may be the genome of an organism, a collection of several genomes, a collection of protein sequences or coding sequences, or even a broader collection of sequences such as the complete GenBank database.

Homology search nowadays is a routine task performed on an everyday basis with numerous applications in evolutionary or functional genomics and many other fields. Making homology suitable for routine use yet took substantial amount of efforts for the bioinformatics research community during the 1980s, when in particular the availability of sequenced DNA started mushrooming. The first landmark contribution to homology search was the idea initiated by Smith and Waterman (1981) to carry the idea behind the Needleman-Wunsch algorithm to local alignments. Rather than aligning two sequences from beginning to the end, the Smith-Waterman algorithm aims to identify subsequences of each of the two sequences that can be aligned under a small number of insertions, deletions and substitutions. This is particularly attractive for searching genes within genomes, as the alignable subregion of the genome is explicitly determined in a local sequence alignment. Computationally, the Smith-Waterman algorithm is a simple yet clever modification of the Needleman-Wunsch algorithm. This means that indeed the complete "dot plot matrix" needs to be considered during the alignment procedure. Note that the size of the matrix is the product of the length of the query and the size of the database. In a quite common scenario where the query comprises 1000 nucleotides and is searched against the human genome with 3 billion nucleotides, this quite obviously results in running times that too easily become infeasible for a routine task, even if using substantially fast and powerful computer hardware.

While the Smith-Waterman algorithm indeed provided a highly appropriate concept to address the problem of homology search, it was largely hindered by its slow running time. In subsequent years, substantial efforts of the bioinformatics community were spent on improving the performance of Smith-Waterman-like approaches without sacrificing the accuracy of their results. It took indeed almost ten years before Altschul, Gish, Miller, Myers, and Lipman (1990) came up with the concept behind the Basic local alignment search tool (BLAST) that since its publication in 1990 and until today constitutes standard approach to homology search.

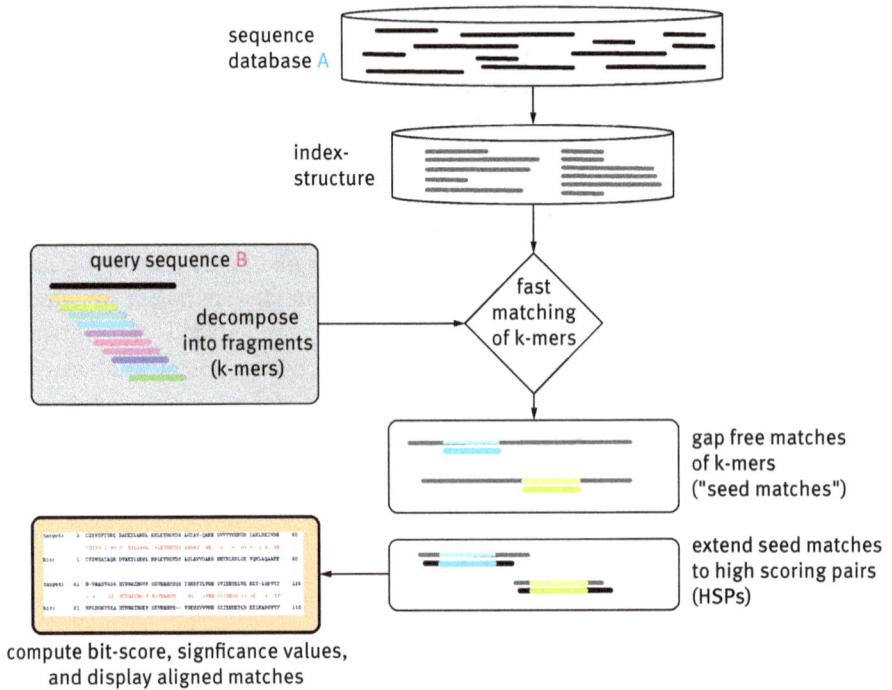

Fig 14.5: Index generation, decomposition into k-mers, seed matching, and extension to matches in a BLAST search.

The key idea behind blast illustrated in Figure 14.5 is to first decompose both the query and the database into short fragments of fixed length, i.e. extract all subsequences of either 9 nucleotides (DNA) or 3 peptides (protein). Using these so-called k-mers, where k denotes the fragment length, seed matches of the query k-mers in the database sequence are determined. Seed matches may involve a few mismatches, but no insertions or deletions. The seed matches are then extended to the front and the back using a Smith-Waterman approach into longer local alignments, which are then reported as a result of the BLAST search is their alignment score exceeds a threshold that is typically specified by the user along with the query.

BLAST indeed allows to search very large databases within running times that are usually in the order of few seconds. The reason for the efficiency lies in searching for the seed matches, which utilizes a precomputed index structure for the database sequence. With this index at hand, a sophisticated algorithm allows identifying seed matches in a way that the running time is not affected by the size of the database. The only limitation of BLAST compared to the much slower Smith-Waterman algorithm is that it may only find those local alignments that contain a gap-free seed-match of either 9 nucleotides or 3 peptides. While for many scenarios, this condition tends to be satisfied, it reaches its limits when searching across longer evolutionary time gaps or when dealing with sequences such as many non-

Table 14.4: Types of homology searches using BLAST.

blastn	Search a nucleotide database using a nucleotide query
blastp	Search a protein database using a protein query
blastx	Search a protein database using a translated nucleotide query
tblastn	Search a translated nucleotide database using a protein query
tblastx	Search a a translated nucleotide database using a translated nucleotide query

coding RNA genes (see Section 14.4.1) evolution at sequence level appears to follow a more fragmented pattern.

According to the different variants of homology search involving either protein or (translated) DNA sequences, BLAST supports all possible combinations of sequences, which are summarized in Table 14.4.

Homology search across larger evolutionary time scales, or queries involving rapidly evolving sequences, may result in either no matching sequences, or a potentially large number of ambiguous matches, where none of the alignment scores stands out significantly to identify a true homologue. Indeed, such matches may occur just by chance, as the chance for a subsequence that randomly matches the query sequence rises proportionally with the size of the database. The question of what is an alignment score that identifies a statistically significant homologue naturally arises. For local alignments as in BLAST, this questions has been answered through work by Karlin and Altschul (1990), who introduced corresponding statistical measures along with the original publication of BLAST in 1990. The so-called Karlin-Altschul statistics can be explained in terms of searching 561 peptides long *C. elegans* TERT against the known coding repertiore of the *A. mellifera* genome, which in total comprises 12,154,792 peptides. The highest scoring match reported in this search achieves a score of $S = 32.3$, which is the sum of all BLOSUM scores of aligned peptides (omitting positions that involve a gap in either sequence). To judge whether a score of $S = 32.3$ is significant or not, Karlin-Altschul statistics consider what would happen in a random sequence: Assume we generate a database of random protein sequences, comprising the same number of 12,154,792 peptides (and following an identical distribution of the 20 amino acids). Assume also we generate a random query of 561 peptides. Now, in this random scenario, what is the probability that we obtain a BLAST hit whose score exceeds 32.3? The answer to this question indeed defines the p-value. Obviously, if this probability is very low (say 0.001), the hit we obtained will be statistically very significant. In this random sequence scenario we may also ask how many matches we are expected to find whose score exceeds $S = 32.3$. The answer to this question defines the E-value, where again small E-values close to 0 indicate significant matches.

In principle, the p-value and E-value could be computed in an empirical manner: One could actually generate say 100 random query and database sequences, and count the matches whose score exceeds 32.3, which would allow us to compute the significance scores immediately according to their definition. As this means multiplying the running time by a large factor just to obtain significance scores, it obviously defeats the purpose of BLAST as a highly running-time optimized procedure. Luckily, this is indeed not necessary, and the p-value can be computed according to the surprisingly simple formula

$$E = m \times n \times 2^{-S'},$$

where S' is a normalized version of the actual score S, and m and n are the size of query and database, respectively. The p-value can be obtained from the E-value using

$$P = 1 - e^{-E}.$$

the formulas for p-value and E-value result from the observation that BLAST hits fall into the category of *rare events* and thus follow a Poisson distribution.

There are several aspects of p-values and E-values to be aware of when interpreting the outcome of BLAST homology searches. First, the E-value is directly proportional to the size of the database. It is an obviously prohibitive shortcut to "improve" the E-value by reducing the size of the database, e.g., dividing homology search in a genome into separate searches for each chromosome. Also, low complexity parts of sequences, e.g. several consecutive repetions of one and the same nucleotide or peptide, are unlikely due to true homology, but will generally achieve over-optimistic significance scores if such low-complexity regions are not filtered out in advance (which BLAST can do automatically). The same phenomenon occurs even more pronounced when the hit alignment involves repeat regions, so that BLAST searches should in general be performed against repeat masked genomes.

There is no universal threshold for either p-value or E-value to distinguish true homologs from false positive ones. The E-value is more commonly used, and some often applied cut-offs are 10, 10^{-3} or 10^{-9}. Yet, beside the size of the database and the query, there are other criteria that are impossible to take into account by BLAST itself. For example, the family of G-protein coupled receptor (GPCR) proteins, a family of seven-transmembrane receptors of early eucaryotic origin, is one the one hand highly conserved, and on the other hand their genomic repertoire underwent rapid gene duplications and losses for instance within the mammals. The repertoire of human GPCRs involves less than 900 receptors, whereas in mouse, there are more than 1600 known GPCRs. Taking any individual receptor from human or mouse and blasting it against the other genome will certainly yield a large subset of all GPCRs, or at least one GPCR-subfamily, as hits with seemingly very low E-values. Yet, identifying the "true homologue" of a GPCR (or rather all potential isoforms of it) obviously requires a more thoughtful approach than merely relying BLAST significance scores.

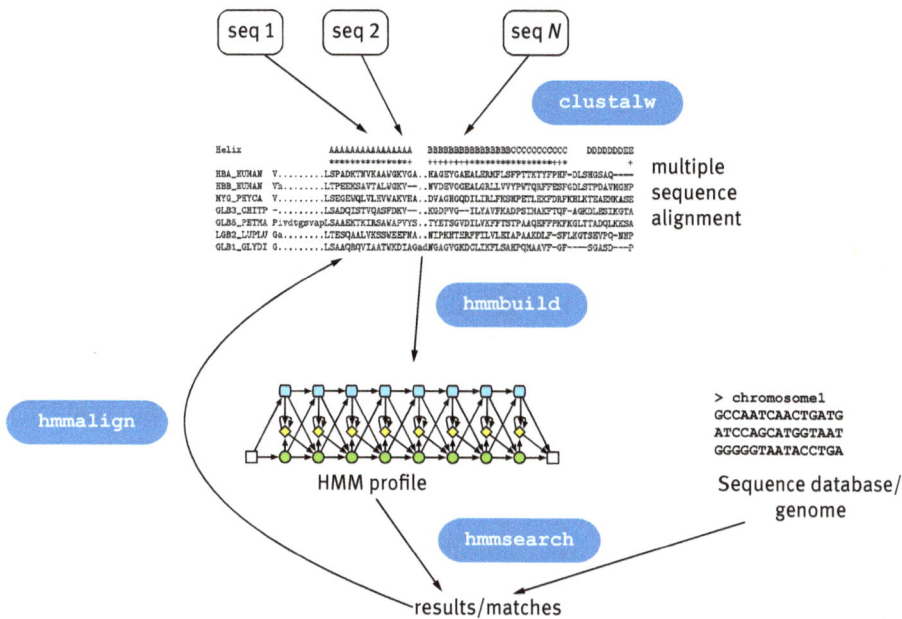

Fig 14.6: Principle workflow of homology search using profile HMMs.

When trying to identify members of a well-characterized family of proteins in a newly sequenced genome, one can also rely on the known evolutionary patterns of the family in order to obtain more reliable homologues. For instance, the TERT protein is known to contain nine highly conserved motifs, each of which is few dozens of amino acids in length and where only few positions vary between different vertebrates. During homology search, the regions belonging to these motifs are obviously much more relevant than positions outside these motives. For a BLAST search, this is however difficult to take into account. A much more systematic way of utilizing such conservation patterns are profile-based approaches to homology search. For vertebrate TERT, such profile based approach in fact will utilize a multiple sequence alignment of all known vertebrate TERT proteins as an input, and infer a statistical search pattern from it. This search pattern incorporates the distribution of amino acids in each column, and also the probability of insertions or deletions. Doing this in a systematic manner in fact leads to the concept of Hidden Markov Models (HMMs), which is illustrated in Figure 14.6. An advantage of HMMs is that the statistical model allows matching sequences to be integrated into an existing alignment in a straight-forward manner, so that a gene family can be traced by searching occurrences from a small core repertoire in few closely related species, and the repertoire can be gradually extended to more and ever more evolutionarily distant species by integrating newly identified homologues into the search pattern. Due to these characteristics, HMMs constitute the methodological core underlying the Rfam database, which provides alignments of known and well-characterized families of proteins.

14.3.3 Whole-Genome Comparison

As can be seen from the aforementioned example of identifying GPCR homologues between human and mouse, taking an isolated look on an individual gene or a small family of genes easily hits limitations, even when utilizing the more advanced approach of HMMs. A broader picture can be obtained when comparing two genomes as a whole. Doing so on the basis of all ORFs annotated in a genome leads to the concept of orthology annotations, which aim to resolve the evolutionary relationships between all genes in two (or more) genomes in a systematic manner. The key concept behind orthology annotations is obviously orthology, where two genes are termed orthologous if they descend from the same ancestral gene separated by a speciation event. Difficulties to identify orthologues arise in the presence of paralogues in one or both of the genomes, i.e., genes that are duplicated in one genome, but not in their most recent common ancestor species. Untangling these relationships is a delicate and computationally demanding task, which involves a complete all-against-all comparison of the annotated genes in both genomes, typically using complete Smith-Waterman or Needleman-Wunsch style alignments rather than BLAST scores, which are not accurate enough for untangling some more obfuscated evolutionary relationships. The resulting "all-vs-all" scoring matrix is used to identify unique orthologue pairs using combinatorial assignment algorithms. Some recent variants even take into account syntheny, i.e., if two genes occur in direct neighborhood to each other in both genomes, both gene pairs receive preference for being identified as orthologue pairs. for many genes, it is not possible to infer one-to-one orthology relationships. Some approaches then infer many-to-many orthologue assignments, indicating that one group of genes in genome *A* contains orthologues to another group of ortholgues in genome *B*, without being able to resolve the exact one-to-one orthologue pairs. In most cases, these orthology groups are small and comprise 2–10 genes only for common vertebrate orthology annotations. Pairs of genes within one group of a many-to-many assignment obviously constitute potential candidates for paralogues.

While orthology annotations are largely useful, they are inherently limited to the protein coding part of genomes. Often, however, it will be useful or necessary to identify evolutionary conservation patterns of the remaining non-protein coding part of the genome. For example, if a transcript is identified using PCR or sequencing of reversely transcribed RNA that originates from a genomic location without any potential to code for protein, one of the first questions to be raised will be whether the transcribed sequence is evolutionarily conserved, and if so how well and across what species. Such information is systematically acquired through so-called genome wide alignments. The goal of genome wide alignments is quite simple, namely to provide multiple sequence alignments of orthologous genomic loci. In practice, obtaining such alignments is a complex task – note that the evolution of genomes follows complex and not fully understood patterns, where for instance long stretches of sequence change their positions across different chromosomes in so-called rearrangement events.

When dealing with two genomes only, a variant of the BLAST algorithm, BLASTZ proposed in 2002 can efficiently find alignable regions between two genomes and compute the pairwise alignments between these regions. Things get more involved in the presence of three or more genomes, for which a program called MULTIZ in combination with a so-called threaded blockset aligner has been proposed. The resulting alignments are linked in popular genome browsers for broad ranges of taxa, and beside the alignments themselves the genome wide alignments can be used for computing conservation scores, which provide an immediate insight how well a given locus e.g. in the human genome is conserved throughout other mammals or vertebrates. An example of these conservation scores can be seen in Figure 14.2, where the genomic context of the C. elegans TERT gene is displayed within the UCSC genome browser, where the annotation track labeled Conservation is obtained from MULTIZ alignments with related species. The alignable regions in closely related species (*P. pacificus*, *C. japonica*, *C. brenneri* and *C. briggsae*) obtained from these MULTIZ alignments are also displayed.

Software Tools and Implementations

Table 14.5: Sequence alignment software.

ClustalW	Multiple sequence alignments	www.ebi.ac.uk/Tools/msa/clustalw2
TCoffee	Multiple sequence alignments	www.tcoffee.org
muscle	Multiple sequence alignments	www.drive5.com/muscle
BLAST	Local alignment based homology search	blast.ncbi.nlm.nih.gov
ssearch	Implementation of the Smith-Waterman algorithm	www.biology.wustl.edu/gcg/ ssearch.html
hmmer	Implementation of profile HMMs	hmmer.janelia.org
Pfam	Sequences and alignments of protein families	pfam.xfam.org
OrthoDB	Orthology annotation database	www.orthodb.org
OrthoMCL	Software for computing orthology annotations	orthomcl.org
InParanoid	Software for computing orthology annotations	inparanoid.sbc.su.se
BLASTZ	Pairwise whole genome alignments	www.bx.psu.edu/~rsharris/lastz/
MULTIZ	Multiple whole genome alignments	www.bx.psu.edu/miller_lab/

14.4 Transcriptome analysis and non-coding RNAs

The bioinformatics approaches described in this chapter so far are largely focused on the protein coding part of the genome. Yet, not just in human, but most other eukaryotic genomes, only a small fraction of the genome codes for protein. In fact, tremendous efforts have been spent to development approaches for analyzing the non-protein coding sequences, and understand the origins of those parts of the genome that have often been referred to as junk DNA, albeit with the connotation that garbage you throw away, junk you keep (which is commonly attributed to Sydney Brenner[1]). Looking at the conservation score curve of the C. elegans TERT gene in Figure 14.2, one can clearly see that some parts of intronic sequence are as highly conserved as exonic sequences, suggesting a sequence-related function of the intronic areas that is evolutionarily conserved on the level of DNA sequence. While for intronic sequences it is not too surprising to observe a high degree of conservation, highly conserved areas can be found throughout all eukaryotic genomes (typically identified through MULTIZ genome wide alignments) even in genomic "deserts" located far away from any protein coding gene. As evolutionary conservation of a genomic region is unlikely to occur by chance, this provides evidence for functional roles of non-protein coding regions in the genome.

14.4.1 Non-coding RNA secondary structure

As findings from the ENCODE project (see Box 14.2) suggest, up to 90 % of the non-repetitive part of the genome are transcribed in some cellular context. An important contribution to answering the imminent question which of these transcripts are functional and in which way lies in the structures into which these single-stranded RNA transcripts fold.

It is commonly known that in transcribed single-stranded RNA, complementary nucleotides base-pair with each other within the transcripts. These base-pairing patterns lead the RNA molecule to fold into a spatial structure, which in many cases is known to be relevant for exhibiting certain functions inside cells. The most prominent example presumably is the cloverleaf structure of transfer RNA displayed in Figure 14.7, which takes the role of translating the genetic code into amino acids.

The secondary structure of an RNA sequence can be represented in different ways. First, it can be drawn as arc diagram (Figure 14.7), where the sequence is written in linear order, and an arc is drawn between any two positions in the se-

---

1 Sydney Brenner is a South African born British biologist. Beside his Nobel prize awarded work on establishing C. elegans as a model organism for developmental biology, he was a main contributor to other key discoveries such as the unveiling of the genetic code.

Box 14.2: The ENCODE Project.

The sequencing of the human genome was famously referred to as a race to the start line ten years onwards, as it may have raised more questions regarding what parts of the genome are functional and it what way than the availability of the genome sequence has answered. Thus, as a follow-up to the human genome project, the ENCODE (Encyclopedia of DNA Elements) consortium was founded aiming to identify all functional elements in the human genome. This ambitious enterprise is investigated within several subprojects, which study aspects of mRNA and ncRNA expression profiling, transcription factor binding, histone modifications and replacement, chromatin structure, DNA replication initiation and timing, as well as genomic variability in terms of SNPs and and CNVs. The corresponding large-scale experimental efforts cover a large range of cells types, in its current phase involving 147 different cell types in total. While one of the birds-eye conclusions that 80 % of the genome are functional may be somewhat arguable due to the quite liberal definition of what is functional, the ENCODE project has contributed many insights into functional elements and how they affect cells and phenotypes in different contexts, for example a preliminary sketch of the architecture of the network of human transcription factors. The concepts and protocols of the ENCODE project have been carried to related projects on other model organisms in the modENCODE project, where *Drosophila melanogaster* and *Caenorhabditis elegans* are under investigation using an analogous approach. Just as for the ENCODE project, all data produced are released for public use.

quence that base pair with each other. Second, one can list all pairs of positions that base pair with each other (Figure 14.7). Third, one can represent the secondary structure as a sequence of parentheses and dot characters. Base pairing positions are indicated by matching pairs of parentheses, while nucleotide positions not involved in any base pairing are indicated by a dot. The structurally possibly most intuitive representation are planar representations as a fourth possible graphical representation, where the sequence is drawn following a non-linear path so that pairing bases are drawn immediately next to each other (Figure 14.7).

The base-pairings in a non-coding RNA molecule lead to higher order structures that are best characterized in the planar representation, where the base-pairing patterns delimit different types of areas in the plane. Looking at the four "leaves" in the tRNA cloverleaf structure, the corresponding areas are adjacent to only one base pairing, and one consecutive stretch of unpaired nucleotides. Stretches of unpaired nucleotides are referred to as loop regions, and those adjacent to only one base pairing are termed hairpin loops due to their obvious similarity to a hairpin. The central "branching point" in the cloverleaf structure, on the other hand, is adjacent to three base-pairings. The corresponding unpaired nucleotides in the structure are called a multi loop. Another loop type not commonly part of the tRNA cloverleaf structure, but found in many other structured RNA molecules, is the interior loop, which is adjacent to precisely two base pairings. A special form of the interior loop is the bulge loop, where one stretch of adjacent unpaired nucleotides has length zero. If both stretches have length zero, no unpaired nucleotide is adjacent to the area, which is called a stacking pair. Sequentially stacked stacking pairs are also referred to as stems. Finally, stretches of unpaired nucleotides at either end of the sequence define so-called dangling ends.

(a)

GGGCUAUUAGCUCAGUUGGUUAGAGCGCACCCCUGAUAAGGGUGAGGUCGCUGAUUCGAAUUCAGCAUAGCCCA

| 1 | 5 | 1 0 | 1 5 | 2 0 | 2 5 | 3 0 | 3 5 | 4 0 | 4 5 | 5 0 | 5 5 | 6 0 | 6 5 | 7 0 |

(b) ((((((((..((((.........)))).(((((.......))))).....(((((.......)))))))))))))

(c) {{1, 73}, {2, 72}, {3, 71}, ... ,

    {53, 63}, {54, 62}}

(d)

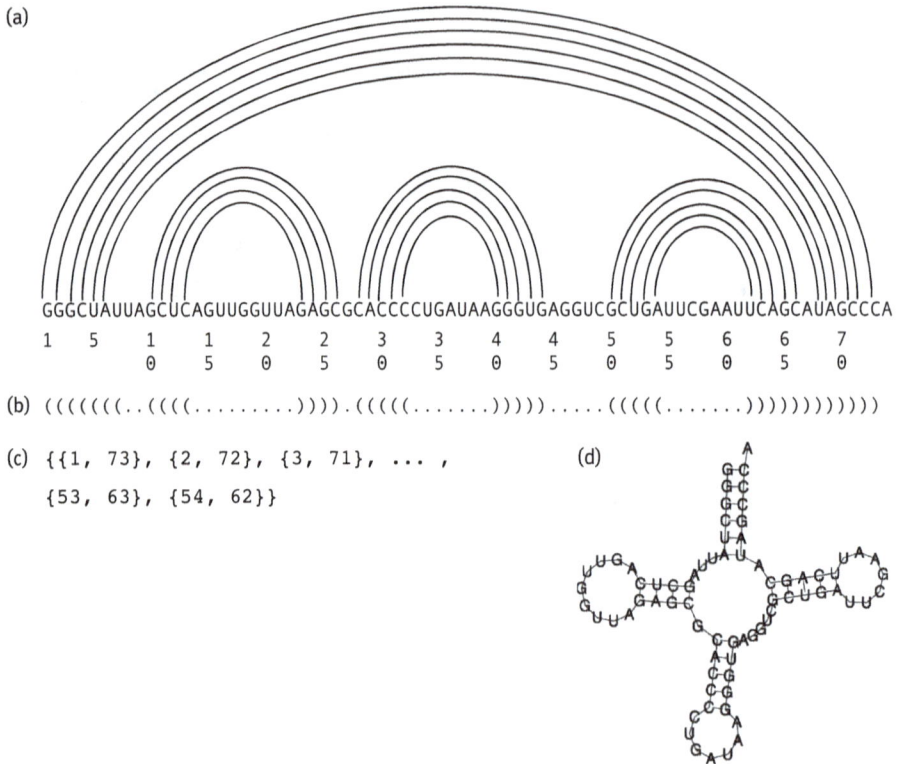

Fig 14.7: Transfer RNA secondary structure in different representations. (a) Arc diagrams, (b) parentheses string, (c) set representation and (d) planar representation.

The aforementioned structural elements are displayed in Figure 14.8, and every RNA secondary structure can be decomposed into precisely these components – as long as the structure does not involve pseudoknots. Pseudoknots result from base pairings that lead to crossing arcs in the arc diagram, i.e., from a base pairing between positions i and j and another base pairing between k and l such that $i<k<j<l$. While many known non-coding RNA genes such as tRNA do not involve pseudoknots, important examples such as telomerase RNA exist where pseudoknots are evolutionarily conserved and take an important functional role. In telomerase RNA, a pseudoknotted region hosts the template for the telomeric repeat, which is a structural feature present in all known instances of telomerase RNA throughout the eucaryotes.

Following seminal work by Matthews and Turner (1999), an energy model exists that can assign stabilizing or destabilizing energy terms to each structural element in a given un-pseudoknotted secondary structure. Thus, given a fixed secondary structure, one can determine its free energy by summing up all free energy contributions from the decomposition into structural elements. It somewhat suggests itself to use the availability of an energy model to actually predict a minimum

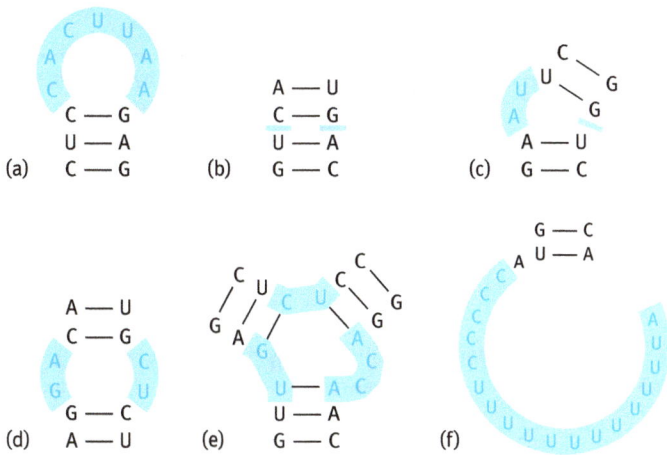

Fig 14.8: Loop elements in RNA secondary structures. (a) Hairpin loop, (b) stacking pair, (c) bulge loop, (d) interior loop, (e) multi loop and (f) dangling end.

free energy structure from the sequence of a non-coding RNA. A naive way of doing so would be to enumerate all possible base pairing patterns, and determine the decomposition into structural elements along with the free energy, while keeping track of which structure achieved the minimal free energy. As however the number of possible base pairings patterns grows exponentially with the length of the sequence, this approach would be prohibitively slow even for short RNA sequences. A better way was found by Michael Zuker (1980), who proposed an algorithm to predict the minimum free energy secondary structure of a given sequence that manages to avoid exponential running time, building on a so-called dynamic programming approach introduced earlier by Ruth Nussinov (1980).

While the Matthews-Turner model is quite realistic for RNAs in aqueous solution, it is usually far from precise in the presence of a cellular surrounding, where RNAs form ribonuclear complexes with proteins and other external factors affect the base pairing patterns. Thus, structure prediction based on running the Zuker algorithm on an individual RNA sequence leads to results of limited use in general. For example, the Zuker algorithm reports the well known cloverleaf structure only for a small fraction of known tRNAs; the cloverleaf structure is realized by a structure that in energetically not minimal in the Matthews Turner model.

To obtain realistic predictions of RNA secondary structure despite these effects, it has turned out that taking an evolutionary point of view is often helpful in the form of consensus folding. If beside a single non-coding RNA sequence, several homologues from closely related species are known, one can first align the corresponding sequences using a conventional multiple sequence alignment program. Now, the Matthews-Turner model can be enhanced to provide free energies of structures of aligned sequences rather than of individual sequences. An adaptation of

Table 14.6: Selected families of non-coding RNA.

Acronym	Class	Length	Function	Phylogenetic Distribution
tRNA	transfer RNA	70–80	translation	ubiqitous
16S/18S rRNA	ribosomal RNA	~1500	translation	ubiquitos
28S rRNA	ribosomal RNA	~3000	translation	ubiquitos
5S rRNA	ribosomal RNA	130	translation	ubiquitos
RNAse P		220–440	tRNA maturation	ubiquitos
RNAse MRP		250–350	endonuclease, rRNA maturation	eucarya
telRNA (TERC)	telomerase RNA	150–1800	replication termination	eucarya
miRNA	microRNA	~22	post-transcriptional regulation	multi-cellular organisms
snRNA	small nuclear RNA	100–160	spliceosomal	eucarya
snoRNA	small nulcleolar RNA	60–130	rRNA modification	eucarya
7SK RNA		~300	transcription regulation	animalia
Y RNA		80–100	part of Ro particle	metazoa
vRNA	vault RNA	80–100	part of vault particle	vertebrata

the Zuker algorithm then allows to infer a minimum energy structure that the aligned sequences fold into. A major advantage of consensus folding is that the resulting consensus structure may unveil compensatory mutations – if for instance a base pairing is realized by an AU pairing in one species has been mutated into a GC base pairing in another species. Such compensatory mutations are very unlikely to occur by chance. If a structure obtained from consensus folding involves many such compensatory mutations, this is a clear sign of evolutionarily conserved secondary structure. Compensatory mutations are indeed commonly observed, for instance in the Rfam database as the major resource of evolutionarily conserved families of non-coding RNA genes, which contains sequences, structures and alignments of more than 2000 known families of ncRNAs. Some prominent families of ncRNAs are listed in Table 14.6. Note that these classes are only the most prominent among the more than 2000 families of ncRNAs, and in particular through the use sequencing technology, novel functional classes have been discovered in recent years.

Beside unveiling compensatory mutations, consensus folding yields other signals of conserved secondary structure. One such signal is obtained from the minimum free energy of the consensus secondary structure in relation to the average minimum free energies obtained from folding each of the aligned sequences individually using the conventional Zuker algorithm. If the ratio between the consensus MFE and the average individual MFE is close to one, a strong evidence of structure conservation is given. Approaches such as RNAZ, EVOFOLD or PFOLD capture structural conservation indicators in a statistical manner. Applied to genome wide alignments, they have predicted thousands of candidates for conserved mammal or vertebrates, many of which have been validated experimentally.

14.4.2 Homology patterns of non-coding RNAs

Due to the selection pressure on the level of secondary structure, the evolution of ncRNAs sequences follows substantially different patterns than protein coding sequences. This has two consequences. First, ncRNAs are generally more difficult to align. While protein sequences and protein coding sequences can be considered well alignable if sequence identity exceeds 30 % – often even 10 % sequence identity is sufficient – it has been found that trustworthy alignments of ncRNAs require a level of 60 % sequence identity. Moreover, open reading frames are relatively easy to spot in a genome due to the presence of start and stop codons and a well understood distribution of the amino acids the codons code for. Non-coding RNAs exhibit no such universal features, so the only signals left to recognize them are either signals of evolutionary conservation – inherently prohibiting studies focused on a single species – or relying on features specific for individual ncRNA gene families – inherently prohibiting studies involving the complete repertoire of ncRNAs. It should be noted that simple BLAST search generally fails to report true homologs of ncRNAs, where the true homologs will either not occur in the list of BLAST hits, or will be hidden among a long list of equally significantly scoring false positives. Performing a BLAST search for an arbitrary human tRNA in the chimp will be a curing exercise to experience the limitations of BLAST in relation to ncRNAs.

In order to reliably annotate the repertoire of known families of ncRNAs, specific computational approaches have been developed. Considering the importance of evolutionary conservation in ncRNAs, profile approaches such as HMMs are generally relevant for ncRNA homology search. However, HMM are not capable of capturing the conservation patterns behin compensatory mutations, which can affect distantly located sites, whereas HMMs only have a "local" view on individual or immediately consecutive sites in an aligned family profile. To facilitate recognition of conserved base pairings, so-called covariance models have been successfully utilized to generate homology patterns of aligned profiles of ncRNAs. Mathematically, covariance models are indeed a generalization of HMMs, and as such constitute the algorithmic foundation behind the Rfam database, just as HMMs constitute the methodological foundation behind the Pfam database.

Some families of ncRNAs such as TERC, are resilient even to covariance models due to their obfuscated evolutionary patterns. Indeed, a covariance model of mammalian TERC is sufficient to identify TERC in non-mammalian vertebrates such as the teleosts, but fails to identify invertebrate TERC. In such cases, family-specific homology modeling combining several approaches and other evidence such as potentially conserved flanking genomic context may help to reliably annotate these families.

A special case must be made for microRNAs as a functionally very important class of ncRNAs, whose functional core unit is the only 20–25 nt long mature microRNA, which hybridizes with mRNAs and thus alters their translation into proteins. The mature microRNA is obtained from the 60–150 nucleotides long microRNA precursor (pre-miRNA), whose main structural component is a long and energetically stable hairpin, which is processed into the mature microRNA by a protein complex involving the Dicer protein. The original transcript of a microRNA, the so-called primary microRNA (or pri-miRNA) is even longer than the precursor, and is processed by a complex including the Drosher protein as its main component into the precursor. Sequence and structural conservation is relatively weak for miRNAs in general, often limited to a certain degree of sequence conservation of the mature microRNA residing within an energetically stable hairpin loop in the precursor. While some subfamilies of miRNAs are known to be evolutionarily relatively old, their overall repertoire evolves quite rapidly, and the evolutionary mechanism behind the generation and loss of miRNAs is still enigmatic. Detection of miRNAs from a genome sequence alone can be considered impossible in general, but successful bioinformatics approaches exist in combination with transcriptome sequencing and will be discussed in Section 14.4.3.

14.4.3 Gene expression analysis

Beyond computational predictions of either protein coding or non-coding genes, it is obviously of high relevance to link these predictions to experimental data obtained from so-called gene expression studies, which experimentally characterize what has been transcribed. The study of the entirety of all transcripts and their relations to each other and other entities is also commonly referred to as transcriptomics.

The first experimental basis to study the transcriptome was provided by microarray technology. On a microarray, single-stranded RNA (or in some cases DNA) oligonucleotides, i.e., short fragments of a limited length of usually 20–30 nucleotides, can be applied to a microchip-like surface. Oligonucleotides with a specific sequence can be fixed at a designated *spot* on the chip. If now single-stranded and fluorescently labeled RNA from a cell or tissue sample is applied to this chip, the sample RNAs will hybridize with reversely complementary counterparts on the chip and produce a fluorescence signal on the respective spot. Capturing a fluorescence image of the complete microarray after hybridization, and assigning fluorescence signals to the spots with known oligosequences now allows to identify what has been transcribed from spots exhibiting a fluorescence signal, or has not been transcribed from spots lacking a fluorescence signal. A major limitation of microarrays is the number of spots, where for instance the limit lies within roughly a few spots (i.e., oligonucleotides) per protein-coding gene on the human genome. Microarrays are thus well suitable for studying expression of annotated mRNAs. However, microarrays for obtaining expression information on non-coding parts of the genome, so-called tiling arrays, require many spots that need to be distributed across several chips are correspondingly costly. Another difficulty of microarrays results from correlating the fluorescence intensities of spots to the abundance of the corresponding transcript, which in general does not follow immediate proportionality, but may be distorted by artifacts resulting from the hybridization and fluorescent labeling procedure.

Due to the rapidly falling cost of sequencing DNA, microarray technology has become more or less overtaken by sequencing reversely transcribed cDNA, which experimentally is based on the RNA-Seq protocol. A major influence on what can be seen in the resulting sequencing data depends on filtering steps and the read length under consideration. A main challenge in filtering RNA is to eliminate ribosomal RNAs, which represent up to 90 % of all RNA in a cell, but would overshadow the actual RNA of interest if not filtered out. Beside filtering steps aiming to immediately filter our rRNAs, some protocols extract polyadenelated transcripts, which however will mainly extract mRNAs and eliminate many ncRNAs which are not polyandenelated. A further common filtering is size filtering, which has a major influence on what can be seen from the resulting sequencing data. If only small RNAs (e.g. within the typical length range of mature miRNAs) are filtered out, the expression of those small RNAs can be investigated; their signal may yet be overshadowed by degredation products of other longer RNAs. For lbanyy the transcription of protein coding genes, either no size filtering or a much longer size range will be selected.

The sequenced RNA fragments do not immediately yield expression information. Rather, they need to be mapped to the genome to identify the genomic loci from which they have been transcribed. For this read mapping task, dedicated software packages are available that work highly efficient and can take into account typical sequencing errors, where roughly one mismatch per 25 nucleotides is a reasonable fault tolerance to be taken into account. For reads exceeding a length of 70 nucleotides, few insertions or deletions may need to be allowed for.

A further challenge in read mapping lies in the architecture of eukaryotic genes and is illustrated in Figure 14.3. As mRNAs, as well as a number of known ncRNAs, are produced by the splicing of several exons, reads crossing exon boundaries cannot be mapped to the genome directly. Read mappers offer different strategies to deal with this. One possibility is to equip the mapping process with an annotation of all exons in the genome to be mapped against, so that the mapper is aware of exon ends and continue mapping at the start of next exon. This procedure works well for well annotated genes, but has limitations when dealing with splice variants or even unannotated genes. In this case, a shotgun assembly of the transcriptome can be performed, which however is computationally very demanding. However, long transcripts assembled from many shorter reads can usually be mapped to unique loci on the genome (taking into account exon structure), and provide full information on splicing variation.

If no major bias has been introduced during sample preparation, RNA-Seq data also allow to quantify the expression of specific transcripts. To this end, the average number of read nucleotides mapping onto a nucleotide in an annotated region can be determined. As this number statistically grows proportionally with the amount of sequenced RNA, it needs to be normalized by the amount of sequenced nucleotides. The commonly used unit to measure expression of an annotated region is thus reads per kilobase of transcript per million reads mapped (RPKM), which for gene g is formally defined as

$$\text{RPKM}_g = \frac{r_g \times 10^9}{L_g \times R}$$

where r_g is the number of reads mapping to the region annotated for g, L_g is the number of regions in the mapable are of g, and R is the total number of reads in the sample. Note that for rare transcripts such as miRNA precursors, it may easily be the case that no mappable read is observed, leading to an expression intensity of 0. The only general cure to see rare transcripts is to increase sequencing depth, i.e., produce more reads at proportionally higher sequencing costs.

Given expression intensities in the form of RPKM values for annotated genes allows to compare the expression of genes to be compared across samples. Typical scenrioes are the comparison of "treatment versus control" such as wild type vs. mutant, disease-affecte vs. control, or different time points in a biological process such as the cell cycle. One typical question regards the identification of differentially expressed genes (sometimes also more simply but less accurately referred to as differential genes). The most straightforward approach to identify differentially expressed genes is to consider the RPKM ratios for each gene between the two condition, and consider genes differentially expressed if this ratio falls out of a certain lbanyy, such as being less than .1 (or .5) or exceeding 10 (or 2, respectively). The identification of differential genes can be made more reliable at the cost of more experimental effort by considering several replicates for each of the two conditions. As each replicate yields one expression intensity, one can ask for each gene whether the expression intensities in the treatment group replicates significantly differ from the intensities in the control replicates, which leads to a statistical test (e.g., a chi-squared test) to be performed for each gene, where cutoffs can be defined for the significance scores yielded by the statistical test used.

The analysis of gene expression data gets more involved if more than two conditions are involved. For example, one may study the expression of genes through the cell cycle by obtaining gene expression profiles at 10 different time points in cell cycle synchronized cellular samples. In this setting, each gene is associated with an intensity vector consisting of 10 components, so that the complete expression profile will comprise roughly 20,000 data points (one for each gene) in a 10-dimensional vector space. To handle these high-dimensional data points, techniques of high-dimensional statistics and data mining can be applied. For example, clustering algorithms will partition the data points into a small number of similarity classes, where the genes within each class may be associated with certain properties regarding their behavior in the cell cycle (e.g., "genes that have a sharp expression peak in S-phase").

Software Tools and Implementations

Table 14.7: Software for RNA Bioinformatics.

Vienna RNA	RNA folding and structure comparison	rna.tbi.univie.ac.at
mfold	RNA folding	mfold.rna.albany.edu
RNAz	identifying conservation in consensus folding	tbi.univie.ac.at/~wash/RNAz/
evofold	identifying conservation in consensus folding	users.soe.ucsc.edu/~jsp/EvoFold/
infernal	covariance models	infernal.janelia.org
bowtie, tohat	read mapping	ccb.jhu.edu/software/tophat
cufflinks	differential gene identification	cufflinks.cbcb.umd.edu
segemehl	read mapping	www.bioinf.uni-leipzig.de/Software/segemehl/
bwa	read mapping	bio-bwa.sourceforge.net
soap	read mapping, genome and transcriptome assembly	soap.genomics.org.cn
Bioconductor R	platform for statistical analysis	bioconductor.org

14.5 Other topics in bioinformatics biotechnolgy

Although a relatively young discipline, bioinformatics has branched into numerous subfields, in fact too many to be covered within one single chapter. In the final section of this chapter, some other key areas in bioinformatics will be sketched in terms of the key questions that these areas deal with.

14.5.1 Evolutionary bioinformatics

A central role in bioinformatics is often taken by evolutionary aspects, even when dealing with questions that may not have an immediately apparent evolutionary component, as we have seen when dealing with RNA secondary structures or substitution matrices for amino acids. In fact, evolutionary approaches act behind the scenes of many of the approaches and software tools discussed in this chapter. One of the key topics in evolutionary bioinformatics is the reconstruction of phylogenetic trees, typically from aligned DNA or protein sequences. Many approaches have been developed for this essential task, and led to other important ideas and questions. For example, evolutionary bioinformatics has brought forth evolution rate models that immediately model evolution at the level of DNA. Also, topics such as the study evolution at the level of genomes, where rearrangements, duplications and deletions of large genomic areas or complete chromosomes rather than individual nucleotide mutation constitute the basic events, has attracted significant attention. Further topics regard the study of coevolution, e.g. identifying correlated events in the evolution of sequences in a host and a symbiont (or parasite).

14.5.2 Regulatory genomics and systems biology

In Section 14.4.3, we introduced methods to identify genes whose transcription is differentially regulated in different contexts. The question *how* these genes are differentially regulated is the topic of regulatory genomics. One main subject of research in regulatory genomics are promoter regions which are constituted by the sequence upstream to the transcription start site, typically a few thousands of nucleotides. In these regions, one can identify different regulatory relevant regions such as CpG islands and transcription factor binding sites (TFBSs), where bioinformatics plays an important role in their characterization and identification. Also, in the characterization and detection of microRNAs and their targets, bioinformatics approaches play an important role.

Regulatory genomics is not limited to identifying individual regulators and their specific targets. Questions such as "who regulates the regulators" leads to network representations of regulatory interactions, whose study is usually considered to be the subject of systems biology. Another systems biological question arising from the study of differential genes is which pathways are mainly affected by the genes that have been observed to be altered. As a resource to answer such questions, the bioinformatics community has brought forth tools such as gene ontology databases, which systematically capture terms (such as pathways or subcellular location) associated with genes, in some cases also using automated analysis of scientific literature using text mining approaches.

14.5.3 Proteomics data analysis

Beside at the level of transcription, gene expression can also be studied at the level of translation, where modern proteomics approaches provide means to characterize all or a large number of proteins present in a sample, most notably using mass spectrometry. A large portion of proteom bioinformatics thus deals with the analysis of such mass spectrometric data, also utilizing the concepts introduced in Section 14.4.3 to identify differential genes.

14.5.4 Bioimage informatics

Biological imaging technology has been developing at rapid pace in recent years and was a driving force behind numerous insights into biological systems. Imaging technology has been driven by diverse enhancements of conventional microscopy in terms of spatial and temporal resolution, dimensionality, or combinatorial label-

Box 14.3: Glossary for Section 14.5.

Transcription factor binding site. Transcription factors bind to sequence specific sites of DNA in promoter regions. The binding sites are 8–15 nucleotides long, and while not always being identical, they usually differ by few nucleotides only. Their similarity patterns are modeled through sequence logos, which are obtained using information theoretic concepts.

Systems biology. Many biological processes involve a large number of players, encompassing proteins as well as RNA or other small molecules. Systematically studying how functions emerge from the interplay of many components is the subject of Systems Biology, which utilizes a broad range of concepts from areas such as control theory or mathematical graph theory.

Gene ontology. Genes are associated with different terms at different levels such as their function (e.g., transcription factor, membrane protein), their location (e.g. nuclear, mitochondrial), or a cellular process (such as cell cycle, translation). Maintaing corresponding categorized vocabularies and their systematic association with genes is what constitutes a gene ontology. The gene onotology consortium (www.geneontlogy.org) is the main resource for accessing and contributing to ontologies.

Text mining. Beside the systematic study of biological data, systematically and automatically analyzing the scientific literature has proven a surprisingly useful tool to understand biological phenomena, for instance based on the occurrence and identification of relevant key words, the citation patterns of publications, and linking with other databases.

Live cell imaging. Using a microscope to study cells alive rather than in a fixated state is of importance for many cell types, in particular when studying the cell cycle or cell types where motility is of relevance such as immune cells. Such live cell imaging requires dedicated protocols that are sufficiently fast on the one hand, and on the other hand are not lethal to cells through either temperature, or through the use of labels, light, or reagents.

Fourier-transform infrared microscopy. Different molecules absorb the energy of different wavelengths of light in different ways. Capturing the absorbance of a cell or tissue sample across a large bandwidth of wavelengths, in particular in the infrared range of light, thus provides a characteristic fingerprint of the molecular composition of the sample, and thus allows to differentiate different states or phenotypes of cell or tissue samples. Using state-of-the-art vibrational spectroscopic techniques, these spectra can be captured at high spatial resolution of roughly 5 μm, yielding images where each pixel is represented by an absorbance spectrum. This spectral image can be used to resolve tissue structure and identify disease-associated regions.

ing. While novel technologies of super resolution microscopy pushed the limits of resolution close to single-protein level, combinatorial labeling approaches contributed to a highly resolved understanding of co-localization patterns within cells and tissues. Moreover, live cell imaging has become popular for studying 2D or 3D cellular motion patterns over time. Beyond the scope of light microscopy, spectral imaging approaches recently proved to be useful tools for analyzing and understanding tissue structure. Among those, Fourier Transform Infrared (FTIR) and Raman spectroscopic imaging as well as mass spectrometry imaging received major attention.

The widespread availability of novel imaging technologies imposes major challenges to turn illustrative images into quantifiable scientific measurements; hence, bioimage informatics has gained significance as a discipline within the field of computational biology. As quantifying observations in 2D, 3D, or 4D image data is significantly more demanding than the analysis of sequencing or microarray data, many developments in bioimage informatics are still in their infancy. In fact, many of the computational challenges that arise from image data require fundamentally new algorithmic approaches. Beyond providing quantitative analysis of image data, the long-term perspective for bioimage informatics will be to establish computational approaches as an "additional lense" for imaging technology, facilitating insights into biological systems beyond the reach of naive image analysis.

The increasing availability of bioimaging devices led to a number of computational methods for image data analysis developed in recent years. These methods cover applications such as building atlases for model organisms, reconstructing neural wiring diagrams and comparing them between individuals, or understanding dynamic cellular processes through time-series imaging. A common theme in these applications is the extraction of quantitative information from the underlying images, which in many cases requires the development of substantially new and non-trivial algorithms. In some cases, such quantitative analysis approaches helped to answer systems biological question. For example, observing a cell division under the knock-out of a key gene in the cell cycle using live cell imaging will much better reveal its cell cycle role than identifying it as differentially expressed together with hundreds or thousands of other, less cell cycle relevant, genes.

Key-terms

shotgun sequencing, whole genome shotgun sequencing, read length, genomic repeats, pairwise end sequencing, contigs, scaffolds, coverage, CpG islands, genome browser, genome annotation, single nucleotide polymorphism, copy number variation, SNP calling, resequencing, exome, dot plot, Needleman-Wunsch, Smith-Waterman, multiple sequence alignment, agglomerative alignment, substitution matrix, homology search, local alignment, Basic Local Alignment Search Tool, seed match, p-value E-value, hidden Markov model, orthology annotation, orthologue, paralogue, synteny, genome wide alignment, genomic rearrangement, RNA secondary structure, hairpin loop, internal loop, bulge loop, stacking pair, multi loop, pseudoknot, consensus folding, compensatory mutation, covarianve model, microRNA, gene expression, transcriptome, read mapping, reads per kilobase of transcript per million reads mapped, differentially expressed gene

Questions

- Repeat the blast search for the C. elegans TERT query against the genome sequences of D. melanogaster and A. mellifera by querying directly against the DNA sequence rather than against the RefSeq protein sequences. What variant of blast do you need to use? Is the E-value for the honey bee TERT higher, lower, or equal to the E-value obtained by the blastp search sketched in Figure 14.2?
- Identify whether the genome of the bean beetle (E. varivesits) has a telomeric repeat in each of its chromosomes. Note: the bean beetle genome is available from www.beanbeetles.org
- Identify whether the bean beetle has a copy of the TERT protein. Note: a blast interface to the bean beetle genome is available from www.beanbeetles.org
- Retrieve the tRNA sequence with ID B0285.t1 from the WormBase database (www.wormbase.org)
- Use the blast interface from http://metazoa.ensembl.org/Caenorhabditis_briggsae/ to blast the tRNA sequence with WormBase ID B0285.t1 (Problem 4) against the genomes of C. briggsae. What variant of blast should you use? For the 5 highest ranking blast hits, investigate whether they are annotated as tRNA in the C. briggsae genome. What is your conclusion regarding the suitability of blast for identifying homologues of non-coding RNA genes?
- Go to the Rfam website rfam.janelia.org and download the tRNA seed sequences.
- Take the first five tRNA sequences from the tRNA seed sequences obtained from Problem 6 and predict an RNA secondary structure using the RNAfold server from rna.tbi.univie.ac.at. How many of the structures resemble the tRNA cloverleaf structure?
- Align the first 20 of the tRNA seed sequences obtained from Problem 6 using the ClustalW web server (www.ebi.ac.uk/Tools/msa/clustalw2/), and save the resulting alignment file.
- Use the alignment of the 20 tRNA sequences from Problem 8 to compute a consensus structure using the RNAalifold server from rna.tbi.univie.ac.at. Does the consensus structure resemble the typical tRNA cloverleaf structure?

Further readings

Durbin, R., Eddy, S., Krogh, A., et al. 1998. Biological Sequence Analysis: Probabilistic Models of Proteins and Nucleic Acids, *Cambridge University Press*.

Mount, D. W. Bioinformatics: Sequence and Genome Analysis, 2nd ed, *Cold Spring Harbor Laboratory Press*.

References

Altschul, S. F., Gish, W., Miller, W., et al. 1990. Basic local alignment search tool. *J Mol Biol* 215 (3): 403–410.

Dayhoff, M. O. & Schwartz, R. M. 1978. A model of evolutionary change in proteins. *In Atlas of protein sequence and structure*.

Henikoff, S. & Henikoff, J. G. 1992. Amino acid substitution matrices from protein blocks. *Proc Natl Acad Sci* 89(22): 10915–10919.

Karlin, S. & Altschul, S. F. 1990. Methods for assessing the statistical significance of molecular sequence features by using general scoring schemes. *Proc Natl Acad Sci* 87(6): 2264–2268.

Mathews, D. H., Sabina, J., Zuker, M., et al. 1999. Expanded sequence dependence of thermodynamic parameters improves prediction of RNA secondary structure. *J Mol Biol* 288(5): 911–940.

Needleman, S. B. & Wunsch, C. D. 1970. A general method applicable to the search for similarities in the amino acid sequence of two proteins. *J Mol Biol* 48(3): 443–453.

Nussinov, R. & Jacobson, A. B. 1980. Fast algorithm for predicting the secondary structure of single-stranded RNA. *Proc Natl Acad Sci* 77(11): 6309–6313.

Shannon, C. E. 1957. A universal Turing machine with two internal states. *Automata studies* 34: 157–165.

Smith, T. F. & Waterman, M. S. 1981. Identification of common molecular subsequences. *J Mol Biol* 147(1): 195–197.

Staden, R. 1982. Automation of the computer handling of gel reading data produced by the shotgun method of DNA sequencing. *Nucl Acids Res* 10(15): 4731–4751.

Turing, A. M. 1936. On computable numbers, with an application to the Entscheidungsproblem. *J Math* 58: 345–363.

Weber, J. L. & Myers, E. W. 1997. Human whole-genome shotgun sequencing. *Genome Res* 7(5): 401–409.

Glossary

AAV	Adeno-associate viruses developed for gene therapy. AAVs have a small genome, which can easily be genetically modified, infect many different dividing and non-dividing cells and cell-types and mediate long-term transgene expression, without integration into the genome. AAVs are also used to deliver optogenetic tools to organs, tissues, and cell-types.
Actinomycetes	Heterogeneous group of aerobic, Gram-positive soil bacteria able to grow in branched cellular networks called mycelium. Most actinomycetes form spores.
Active ingredient	Molecule present in a medicinal formulation that is responsible for the action of the medicine.
Adenylyl cyclase	An enzyme which catalyzes the cyclization of adenosine triphosphate (ATP) into cyclic adenosine monophosphate (cAMP).
Adverse drug reactions	Any response to a drug which is noxious and unintended, including lack of efficacy.
Affinity purification	Biochemical mainly chromatographic purification procedure that is based on specific ligand –protein interactions, often conveyed by affinity tags (q. v. tags)
Agglomerative alignment	Technique to compute multiple sequence alignments based on pairwise alignments along a tree.
Agrobacterium tumefaciens	A soil bacterium, which infects plants and transfers a part of its own genome (the T-DNA of the T_i plasmid) into the genome of the infected plant cell
Alcohol fermentation	Biological process, also known as ethanol fermentation, by which sugars are converted into cellular energy (ATP) with the generation of ethanol and carbon dioxide as waste products. Performed by yeast and certain bacteria in the absence of oxygen (anaerobic).
Ames test	Method that uses bacteria to test whether a given chemical can cause mutations in the DNA of the test organism.
Amflora	A genetically modified potato, whose starch consists only of amylopectin and does not contain amylose. Amylopectin is used in several industries as thickener.
Amylase	An enzyme that catalyzes the conversion of starch into sugars.
Angiogenesis	Growth of blood vessels.
Angkak	Also known as red (yeast) rice. Produced by fermentation of rice with the mold Monascus purpureus.
Antibiotic selection marker	A genetic feature imparting the resistance of an organism to an antibiotic
Anti-fouling agent	An agent that prevents fouling.
Aquaculture	Farming of aquatic organisms (freshwater or saltwater) under controlled conditions.
Aquafarming	see Aquaculture.
Archaea	Prokaryotic microorganisms forming, besides the eukaryotes and bacteria, the third "urkingdom" of life.

Archaerhodopsin (Arch)	A proton pump from the archaeon *Halorubrum sodomense*. A proton (H$^+$) gradient over the membrane is commonly used in bacteria for energy (ATP) production. As an optogenetic tool, Arch is used for light induced hyperpolarization of cell membranes with peak excitation around 575 nm.
Astaxanthin	A keto-carotenoid that belongs to the class of xanthophylls. Astaxanthin has antioxidant properties and gives red salmon its red color.
Asymmetric cell division	Keeps the number of stem cells constant by generating one stem cell with its mother's properties and one more differentiated cell.
Asymmetric synthesis	A prochiral substrate is converted selectively to one of the two possible enantiomers of a chiral product.
Bacteriorhodopsin	A retinal protein from the archaeon *Halobacterium salinarum* acting as a light-driven proton pump
Basta	The trade name of an herbicide. The active ingredient is phosphinothricin, which inhibits the process of nitrogen assimilation in plants. Resistance against phosphinothricin can be achieved in genetically modified plants by expressing an enzyme that catabolizes the compound.
Batch cultivation	Type of cultivation, in which the culture grows until the limitation of one or more components (in case of photobioreactors also light) is reached. The culture remains in the bioreactor until the end of the run.
Beer	Alcoholic beverage with the original ingredients barley malt, hops and yeast. Saccharification of barley starch and subsequent alcohol fermentation by yeast.
Bicistronic gene expression	Expression at the same time of two different genes on the same vector.
Biocatalysis	The use of natural catalysts, such as enzymes, to perform chemical transformations.
Biofuels	Chemical entities that store primarily energy and that are derived from a biological process (excluding fossil fuels).
Biohydrogen	Typical biofuel, since it is produced by an enzyme (hydrogenase) that reduces protons with electrons derived from cellular metabolism and in some cases from photosynthesis.
Biolistics	A method to transform (genetically modify) a plant or algal cell, where small gold or tungsten particles, which are coated with the DNA to be transformed, are shot into plant cells by the use of a gene gun.
Biological control circuit	Artificially assembled network of genes and their products which can perform a set of logical functions. They can act as switches or react in a defined manner to an occurring stimulus.
Bioluminescence	Emission of light by a living organism. It depends on an enzyme (luciferase), a substrate (luciferin), and energy (typically ATP).
Biomass	Organic material organisms are made of.
Biotransformation, microbial	Defined chemical alteration of a given compound (e.g., a drug precursor) by microbes or isolated cells. Occurs due to the catabolic versatility of microbial organisms and is often stereospecific.

Blastocoel	The cavity of a blastocyst.
Blastocyst	Early preimplantation stage of an embryo before it implants into the placenta. The blastocyst has already the first specialized cells, the trophoectoblast cells, which later form the embryonic part of the placenta.
Blending	By mixing several components of different origin or different characteristics, a product with defined properties can be achieved.
BLUF domain	BLUF is a photoreceptor protein domain that uses flavin adenine dinucleotide (FAD) as a chromophore to sense blue light. For optogenetic applications, BLUF domains are coupled to catalytic, enzymatic like, protein domains involved in for example cyclic nucleotide metabolism.
bp	Base pair of DNA
Broad host range vector	A plasmid vector that can be transferred and replicated in a number of different non-related bacteria.
Bt toxin	Crystal proteins from spores of *B. thuringensies*, which are toxic to certain insect larvae.
CAL-A	A lipase that was isolated from the yeast *Candida antarctica*. It is a widely-used industrial biocatalyst.
Carotenoids	Pigments belonging to the chemical group of isoprenoids found mostly in photosynthetic organisms, where they are mainly used as antioxidants or for light absorption.
Casein	Milk protein and a mixture of different proteins: αS1, αS2, β and κ-casein.
Catalyst	It lowers the activation energy and increases the rate of a reaction. Catalysts are not consumed by the reaction but are recycled.
Cellulase	An enzyme that catalyzes the decomposition of cellulose and polysaccharides. Cellulases are produced by fungi, bacteria, and protozoans.
Central metabolic pathways	Pathways that provide precursor metabolites to all other pathways. For the aerobic metabolism of carbohydrate, the central metabolic pathways are glycolysis (Emden-Meyerhoff-Parnas pathway), the Entner-Doudoroff-pathway, pentose phosphate pathway, citric acid cycle, and the respiratory chain.
Channelrhodopsin 2 (ChR2)	A light-gated ion channel from the unicellular green algae *Chlamydomonas reinhardtii*. In the green algae, ChR2 is used for visible light perception and phototaxis. As research tool, ChR2 causes an influx of positively charged ions such as sodium and protons to depolarize membranes of excitable cells with light.
Chaperone	Proteins that assist in correct folding of proteins.
Cheese	Fermentation of milk by bacteria or by fungi results in the formation of this food.
Chemical space	Space spanned by a group of chemical compounds.
Chimera	An organism that consist of cells from more than one organism. For the knock out technologies, the chimera are usually generated as a result from the blastocyst injection.

Chirality	Organic molecules that cannot be superimposed with their mirror-image are referred to as chiral. Molecules that can be transformed by a symmetry operation (e.g., a rotation) into their mirror image, are achiral. Consequently, all molecules lacking a symmetry element such a symmetry axis or a rotational center are chiral.
Chitinase	An enzyme that catalyzes the degradation of chitin present in the cell walls of fungi and in the exoskeleton of some animals.
Chromatophores	Pigment-containing endomembranes of bacteria containing the photosynthetic apparatus, such as in *Rhodobacter spp.*
Citric acid	Weak organic acid with the chemical formula $C_6H_8O_7$. In biotechnology, millions of tons of citric acid are produced by the fungus *Aspergillus niger.*
Classical breeding	Also traditional breeding, terms that are used to describe breeding which only makes use of selection of plants with desirable traits, usually in combination with crossing.
Cloning	Is a process of producing similar genetically identical individuals. In biotechnology, cloning is a general term that refers to all operations where genes and other DNA fragments are amplified or transferred between organisms.
Codon usage	Describes the phenomenon during protein biosynthesis that different species tend to prefer one or few of several codons of the general genetic code that encode the same amino acid.
Common Technical Document	Application dossier for the registration of medicines across Europe, Japan, and the United States.
Compensatory mutation	Simultaneous substitution of distant nucleotides within a non-coding RNA gene that preserves base pairing between the two sites.
Competence	Capability of bacteria to take up genetic material from the surrounding medium.
Complementation	Compensation of a genetic defect in a host strain, usually conferred by a transformed vector encoding the required gene.
Conidia	Asexual spores that are generated by most molds. Conidia are generated by mitosis and are thus mitospores.
Conjugation	Transfer or DNA fragments or plasmids between bacteria of the same species or different genera by cell-cell contacts.
Contig	Stretch of DNA reconstructed from overlap information of individual (single end) reads based on overlap information in whole genome shotgun sequencing.
Continuous cultivation	Type of cultivation, which keeps the culture at constant density by removal of cells and addition of fresh media in an automated system.
Conventional breeding	All breeding techniques that do not use transgenic methods. These include classical breeding techniques but also other techniques like micropropagation and mutation breeding.
Copy number variation	Genomic site involving a large number of nucleotides that are variable within individuals of a species or a population, and lead to specific genes having a larger or smaller number of copies in different individuals.

Cotton bollworm	A caterpillar which attacks cotton plants and which can effectively be controlled by expression of certain Bt toxins in genetically modified cotton plants.
Coverage	Number of nucleotides sequenced in a given experiment divided by the number of nucleotides in a given genomic reference sequence.
CpG island	Stretches of genomic DNA with a high frequency of CpG sites, often found in promoter regions and associated with DNA methylation.
Crabtree effect	Describes the phenomenon that yeast can undergo fermentation, even under aerobic conditions, if high glucose concentrations are present.
Cre/loxP	A site-specific recombination system utilizing a type I topoisomerase (cre-recombinase) derived from bacteriophage P1, which catalyzes recombination events between short palindromic DNA sequences, so called loxP sites.
Cryptochromes (CRY)	In optogenetic applications, CRYs can be fused to proteins of choice to control protein-protein interactions involved in for example gene transcription and membrane recruitment of proteins.
Cyanobacteria	Formerly termed "blue-green algae", but cyanobacteria is a phylum in the kingdom of bacteria. Cyanobacteria are the only bacteria that perform oxygenic photosynthesis. According to the endosymbiotic theory, cyanobacteria are the ancestors of the chloroplasts found in plants and eukaryotic algae.
Derivatization	Process by which a molecule is modified to optimize its properties.
Desaturase	An enzyme that introduces a double bond in a fatty acid.
Diastereomer	Molecules with two or more stereocenters form diastereomers. In contrast to enantiomers, diastereomers have different chemical and physical properties.
Dihydrofolate reductase	An enzyme required for the synthesis of nucleic acids.
Directed evolution	An experimental algorithm for the optimization of enzymes by randomized variation of the gene and identification of improved variants by selection or screening.
Disinfectant	Antimicrobial product that destroys or inhibits the growth of harmful microorganisms.
Dot plot	Visualization of sequence similarity based on a matrix obtained from writing one sequence in rows and the other in columns, and indicating similar sites by black or gray dots.
Downstream processing	Includes all steps that are necessary to isolate fermentation products (i.e., harvesting, extraction etc.) from microbial sources in a biotechnological workflow.
DREADDs	Designer receptors exclusively activated by designer drugs. Developed by a random mutagenesis approach.
Drug Target	The molecular structure (in most instances a protein) a medicinal compound is binding to.
Drugability (of a drug target)	Term used to describe the likelihood of finding chemicals selectively binding to the target.

Electron transfer reactions	Reactions that involve transfer of electrons by the oxidation of one partner and reduction of the other (i.e. redox reactions). The redox potential difference (pE) of each partner determines the driving force of the reaction.
Embryonic stem cells (ESCs)	Stem cells derived from early embryos with fundamental properties. Most ESC lines are derived from eggs that have been fertilized *in vitro* and donated for research purposes.
Enantiomer	The two different configurations of a chiral molecule are referred to as enantiomers. Enantiomers have equal chemical and physical properties.
Enantiomeric excess (*ee*)	Describes the optical purity of an organic compound.
Enantioselectivity	The enantioselectivity of a reaction indicates by how much an enantiomer is faster converted than the other. It is defined as the ratio of the catalytic efficiencies k_{cat}/K_M of both enantiomers.
Entner-Doudoroff-pathway	Metabolic pathway only found in prokaryotes for the oxidation of glucose to pyruvate via the intermediate 2-keto-3-deoxy-6-phosphogluconate (KDPG). Yields 1 ATP, 1 NADH and 1 NADPH per molecule of glucose.
Enzyme unit U	Is defined as the molar product formation per min.
Enzyme	Enzymes catalyze all molecular transformations in a cell. While they are mostly polypeptides, several RNA-based enzymes or riboproteins are also known.
Esterase	An enzyme that cleaves an ester into an alcohol and an acid.
Esterification	A reaction forming an ester out of an alcohol and an organic acid. The reaction requires a removal of water to shift the equilibrium to the product side.
European corn borer	A caterpillar which attacks maize plants and which can effectively be controlled by expression of certain Bt toxins in genetically modified maize plants.
Eutrophication	Oversupply of nutrients causes excessive growth of plants and algae and leads to oxygen depletion.
Exome	Entirety of all protein-coding parts of the genome.
Extracellular matrix (ECM)	A collection of proteins (e.g. glycoproteins and proteoglycans) secreted by cells into the extracellular space. The ECM represents a supportive constituent of the stem cell niche.
Fatty acid	Carboxylic acid with a long hydrocarbon chain (an aliphatic chain). The C-atoms of the chain can be connected by single bonds (saturated) or double bonds (unsaturated).
Fatty acid synthase	A multi-enzyme protein that catalyzes fatty acid synthesis.
FDA (Federal Drug Administration)	The federal agency in the USA dealing with drug safety and approvals. The FDA has established itself as leading medical government agency in the world.
Filamentous fungi	Typical for these fungi is the formation of hyphae. These are tubular, elongated, and thread-like structures. Hyphae grow usually at their tips and contain multiple nuclei.
Flat panel photobioreactor	Closed reactor system for the cultivation of microalgae. Due to a high surface to volume ratio, flat panel reactors enable an efficient use of light.

Flavr Savr	A genetically modified tomato plant which has a strongly reduced level of polygalacturonase enzyme, and thus shows delayed softening of the tomato fruit. The Flavr Savr tomato was the first genetically modified plant that was released as food.
Fluorescence	Emission of light by a protein (or other material) that has absorbed light of a different (usually shorter) wavelength.
Flux balance analysis	An approach to predict fluxes in a metabolic network based on stoichiometric models. These models are constrained by several factors, as assumption of a steady state, definition of irreversible reactions and magnitudes of fluxes.
Fluxomics	Metabolic flux analysis aims at quantitative representation of a cell's metabolic state. It measures the metabolite fluxes of a reaction network occurring in an organism.
Food biotechnology	A branch of food science in which modern biotechnological techniques are applied to improve food production or food itself.
Formulation	Pharmaceutical formulations contain different chemical substances, including the active ingredient, that are combined to produce a final medicinal product.
Fouling	Accumulation of unwanted material (living organisms, organic or inorganic matter) on a solid surface.
Fruiting body	A multi-cellular structure with spore-producing tissues found in many fungi. The fruiting body is always part of the sexual cycle of a fungus.
G protein-coupled receptor (GPCR)	Integral plasma membrane proteins spanning seven transmembrane helices. GPCRs transduce extracellular signals to intracellular relay proteins (G proteins) to activate downstream signaling cascades.
Galenics formulations	They deal with the compositions of medicines that optimize their absorption.
Gene gun	An apparatus that is used in biolistics to shoot DNA coated particles into a plant or algal cell.
Gene therapy	An experimental technique for gene delivery into individual cells to treat or prevent genetic diseases.
Genetic marker	Genetic property that facilitates to distinguish different bacterial strains from another, e. g. the capability to synthesize an essential amino acid or an antibiotic resistance factor.
Genome annotation	Identification of functional associations of genomic sites, such as identification of protein coding genes, non-coding RNA genes, or regulatory sites.
Genome browser	Software to visualize annotations of genome sequences, often implemented as online applications in web browsers.
Genome engineering	Rational remodeling of an organism's genome and alteration of its genomic structure.
Genomic rearrangement	Major changes of DNA involving relocation of fragments, whose size may range from few hundred to several millions of base pairs.

Golden Rice	A genetically modified rice plant that produces provitamin A in the grain, which makes the grain looking yellow or golden. Golden Rice was produced to fight vitamin A deficiency in countries, where rice is a staple food.
Good manufacturing practice	Set of regulations that need to be applied in pharmaceutical manufacturing.
Gram stain	A dye staining procedure, which allows to differentiate bacteria according to their thick peptidoglycan cell wall and a single membrane (Gram-positives, e. g. *Bacillus subtilis*) from those containing no or peptidoglycan cell wall or have two membranes (Gram-negatives, e.g., *Escherichia coli* and *Rhodobacter spp.*). The term relates to the name of its inventor, the scientist H. C. Gram.
Green fluorescent protein	A protein with an intrinsic chromophore that emits green fluorescent light when exposed to UV light.
Halorhodopsin	A chloride pump from *Natromonas pharaonis* (NpHR). Cl⁻-transport is used by halophilic prokaryotes for osmotic balance and to maintain Cl⁻-dependent cellular processes, such as growth and motility. NpHR is used as optogenetic tool to hyperpolarize neurons with yellow light.
Hamanatto	Black soy beans fermented with molds.
hERG activity	Activity of a compound on the product of the "human Ether-à-go-go-Related Gene", a potassium channel predictive of irregularities in heartbeat.
Heroin	Chemical derivative of morphine.
Heterologous expression	Expression of "foreign" genes, which stem from another source than the host organism.
Heterosis	Also known as hybrid vigor, describes the phenomenon that offspring of pure-line parents often have superior characteristics compared to their parents.
Hexokinase	An enzyme that phosphorylates hexose sugars.
Hidden Markov model	Generative statistical model for sequences of events (e.g., for nucleotide or protein sequences).
High throughput screening	Screening process by which thousands or even millions of compounds (e.g., enzyme variants) are tested with high speed (often in an automated proceeding) against predefined targets.
High value products	Commercial products that have a high financial value and thus realize high profits.
Homologous expression	Expression of endogeous genes in a host, called "overexpression" if the protein level is above normal physiological level.
Homology search	Identification of sequences within a database of sequences that are similar (and putatively evolutionarily related) to a given query sequence.
Hybrid breeding	A breeding technique that makes use of heterosis by crossing two (nearly) pure-line parents. Hybrid breeding is standard in maize breeding but is also often used for other plants.
Hybrid vigor	see heterosis.

Hydrogen	In its molecular form (H_2), hydrogen has a high energy content and is combusted with oxygen to water. It is regarded as clean energy carrier, but is also needed in many chemical syntheses.
Hydrolases	Group of enzymes that catalyze the reversible hydrolytic cleavage of a chemical bond. For example, ester bonds are cleaved by nucleases, phosphodiesterases, lipases or phosphatases.
Inclusion bodies	Electron microscopically visible compartments in bacteria that are formed by aggregation of unfolded protein inside the cytoplasm of bacteria.
Induced pluripotent stem cells (iPSCs)	A type of pluripotent stem cell that can be generated from adult cells by re-programming. So iPSCs might bypass the need for embryonic stem cells and can be made in a patient-matched manner. The iPSC technology was pioneered by Yamanaka's team in Kyoto, Japan.
Inducible promotor	q. v. promotor. A promoter whose transcription activity is controlled by externally added factors, e.g., via derepression by release of a complex of a repressor protein bound to an inducer molecule.
Inducible protein expression	Translation of a gene resulting after transcription of the respective gene controlled by an inducible promotor.
Input trait	Trait that influence the (financial) input of the farmer. Insect resistance, for example, is an input trait because plants with this trait will have a higher yield and need less spraying with insecticides.
Introgression	The introduction of a chromosomal fragment from one organism into the genome of another organism by crossing and repeated backcrossing.
Inulin	Polysaccharide found in many plant types and belongs to the dietary fibers and the class of fructanes.
Investigator brochure	Document informing clinical researchers about the background and the details of a clinical trial.
Isoprenoids	Natural molecules composed of a varying number of isoprene units. Isoprene (2-methyl-1,3-butadiene) consists of five carbon atoms according to the chemical formula $CH_2=C(CH_3)CH=CH_2$
Kinetic resolution	The separation of two enantiomers by an enantioselective reaction. Principal is a different reaction rate of the enzyme towards the two enantiomers.
Knock out	Also known as gene deletion. As a consequence, the corresponding organism lacks expression of the corresponding gene.
Lactic acid fermentation	Biological process by which sugars are converted to cellular energy (ATP) with the generation of lactic acid (lactate) as a waste product. Performed under anaerobic conditions by certain bacteria, muscle cells, plants or lower eukaryotes. In bacteria, homofermentative and heterofermentative lactic acid fermentation can be distinguished (dependent of product).
Lactoferrin	Iron-transporting protein present in milk and blood plasma.

Lead compound	Molecule with pharmacological activity likely to be therapeutically useful, which may still have a suboptimal structure that requires modification.
Light stress	A condition under which a photosynthetic organisms absorbs light energy that exceeds the capacity of the photosynthetic electron transport chain and its electron sinks, respectively. Light stress can occur in high light or when the electron flow capacity is limited, e.g. under nutrient deprivation, and typically results in the production of reactive oxygen species.
Light-oxygen-voltage protein (LOV) domain	Flavin binding parts of phototropins. LOV domain-based photoswitches are used to control protein-protein interactions by blue light.
Lipase	An enzyme that catalyzes the hydrolysis of lipids (fats). Lipases belong to a subclass of esterases, they hydrolyse ester bonds.
Lipids	Hydrophobic natural compounds that usually consist of long hydrophobic hydrocarbon chains. Common lipids are membrane lipids or triacylglycerols, in which two or three fatty acids are bound to a hydrophilic head group so that the whole molecule is amphiphilic.
Lysozym	An enzyme that cleaves the backbone of the bacterial cell wall (peptidoglycan).
Malolactic fermentation	Microbiological process (performed by lactic acid bacteria) in winemaking in which tart-tasting malic acid, naturally present in grape must, is converted to softer-tasting lactic acid.
Marker gene	Not to be confused with genetic marker. A marker gene is a gene, whose presence can easily be accessed, e.g., genes for resistance against antibiotics or herbicides. In transformation, marker genes are coupled to the gene of interest to select those cells in which gene transfer was successful.
Marker	Also genetic marker, a known DNA sequence that differs between two parents and which can be identified by certain techniques. A set of markers allows the mapping of a trait to the genome by analyzing the offspring with regard to the distribution of the trait and the markers.
Medicinal chemistry	Chemistry directed at identification and optimization of chemical compounds for medical use.
Meiosis	Special cell division resulting in animals in either sperm cells or eggs, in plant in either pollen or eggs and in lower eukaryotes in meiospores.
Meiospores	Haploid spores that are generated by meiosis during the sexual cycle of a fungus.
Melanopsin	A light-activated $G_{q/11}$ protein-coupled receptor located on intrinsically photosensitive retinal ganglion cells (ipRGSCs). Photoentrainment of ipRGSCs is critically for the control of circadian rhythms and the pupillary light reflex. Melanopsin is used as optogenetic tool for controlling the Gq pathway by blue light.

Metabolic control analysis	Provides a framework which quantitatively connects properties of the metabolic network with properties of each reaction node. This is achieved by calculating change of steady state rates in response to the change of a metabolic network parameter.
Metabolic engineering	Applying gene technology (i.e. heterologous gene expression, altered expression of host genes) to optimize e.g. microbial production systems in order to generate a certain product of commercial value.
Metabolic network	A given set of physical and chemical processes occurring in a cell, determining its biochemical and physiological properties.
Metabolic streamlining	Metabolic engineering with focus on the elimination of competing pathways and an increased performance of the target pathway for enhanced product yield.
Metabolomics	Analysis of the complete set of small molecules involved in the metabolic pathways of an organism.
Metagenomics	Culture-independent sequencing of genetic material from environmental samples.
Microalgae	A term generally applied for microscopic algae both of the eukaryotic and prokaryotic type.
Micropropagation	Clonal propagation of plant cells or tissues in sterile culture. Finally, whole plants can be regenerated from them. Thus, several thousand clones can be produced from a single plant.
MicroRNA (miRNA)	Family of short non-coding RNA genes that are processed by the Dicer protein from a structured precursor transcript. miRNAs control often developmental processes.
Miso	Is produced by fermenting soy beans by the fungus *Aspergillus flavus* var. *oryzae*. It is used as a traditional Japanese seasoning.
Mitospores	Haploid or diploid spores in eukaryotes that occur as products of mitosis.
Molar yield	Defines how many moles of product are produced per mole substrate.
Morula	Early preimplantation stage of an embryo with no specialized cells.
Müller glia cells	Intrinsic glia of the vertebrate retina which provide structural support, contribute to cellular homeostasis, and act as living optical fibers. In lower vertebrates or under defined conditions, Müller glia exhibit also de-differentiation capacity and can become retinal stem cells that divide and differentiate into multiple retinal cell types.
Multipotency	Capacity of stem cells from specific tissues, where only a limited amount of different cell types can appear in the lineage.
Mutagenesis	The generation of mutations within the genetic information of an organism.
Mutation breeding	Use of mutagenizing agents (often radioactive radiation) to invoke mutations that produce plants with new traits.
New chemical entity	A molecule that has not been used in a medicinal product before.
New Drug Application	Application process for drug approval in the USA.

Next Generation Sequencing	Gel free DNA sequencing methods that led to a tremendous throughput increase in DNA sequencing and are not based on the commonly used Sanger technology.
Non-ribosomal peptide synthetase	Enzymes that synthesize non-ribosomal peptides with cyclic or branched structures. Both non-proteinogenic and proteinogenic amino acids are used as substrates.
Nutrient deficiency	A condition under which a nutrient is absent or present in too low amounts to sustain normal cell metabolism.
Opsins	Light-sensitive proteins that convert light into electro-chemical signals.
Optogenetics	A technique combining genetic and optic approaches to control well-defined neuronal populations within a millisecond timescale.
Organic acids	Weak acids that unlike mineral acids do not dissociate completely in water. Typical biotechnical applications are the production and preservation of food.
Output trait	Trait that influences the product (output), e.g. different flower color, increased nutritional value, longer shelf life etc.
Pairwise end sequencing	Technique of sequencing the 5′ end and the 3′ end of a long fragment of DNA. While the length of the sequenced part of the 5′ and the 3′ end is limited by the read length, the fragment itself can be much longer, e.g. tens or hundreds of thousands of nucleotides. P.e.s. is commonly used to overcome the problem of sequencing repetitive regions in whole genome shotgun sequencing.
Particle bombardment	See biolistics.
Pasteur effect	Describes the phenomenon that yeasts grown under anaerobic conditions consume more glucose than under aerobic conditions.
Pectinases	Enzymes that breaks down pectin, a polysaccharide of plant cell walls. Pectinases are usually applied for the extraction of fruit juice or for the production of wine.
Personalized medicine	Based on cells from the patient himself, isolated from biopsies or easily available tissues and cultivated and manipulated in a disease-specific manner. Upon transplantation of treated cells or tissue components developed there from a curing of the specific disease or selective tissue replacement can be achieved without immunosuppression.
Pesticide	Agent that controls any unwanted organism. It includes insecticides, herbicides, fungicides, and antimicrobial compounds.
Pharmacophore	Ensemble of steric and electronic features necessary to ensure optimal supramolecular interactions with a specific biological target.
Pharmacovigilance	Drug Safety, i.e., the pharmacological science relating to the collection, detection, assessment, monitoring, and prevention of adverse effects of pharmaceutical products
Phosphatase	An enzyme that removes a phosphate group from its substrate.
Photobioreactor	Bioreactor (or fermenter) specifically designed to house photosynthetic microalgae, e.g., it provides a light source.
Photosynthesis	The biological process by which photosynthetic organisms convert light energy to chemical energy and finally biomass.

Photosynthetic efficiency	Percentage of supplied light energy that is converted into chemical energy.
Phycology	The science of studying algae.
Phytase	A phosphatase enzyme that releases phosphorus from organic phosphor-containing compounds.
Phytochromes	Light sensing chromoproteins in plants. Plants use phytochromes to detect light in the red and far-red region of the visible spectrum. The light-sensing molecule is a covalently bound linear tetrapyrrole, bilin chromophore. As optogenetic tool, phytochromes can be used to control interactions between two protein partners.
Plantibody	An antibody that was produced by expression in plants or plant cell cultures.
Plasmid	Non-chromosomal, circular DNA that is replicated independently. Mostly found in bacteria, and in some Archaea.
Pluripotency	Ability of cells to generate derivatives of all three germ layers, ectoderm, mesoderm, and endoderm. ESCs and iPSCs are pluripotent.
Polyketide synthase	A multi-domain enzyme that produces polyketides. These secondary metabolites are usually synthesized through the decarboxylative condensation of malonyl-Co and can be found in species from all kingdoms of life.
Polymerase	Enzyme that catalyses the production of nucleic acids. DNA polymerases are required for replication and RNA polymerases for transcription of genes.
Polyunsaturated fatty acids (PUFAs)	Fatty acids with several double bonds. Some PUFAs, such as α-Linolenic acid with three double bonds, are essential for humans.
Primary embryonic fibroblast (PMEF)	Cells isolated from an embryo and taken into culture.
Productivity	Reflects the amount of product formation per amount of catalyst.
Progenitor cells	Cells which differentiate into a specific cell type and divide only a limited number of times.
Promoter	A genetic element located upstream of a structural gene that controls the transcription of the respective gene. In bacteria, it contains the −10 bp and −35 bp region (position 0 bp is the site of transcription initiation), and is typically bound by transcription factors and RNA Polymerases.
Pronucleus	Exists in the fertilized egg for the first hours after fertilization. After finishing the last division steps for the egg, the male and the female pronuclei fuse.
Protease	An enzyme that hydrolyses peptide bonds and therefore proteins. Proteases are found in organisms from all kingdoms of life.
Protein engineering	A method for the generation of enzyme variants with tailor-made catalytic properties for technical applications.
Protein folding	Highly regulated process inside a cell that ensures the production of functionally active proteins. Protein folding starts immediately after protein synthesis at the ribosome and is dependent on the function of numerous chaperones.

Protein toxicity	Recombinantly expressed proteins can often be toxic for a host cell, either by causing non-physiological catalytic activity or by blocking vital host functions due to the overproduced protein mass (e.g., production of membrane proteins).
Proteobacteria	A phylum of bacteria that only includes Gram-negative bacteria such as *Escherichia coli* and *Rhodobacter spp*. The phylum is subdivided into classes named with greek letters (e.g., Gamma proteobacteria).
Proteomics	This is the analysis of the entire set of proteins, which can be expressed by the genome of a cell, in a tissue or in a body liquid. Proteomes can be elucidated using gel based methods as 2D gels or gel-free MS-based methods.
Protoplast	A plant, algal or fungal cell whose cell wall was removed by certain hydrolytic enzymes. Protoplasts of different plants can be fused to produce a chimera. They are also used for the DNA-mediated transformation.
Protoplast fusion	The cell wall of microorganisms is lysed only leaving the plasma membrane. Transfer of genes and fusion of protoplasts is induced by chemical and physical processes.
P-value	Probability that the value of a variable (e.g., alignment score) is greater than or equal to the observed value strictly by chance.
Pyridoxal-phosphate	An organic cofactor that catalyzes several transformations of molecules with an amino a group, including transamination, racemization, and dehydration reactions.
Qualified person	Person responsible for controlling the quality of a drug and releasing manufacturing batches for commercial use.
Radial glia cells	Radial-shaped stem cells that span the width of the developing cortex. Radial glia cells are multipotent, self-renew, and generate neurons and glial cells of the central nervous system.
Random mutagenesis	Cells are subjected to mutagenic radiation or a mutagenic chemical so that indiscriminate mutations occur. Mutants with characteristics advantageous for biotechnology are then screened for.
Rare codons	Amino acid coding base triplets which are used at very low frequency in the host organism. Occurs when the codon usages of the host and the genetic source organism are very different from each other.
RASSLs	Engineered receptors activated solely by synthetic ligands. These receptors lost the ability to respond to their endogenous ligands and are instead activated by synthetic compounds.
Rational protein design	Rationally developed assumptions on catalytic properties of enzymes are used for predictions of the effect of amino acid substitutions. Site-directed mutagenesis and characterization then verifies the predicted variants.
Read length	Number of nucleotides that is sequenced in one consecutive stretch in a given experiment. Typical read lengths are 1000 nucleotides when using conventional Sanger sequencing, and few hundred for next generation sequencing approaches.

Read mapping	Identification of the genomic site of origin of sequenced DNA or RNA fragments.
Recombinant proteins	Proteins that are synthesized by expression of recombinant genes, often in heterologous hosts. Usually, these genes are generated by genetic engineering.
Regioselectivity	Is the ability of a catalyst to distinguish between chemically similar functional groups that are on different positions of a molecule.
Replicon	A DNA molecule replicating from a single origin such as a plasmid. The "origin of replication" (*ori*) is the genetic element that host specifically controls this process.
Retinogenesis	Comprehensive phase of cellular proliferation, differentiation, maturation, and "synaptic fine tuning" in the developing retina.
Reverse pharmacology	Search for bioactive molecules by screening against predefined molecular targets.
Rhodopsin	30–60 kDa membrane-bound light sensitive protein that consists of two parts: The apoprotein, termed opsin and the chromophore, which constitutes the prosthetic group of the opsin.
Roundup	Trade name for glyphosate, the most sold herbicide today. Glyphosate inhibits the biosynthesis of aromatic amino acids. Resistance in plants can be achieved by expressing a tolerant version of the targeted enzyme or by expressing an enzyme that catabolizes the compound, or both.
Sauerkraut	Fine cut cabbage that has been fermented by lactic acid bacteria.
Scaffold	Stretch of DNA reconstructed from distance information between contigs in pairwise end reads.
Secondary metabolites	Organic compounds produced by living organisms that are not directly involved in the normal growth, development, or reproduction of an organism.
Secretion	Transport of synthesized proteins from the cytoplasma to the external medium.
Selection	Keeping alive only those members of the subgroup that contain a selective marker from a background of non-bearing organisms. E. g., bacteria bearing plasmid vectors equipped with antibiotic resistance factor genes are selected against all antibiotic sensitive bacteria without plasmids.
Sequence alignment	Writing two (*pairwise* S.a.) or more (*multiple* S.a.) sequences in a way that homologous positions occur within the same columns, while inserted or deleted sites are indicated by gap ('-') characters.
Shellfish	Water animals bearing an exoskeleton; e.g., mussels, oysters, shrimp, lobster.
Shotgun sequencing	Reconstruction of long stretches of DNA (or sometimes RNA) sequences based on overlap information between short sequenced fragments.
Single nucleotide polymorphism	Genomic site involving only a single nucleotide that is variable within individuals of a species or a population.

SMART breeding	"Selection with markers and advanced reproductive technologies". SMART breeding makes use of genetic markers which allow selecting promising offspring from a cross before the desired trait becomes visible and to characterize their genomic background. Thus, breeding programs can be shortened by several generations.
SNP calling	Identification of single nucleotide polymorphisms, typically based on DNA sequencing approaches.
Solar irradiance	The irradiance (flux of electromagnetic radiation per area of a surface, SI unit W/m^2) of the sun received on the surface of the Earth.
Soy sauce	Typical product of Asian cuisine. This sauce is made from boiled soy beans and wheat (only Japan) that were fermented with *Aspergillus oryzae* or *Aspergillus sojae* and several yeasts and bacteria.
Space time yield (STY)	This parameter indicates the amount of product that is formed in a certain time and volume.
Specific activity	Enzyme activity U per mg enzyme.
Spores	Propagation units that are produced by microbes or plants, mostly unicellular. Spores are usually the product of asexual or sexual reproduction.
Standard Operating Procedures	Controlled documents describing processes within a company that can only be adapted according to rules described in a more fundamental standard operating procedure.
Stem cell factory	Standardized protocols for the high-throughput treatment of cell cultures. Automated processes and computer-aided control mechanisms can ensure effective cultivation of high numbers of samples under GMP standards.
Stem cell niche	Microenvironment of the stem cells themselves, surrounding cells, blood vessels, trophic support, diffusible factors, like growth factors, and the extracellular matrix. The composition of these factors is specific for the distinct stem cell compartments.
Stem cells	Cells with the potential to develop into different cell types and exhibit self-renewal and multipotency.
Strain design	Process of strain selection or optimization to achieve a defined alteration of a strain.
Substitution matrix	Matrix to assign scores to one protein (or nucleotide) being substituted by another protein (or nucleotide) in a sequence alignment.
Symmetric cell division	Division of a stem cell leading to two identical daughter cells. Expansive, when both daughter cells are stem cells; consumptive, when the daughter cells are committed precursors or differentiated cells.
Synthetic biology	Biological approach where biological networks or cells are planned and constructed from scratch, often using artificial gene sequences.
Systems Biology	Field of biology which aims at quantitatively understanding the entire set of dynamic processes occurring in an organism. Transcriptome, proteome, metabolome, and fluxome data is integrated and used to create a quantitative mathematical model of cellular processes.
T7 polymerase	RNA polymerase from the bacteriophage T7, which specifically initiates transcription after binding to the T7 promoter. The enzyme is typically used for genetic engineering.

Tag	Part of a polypeptide that is used for affinity purification (*q.v.*). Typically, a tag is a short peptide (e.g., 6x Histidine) or a protein (e.g., Maltose-binding-protein) located N- or C-terminally of the target protein.
T-DNA	"Transfer DNA", that part of the Ti plasmid of *A. tumefaciens*, which is transferred to the plant genome. The T-DNA is defined by two border sequences.
Tempeh	A traditional soy product. Usually the fungi *Rhizopus oligosporus* or *Rhizopus oryzae* are used for fermentation of soy beans.
Teratomas	Tumors that inherit cells from all three germ layers, namely the ecto-, endo-, and mesoderm. They can develop when pluripotent cells are transplanted and are, therefore, a good proof for pluripotency.
Tetraploid	A cell carrying 4 sets of each chromosome.
Theka interna cells	Cells within the ovaries helping the egg to develop by providing hormones.
Thermophile	Organism living at high temperatures, typically with a growth optimum between 45 and 80 °C.
TILLING	"Targeted induced local lesions in genomes". A technique to rapidly identify a desired mutation in a gene of interest in a high number of plants.
Ti-plasmid	The tumor-inducing plasmid from *A. tumefaciens* carries the T-DNA and other genes (e.g. virulence genes), which are necessary for the transfer of the T-DNA into the plant genome.
Totipotency	Capacity of cells to generate into different cell types (e.g. embryonic and extraembryonic tissues). The fertilized egg cell, the zygote is totipotent as well as cells until the third round of cell division in the 8-cell-stage of the embryo. Since plant cells, in contrast to animal cells, generally stay totipotent, a whole plant may be regenerated from a single cell.
Transaminase	A pyridoxal-dependent enzyme that catalyzes the conversion of a ketone into an amino acid and vice versa.
Transcription	Synthesis of messenger RNA according to the gene encoded by DNA
Transcriptome	Entirety of all RNA transcribed from genomic DNA.
Transcriptomics	Sequencing and quantification of all RNA transcripts extracted from a cell. Aim of transcriptomics is to analyze the global dynamics of gene expression in an organism.
Transdermal formulation	Drug formulation that allows application of the active ingredient through the skin.
Trans-fats	The catalytic hardening of fats causes the formation of the *trans*-isomers of fatty acids.
Transformation	Uptake of unprotected genetic material such as DNA fragments or plasmid vectors by transformation of competent cells.
Transgene	A gene (or genetic material) that does not naturally exist in a given cell but has been transferred to recipient strains/cells by genetic techniques. This can occur naturally, e.g. by viruses, or by applying genetic engineering.
Transient amplifying precursors (TAPs)	Progeny of stem cells that exhibit a high proliferation rate and give rise to a high number of offspring cells.

Transposon	Mobile genetic element, which can be transferred from one position in the genome to another position.
Triacylglycerols (TAGs)	Lipids, in which three fatty acids are esterified to a glycerol molecule. In organisms, TAGs serve as fatty acid reservoirs as an energy storage or for remodeling processes. TAGs are the basis of biodiesel.
Triglycerid	Most frequent molecular structure of neutral lipids, in which three fatty acids are bound to a glycerol molecule.
Triticale	Progenies of a cross of wheat (*Triticum sp.*) and rye (*Secale cereale*).
Trophoectoblast	Cells of a blastocyst that generate the embryonic part of the placenta.
Vector	Artificial plasmid, which is used for cloning and/or recombinant expression of genes in molecular biology. Vectors normally contain antibiotic resistance factor genes for selection and stable maintenance in host cells.
Vitamins	Organic compounds that are needed only in small amounts, but that are essential for a certain function (for example as co-factor for enzymes, such as vitamin B_{12}). The definition of vitamin includes that it cannot be synthesized by the organism that needs it.
White Biotechnology	Part of the biotechnological industry which produces compounds needed in nutritional and chemical industries.
Whole genome shotgun sequencing	Reconstruction of complete genome sequences using shotgun sequencing, typically involving paired end reads.
Wine	Alcoholic beverage made from fermented grapes or other fruits. During alcohol fermentation, yeasts convert the sugar of the grapes into ethanol and carbon dioxide.
Yamanaka cocktail	Combination of the transcription factors Sox2, Oct4, Klf4 and c-Myc for the reprogramming of differentiated cells to induce pluripotency, developed by Yamanaka's team in Kyoto, Japan.
Yeasts	Unicellular eukaryotic organisms that are classified in the fungal phyla ascomycota or basidiomycota. The most prominent yeast species is *Saccharomyces cerevisiae*, which is used for baking and alcohol production.
Yoghurt	Food produced during lactic acid fermentation of milk. Lactose in milk is converted to lactic acid which precipitates the milk protein to give yogurt its texture and its characteristic taste.
Zona pellucida	A glycoprotein matrix of the egg that is generated by the follicular cells.

Index

In cases, where key words occur more than once in a chapter, only the first page of the appropriate chapter is given.

www.ingramcontent.com/pod-product-compliance
Lightning Source LLC
Chambersburg PA
CBHW072009230326
41598CB00082B/6873

9 783110 341102